Lecture Notes
in Computational Science
and Engineering

82

Editors

For further volumes:
http://www.springer.com/series/3527

Björn Engquist • Olof Runborg
Yen-Hsi R. Tsai
Editors

Numerical Analysis of Multiscale Computations

Proceedings of a Winter Workshop at the Banff International Research Station 2009

Springer

Editors
Björn Engquist
Yen-Hsi R. Tsai
The University of Texas at Austin
Department of Mathematics
Institute for Computational
Engineering and Sciences
University Station C 1200 1
78712-0257 Austin, Texas
USA
engquist@ices.utexas.edu
ytsai@ices.utexas.edu

Olof Runborg
Royal Institute of Technology (KTH)
Dept. of Numerical Analysis, CSC
Lindstedtsvägen 3
100 44 Stockholm
Sweden
olofr@nada.kth.se

ISSN 1439-7358
ISBN 978-3-642-21942-9 e-ISBN 978-3-642-21943-6
DOI 10.1007/978-3-642-21943-6
Springer Heidelberg Dordrecht London New York

Library of Congress Control Number: 2011938020

Mathematics Subject Classification (2010): 65Cxx, 65P99, 65Zxx, 35B27, 65R20, 70Hxx, 70Kxx, 70-08, 74-XX, 76S05, 76Txx

Cover photo: By courtesy of Jon Häggblad

Printed on acid-free paper

Springer is part of Springer Science+Business Media (www.springer.com)

Preface

The recent rapid progress in multiscale computations has been facilitated by modern computer processing capability and encouraged by the urgent need to accurately model multiscale processes in many applications. For further progress, a better understanding of numerical multiscale computations is necessary. This understanding must be based on both theoretical analysis of the algorithms and specific features of the different applications.

We are pleased to present 16 papers in these proceedings of the workshop on Numerical Analysis and Multiscale Computations at the Banff International Research Station for Mathematical Innovation and Discovery, December 6–11, 2009. The papers represent the majority of the presentations and discussions that took place at the workshop. The goal of the workshop was to bring together researchers in numerical analysis and applied mathematics with those focusing on different applications of computational science. Another goal was to summarize recent achievements and to explore research directions for the future. We feel that this proceeding lives up to that spirit with studies of different mathematical and numerical topics, such as fast multipole methods, homogenization, Monte Carlo techniques, oscillatory solutions to dynamical systems, stochastic differential equations as well as applications in dielectric permittivity of crystals, lattice systems, molecular dynamics, option pricing in finance and wave propagation.

Austin and Stockholm

Björn Engquist
Olof Runborg
Yen-Hsi Richard Tsai

Acknowledgements

We thank Banff International Research Station (BIRS) and its excellent staff for hosting the workshop which formed the basis for the present volume. We also thank the Institute for Computational Engineering and Sciences (ICES), the University of Texas at Austin and the Swedish e-Science Research Center (SeRC) for support that resulted in the inception and the completion of this book.

Contents

Explicit Methods for Stiff Stochastic Differential Equations

Assyr Abdulle

Abstract Multiscale differential equations arise in the modeling of many important problems in the science and engineering. Numerical solvers for such problems have been extensively studied in the deterministic case. Here, we discuss numerical methods for (mean-square stable) stiff stochastic differential equations. Standard explicit methods, as for example the Euler-Maruyama method, face severe stepsize restriction when applied to stiff problems. Fully implicit methods are usually not appropriate for stochastic problems and semi-implicit methods (implicit in the deterministic part) involve the solution of possibly large linear systems at each time-step. In this paper, we present a recent generalization of explicit stabilized methods, known as Chebyshev methods, to stochastic problems. These methods have much better (mean-square) stability properties than standard explicit methods. We discuss the construction of this new class of methods and illustrate their performance on various problems involving stochastic ordinary and partial differential equations.

1 Introduction

The growing need to include uncertainty in many problems in engineering and the science has triggered in recent year the development of computational methods for stochastic systems. In this paper we discuss numerical methods for stiff stochastic differential equations (SDEs). Such equations are used to model many important applications from biological and medical sciences to chemistry or financial engineering [16,32,39]. A main issue for practical application is the problem of stiffness. Various definitions of stiff systems for ordinary differential equations (ODEs) are proposed in the literature [19] (see also [26, Chap. 9.8] for a discussion in the

A. Abdulle (✉)
Section of Mathematics, Swiss Federal Institute of Technology (EPFL), Station 8, CH-1015, Lausanne, Switzerland
e-mail: assyr.abdulle@epfl.ch

B. Engquist et al. (eds.), *Numerical Analysis of Multiscale Computations*, Lecture Notes in Computational Science and Engineering 82, DOI 10.1007/978-3-642-21943-6_1,

stochastic case). Central to the characterization of stiff systems is the presence of multiple time scales the fastest of which being stable. The usual remedy to the issue of stiffness (in the deterministic case) is to use implicit methods. This comes at the cost of solving (possibly large and badly conditioned) linear systems. For classes of problems (dissipative problems), explicit methods with extended stability domains, called Chebyshev or stabilized methods, can be efficient [2, 3, 24, 27] and have proved successful in applications (see for example [4, 14, 18, 21] to mention but a few). In this paper we review the recent extensions [5–8] of Chebyshev methods to mean-square stable stochastic problems with multiple scales.

We close this introduction by mentioning that the stability concept considered in this paper, namely the mean-square stability, does not cover some classes of interesting multiscale stochastic systems. Indeed, adding noise to a deterministic stiff system (where Chebyshev or implicit methods are efficient) may lead to stochastic problems for which the aforementioned methods are not accurate. Adding for example a suitably scaled noise ($\epsilon^{-1/2} dW(t)$) to the fast system of the following singular perturbed problem

$$dx = f(x,y)dt, \qquad x(t_0) = x_0, \tag{1}$$

$$dy = \frac{1}{\epsilon} g(x,y)dt, \qquad y(t_0) = y_0, \tag{2}$$

where $\epsilon > 0$ is a small parameter, can lead to a fast system with a non-trivial invariant measure. To capture numerically the effective slow variable, requires to correctly compute the invariant measure of the fast system. This might not be possible for implicit[1] or Chebyshev methods, if one uses large stepsize for the fast process. Even though such problems are not mean-square stable, the stability properties of implicit or Chebyshev methods still allow to compute trajectories which remain bounded. But the damping of these methods may prevent the capture of the right variance of the invariant distribution (see [7, 28] for examples and details). In such a situation one should use methods relying on averaging theorems as proposed in [37] and [15].

The paper is organized as follows. In Sect. 2 we discuss stiff stochastic systems and review the mean-square stability concept for the exact and the numerical solution of an SDE. Next, in Sect. 3 we introduce the Chebyshev methods for stiff ODEs. The extension of such methods to SDEs (called the S-ROCK methods) are presented in Sect. 4. In Sect. 5, we study the stability properties of the S-ROCK methods. Numerical comparison illustrating the performance of the S-ROCK methods and comparison with several standard explicit methods for SDEs are given in Sect. 6.

[1] There is one exception, namely the implicit midpoint rule, which works well for (1)-(2) when the fast process is linear in y. This is due to the lack of damping at infinity [28].

2 Stiff Stochastic Systems and Stability

As an illustrative example, consider the following stochastic partial differential equation (SPDE), the heat equation with noise (see [6]):

$$\frac{\partial u}{\partial t}(t,x) = D\frac{\partial^2 u}{\partial x^2}(t,x) + \mu u(t,x)\dot{W}(t), \qquad t \in [0,T], \quad x \in [0,1], \tag{3}$$

where we choose the initial conditions $u(0,x) = 1$, and mixed boundary conditions $u(t,0) = 5$, $\frac{\partial u(t,x)}{\partial x}|_{x=1} = 0$ and $D = 1$. Here $\dot{W}(t)$ denotes a white noise in time.[2] To solve numerically the above system, we follow the method of lines (MOL) and discretize first the space variable

$$dY_t^i = \frac{Y_t^{i+1} - 2Y_t^i + Y_t^{i-1}}{h^2} + \mu Y_t^i \, dW_t, \qquad i = 1,\ldots,N, \tag{4}$$

to obtain (a large) system of N SDEs, where $N = \mathcal{O}(1/h)$ (Fig. 1).

Remark 1. Notice that we used finite differences (FDs) to perform the spatial discretization. We emphasize that finite element methods (FEMs) could have been used as well. In a first step one would obtain a system $MY' = \ldots$, where M is the mass matrix. For low order FEs a cheap procedure, called mass lumping, allows to transform M into a diagonal matrix without loss of accuracy for the numerical method [36].

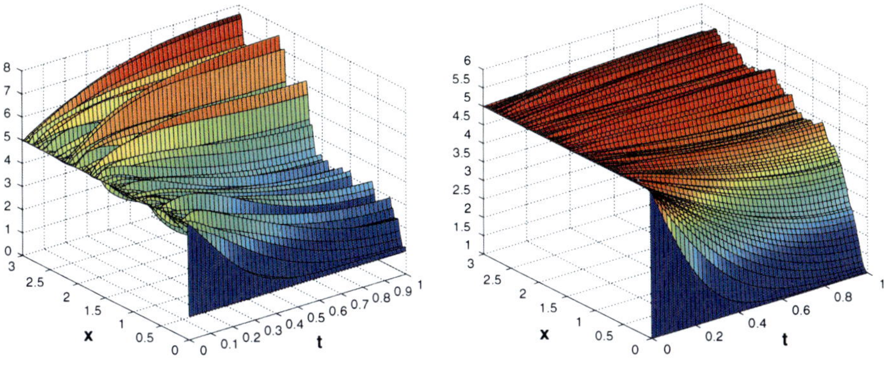

Fig. 1 One realisation of the system (4) with the Euler-Maruyama method (*left figure*); average over 100 realizations (*right figure*). Parameters values: $D = k = 1, N = 50, \Delta t = 2^{-14}$, $t \in [0,3]$

We first write the system (4) in the form $dY = (AY + B(Y))dt + GYdW_t$, where A is a tridiagonal matrix (approximation of the second order partial differential

[2] We will not discuss the precise meaning of (3), whose rigorous definition involves an integral equation [13, 38].

operator), $B(Y)$ is a vector accounting for the boundary conditions, and G is a (diagonal) matrix accounting for the multiplicative noise. When then obtain after (simultaneous) diagonalization, the system of SDEs (with appropriate boundary conditions omitted here) reads

$$dY_t^i = \lambda_i Y_t^i dt + \mu Y_t^i \, dW_t, \qquad i = 1, \ldots, N, \tag{5}$$

where $\lambda_i \in [-\mathcal{O}(N^2), 0]$ (see [6] for details). As for (3), the rigorous interpretation of (5) is an integral form involving a stochastic integral for which various "calculus" can be used, most often the Itô or the Stratonovich calculus [9]. The numerical methods described in this paper have been derived for both calculus. For the time being, we will consider Itô form. The simplest numerical scheme to solve (5) (assuming Itô form) is the Euler-Maruyama method, a generalization of the Euler scheme for ordinary differential equations (ODEs) introduced in [30]

$$Y_{n+1} = Y_n + \Delta t \lambda Y_n + I_n \mu Y_n, \tag{6}$$

where $I_n = W(t_{n+1}) - W(t_n)$ are independent normal $\mathcal{N}(0, \Delta t)$ random variables.

As for ODEs, two important issues arise when deriving numerical methods for SDEs, namely the accuracy and the stability of the approximation procedure.

Accuracy. Consider

$$dY = f(t,Y)dt + \sum_{l=1}^{M} g_l(t,Y) dW_l(t), \qquad Y(0) = Y_0, \tag{7}$$

where $Y(t)$ is a random variable with values in $\mathbb{R}^d$, $f : [0,T] \times \mathbb{R}^d \to \mathbb{R}^d$ is the drift term, $g : [0,T] \times \mathbb{R}^d \to \mathbb{R}^d$ is the diffusion term and $W_l(t)$ are independent Wiener processes. Assuming that f and g are continuous, have a linear growth and are uniform Lipschitz continuous with respect to the variable Y, that Y_0 has finite second order moment and is independent of the Wiener processes, one can show the existence and uniqueness of a (mean-square bounded) strong solution of (7) (see for example [31, Chap. 5.2] for details). Consider for the numerical approximation of (7) the one-step method of the form

$$Y_{n+1} = \Phi(Y_n, \Delta t, I_{n_1}, \ldots, I_{n_M}), \tag{8}$$

where $I_{n_l} = W_l(t_{n+1}) - W_l(t_n)$ are independent Wiener increments drawn from the normal distributions with zero mean and variance $\Delta t = t_{n+1} - t_n$. The numerical method (8) is said to have a strong order ρ, respectively weak order of ρ, if there exists a constant C such that

$$\mathbb{E}(|Y_n - Y(\tau)|) \leq C(\Delta t)^{\rho} \text{respectively} \left|\mathbb{E}(G(Y_n)) - \mathbb{E}(G(Y(\tau)))\right| \leq C(\Delta t)^{\rho}, \tag{9}$$

for any fixed $\tau = n\Delta t \in [0,T]$ (Δt sufficiently small) and for all functions $G : \mathbb{R}^d \to \mathbb{R}$ that are $2(\rho+1)$ times continuously differentiable with partial derivatives having polynomial growth.

Remark 2. In general, for numerical methods depending only on the first Wiener increment $W_l(t_{n+1}) - W_l(t_n)$ the highest strong and weak order that can be obtained are $1/2$ and 1, respectively. Strong order one can be obtained for 1-dimensional problems or if commutativity conditions hold for the diffusion functions g_l [12, 26, 34].

Stability. We have to investigate for what Δt does a numerical method $Y_{n+1} = \Phi(Y_n, \Delta t, I_{n_1}, \ldots, I_{n_M})$ applied to (7) share the stability properties of the exact solution Y_t. Widely used measures of stability for SDEs are mean-square stability, which measures the stability of moments, and asymptotic stability (in the large), which measures the overall behavior of sample functions [20]. We will focus here on mean-square stability. For linear autonomous system of SDEs, this concept of stability is stronger than asymptotic stability (see [9, Chap. 11]) or [20]). Consider the SDE (7) with $f(t,0) = g_l(t,0) = 0$ and with a nonrandom initial value Y_0. The steady solution $Y = 0$ of (7) is said to be mean-square stable if there exists δ_0 such that

$$\lim_{t\to\infty} \mathbb{E}\big(|Y(t)|^2\big) = 0, \quad \text{for all } |Y_0| < \delta_0. \tag{10}$$

In order to analyze the stability of numerical methods one has to restrict the class of problems considered. Inspired by (5) and following [22, 35] we consider the scalar linear test equation

$$dY = \lambda Y dt + \mu Y dW(t), \qquad Y(0) = Y_0, \tag{11}$$

where $\lambda, \mu \in \mathbb{C}$. For $\mu = 0$ one recovers the Dahlquist test equation, which is instrumental in developing the linear A-stability theory for ODEs [19, Chaps. 4.2, 4.3].

Remark 3. We note that for SDEs, it is at first not clear to which extend the study of a scalar linear test problem is relevant to systems of linear equations or fully nonlinear equations. Recent work, however, suggest that stability analysis for the scalar test equation is relevant for more general systems [10].

The test equation (11) can be solved analytically and the solution reads

$$Y(t) = Y_0\, e^{((\lambda - \frac{\mu^2}{2})t + \mu W(t))} \text{ (Itô)}, \quad Y(t) = Y_0\, e^{(\lambda t + \mu W(t))} \text{ (Stratonovich)}, \tag{12}$$

and we have for the mean-square stability

$$\lim_{t\to\infty} \mathbb{E}\big(|Y(t)|^2\big) = 0 \iff \begin{cases} \{(\lambda,\mu) \in \mathbb{C}^2;\ \Re\lambda + \frac{1}{2}|\mu|^2 < 0\} \text{ (Itô)}, \\ \{(\lambda,\mu) \in \mathbb{C}^2;\ \Re\lambda + (\Re\mu)^2 < 0\} \text{ (Stratonovich)}. \end{cases} \tag{13}$$

If we apply the Euler-Maruyama method (6) to (11) we obtain

$$\mathbb{E}(|Y_{n+1}|^2) = (|1+p|^2 + q^2)\mathbb{E}(|Y_n|^2), \tag{14}$$

where $p = \Delta t \lambda, q = \sqrt{\Delta t}\mu$ and thus, the method is mean-square stable if and only if $|1+p|^2 + q^2 < 1$. More generally, if we apply the numerical scheme (8) to the test problem (11), square the result and take the expectation, we obtain

$$\mathbb{E}(|Y_{n+1}|^2) = R(p,q)\mathbb{E}(|Y_n|^2), \tag{15}$$

where $p = \Delta t\lambda, q = \sqrt{\Delta t}\mu$ and where $R(p,q)$ is a function in $\Re(p)$, $\Im(p)$, $\Re(q)$, $\Im(q)$ (a polynomial in these variables if the method is explicit). We say that a numerical method is mean-square stable for the test problem (11) if and only if

$$\lim_{n\to\infty} \mathbb{E}\big(|Y_n|^2\big) = 0 \iff (\Delta t\lambda, \sqrt{\Delta t}\mu) \in \mathscr{S} := \{p,q \in \mathbb{C};\ R(p,q) < 1\}. \tag{16}$$

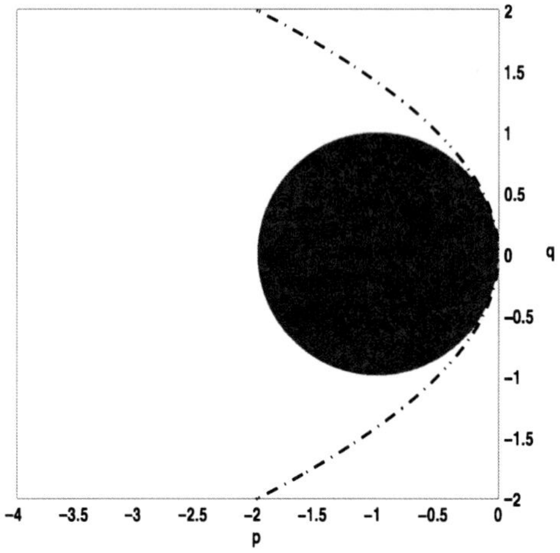

Fig. 2 Stability domain of the Euler-Maruyama method (*black disk*) for $\lambda, \mu \in \mathbb{R}$. The dashed curve represent the boundary of the exact stability domain (the left part of the curve lies in the stability domain)

In order to be able to visualize the stability region, we restrict ourself to the case $\lambda, \mu \in \mathbb{R}$. We see in Fig. 2 that the stability domain of the Euler-Maruyama method is a disk of radius 1 centered at $p = -1$, while the stability domain of the exact test problem is the unbounded region on the left of the dashed curve. The Euler-Maruyama has thus a restricted stability region. For the problem (3) (see also (5)) this explicit method will thus face a severe time step restriction due to stability constraint (see Fig. 7 in Sect. 6). One could use semi-implicit methods (implicit method in the drift term) to obtain method with much better stability properties. This comes however with the cost of solving nonlinear equations at each stepsize. This can be numerically expensive for large systems (see e.g. (4)), specially if one needs to simulate many realizations. We will explain in the next section how mean-square stability can be improved without giving up the explicitness of the numerical method.

3 Chebyshev Methods

Chebyshev methods are a class of explicit one-step methods with extended stability domains along the negative real axis. The basic idea for such methods goes back to the 1960s with Saul'ev, Franklin and Guillou and Lago (see [19, Sect. IV.2] and the references therein). It can be summarized as follows: consider a sequence of forward Euler methods $\Psi_{h_1},\ldots,\Psi_{h_m}$ with a corresponding sequence of timesteps $h_1,\ldots,h_m$ and define a one-step method as the composition $\Psi_{\Delta t} = (\Psi_{h_m} \circ \ldots \circ \Psi_{h_1})(y_0)$ with stepsize $\Delta t = h_1 + \ldots + h_m$. Next, given m, optimize the sequence $\{h_i\}_{i=1}^m$, so that

$$|R_m(x)| = \left|\prod_{i=1}^{m}\left(1+\frac{h_i x}{\Delta t}\right)\right| \leq 1 \quad \text{for } x \in [-l_m, 0],$$

with $l_m > 0$ as large as possible. The resulting numerical method will thus be a *m-stage method*. The solution of the above optimization problem is given by shifted Chebyshev polynomials

$$R_m(x) = T_m(1 + x/m^2) = 1 + x + a_2 x^2 + \cdots + a_m x^m,$$

where $\{T_j(x)\}_{j\geq 0}$ are the Chebyshev polynomials given recursively by

$$T_0(x) = 1,\ T_1(x) = x,$$

and

$$T_j(x) = 2xT_{j-1}(x) - T_{j-2}(x), \quad j \geq 2.$$

We see that the optimal sequence of $\{h_i\}_{i=1}^m$ is given by $h_i = (-1/x_i)\Delta t$, where x_i are the zeros of $R_m(x)$ and the maximal stability domain on the negative real axis increases *quadratically* with the number of stages m and is given by $l_m = 2m^2$. The property $R_m(z) = 1 + x + \mathcal{O}(x^2)$ ensure the first order convergence of the numerical method. Besides the stability of the "super stepsize" Δt, one has also to care about the internal stability (accumulation of errors within one step) of the method as m can be large. This can be achieved either by a proper ordering of the Euler steps h_i [27] or by exploiting the three-term recurrence relation of the orthogonal polynomials [24]. Following the second strategy we consider a m-stage numerical method given by

$$\begin{aligned}
k_0 &:= y_0\\
k_1 &:= y_0 + \frac{\Delta t}{m^2} f(k_0)\\
k_j &:= \frac{2\Delta t}{m^2} f(k_{j-1}) + 2k_{j-1} - k_{j-2}, \quad 2 \leq j \leq m\\
y_1 &:= k_m.
\end{aligned} \tag{17}$$

Applied to the test problem $y' = \lambda y$, this method gives for the internal stages

$$k_j = T_j\left(1 + \Delta t \lambda / m^2\right) y_0, \qquad j = 0,\ldots,m, \tag{18}$$

and produces after one step $y_1 = R_m(\Delta t\lambda)y_0$, where $R_m(x) = T_m(1+x/m^2)$, is the shifted Chebyshev polynomial of degree m ($x = \Delta t\lambda$).

These methods have been originally developed for deterministic problems with eigenvalues along the negative real axis. A typical (deterministic) stability domain $\mathcal{S}_m$ of a Chebyshev method is sketched in Fig. 3 (left figure), where

$$\mathcal{S}_m := \{z \in \mathbb{C}; |R_m(z)| < 1\}.$$

Recall that for the linear stability of deterministic ODE solvers, one considers (11) with $\mu = 0$ [19, Chaps. 4.2, 4.3]. It can be seen in Fig. 3 that the boundary of the

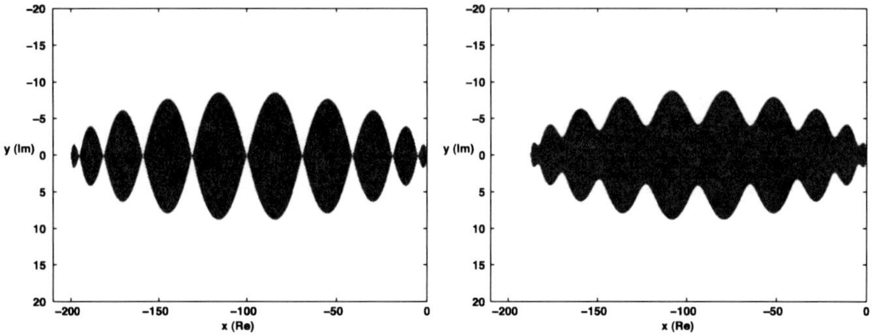

Fig. 3 Stability domain of first order Chebyshev method (degree $m = 10$) with variable damping $\eta = 0$ (*left figure*), $\eta = 0.1$ (*right figure*)

stability domain along the negative real axis is 200, for $m = 10$. However, there are regions in $[0, 200]$, precisely when $T(1 + x/m^2) = 1$, with no stability in the direction of the imaginary axis.

To overcome the aforementioned issue, it has been suggested by Guillou and Lago [17] to replace the requirement $|R_m(x)| \leq 1$ in $[-l_m, 0]$ by $|R_m(x)| \leq \eta < 1$ in $[-l_m^\eta, -\epsilon]$, where ϵ is a small positive number. The number η is called the damping parameter or sometimes just the "damping". This can done for the polynomials $T_m(1+x/m^2)$ by a division with $T_m(\omega_0) > 1$, where $\omega_0 = 1 + \eta/m^2$. To obtain the right order of accuracy with this modified stability function, one does a change of variables and obtains $R_{m,\eta}(x) = T_m(\omega_0 + \omega_1 x)/T_m(\omega_0)$, where $\omega_1 = T_m(\omega_0)/T'_m(\omega_0)$ (see [19, Sect. IV.2]). By increasing the parameter η the strip around the negative real axis included in the stability domain can be enlarged as can be seen in Fig. 3 (notice that this reduces the value of l_m as $l_m^\eta < l_m$ for $\eta > 0$). The formula (18) can be modified appropriately to incorporate damping.

Higher order quasi-optimal Chebyshev methods: the ROCK methods. Higher order methods, called ROCK, for orthogonal Runge-Kutta Chebyshev methods, based on orthogonal polynomials have been developed in [2, 3]. The stability functions are given by polynomials $R_m(x) = 1 + x + \ldots + x^p/p! + \mathcal{O}(x^{p+1})$ of order p (i.e., $R_m(x) - e^x = \mathcal{O}(x^{p+1})$) and degree m with quasi optimal stability domains along the negative real axis. These polynomials can be decomposed as [1]

$$R_m(x) = w_p(x)P_{m-p}(x),$$

where $P_{m-p}(x)$ is a member of a family of polynomials $\{P_j(x)\}_{j\geq 0}$ orthogonal with respect to the weight function $(1-x^2)^{-1/2}w_p(x)^2$. (The function $w_p(x)$ is a polynomial of degree p with only complex zeros when p is even and with only one real zero when p is odd.[3]) The idea for the construction of a numerical method is then as follows: the 3-term recurrence relation of the orthogonal polynomials $\{P_j(x)\}_{j\geq 0}$

$$P_j(x) = (\alpha_j x - \beta_j)P_{j-1}(x) - \gamma_j P_{j-2}(x),$$

is used to define the internal stages of the method

$$K_j = \Delta t \alpha_j f(K_{j-1}) - \beta_j K_{j-1} - \gamma_j K_{j-2}, \qquad j = 2, \ldots, m-p.$$

This ensures the good stability properties of the method. A p-stage finishing procedure with the polynomial $w_p(z)$ as underlying stability function ensures the right order of accuracy of the method.

Gain in efficiency. Assume that Δt is the stepsize corresponding to the desired accuracy to solve an initial value problem $y' = f(t, y)$ in the interval $[0, T]$. Let ρ be the spectral radius of the Jacobian $\partial_y f$. A standard explicit method, as the Euler method, must satisfy $\delta t = C/\rho$ (for stability) and thus needs $\Delta t\rho/C$ function evaluations in each interval Δt. For a Chebyshev method, we can select a stage number $m = \sqrt{\Delta t\rho/C}$. As the number of function evaluations is equal to the stage number of the Chebyshev method[4] only the square root of the function evaluations needed for standard explicit method are required for each stepsize (notice that the constant C can be different for the two methods but is in both cases of moderate size).

4 The S-ROCK Methods

We now present the Stratonovich and the Itô stochastic ROCK (S-ROCK) methods derived in [5–7]. When modeling physical systems with SDEs, the question of the choice of the stochastic integral arises. SDEs with Stratonovich integrals are stable with respect to changes in random terms and are often used for systems where the noise is "added" as fluctuation of a deterministic system. SDEs with Itô integrals are preferred for systems with internal noise where the fluctuation is due to the systems itself as for example in chemical reactions due to the property of "not looking into the future" of the Itô integral (i.e., the martingale property)

---

[3] The ROCK methods have been developed for p even ($p = 2, 4$). They could be obtained for p odd provided a proper treatment of the real zero of $w_p(x)$.

[4] Strictly speaking this is true for first order Chebyshev methods. For higher order methods, as the ROCK methods, the number of function evaluations is not equal but still close to the stage number m (see [2, 3]).

[26, 31]. Of course, there are conversion rules from one calculus to the other. However, these rules involve the differentiation of the diffusion term which can be cumbersome and costly. It is thus preferable to derive genuine formulas for both calculus. Furthermore, it is sometimes desirable to have stabilized explicit methods for discrete noise. This has been considered in [8], where the τ-ROCK methods have been developed and we briefly comment on these methods as well in what follows.

4.1 Construction of the S-ROCK Methods

Inspired by the ROCK methods, we consider methods based on:

- Deterministic Chebyshev-like internal stages to ensure good stability properties (stages $1,2,\ldots,m-1$).
- A finishing stochastic procedure to incorporate the random process and obtain the desired stochastic convergence properties.

As for deterministic methods, the use of damping plays a crucial role and allows to enlarge the width of the stability domains in the direction of the "stochastic axis" (e.g, the q axis in Fig. 2). This is discussed in Sect. 5.

Deterministic Chebyshev stages. Define the $m-1$ stages of the S-ROCK method by

$$\begin{aligned}
K_0 &= Y_n,\\
K_1 &= Y_n + \Delta t \frac{\omega_1}{\omega_0} f(K_0),\\
K_j &= 2\Delta t \omega_1 \frac{T_{j-1}(\omega_0)}{T_j(\omega_0)} f(K_{j-1}) + 2\omega_0 \frac{T_{j-1}(\omega_0)}{T_j(\omega_0)} K_{j-1} - \frac{T_{j-2}(\omega_0)}{T_j(\omega_0)} K_{j-2},
\end{aligned}$$

for $j=2,\ldots,m-1$, where $\omega_0 = 1+\eta/m^2$ and $\omega_1 = T_m(\omega_0)/T_m'(\omega_0)$. Recall that η is the damping parameter which will be optimized (see Sect. 5).

Stochastic stages. We have now to incorporate the noise in an appropriate way. While the deterministic stages are the same for the various S-ROCK methods, the finishing procedure will be different to take into account the various stochastic calculus of the underlying SDE and the desired accuracy of the methods.

Itô S-ROCK methods (multi-dimensional SDEs). We define the finishing procedure as

$$\begin{aligned}
K_m &= 2\Delta t \omega_1 \frac{T_{m-1}(\omega_0)}{T_m(\omega_0)} f(K_{m-1}) + 2\omega_0 \frac{T_{m-1}(\omega_0)}{T_m(\omega_0)} K_{m-1} - \frac{T_{m-2}(\omega_0)}{T_m(\omega_0)} K_{m-2}\\
&\quad + \sum_{l=1}^{M} I_{n_l} g_l(K_{m-1}),\\
Y_{n+1} &= K_m.
\end{aligned} \tag{19}$$

Itô S-ROCK methods (commutative noise[5] or one dimensional Wiener process). In that special case, one can improve the strong convergence of the method by considering the finishing procedure

$$
\begin{aligned}
K_{m-1}^{*} &= K_{m-1} + \sum_{r=1}^{M} g_r(K_{m-1}) I_{n_r},\\
K_{m-1}^{**,l} &= K_{m-1} + \sqrt{\Delta t}\, g_l(K_{m-1}), \quad l = 1,2,\ldots,M,\\
K_m &= 2\Delta t \omega_1 \frac{T_{m-1}(\omega_0)}{T_m(\omega_0)} f(K_{m-1}) + 2\omega_0 \frac{T_{m-1}(\omega_0)}{T_m(\omega_0)} K_{m-1} - \frac{T_{m-2}(\omega_0)}{T_m(\omega_0)} K_{m-2}\\
&\quad + \sum_{l=1}^{M} I_{n_l} g_l(K_{m-1}) + \frac{1}{2}\sum_{l=1}^{M} I_{n_l}\Big(g_l(K_{m-1}^{*}) - g_l(K_{m-1})\Big)\\
&\quad - \frac{1}{2}\sum_{l=1}^{M} \sqrt{\Delta t}\Big(g_l(K_{m-1}^{**,l}) - g_l(K_{m-1})\Big),\\
Y_{n+1} &= K_m.
\end{aligned} \tag{20}
$$

Remark 4. For $M = 1$ the above formula can be further simplified and written as

$$
\begin{aligned}
K_{m-1}^{*} &= K_{m-1} + \sqrt{\Delta t}\, g(K_{m-1}),\\
K_m &= 2\Delta t \omega_1 \frac{T_{m-1}(\omega_0)}{T_m(\omega_0)} f(K_{m-1}) + 2\omega_0 \frac{T_{m-1}(\omega_0)}{T_m(\omega_0)} K_{m-1} - \frac{T_{m-2}(\omega_0)}{T_m(\omega_0)} K_{m-2}\\
&\quad + I_n g(K_{m-1}) + \frac{I_n^2 - \Delta t}{2\sqrt{\Delta t}}\big(g(K_{m-1}^{*}) - g(K_{m-1})\big),\\
Y_{n+1} &= K_m.
\end{aligned} \tag{21}
$$

Stratonovich S-ROCK methods (multi-dimensional SDEs). We define the finishing procedure as

$$
\begin{aligned}
K_{m-1}^{*} &= K_{m-1} + \frac{T_m(\omega_0)}{2\omega_0 T_{m-1}(\omega_0)} \sum_{l=1}^{M} I_{n_l} g_l(K_{m-2}),\\
K_m &= 2\Delta t \omega_1 \frac{T_{m-1}(\omega_0)}{T_m(\omega_0)} f(K_{m-1}) + 2\omega_0 \frac{T_{m-1}(\omega_0)}{T_m(\omega_0)} K_{m-1} - \frac{T_{m-2}(\omega_0)}{T_m(\omega_0)} K_{m-2}\\
&\quad + \frac{\omega_0 T_{m-1}(\omega_0)}{T_m(\omega_0)} \sum_{l=1}^{M} I_{n_l}\big(g_l(K_{m-1}) - g_l(K_{m-2})\big),\\
Y_{n+1} &= K_m.
\end{aligned} \tag{22}
$$

Notice that this method has order one when solving SDEs with commutative noise or with only one Wiener process [5].

[5] Consider $L^l = \sum_{k=1}^{d} g_l^k \frac{\partial}{\partial y^k}$, $l = 1,2,\ldots,M$. Commutative noise means that the condition $L^l g_r^k = L^r g_l^k\ \forall l,r = 1,\ldots,M;\ k = 1,\ldots,d$ holds for the diffusion functions [26].

S-ROCK methods for discrete noise. The procedure explained above can be generalized to stochastic problems with other types of noise. In [8], the approximation of SDE for chemical kinetic systems has been considered. The SDE is of the form[6]

$$dY_t = \sum_{j=1}^{M} \nu_j \mathscr{P}(a_j(Y_{t}-)dt),$$

where Y_t is a $N-$dimensional state vector (corresponding to the N species of the reaction) with components in $\mathbb{N}$, ν_j is a state-change vector, a_j is a propensity function (the number of possible combination of reactant molecules involved in the jth reaction, times a stochastic reaction rate constant) and $\mathscr{P}(a_j(Y_t-)dt)$ is a state-dependent Poisson noise. We now make the decomposition

$$\begin{aligned} dY_t &= \sum_{j=1}^{M} \nu_j a_j(Y_{t-})dt + \sum_{j=1}^{M} \nu_j \Big(\mathscr{P}(a_j(Y_{t-})dt) - a_j(Y_{t-})dt \Big) \\ &= f(Y_{t-})dt + dQ_t, \end{aligned} \tag{23}$$

where f and Q are called the drift part and jump part, respectively (see [29]). This form is similar with SDEs driven by Wiener processes, except for the different noise. Similarly as for the Itô or the Stratonovich S-ROCK methods, the $m-1$ deterministic Chebyshev stages can be applied to the drift part of (23), and the noise term can be incorporated in the finishing procedure in an appropriate way to solve (23) (we refer to [8] for details).

4.2 Accuracy of the S-ROCK Methods

Before considering the stability properties of our methods (the main motivation to consider the formulas introduced in Sect. 4.1) we briefly discuss their accuracy. As mentioned in Sect. 1, by considering numerical methods depending only on the first Wiener increment, strong accuracy higher than $\rho = 1/2$ or weak accuracy higher than $\rho = 1$ cannot be obtained. Only in the special case of commutative, diagonal or one dimensional noise, strong order $\rho = 1$ is possible. The theorems below show that the S-ROCK methods enjoy the highest possible accuracy for numerical methods involving only the first Wiener increment.

Theorem 1 ([5–7]). *For $m \geq 2$, the methods* (19) *(Itô) and* (22) *(Stratonovich) applied to* (7) *(with f and g_l sufficiently smooth) satisfy*

$$\mathbb{E}(|Y_N - Y(\tau)|) \leq C(\Delta t)^{1/2}, \qquad |\mathbb{E}(G(Y_N)) - \mathbb{E}(G(Y(\tau))| \leq C\Delta t \tag{24}$$

for any fixed $\tau = N\Delta t \in [0, T]$ and Δt sufficiently small and for all functions $G : \mathbb{R}^d \to \mathbb{R}$, 4 times continuously differentiable and for which all partial derivatives have polynomial growth.

[6] See [29] for a rigorous description of the problem.

Theorem 2 ([5–7]). *Assume that* (7) *(with f and g_l sufficiently smooth) has commutative noise or that $M = 1$. Then, for $m \geq 2$, the methods* (20),(21) *(Itô) and* (22) *(Stratonovich) applied to* (7) *(with f and g_l sufficiently smooth) satisfy*

$$\mathbb{E}(|Y_N - Y(\tau)|) \leq C\Delta t \tag{25}$$

for any fixed $\tau = N\Delta t \in [0,T]$ and Δt sufficiently small.

For the proofs of these theorems we refer to [5,6] (Stratonovich S-ROCK methods) and [7] (Itô S-ROCK methods).

5 Extended Mean-Square Stability and Damping

We study here the mean-square stability property of the S-ROCK methods. By applying any of the methods (19),(20),(21) or (22) to the scalar test problem (11), squaring the results and taking the expectation we obtain the mean-square stability function (see (15))

$$R_m(p,q) = \frac{T_m^2(\omega_0 + \omega_1 p)}{T_m^2(\omega_0)} + Q_{m-1,r}(p,q), \tag{26}$$

where $Q_{m-1,r}(p,q)$ is a polynomial of degree $2(m-1)$ in p and of degree $2r$ in q. The precise form of $Q_{m-1,r}(p,q)$ depends on the specific numerical method considered. Define $\Theta_j = \frac{T_j(\omega_0+\omega_1 p)}{T_j(\omega_0)}$. For the method (19) we have $r = 1$ and

$$Q_{m-1,1}(p,q) = q^2\Theta_{m-1}.$$

For the method (21) $r = 2$ and

$$Q_{m-1,2}(p,q) = q^2\Theta_{m-1} + \frac{q^4}{2}\Theta_{m-1}.$$

Finally, $r = 2$ for the method (22) and

$$Q_{m-1,2}(p,q) = q^2\left(\Theta_m\Theta_{m-2} + \left[\Theta_{m-2}\left(\frac{\omega_1}{\omega_0}p + 1\right)\right.\right.$$
$$\left.\left. + \omega_0\frac{T_{m-1}(\omega_0)}{T_m(\omega_0)}(\Theta_{m-1} - \Theta_{m-2})\right]^2\right) + \frac{3}{4}q^4\Theta_{m-2}^2.$$

In Fig. 4, we plot the mean-square stability domains for the method (19) with various values of damping for $m = 5$. We observe that without damping, the stability along the p axis (the "deterministic axis") is optimal (i.e., $2\cdot 5^2$). But there are points (close to the p axis) with no stability in the direction of the q axis (see Fig. 4, (left)). As $Q_{m-1,r}(p,0) =$ these points are exactly the points where $T_m^2(1 + p/m^2) = 1$. For

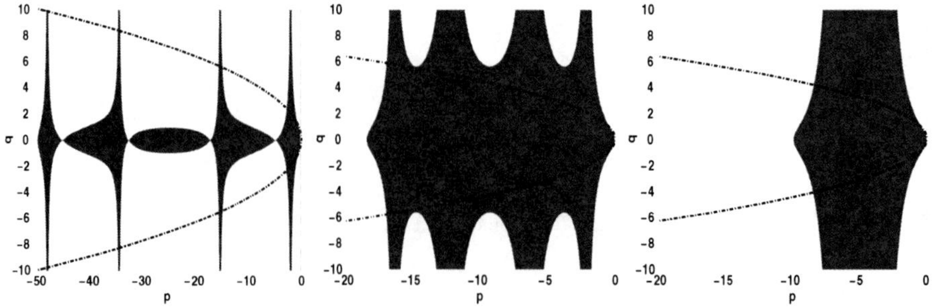

Fig. 4 Mean-square stability regions for the method (19) with various values of damping ($m = 5$). *Left figure* (no damping, $\eta = 0$), *middle figure* (optimal damping, $\eta = 4.7$), *right figure* (infinite damping)

infinite (or very large) damping the mean-square stability domain covers a portion of the stability domain of the test equation (11), but the stability domain along the p axis becomes linear in m (i.e., $2 \cdot 5$, see Fig. 4 (right)). The mean-square stability domain for what will be called the optimal damping value covers a "large" portion of the stability domain of the test equation (see Fig. 4, (middle)). In order to quantify these observations we define a "portion" of the stability domain (13) by

$$\mathcal{S}_{\text{SDE},s} = \{(p,q) \in [-s,0] \times \mathbb{R}; |q| \leq \sqrt{-p}\} \text{ (Stratonovich)}, \tag{27}$$

or

$$\mathcal{S}_{\text{SDE},s} = \{(p,q) \in [-s,0] \times \mathbb{R}; |q| \leq \sqrt{-2p}\} \text{ (Itô)}, \tag{28}$$

where $s > 0$. We then consider two parameters l and d related to a numerical stability domain $\mathcal{S}$ by

$$l = \max\{|p|; p < 0, [p,0] \subset \mathcal{S}\}, \qquad d = \max\{r > 0; \mathcal{S}_{\text{SDE},s} \subset \mathcal{S}\}. \tag{29}$$

Clearly, $d \leq l$, and for mean-square stability, it is the parameter d which has to be optimized. For the S-ROCK methods, as can be seen in Fig. 4, l and d depend on the stage number m and the value of the damping parameter η. We thus denote these parameters by $l_m(\eta)$ and $d_m(\eta)$. The following lemmas give important information on the value of $l_m(\eta)$ and a bound of the possible values for $d_m(\eta)$, the parameter which characterizes the stability domains of our methods.

Lemma 1 ([6,7]). *Let $\eta \geq 0$. For all $m \geq 2$, the m-stage numerical method (22) has a mean-square stability region $\mathcal{S}_m^\eta$ with $l_m(\eta) \geq c(\eta)m^2$, where $c(\eta)$ depends only on η.*

Lemma 2 ([6,7]). *For all $m \geq 2$*

$$l_m(\eta) \to 2m \quad \textit{for} \quad \eta \to \infty. \tag{30}$$

In view of the above two lemmas we make the following important observation: for any fixed η, the stability domain along the p axis increases quadratically

(Lemma 1), but for a given method, i.e., a fixed m, increasing the damping η to infinity reduces the quadratic growth along the p axis into a linear growth (Lemma 2). Since $d_m(\eta) \le l_m(\eta)$ there is no computational saving compared to classical explicit methods for this limit case.

Optimized methods. Our goal is now for a given method to find the value of η, denoted η^* which maximize $d_m(\eta)$, i.e.,

$$\eta^* = \text{argmax}\{d_m(\eta); \eta \in [0, \infty)\}. \quad (31)$$

The corresponding optimal values $d_m(\eta^*)$ for $m \le 200$ have been computed numerically and are reported in Fig. 5 for the Itô S-ROCK methods (19) and in Fig. 6 for the Stratonovich S-ROCK methods (22). We also report in the same figures the values of $l_m(\eta^*)$ and η^*. We see that for $\eta = \eta^*$, $d_m(\eta^*) \simeq l_m(\eta^*)$. The dashed and the dash-dotted lines in the plots reporting the values of $d_m(\eta^*)$, represent a quadratic and a linear slope, respectively. We clearly see that the portion of the true

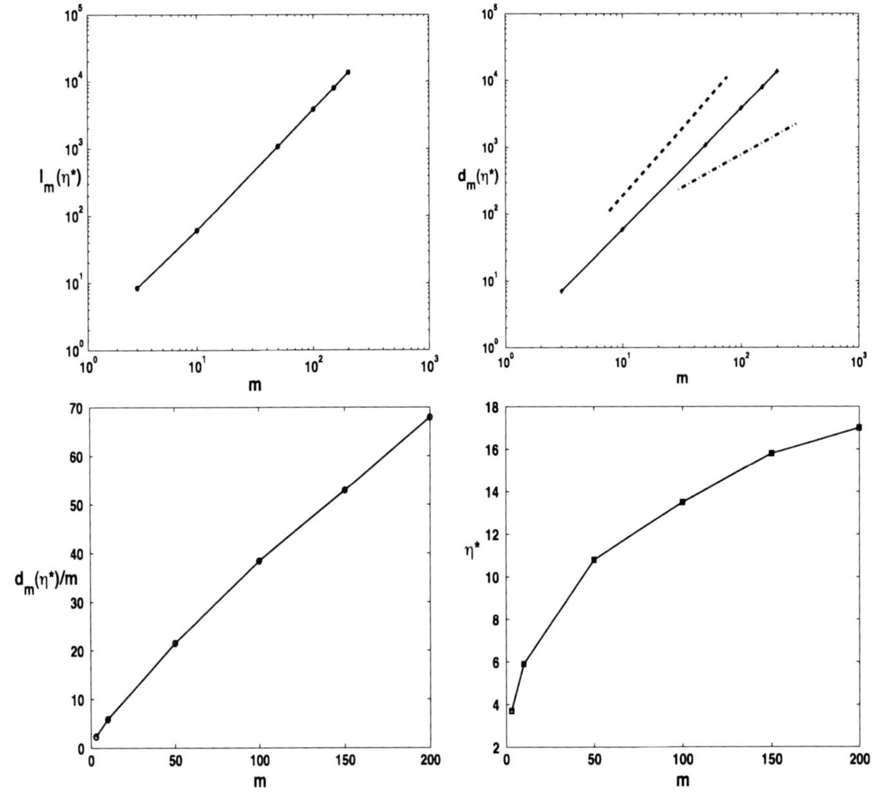

Fig. 5 Values of $\eta^*, l^{\eta^*}, d^{\eta^*}$ as a function of m and the ratio $d_m^{\eta^*}/m$ (stability versus work) for the Itô S-ROCK methods (19). The *dashed* and the *dash-dotted lines* in the upper-right figure represent a quadratic and a linear slope, respectively

stability domain included in the stability domain of our numerical methods grows super-linearly (close to quadratically) for both the Itô and the Stratonovich S-ROCK methods. Finally we study the efficiency of the methods by reporting the quantity

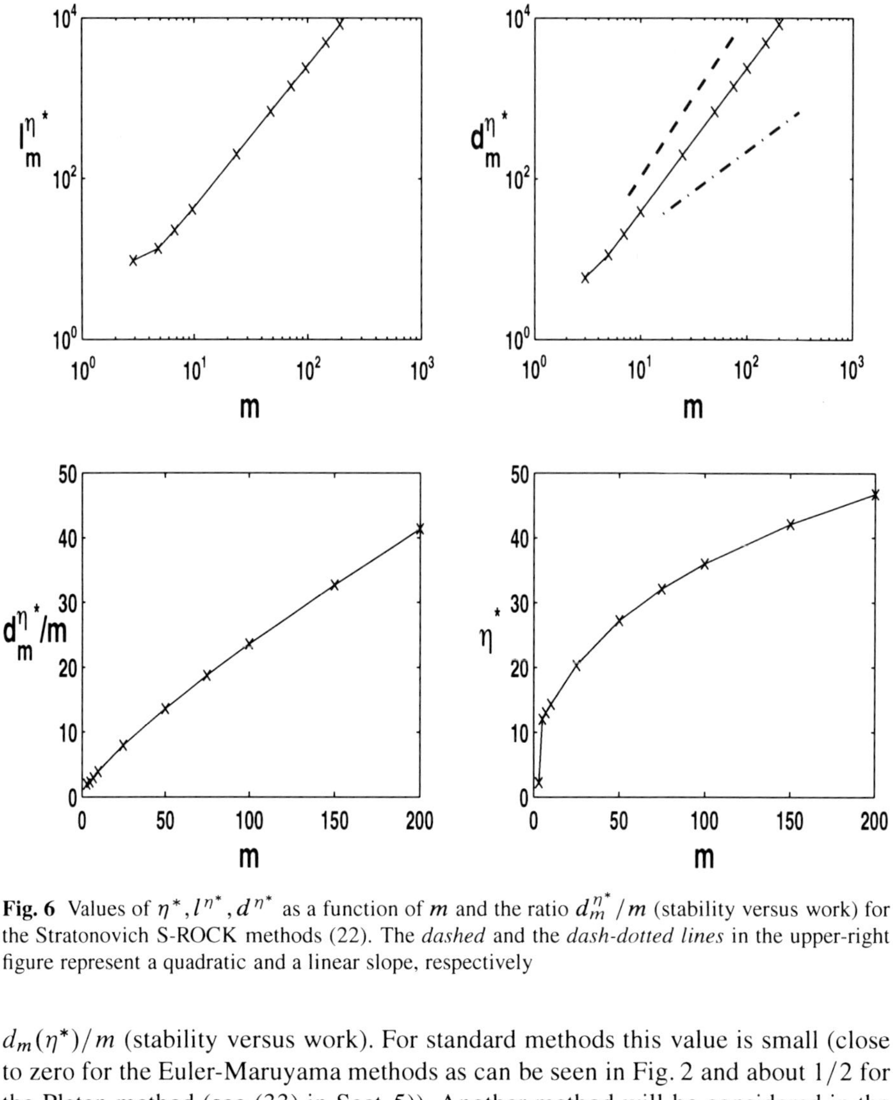

Fig. 6 Values of $\eta^*, l^{\eta^*}, d^{\eta^*}$ as a function of m and the ratio $d_m^{\eta^*}/m$ (stability versus work) for the Stratonovich S-ROCK methods (22). The *dashed* and the *dash-dotted lines* in the upper-right figure represent a quadratic and a linear slope, respectively

$d_m(\eta^*)/m$ (stability versus work). For standard methods this value is small (close to zero for the Euler-Maruyama methods as can be seen in Fig. 2 and about $1/2$ for the Platen method (see (33) in Sect. 5)). Another method will be considered in the numerical experiments, namely the RS method [11, p. 187] developed with the aim of improving the mean-square stability of the Platen method. This method has a larger l value than the Platen method but a smaller d value and the efficiency of this method (as measured here) is about 0.3. We see that S-ROCK methods are orders of magnitude more efficient (for the aforementioned criterion of efficiency related to stability) than standard explicit methods for SDEs.

6 Numerical Illustrations

In this section we illustrate the efficiency of the S-ROCK methods. As mentioned in the beginning of Sect. 4, different applications require different stochastic integrals and we will consider both Itô and Stratonovich SDEs in the following examples. The first example is the heat equation with noise mentioned in the introduction. For this problem we consider the Stratonovich S-ROCK methods. The second example is a chemical reaction modeled by the chemical Langevin equation. The Itô S-ROCK methods will be used for this latter problem. For both examples, we compare the S-ROCK methods with standard explicit methods.

Example 1: heat equation with noise. We consider the SPDE (3), where we choose this time the Stratonovich modeling for the noise. We follow the procedure explained in Sect. 2 and transform the SPDE in a large system of SDEs

$$dY_t^i = \frac{Y_t^{i+1} - 2Y_t^i + Y_t^{i-1}}{h^2} + \mu Y_t^i \circ dW_t, \qquad i = 1, \ldots, N, \tag{32}$$

where the symbol $\circ$ denotes the Stratonovich form for the stochastic integral. In our numerical experiments, we compare the Stratonovich S-ROCK methods (22) with two other methods, the method introduced by Platen [33] (denoted PL) given by the two-stage scheme

$$\begin{aligned} K_n &= Y_n + \Delta t f(Y_n) + I_n g(Y_n), \\ Y_{n+1} &= Y_n + \Delta t f(Y_n) + I_n \frac{1}{2}(g(Y_n) + g(K_n)), \end{aligned} \tag{33}$$

and the RS method, introduced by P.M. Burrage [11, p. 187]. This is a 2-stage method constructed with the aim of improving the mean-square stability properties of the Platen method and is given by

$$\begin{aligned} K_n &= Y_n + \frac{4}{9}\Delta t f(Y_n) + \frac{2}{3} J_n g(Y_n), \\ Y_{n+1} &= Y_n + \frac{\Delta t}{2}(f(Y_n) + f(K_n) + \frac{1}{4}(g(Y_n) + g(K_n))J_n. \end{aligned} \tag{34}$$

Both methods have strong order 1 for one-dimensional systems or systems with commutative noise as (4). This is also the case for the Stratonovich S-ROCK methods (22). We have seen, at the end of Sect. 5, that the stability domains of both methods, PL and RS, cover only a small portion of the stability domain corresponding to the stochastic test equation and this is in contrast with the S-ROCK methods. In Fig. 7 we monitor the number of function evaluations (cost)[7] needed by the various methods to produce stable integrations when increasing the value of N,

[7] By number of function evaluations we mean here the total number of drift and diffusion evaluations.

i.e., the stiffness of the problem. For the S-ROCK methods we vary the number of stages to meet the stability requirement (this value is indicated in Fig. 7).

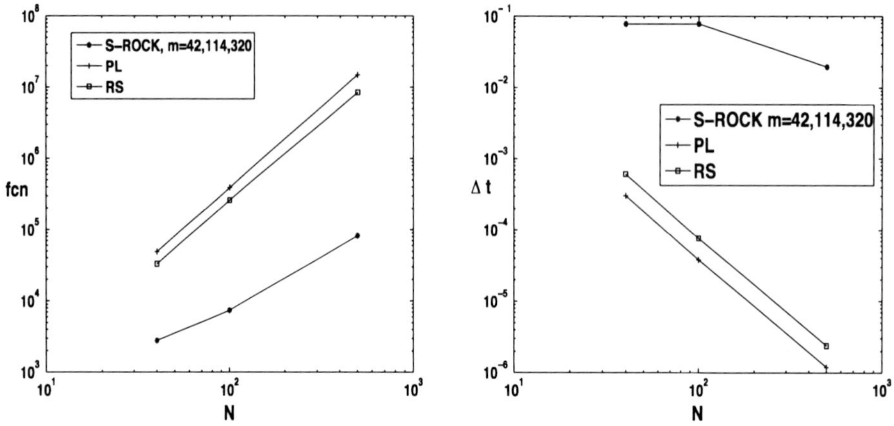

Fig. 7 Function evaluations and stepsize as a function of N. For PL and RS, we choose the largest stepsize to have a stable integration of (32) (strong error $< 10^{-1}$). For the S-ROCK methods, we can vary the stage number m to meet the stability requirement (we fixed the highest stage number at $m = 320$)

We see that the S-ROCK method reduces the computational cost by several orders of magnitude as the stiffness increases. In the same figure we see the value of the stepsize needed for the different methods, again as a function of N. As expected, the standard explicit methods, as PL or RS face severe stepsize restriction as the stiffness increases. This example demonstrates that for classes of SPDEs there is a real advantage in using explicit stabilized methods such as S-ROCK methods. We notice that the stepsize is reduced for the highest value of N for the S-ROCK methods (see Fig. 7 (right)). We could have kept the same stepsize but the stage number would then have become quite large. It is well-known for Chebyshev methods that in order to control the internal stability of the method one should avoid computation with a very high stage number [3]. Here we fixed the highest stage number at $m = 320$.

Example 2: a chemical reaction. We know illustrate the use of the Itô S-ROCK methods. Following [7] we consider a stiff system of chemical reactions given by the Chemical Langevin Equation (CLE). We study the Michaelis-Menten system, describing the kinetics of many enzymes. This system has been studied in [23] with various stochastic simulation techniques. The reactions involve four species: S_1 (a substrate), S_2 (an enzyme), S_3 (an enzyme substrate complex), and S_4 (a product) and can be described as follows: the enzyme binds to the substrate to form an enzyme-substrate complex which is then transformed into the product, i.e.,

$$S_1 + S_2 \xrightarrow{c_1} S_3 \tag{35}$$

$$S_3 \xrightarrow{c_2} S_1 + S_2 \tag{36}$$

$$S_3 \xrightarrow{c_3} S_2 + S_4. \tag{37}$$

The mathematical description of this kinetic process can be found in [25]. For the simulation of this set of reactions we use the CLE model

$$dY(t) = \sum_{j=1}^{3} \nu_j a_j(Y(t))dt + \sum_{j=1}^{3} \nu_j \sqrt{a_j(Y(t))}dW_j(t), \tag{38}$$

where $Y(t)$ is a 4 dimensional vector describing the state of each species $S_1, \ldots, S_4$. The Itô form used in (38). The functions $a_j(Y(t))$, called the propensity functions, give the number of possible combinations of molecules involved in each reaction j. For the above system they are given by

$$a_1(Y(t)) = c_1 Y_1 Y_2, \quad a_2(Y(t)) = c_2 Y_3, \quad a_3(Y(t)) = c_3 Y_3.$$

The vectors ν_j, called the state-change vectors, describe the change in the number of molecules in the system when a reaction fires. They are given for the three reactions of the above system by $\nu_1 = (-1,-1,1,0)^T$, $\nu_2 = (1,1,-1,0)^T$, $\nu_3 = (0,1,-1,1)^T$. We set the initial amount of species as (the parameters are borrowed from [39, Sect. 7.3])

$$Y_1(0) = [5 \times 10^{-7} n_A vol], \quad Y_2(0) = [5 \times 10^{-7} n_A vol], \quad Y_3(0) = 0, \quad Y_4(0) = 0,$$

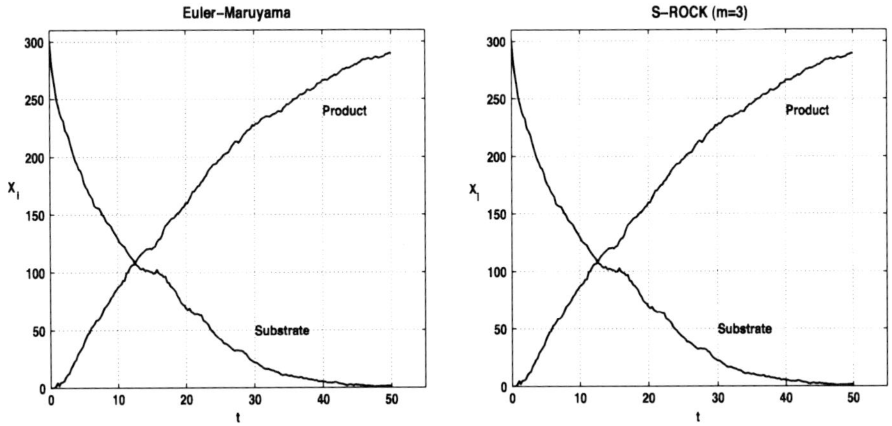

Fig. 8 One trajectory of the Michaelis-Menten system solved with the Euler-Maruyama method (*left figure*) and the S-ROCK method (*right figure*) for $c_1 = 1.66 \times 10^{-3}, c_2 = 10^{-4}, c_3 = 0.10$ (the stepsize is $\Delta t = 0.25$ and the same Brownian path is used for both methods; $m = 3$ for the S-ROCK method)

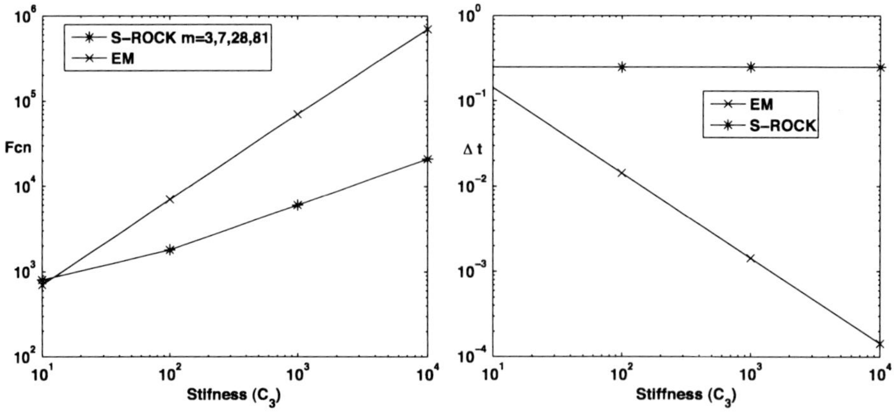

Fig. 9 Numerical solution of (38) with the Euler-Maruyama and the S-ROCK methods. Number of function evaluations as a function of c_3 for both methods (*left figure*). Size of the timestep Δt as a function of c_3 (Euler-Maruyama); $\Delta t = 0.25$ for the S-ROCK method and the stage number m is adapted to the stiffness (*right figure*)

where $[\,\cdot\,]$ denotes the rounding to the next integer and $n_A = 6.023 \times 10^{23}$ is the Avagadro's constant (number of molecules per mole) and vol is the volume of the system.

In the following numerical experiments, we solve numerically the SDE (38) with the Itô S-ROCK methods and the Euler-Maruyama method (6). This latter method is often used for solving the CLE. As the CLE has multidimensional Wiener processes, we use the S-ROCK methods (19). We first compare the solutions along time for the two methods ($t \in [0, 50]$), with parameters leading to a non-stiff system for (38). As expected, we observe in Fig. 8 a very similar behavior of the two methods.

We next increase the rate of the third reaction in (35)–(37), $c_3 = 10^2, 10^3, 10^4$ corresponding to an increasingly fast production. The resulting CLE becomes stiff and the Euler-Maruyama method is inefficient. In Fig. 9 we report the stepsizes and the number of function evaluations needed for the Euler-Maruyama and the S-ROCK methods. The stepsize is chosen as $\Delta t = 0.25$ for the S-ROCK methods. For the Euler-Maruyama method we select for each value of c_3 the largest stepsize which leads to a stable integration. Thus, for the Euler-Maruyama method, stability is achieved by reducing the stepsize while for the S-ROCK method, it is achieved by increasing the stage number ($m = 3, 7, 28, 81$). Notice that for both methods, one evaluation of "$g(Y)dW(t)$" is needed per stepsize. Thus, by keeping a fixed stepsize, the number of generated random variables remains constant as the stiffness increases for the S-ROCK methods, while this number increases linearly (proportional to the stepsize reduction) for the Euler-Maruyama method. Taking advantage of the quadratic growth of the stability domains, we see that the number of function evaluations is reduced by several orders of magnitude when using the S-ROCK methods instead of the Euler-Maruyama method.

References

1. A. Abdulle, *On roots and error constant of optimal stability polynomials*, BIT 40 (2000), no. 1, 177–182.
2. A. Abdulle and A.A. Medovikov, *Second order Chebyshev methods based on orthogonal polynomials*, Numer. Math., 90 (2001), no. 1, 1–18.
3. A. Abdulle, *Fourth order Chebyshev methods with recurrence relation*, SIAM J. Sci. Comput., 23 (2002), no. 6, 2041–2054.
4. A. Abdulle and S. Attinger, *Homogenization method for transport of DNA particles in heterogeneous arrays,* Multiscale Modelling and Simulation, Lect. Notes Comput. Sci. Eng., 39 (2004), 23–33.
5. A. Abdulle and S. Cirilli, *Stabilized methods for stiff stochastic systems,* C. R. Acad. Sci. Paris, 345 (2007), no. 10, 593–598.
6. A. Abdulle and S. Cirilli, *S-ROCK methods for stiff stochastic problems,* SIAM J. Sci. Comput., 30 (2008), no. 2, 997–1014.
7. A. Abdulle and T. Li, *S-ROCK methods for stiff Itô SDEs*, Commun. Math. Sci. 6 (2008), no. 4, 845–868.
8. A. Abdulle, Y. Hu and T. Li, *Chebyshev methods with discrete noise: the tau-ROCK methods*, J. Comput. Math. 28 (2010), no. 2, 195–217
9. L. Arnold, *Stochastic differential equation, Theory and Application,* Wiley, 1974.
10. E. Buckwar and C. Kelly, *Towards a systematic linear stability analysis of numerical methods for systems of stochastic differential equations*, SIAM J. Numer. Anal. 48 (2010), no. 1, 298–321.
11. P.M. Burrage, *Runge-Kutta methods for stochastic differential equations*. PhD Thesis, University of Queensland, Brisbane, Australia, 1999.
12. K. Burrage and P.M. Burrage, *General order conditions for stochastic Runge-Kutta methods for both commuting and non-commuting stochastic ordinary differential equation systems,* Eight Conference on the Numerical Treatment of Differential Equations (Alexisbad, 1997), Appl. Numer. Math. 28 (1998), no. 2-4, 161–177.
13. P. L. Chow, *Stochastic partial differential equations*, Chapman and Hall/CRC, 2007.
14. M. Duarte, M. Massota, S. Descombes, C. Tenaudc , T. Dumont, V. Louvet and F. Laurent, *New resolution strategy for multi-scale reaction waves using time operator splitting, space adaptive multiresolution and dedicated high order implicit/explicit time integrators*, preprint available at hal.archive ouvertes, 2010.
15. W. E, D. Liu, and E. Vanden-Eijnden, *Analysis of multiscale methods for stochastic differential equations*, Comm. Pure Appl. Math. 58 (2004), no. 11, 1544–1585.
16. D.T. Gillespie, *Stochastic simulation of chemical kinetics*, Annu. Rev. Phys. Chem. 58 (2007), 35–55.
17. A. Guillou and B. Lago, *Domaine de stabilité associé aux formules d'intégration numérique d'équations différentielles à pas séparés et à pas liés. Recherche de formules à grand rayon de stabilité,* in Proceedings of the 1er Congr. Assoc. Fran. Calcul (AFCAL), Grenoble, (1960), 43–56.
18. E. Hairer and G. Wanner, *Intégration numérique des équations différentielles raides*, Techniques de l'ingénieur AF 653, 2007.
19. E. Hairer and G. Wanner, *Solving ordinary differential equations II. Stiff and differential-algebraic problems*. 2nd. ed., Springer-Verlag, Berlin, 1996.
20. R.Z. Has'minskiĭ, *Stochastic stability of differential equations*. Sijthoff & Noordhoff, Groningen, The Netherlands, 1980.
21. M. Hauth, J. Gross, W. Strasser and G.F. Buess, *Soft tissue simulation based on measured data*, Lecture Notes in Comput. Sci., 2878 (2003), 262–270.
22. D.J. Higham, *Mean-square and asymptotic stability of numerical methods for stochastic ordinary differential equations*, SIAM J. Numer Anal., 38 (2000), no. 3, 753–769.
23. D.J. Higham, *An algorithmic introduction to numerical simulation of stochastic differential equations*, SIAM Review 43 (2001), 525–546.

24. P.J. van der Houwen and B.P. Sommeijer, *On the internal stage Runge-Kutta methods for large m-values,* Z. Angew. Math. Mech., 60 (1980), 479–485.
25. N.G. van Kampen, *Stochastic processes in physics and chemistry,* 3rd ed., North-Holland Personal Library, Elsevier, 2007.
26. P.E. Kloeden and E. Platen, *Numerical solution of stochastic differential equations*, Applications of Mathematics 23, Springer-Verlag, Berlin, 1992.
27. V.I. Lebedev, *How to solve stiff systems of differential equations by explicit methods.* CRC Pres, Boca Raton, FL, (1994), 45–80.
28. T. Li, A. Abdulle and Weinan E, *Effectiveness of implicit methods for stiff stochastic differential equations,* Commun. Comput. Phys., 3 (2008), no. 2, 295–307.
29. T. Li, *Analysis of explicit tau-leaping schemes for simulating chemically reacting systems*, SIAM Multiscale Model. Simul., 6 (2007), no. 2, 417–436.
30. G. Maruyama, *Continuous Markov processes and stochastic equations,* Rend. Circ. Mat. Palermo, 4 (1955), 48–90.
31. B. Oksendal, *Stochastic differential equations,* Sixth edition, Springer-Verlag, Berlin, 2003.
32. E. Platen and N. Bruti-Liberati, *Numerical solutions of stochastic differential equations with jumps in finance*, Stochastic Modelling and Applied Probability, Vol. 64, Springer-Verlag, Berlin, 2010.
33. E. Platen, *Zur zeitdiskreten approximation von Itôprozessen,* Diss. B. Imath. Akad. des Wiss. der DDR, Berlin, 1984.
34. W. Rümlin, *Numerical treatment of stochastic differential equations*, SIAM J. Numer. Math., 19 (1982), no. 3, 604–613.
35. Y. Saitô and T. Mitsui, *Stability analysis of numerical schemes for stochastic differential equations,* SIAM J. Numer. Anal., 33 (1996), no. 6, 2254–2267.
36. V. Thomee, *Galerkin finite element methods for parabolic problems,* 2nd ed, Springer Series in Computational Mathematics, Vol. 25, Springer-Verlag, Berlin, 2006.
37. E. Vanden-Eijnden, *Numerical techniques for multiscale dynamical system with stochastic effects*, Commun. Math. Sci., 1 (2003), no. 2, 385–391.
38. J.B. Walsh, *An introduction to stochastic partial differential equations*, In: École d'été de Prob. de St-Flour XIV-1984, Lect. Notes in Math. 1180, Springer-Verlag, Berlin, 1986.
39. D.J. Wilkinson, *Stochastic modelling for quantitative description of heterogeneous biological systems,* Nature Reviews Genetics 10 (2009), 122–133.

Oscillatory Systems with Three Separated Time Scales: Analysis and Computation

Gil Ariel, Björn Engquist, and Yen-Hsi Richard Tsai

Abstract We study a few interesting issues that occur in multiscale modeling and computation for oscillatory dynamical systems that involve three or more separated scales. A new type of slow variables which do not formally have bounded derivatives emerge from averaging in the fastest time scale. We present a few systems which have such new slow variables and discuss their characterization. The examples motivate a numerical multiscale algorithm that uses nested tiers of integrators which numerically solve the oscillatory system on different time scales. The communication between the scales follows the framework of the Heterogeneous Multiscale Method. The method's accuracy and efficiency are evaluated and its applicability is demonstrated by examples.

1 Introduction

In this paper we study a few interesting phenomena occurring in oscillatory dynamical systems involving three or more separated time scales. In the typical setting, the fastest time scale is characterized by oscillations whose periods are of the order of a small parameter ϵ. Classical averaging and multiscale methods consider

G. Ariel (✉)
Bar-Ilan University, Ramat Gan, 52900, Israel
e-mail: arielg@math.biu.ac.il

B. Engquist
Department of Mathematics and Institute for Computational Engineering and Sciences (ICES), The University of Texas at Austin, TX 78712, U.S.A
e-mail: engquist@ices.utexas.edu

Y.-H. R. Tsai
Department of Mathematics and Institute for Computational Engineering and Sciences (ICES), The University of Texas at Austin, TX 78712, U.S.A
e-mail: ytsai@ices.utexas.edu

B. Engquist et al. (eds.), *Numerical Analysis of Multiscale Computations*, Lecture Notes in Computational Science and Engineering 82, DOI 10.1007/978-3-642-21943-6_2,

the effective dynamics of such systems on a time scales which is independent of ϵ. However, under this scaling, many interesting phenomena, e.g. the nontrivial energy transfer among the linear springs in a Fermi-Pasta-Ulam (FPU) lattice, occur at the $\mathcal{O}(1/\epsilon)$ or even longer time scales. These kind of interesting phenomena motivates our interest in ordinary differential equations (ODEs) with three or more well separated time scales.

A good amount of development in numerical methods for long time simulations has been centered around the preservation of (approximate) invariances. In the past few years, many numerical algorithms operating on two separated scales have been proposed, see e.g. [1–3, 8, 9, 12–14, 16, 17, 19, 30–34]. To our knowledge, very few algorithms were developed considering directly three or more scales.

For our purpose, it is convenient to rescale time so the slowest time scale of interest is independent of the small parameter ϵ. Accordingly, the basic assumption underling our discussion is that solutions are oscillatory with periods that are of the order of some powers in ϵ: $\epsilon^0, \epsilon^1, \dots, \epsilon^m$. We will study the few issues arising from multiscale modeling and computations for ODEs in the form

$$\dot{\mathbf{x}} = \sum_{i=0}^{m} \epsilon^{-i} f_i(\mathbf{x}), \quad \mathbf{x}(0) = \mathbf{x}_0, \tag{1}$$

where $0 < \epsilon \le \epsilon_0$, $\mathbf{x} = (x_1, \dots, x_d) \in \mathbb{R}^d$. We further assume that the solution of (1) remains in a domain $\mathcal{D}_0 \subset \mathbb{R}^d$ which is bounded independent of ϵ for all $t \in [0, T]$. For fixed ϵ and initial condition $\mathbf{x}_0$, the solution of (1) is denoted $\mathbf{x}(t; \epsilon, \mathbf{x}_0)$. For brevity we will write $\mathbf{x}(t)$ when the dependence on ϵ and $\mathbf{x}_0$ is not directly relevant to the discussion.

We will focus only on a few model problems involving three time scales. Our goal is to compute the effective dynamics of such a system in a constant, finite time interval $[0, T]$, for the case $0 < \epsilon \le \epsilon_0 \le 1$. We will characterize the effective dynamics by some suitable smooth functions $\mathbf{x}$ that change slowly along the trajectories of the solutions, albeit possibly having some fast oscillations with amplitudes that are of the order of ϵ^p, $p \ge 1$. Naturally, the invariances of the system will be of interest.

As a simple example, consider the following linear system

$$\begin{cases} \dot{x}_1 &= \frac{1}{\epsilon} x_2 + x_1, \\ \dot{x}_2 &= -\frac{1}{\epsilon} x_1 + x_2, \end{cases} \tag{2}$$

with initial conditions $(x_1(0), x_2(0)) = (0, 1)$. The solution is readily given by $(x_1(t), x_2(t)) = (e^t \sin \frac{t}{\epsilon}, e^t \cos \frac{t}{\epsilon})$. Taking $I = x_1^2 + x_2^2$, we notice that I has a bounded derivative, i.e., that $\dot{I} := (d/dt) I(x_1(t), x_2(t)) = 2I$ is independent of ϵ. For this particular example one can easily solve for I, $I(t) = I(0)e^{2t}$. In fact, the uniform bound on $\dot{I}$ indicates the "slow" nature of $I(x_1(t), x_2(t))$ when compared to the fast oscillations in $(x_1(t), x_2(t))$. This type of characterization of the effective dynamics in a highly oscillation system is commonly used in the literatures. See for example [1, 2, 14, 18, 23–25]. Other approaches to finding slow variables includes, e.g. [5, 6]. We formalize this notion with the following definition.

Definition 1. We say that the function $\xi : \mathbf{x} \in \mathscr{D}_0 \mapsto \mathbb{R}$ has a *bounded derivative to order* $-k$ *for* $0 < \epsilon \le \epsilon_0$ *along the flow* $\mathbf{x}(t)$ in $\mathscr{D}_0$ if

$$\sup_{\mathbf{x}\in\mathscr{D}_0,\epsilon\in[0,\epsilon_0]} |\nabla\xi(\mathbf{x})\cdot\dot{\mathbf{x}}| \le C\epsilon^{-k}, \tag{3}$$

where $\mathscr{D}_0 \subset \mathbb{R}^d$ is an open connected set and C is a constant, both independent of ϵ. For brevity, we will say that ξ *has a bounded derivative along* $\mathbf{x}(t)$ if (3) holds with $k = 0$. Such functions are commonly referred to as slow variables of the system.

When only two separated time scales are considered, the effective behavior of a highly oscillatory system, $\mathbf{x}(t)$, may be described by a suitably chosen set of variables whose derivatives along $\mathbf{x}(t)$ are bounded. In the literature the time dependent function $x_1 = \sin(t)$ with $|\dot{x}_1| = \mathscr{O}(1)$ is naturally regarded as slow and $x_2 = \sin(t/\epsilon)$ with $|\dot{x}_2| = \mathscr{O}(\epsilon^{-1})$ is fast. Similarly $x_3 = \sin(t) + \epsilon\sin(t/\epsilon)$ is slow.

When more than two time scales are involved, we also need to consider $x_4 = \sin(t) + \epsilon\sin(t/\epsilon^2)$ as slow even if $|\dot{x}_4| = \mathscr{O}(\epsilon^{-1})$. *It will be regarded as slow because* $|x_4 - \sin t| = \mathscr{O}(\epsilon)$ *and* $\sin(t)$ *is slow.* As a further example, consider the linear system

$$\begin{cases} \dot{x}_1 = \frac{1}{\epsilon^2}x_2 + \frac{1}{\epsilon} + x_1, & x_1(0) = x_{10}, \\ \dot{x}_2 = -\frac{1}{\epsilon^2}x_1 + x_2, & x_2(0) = x_{20}, \end{cases} \tag{4}$$

The solution is

$$\begin{pmatrix} x_1(t) \\ x_2(t) \end{pmatrix} = \begin{pmatrix} Ae^t \sin\left(\epsilon^{-2}t + \phi\right) - \frac{\epsilon^3}{1+\epsilon^4} \\ Ae^t \cos\left(\epsilon^{-2}t + \phi\right) - \frac{\epsilon}{1+\epsilon^4} \end{pmatrix}, \tag{5}$$

where A and ϕ are determined by the initial conditions $A = x_{10}^2 + x_{20}^2$ and $\tan\phi = x_{10}/x_{20}$. As above, we look at the square amplitude $I = x_1^2 + x_2^2$. Its time derivative is bounded to order -1 since

$$\dot{I} = 2\epsilon^{-1}x_1 + 2I. \tag{6}$$

However, using (5) we find that $I(t) = A^2e^{2t} + O(\epsilon)$. Hence, even though the derivative of $I(t)$ is not bounded for $0 < \epsilon < \epsilon_0$, $I(t)$ consist of a slowly changing part and a small ϵ-scale perturbation. This example demonstrates that the bounded derivative characterization is not necessary for determining this type of effective property.

Accordingly, Sect. 2 gives a definition for the time scale on which a certain variable $\xi(\mathbf{x})$ evolves under the dynamics of ODEs in the form (1). These ideas are further generalized to describe local coordinate systems. In Sect. 3, the dynamics of the variables is analyzed using the operator formalism for homogenization of differential equations, see for example [29]. We focus the discussion to a few example systems in which the singular part of the dynamics is linear. Our observations are discussed in the settings of integrable Hamiltonian systems that can be written in terms of action-angle variables [4].

The effective behavior for certain class of dynamical systems in the long time scale may be modeled by a limiting stochastic process [21, 29, 34]. This approach has been applied, for example, in climate modeling [26]. However, rigorous analysis of such models has only been established in a few particular cases, for example, discrete rapidly mixing maps [7, 20] and the Lorentz attractor [22, 27]. The operator formalism for homogenization is a useful tool in the determination of stochasticity. In this formalism, by matching the multiscale expansions of the differential operator and a probability density function defined in the phase space, one derives a Fokker-Planck equation (or alternatively the backward equation) in the phase space of the given dynamical system. If the leading order terms in the multiscale expansion contain a diffusion term, then one says that the effective behavior of the given oscillatory dynamical system is "stochastic". Thus the effective behavior is approximated in average. In this paper, we consider systems in which no "stochastic" behavior appear in the effective equations.

Section 4 presents a numerical method that uses nested tiers of integrators which numerically solve the oscillatory system on different time scales. The communication between the scales follows the framework of the Heterogeneous Multiscale Method (HMM) [10, 11]. Section 5 presents a few numerical examples. We conclude in Sect. 6.

2 Effective Behavior Across Different Time Scales

In this section we discuss some of the mathematical notions which we use to study systems containing several well-separated time scales.

2.1 Slowly Changing Quantities

Definition 2. A smooth time dependent function $\alpha : [0, T] \mapsto \mathbb{R}^n$ is said to *evolve on the ϵ^k time scale in* $[0, T]$ for some integer k and for $0 < \epsilon \leq \epsilon_0$, if there exists a smooth function $\beta : [0, T] \mapsto \mathbb{R}^n$ and constants C_0 and C_1 such that

$$\sup_{t \in [0,T]} \left| \frac{d}{dt} \beta(t) \right| \leq C_0 \epsilon^{-k},$$

and

$$\sup_{t \in [0,T]} |\alpha(t) - \beta(t)| \leq C_1 \epsilon.$$

This motivates the following definition for a variable, $\alpha(\mathbf{x})$, that evolves on the ϵ^k time scale along the solutions of (1).

Definition 3. A function $\xi(\mathbf{x})$ is said to *evolve on the* ϵ^k *time scale along the trajectories of (1) in* $[0,T]$ *and in an open set* $\mathscr{D}_0$ if, for all initial conditions $\mathbf{x}_0 \in \mathscr{D}_0$, the time dependent function $\xi(\mathbf{x}(t;\epsilon,\mathbf{x}_0))$ evolves on the ϵ^k time scale in $[0,T]$. For brevity, we will refer to quantities and variables that evolve on the ϵ^0 time scale as slow.

In particular, the above definition suggests that if ξ_0 evolves on the ϵ^k time scale, then the limit

$$\xi_0(s;\mathbf{x}_0) = \lim_{\epsilon\to 0} \xi(\mathbf{x}(\epsilon^k s;\epsilon,\mathbf{x}_0)) \tag{7}$$

exists for all $s \in [0,T]$ and $\mathbf{x}_0 \in \mathscr{D}_0$. For instance, in both examples (2) and (4), the square amplitude $I = x_1^2 + x_2^2$ evolve on the ϵ^0 time scale. The difference is that (according to Definition 3), I has a bounded derivative of order 0 along the flow of (2) but not along the flow of (5). More generally, considering $\alpha(t)$ to be the image of $\xi(\mathbf{x}(t))$, Definitions (2) and (3) allows the inclusion of functions such as $\alpha(t) = \epsilon\sin(\epsilon^{-2}t) + \sin(t)$ (with unbounded derivatives) to be characterized as slowly evolving. In the Appendix, we presents an algorithm to identify slow variables based on (7).

Next, in order to understand what algebraic structure in the ODEs may lead to slow variables such as (6), we consider the following slightly more general system

$$\frac{dx}{dt} = \frac{i}{\epsilon^2}x + f_I(x,y,t), \tag{8}$$

$$\frac{dy}{dt} = \frac{1}{\epsilon}g(x)y + f_{II}(x,y,t). \tag{9}$$

Introducing a new variable $z = \exp(-it/\epsilon^2)x$, we obtain

$$\frac{dz}{dt} = \exp\left(-\frac{it}{\epsilon^2}\right) f_I\left(\exp\left(\frac{it}{\epsilon^2}\right)z, y, t\right), \tag{10}$$

$$\frac{dy}{dt} = \frac{1}{\epsilon}g\left(\exp\left(\frac{it}{\epsilon^2}\right)z\right)y + f_{II}\left(\exp\left(\frac{it}{\epsilon^2}\right)z, y, t\right). \tag{11}$$

Assuming that $|y(t)|$ is $\mathscr{O}(1)$ and $z(t)$ is of the form $z_0 + \mathscr{O}(\epsilon^2)$, then the first term on the right hand side in (11) would be bounded if

$$\int_0^t g(x(s))ds = \mathscr{O}(\epsilon), t > 0.$$

This is possible since the oscillations in x occur on a time scale that is much faster than the ϵ-scale, and they may induce an $\mathscr{O}(\epsilon)$ time averaging in $g(x(t))$. Thus, if $g(x - \bar{x}_0)$ is an odd function for

$$\bar{x}_0 := \lim_{\epsilon\to 0+} \int_0^t x(s;x_0)ds,$$

then for fixed values of z, the singular term in (11) "averages out" and would produce only fast oscillations of $\mathcal{O}(\epsilon)$ amplitude in the trajectories of y. In this case, $y(t)$ is a slowly changing quantity along the trajectory, i.e, it evolves on the ϵ^0 scale. Alternatively, if $g(x - \bar{x}_0)$ is even then y evolves on the ϵ time scale.

This observation suggests that in determining whether y changes slowly in time, we may *test if g is odd around a neighborhood of the averages of x*. If so, one can simply ignore the term containing g in solving for y.

We may generalize the observation above to test potential slow variables. Let $\mathbf{x}$ be a quasi-periodic solution of a highly oscillatory system with $\mathcal{O}(\epsilon^{-2})$ frequencies, and assume that $\mathbf{x}$ has an average $\bar{\mathbf{x}}_0$ as $\epsilon \to 0$. Consider $\alpha(t) := \xi(\mathbf{x}(t))$ with

$$\frac{d}{dt}\alpha(t) = \frac{1}{\epsilon} r(\mathbf{x}(t)).$$

Then α may be slow if $r(x - \bar{x}_0)$ is an odd function.

Finally, we point out that the emergence of a slow variable with unbounded time derivative along the oscillatory trajectories may come from a multiscale series expansion of parts of the solution. Consider again (8) and (9). The leading order term comes out naturally when y has an expansion of the form

$$y(t) = y_0(t) + \epsilon h(x(t)) + \cdots.$$

Hence, we expect that the homogenization approach described in the following section should capture such type of effective behavior of a dynamical system.

2.2 Multiscale Charts

Given an oscillatory dynamical system in $\mathbb{R}^d$, functions such as the slow variables in our previous definitions may be used to analyze the structure of the dynamics. For example, the action and angle variables for a given Hamiltonian system provide a coordinate system in the phase space such that the resulting Hamiltonian dynamics is separated into evolutions on certain invariant tori (oscillations) as described by the angle variables, and non-oscillatory evolutions described by the action variables [4]. For example, the function I defined for (2) together with $\arctan(x_2/x_1)$ corresponds to such a situation in which I is non-oscillatory along the dynamics and provides a coordinate perpendicular to the trajectories. In previous work, we propose the use of a similar strategy for a different class of dynamical systems [1, 2].

Consider the oscillatory dynamical system (1), and a family of trajectories $\mathbf{x}(t;\epsilon,\mathbf{x}_0)$ in a open set $\mathscr{D}_0 \subset \mathbb{R}^d$. Let $\Phi : \mathscr{D}_0 \subset \mathbb{R}^d \to U \subset \mathbb{R}^d$ be a diffeomorphism that is independent of ϵ. Thus Φ is a local coordinate system (chart) for $\mathscr{D}_0 \subset \mathbb{R}^d$. We denote the vector $\Phi(\mathbf{x})$ by $(\phi^1(\mathbf{x}), \phi^2(\mathbf{x}), \ldots, \phi^d(\mathbf{x}))$, where $\phi^i(\mathbf{x})$ is a real valued function defined in $\mathbb{R}^d$. We shall refer to ϕ^i as the ith coordinate. Let $n(\Phi, k; \mathbf{x}(\cdot;\epsilon,\mathbf{x}_0))$ denote the number of coordinates in Φ that evolve along

$\mathbf{x}(t;\epsilon,\mathbf{x}_0)$ on time scales that are smaller or equal to ϵ^k. We have the following definition:

Definition 4. A chart Φ is said to be maximally slow if for any other chart $\tilde{\Phi}$ defined on $\mathscr{D}_0$, $n(\Phi,k;\mathbf{x}(\cdot;\epsilon,\mathbf{x}_0)) \geq n(\tilde{\Phi},k;\mathbf{x}(\cdot;\epsilon,\mathbf{x}_0))$ for all k.

Loosely speaking, the coordinates of Φ are as slow as possible. A numerical method for identifying a maximally slow chart for the case in which the singular parts of the dynamics is linear is describes in the appendix.

Let Φ denote a maximally slow chart with k time scales, i.e.,

$$\Phi = (\phi^0,\ldots,\phi^k),$$

where $\phi^i \in \mathbb{R}^{d_i}$ are the variables evolving on the ith time scale, $i = 0,\ldots,k$ and $\sum d_i = d$. Using the principle of averaging iteratively for each scale, effective equations for each time scale can be constructed and the solutions that approximate the exact dynamics of the coordinates ϕ^i in the corresponding time scale. To obtain these equations, faster time scale components are averaged while keeping the slower ones fixed. Formally, we write

$$\dot{\phi}^i = \epsilon^{-i} F^i(\phi^i;\phi^0,\ldots,\phi^{i-1}) + \mathscr{O}(\epsilon),$$

with appropriate initial conditions. The effective equations hold for a time scale which is of the order of ϵ^i. Furthermore, F^i can be obtained iteratively by averaging over the effective dynamics of the faster ϵ^{i+1} scale. Accordingly, we say that the chart Φ is effectively closed.

3 A Homogenization Approach

The multiscale structure of a system can be analyzed using the operator formalism as presented in [29], which in turn, formally generalizes the work of Papanicolaou [28]. Motivated by perturbed integrable systems, in which the dynamics can be written in terms of action-angle variables, we concentrate on example systems in which the singular part of the dynamics is linear. To make our setting more concrete, we consider the linear operators in the vector space $C^\infty(\mathbb{R}^d)$, and with the usual notion of inner product. We shall denote L^* as the adjoint operator for L.

The analysis motivates a numerical multiscale algorithm along the lines of the HMM framework [1, 10, 13]. The algorithm does not assume that the system is given in the convenient action-angle coordinates, but only that such a transformation exists.

Consider a general ODE system whose right hand side depends on ϵ

$$\frac{dx}{dt} = f_\epsilon(x).$$

The associated Liouville equation takes the form

$$\partial_t u^\epsilon(t,x) + f_\epsilon \cdot \partial_x u^\epsilon(t,x) = 0, \tag{12}$$

with an initial condition $u^\epsilon(0,x) = \psi(x)$. Here, ∂_x denotes partial differentiation with respect to x and similarly for t. This is a linear equation whose characteristics coincide with solutions of the ODE. We begin by matching powers of ϵ in the multiscale expansion of the operator $L^\epsilon := f_\epsilon \cdot \partial_x$ and that of the solution u^ϵ. Formally, we write

$$L^\epsilon = \frac{1}{\epsilon^2} L_2 + \frac{1}{\epsilon} L_1 + L_0, \tag{13}$$

and

$$u^\epsilon = u_0 + \epsilon u_1 + \epsilon^2 u_2 + \ldots \tag{14}$$

Substituting the above expansions into (12) yields

$$\partial_t u_0 = \frac{1}{\epsilon^2} L_2 u + \frac{1}{\epsilon}(L_2 u_1 + L_1 u_0) + (L_2 u_2 + L_1 u_1 + L_0 u_0) + \mathscr{O}(\epsilon).$$

Comparing orders of ϵ, we have

$$\frac{1}{\epsilon^2} : \ L_2 u_0 = 0, \tag{15}$$

$$\frac{1}{\epsilon} : \ L_2 u_1 = -L_1 u_0, \tag{16}$$

$$1 : \ \partial_t u_0 = L_2 u_2 + L_1 u_1 + L_0 u_0. \tag{17}$$

We see that a closed effective equation for u_0 can be derived if both $L_2 u_2$ and $L_1 u_1$ can be approximated by operations on u_0 only. This closure is typically done by averaging over some invariant manifolds. In the following subsections, we apply this procedure to some model problems.

3.1 A Two Scales Example

For completeness, we recall the application of the operator formalism in a simple two-scale highly-oscillatory ODE system. Let

$$\begin{cases} \dot{x} = \frac{1}{\epsilon} y + f(x,y), \\ \dot{y} = -\frac{1}{\epsilon} x + g(x,y), \end{cases} \tag{18}$$

with some non-zero initial condition. Changing into polar coordinates $(r,\theta) \in \mathbb{R} \times S^1$ yields

$$\begin{cases} \dot{r} = (xf(x,y) + yg(x,y))/r, \\ \dot{\theta} = -\frac{1}{\epsilon} + (xg(x,y) - yf(x,y))/r^2. \end{cases}$$

It is clear that the amplitude r is a slow variable while the phase ϕ is fast. Hence, we can naively average the right hand side of the equation for r with respect to the fast phase. This yields an effective equation for the amplitude

$$\begin{aligned}\dot{r} &= F(r),\\ F(r) &= \frac{1}{r}\int_{S^1}[x(r,\theta)f(x(r,\theta),y(r,\theta)) + y(r,\theta)g(x(r,\theta),y(r,\theta))]\,d\theta.\end{aligned} \tag{19}$$

Alternatively, using (12) and (13), we derive the following relations:

$$\begin{aligned}L^{\epsilon} &= \frac{1}{\epsilon}L_1 + L_0,\\ L_1 &= y\partial_x - x\partial_y,\\ L_0 &= f(x,y)\partial_x + g(x,y)\partial_y.\end{aligned} \tag{20}$$

Taking $L_2 = 0$, (15) is trivially satisfied. Noting that $L_1 = \partial_\theta$ and that $L_1 = -L_1^*$, we see that the Null space of L_1 is identical to that of L_1^* and constitutes

$$\text{Null } L_1 = \text{Null } L_1^* = \{\xi(x^2+y^2) : \xi \in C^\infty(\mathbb{R})\}, \tag{21}$$

where L_1^* denotes the dual of the operator L_1. Let P denote projection onto Null L_1 obtained by averaging over the fast angle θ, $P[\cdot] = \int_{S^1}[\cdot]d\theta := \langle\cdot\rangle$. It is a projection in the sense that $P^2 = P$. Substituting the asymptotic expansion for u, (14), into the backwards equation (21) yields (compare with (16)–(17))

$$\begin{cases} L_1 u_0 &= 0,\\ L_1 u_1 &= \partial_t u_0 - L_0 u_0.\end{cases}$$

The equation for u_0 implies that $u_0 \in$ Null L_1, i.e., $u_0 = u_0(t,r)$. Formally, we write

$$P u_0 = u_0. \tag{22}$$

The solvability condition for u_1 is

$$[\partial_t - L_0]u_0 \perp \text{Null } L_1^*.$$

Substituting (22) yields

$$P\,[\partial_t - L_0]\,P u_0 = 0.$$

This gives the effective equation for u_0

$$\partial_t u_0 = P L_0 P u_0,$$

where we used the fact that u_0 does not depend on θ, and can therefore be taken out of the averaging. This can be rewritten as

$$\partial_t u_0 = \langle f(x,y)\partial_x u_0 + g(x,y)\partial_y u_0\rangle.$$

Using the chain rule yields

$$\partial_t u_0 = \frac{1}{r}\langle xf(x,y) + yg(x,y)\rangle \partial_r u_0,$$

which is nothing but the Liouville equation associated with the effective ODE (19).

3.2 Three Scales: Example 1

Consider the following three-scale system which involves slow variables whose derivatives are not bounded.

$$\begin{aligned} \frac{d}{dt}\begin{pmatrix} x_1 \\ x_2 \end{pmatrix}, &= \frac{1}{\epsilon^2}\begin{pmatrix} x_2 \\ -x_1 \end{pmatrix} + f(x_1, x_2, y), \\ \frac{dy}{dt} &= \frac{1}{\epsilon}x_1 + f_{III}(x_1, x_2, y), \end{aligned} \tag{23}$$

where $f = (f^I, f^{II})^T$. To get some intuition, consider the unperturbed case $f_I = f_{II} = f_{III} = 0$ with initial conditions $(x_1, x_2, y) = (1, 0, 1)$. The solution is

$$\begin{aligned} x_1(t) &= -\cos(\epsilon^{-2}t), \\ x_2(t) &= \sin(\epsilon^{-2}t), \\ y(t) &= 1 - \epsilon\sin(\epsilon^{-2}t). \end{aligned}$$

Hence, we can see that the system has two variables which evolve on the $O(1)$ time scale: $I = x_1^2 + x_2^2$ and y. In the unperturbed case, both variables are constants.

From (12) and (13), we derive the following relations:

$$\begin{aligned} L^\epsilon &= \frac{1}{\epsilon^2}L_2 + \frac{1}{\epsilon}L_1 + L_0, \\ L_2 &= x_2\partial_{x_1} - x_1\partial_{x_2}, \\ L_1 &= x_1\partial_y, \\ L_0 &= f_I\partial_{x_1} + f_{II}\partial_{x_2} + f_{III}\partial_y. \end{aligned}$$

The Null space of L_2 is

$$\text{Null } L_2 = \text{Null } L_2^* = \{\xi = \xi(r, y)\},$$

where $r^2 = x_1^2 + x_2^2$. Let P denote projection on Null L_2, which can be performed by averaging over the fast phase $\theta = \arctan x_2/x_1$. Let P denote projection on Null L_2. As before, averaging over the fast phase θ is denoted by $\langle\cdot\rangle$.

Arranging terms with the same order of ϵ prefactors, we have (15)–(17) which are analyzed below.

Leading order equation:

The equation for u_0 implies that $u_0 \in \text{Null } L_2$, i.e., $u_0 = u_0(t, r, y)$. Formally, we write

$$Pu_0 = u_0. \tag{24}$$

Order $1/\epsilon$ equation:

The solvability condition for u_1 in (16) implies

$$L_1 u_0 \perp \text{Null } L_2^*, \tag{25}$$

which is equivalent to

$$PL_1 u_0 = 0.$$

This holds since

$$\begin{aligned} PL_1 u_0 &= P\left[x_1 \partial_y\right] u_0(x_1^2 + x_2^2, y) = \langle x_1 \partial_y u_0(x_1^2 + x_2^2, y)\rangle = \\ &= \langle x_1\rangle \partial_y u_0(x_1^2 + x_2^2, y) = 0. \end{aligned}$$

Hence, we formally write

$$u_1 = -L_2^{-1} L_1 u_0. \tag{26}$$

Order 1 equation:

The solvability condition for u_2 in (17) is

$$[\partial_t - L_1 u_1 - L_0 u_0] \perp \text{Null } L_2^*.$$

Substituting in the formal solution (26) yields

$$P\left[\partial_t + L_1 L_2^{-1} L_1 - L_0\right] Pu_0 = 0.$$

For the example at hand, we have

$$u_1 = -L_2^{-1} x_1 \partial_y u_0.$$

Furthermore, $\partial_y u_0$ has the form $g(x_1^2 + x_2^2, y)$, which implies that $\partial_y u_0 \in \text{Null } L_2^*$. Also, $L_2 x_2 = -x_1$. Hence,

$$L_2^{-1} L_1 u_0 = -u_1 = x_2 \partial_y u_0.$$

We conclude that

$$L_1 L_2^{-1} L_1 u_0 = x_1 \partial_y x_2 \partial_y u_0 = x_1 x_2 \partial_{yy} u_0.$$

This yields the effective equation for u_0

$$\partial_t u_0 = -\langle x_1 x_2 \partial_{yy} u_0 \rangle + \langle f_I \partial_{x_1} u_0 \rangle + \langle f_{II} \partial_{x_2} u_0 \rangle + \langle f_{III} \partial_y u_0 \rangle. \tag{27}$$

As before, $\partial_{yy} u_0 \in \text{Null } L_2$ and

$$\langle x_1 x_2 \partial_{yy} u_0 \rangle = \langle x_1 x_2 \rangle \partial_{yy} u_0 = 0.$$

The effective equation (27) becomes

$$\partial_t u_0 = \langle f_I \partial_{x_1} u_0 \rangle + \langle f_{II} \partial_{x_2} u_0 \rangle + \langle f_{III} \partial_y u_0 \rangle,$$

which can be rewritten as

$$\partial_t u_0 = \frac{1}{r} \langle x_1 f_I + x_2 f_{II} \rangle \frac{\partial u_0}{\partial r} + \langle f_{III} \rangle \partial_y u_0.$$

This equation can be identified as the Liouville equation associated with the ODE

$$\begin{cases} \dot{r} = \langle x_1 f_I + x_2 f_{II} \rangle / r, \\ \dot{y} = \langle f_{III} \rangle. \end{cases} \tag{28}$$

We conclude that, to leading order in ϵ, u_0, and hence r and y, evolve on the $O(1)$ time scale and are deterministic.

3.3 Three Scales: Example 2

We consider a simple system involving three time scales

$$\begin{aligned} \frac{d}{dt}\begin{pmatrix} x \\ y \end{pmatrix} &= \frac{1}{\epsilon^2}\begin{pmatrix} -y \\ x \end{pmatrix} + f(x,y,w,z), \\ \frac{d}{dt}\begin{pmatrix} w \\ z \end{pmatrix} &= \frac{1}{\epsilon}\begin{pmatrix} -z \\ w \end{pmatrix} + g(x,y,w,z), \end{aligned}$$

where $f = (f^I, f^{II})^T$ and $g = (g^I, g^{II})^T$. If $f = g = 0$, then (x, y) and (w, z) are decoupled harmonic oscillators with frequencies $2\pi/\epsilon^2$ and $2\pi/\epsilon$, respectively. From (12) and (13), we have the following operators:

$$\begin{aligned} L_2 &= -y\partial_x + x\partial_y, \\ L_1 &= -z\partial_w + w\partial_z, \\ L_0 &= f^I \partial_x + f^{II} \partial_y + g^I \partial_w + g^{II} \partial_z. \end{aligned} \tag{29}$$

Changing variables into polar coordinates: $(x, y) \mapsto (r, \theta)$ and $(w, z) \mapsto (\rho, \phi)$, we have that $L_2 = \partial_\theta$ and $L_1 = \partial_\phi$. These are also the action-angle variables of this system.

Leading order equation:

Following (15), we have $L_2 u_0 = 0$, which implies that u_0 must be constant in $\theta(x,y)$. Thus u_0 depends only on (r,w,z); i.e. $u_0 = u_0(r,w,z)$. As before, let P denote projection onto Null L_2^* defined by averaging with respect to θ; i.e. and $Pu = \langle u \rangle$.

Order $1/\epsilon$ equation:

The equation takes the form

$$\partial_\theta u_1 = -\partial_\phi u_0.$$

The solvability condition is $\langle -\partial_\phi u_0 \rangle = 0$. However, since u_0 does not depend on θ, the solvability condition implies that $\partial_\phi u_0 = 0$. Thus, u_0 must be a function of only r and ρ.

We conclude that r and ρ are the only slow variables (evolve on the ϵ^0 time scale). Both variables have a bounded derivative (of order 0). Furthermore, the first order perturbation, u_1, vanishes.

Order 1 equation:

Substituting $u_1 = 0$ in (17), the equation for u_2 is formally

$$\partial_\theta u_2 = -\partial_t u_0 + L_0 u_0.$$

We notice that in the example at hand, the dynamics of the two angle variables, ϕ and θ, are decoupled and the invariant measure for both variables (with r and ρ fixed) is uniform over a 2D torus T^2. This occurrence is not incidental, but holds for the class of near integrable systems in which the angle variable in systems which are given in action-angle coordinates undergo uniform rotations on a torus [4]. Earlier, we have concluded that u_0 does not depend on the angle variables θ and ϕ. However, the coefficients of L_0 may depend on the angle variables. Therefore, the solvability condition for (3.3) is

$$\partial_t u_0 = \int_{T^2} L_0 u_0 \, d\theta d\phi. \tag{30}$$

Note that the numerical algorithm presented in Sect. 4 does not assume that the system is written in terms of action-angle variables. A more general case in which the ϵ and ϵ^2 time scales cannot be decoupled in an appropriate coordinate system is beyond the scope of this manuscript and will be presented in a future publication.

3.4 Observations

Following this methodology, we have the following observations:

- Slow variables that have bounded derivatives lie in the Null space of both L_2 and L_1.
- Functions which evolves on the ϵ^0 time scale, see Definition 3, need only lie in the Null space of L_2.
- The homogenization approach picks out the new type of slow variables as defined in Definitions 2 and 3 with $k = 0$. See, for example, (28) which is derived from the system defined in (23).
- The effective slow dynamics in all the examples presented in this manuscript is found to be deterministic. However, for a large class of equations involving chaotic solutions [29], the effective PDEs are diffusive. This means that the effective behaviors of the original dynamical systems could be approximated weakly by solutions of the corresponding stochastic differential equations.

4 Numerical Algorithms

In this section, we discuss an approach which invokes our previous two-scale HMM algorithms [1–3] to multiple (> 2) timescale systems. This is achieved by considering a hierarchy of problems, each involving more than two time scales. Consequently, the numerical integrator is constructed as tiers of two-scale HMM solvers. We consider the time scales $\mathcal{O}(\epsilon^2)$, $\mathcal{O}(\epsilon)$, and $\mathcal{O}(1)$.

Suppose we obtain a maximal slow chart, for example, using the method described in the Appendix for identifying polynomial variables evolving on different time scales. We denote this system of coordinates $\xi = (\xi^1, \xi^2, \xi^3)$, where $\xi^i = (\xi^i_1, \ldots, \xi^i_{d_i})$ are the variables evolving on the ith time scale.

The HMM to be constructed should evaluate the rate of change of ξ along the flow of the original oscillatory system. This typically involves numerically averaging over the fast oscillations in the system. For three-scale problems this requires averaging over the $\mathcal{O}(\epsilon^2)$ scale oscillations as well as the $\mathcal{O}(\epsilon)$ scale oscillations, thus obtaining a numerical approximation for the effective equation for the slow variables ξ^0. The method is schematically illustrated in Fig. 1.

Our goal is to numerically integrate the effective equation for ξ^0, which is not known. Following HMM, this equation is approximated by averaging the dynamics of the system over a short-time calculation of the slower, $\mathcal{O}(\epsilon)$ scale. Hence, whenever the algorithm needs to take a coarse, $\mathcal{O}(1)$ step, we call an auxiliary function whose role is to calculate the dynamics on that time scale. This requires approximating the dynamics of ξ^1 for a time segment of order $\mathcal{O}(\epsilon)$. However, with three scales the dynamics of ξ^1 is given by an effective equation which is itself not known, but can nonetheless be approximated with a second tier of HMM integrator. This second tier approximates the effective dynamics of ξ^1 by numerically averaging the dynamics on the $\mathcal{O}(\epsilon^2)$ scale, namely ξ^2. Note that this is possible because we only require to solve ξ^1 on a time segment of order ϵ. Longer time scales of order 1 are not accessible as the error of using the effective averaged equation rather than the full one becomes large.

In [1], the two-scale HMM solver integrates an approximate averaged equation. The averaged equation, which is not known, is approximated by convoluting the solution of the faster time scale with a smoothing kernel. See [1, 13] for details.

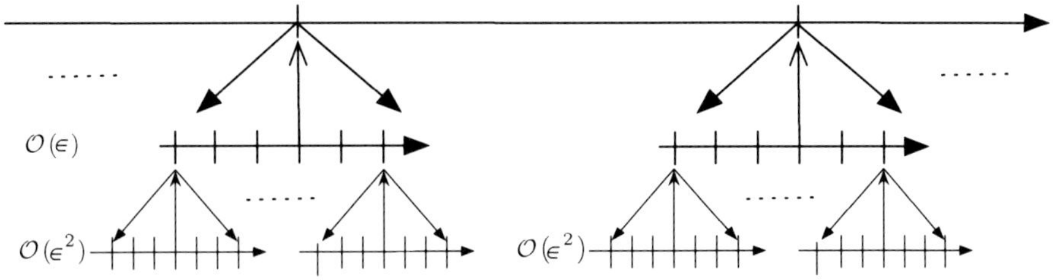

Fig. 1 An illustration of a three scale algorithm

4.1 Accuracy and Efficiency

Consider a three-scale ODE system of the form (1) with a maximally slow chart (ξ_0, ξ_1, ξ_2) in which $\xi_i \in \mathbb{R}^{d_i}$ evolve on the ϵ^i time scale. The system is to be integrated using the three-tier HMM algorithm described above. We will refer to the solver integrating the ϵ^i scale as the i'th tier. The step-size, length of integration and order of accuracy of the integrators at the i'th tier are denoted h_i, η_i and m_i, respectively. For example, on the slowest $O(1)$ time scale we utilize an m_0'th order explicit integrator with step size h_0 and approximate ξ_0 in the range $[0, \eta_0 = T]$. The global error in each run of the i'th tier is denoted E_i. The computational cost of each run of the i'th tier is denoted C_i. The goal is to approximate ξ_0 on a time segment $[0, T]$ with an optimal efficiency C_0 while meeting a prescribed accuracy $E_0 \leq \Delta$.

The numerical analysis is a generalization of the two-scale analysis described in [1]. Recall that in general, the local truncation error in approximating an ODE $\dot{x} = f(x)$ using an m'th order explicit method with step size h is of the order of $M_{m+1} h^{m+1}$, where M_{m+1} is a bound on the $m+1$ time derivative of $f(x(t))$ in the domain of interest. Accordingly, for stiff equations of the form $\dot{x} = \omega f(x)$, the $m+1$ time derivative of $f(x(t))$ is of the order of ω^{m+1}.

- Tier 2: The local error in each $\mathcal{O}(\epsilon^2)$ step of the 2nd tier integrator if of the order of $h_2^{m_2} \epsilon^{-2(m_2+1)}$. Integrating to time η_2, the truncation error of a single run of the 2nd tier integrator is $\eta_2 h_2^{m_2} \epsilon^{-2(m_2+1)}$. Next, the error in approximating the averaged equation using convolution of the approximate numerical solution with a kernel that has q continuous derivatives is [1, 13] $\epsilon^{2q} \eta_2^{-q-1}$. In order to obtain optimal efficiency the two sources of errors need to be the same. Setting $\Delta_2 = \epsilon^{2q} \eta_2^{-q-1} = \eta_2 h_2^{m_2} \epsilon^{-2(m_2+1)}$ yields the optimal scaling of η_2 and h_2 with ϵ and Δ_2.

- Tier 1: The error in each evaluation of $\dot{\xi}_1$, Δ_2, accumulates by taking η_1/h_1 steps of order ϵ to $\Delta_2\eta_1/h_1$. This error needs to be comparable to the truncation error of the tier 1 solver, $\eta_1 h_1^{m_1}\epsilon^{-m_1-1}$ and the averaging error $\epsilon^q\eta_1^{-q-1}$. Equating all terms to equal Δ_1 yields the scaling of η_1, h_1 and Δ_2, and hence η_2 and h_2 with ϵ and Δ_1.
- Tier 0: The error in each evaluation of $\dot{\xi}_0$, Δ_1, accumulates by taking η_0/h_0 steps of order 1 to $\Delta_1\eta_0/h_0$. This error needs to be comparable to the truncation error of the 0th tier solver, $\eta_0 h_0^{m_0}$. Equating all terms to equal Δ yields the scaling of all parameters h_0—h_2 and η_0, η_1 with ϵ and Δ.

We conclude that the overall computational cost

$$C = \frac{\eta_0}{h_0}\frac{\eta_1}{h_1}\frac{\eta_2}{h_2},$$

depends on ϵ, the required accuracy Δ and the orders of the solvers m_0, m_1 and m_2 through

$$C = \Delta^{\left(1-\frac{(m_0+1)(m_1+1)(m_2+1)}{m_0 m_1 m_2}\right)\left(\frac{q+2}{q+1}\right)}\epsilon^{-\frac{(m_2+1)(2+4m_1+q+3m_1 q)}{m_1 m_2 (1+q)^2}}, \tag{31}$$

where, for simplicity we took $\eta_0 = T = 1$. In particular, for a smooth (infinitely differentiable) kernel one may take the limit $q \to \infty$ and the computational cost reduces to

$$C = \Delta^{\left(1-\frac{(m_0+1)(m_1+1)(m_2+1)}{m_0 m_1 m_2}\right)}. \tag{32}$$

We see that for a smooth kernel the cost is independent of ϵ.

5 Examples

In this section we demonstrate the new multi-tier HMM algorithm in a few examples.

5.1 Harmonic Oscillators

Consider the following system describing two coupled harmonic oscillators in resonance

$$\begin{cases} \dot{x}_1 &= -\frac{1}{\epsilon^2}y_1 + \frac{1}{\epsilon}y_2^2 - 3x_1x_2^2, \\ \dot{y}_1 &= \frac{1}{\epsilon^2}x_1 + \frac{1}{2}y_1, \\ \dot{x}_2 &= -\left(\frac{1}{\epsilon^2}+\frac{1}{\epsilon}\right)y_2 - x_2, \\ \dot{y}_2 &= \left(\frac{1}{\epsilon^2}+\frac{1}{\epsilon}\right)x_2 - y_2 + 2x_1^2 y_2. \end{cases} \tag{33}$$

As depicted in Fig. 2, all four state variables oscillate with a period which is of the order of ϵ^2. Hence, x_1, y_1, x_2 and y_2 evolve on the ϵ^2 time scale.

In order to find a maximally slow coordinate system, we change to polar coordinates $(x_i, y_t) \mapsto (I_i, \varphi_i)$, $i = 1,2$ and introduce a polynomial variable θ that describes the 1:1 resonance between the oscillators

$$\begin{aligned} I_1 &= x_1^2 + y_1^2, \\ I_2 &= x_2^2 + y_2^2, \\ \theta &= x_1 x_2 + y_1 y_2, \\ \cos\varphi_1 &= x_1/\sqrt{I_1}. \end{aligned} \tag{34}$$

The corresponding time derivatives are

$$\begin{aligned} \dot{I}_1 &= \frac{2}{\epsilon} x_1 y_2^2 - 6x_1^2 x_2^2 + y_1^2, \\ \dot{I}_2 &= -2I_2 + 4x_1^2 y_2^2, \\ \dot{\theta} &= \frac{1}{\epsilon}(x_2 y_2^2 + y_1 x_2 - x_1 y_2) + (-y_1 y_2/2 - x_1 x_2 - 3x_1 x_2^3 + 2x_1^2 y_1 y_2), \\ \dot{\varphi}_1 &= \frac{1}{\epsilon^2}. \end{aligned} \tag{35}$$

It appears as if $(I_1, I_2, \theta, \varphi_1)$ is a chart in which φ_1 evolves of the ϵ^2 time scale, I_1 and θ evolve on the ϵ time scale while I_2, which has a bounded derivative, evolves on the $\mathcal{O}(1)$ scale. The dynamics of the three slow variables I_1, I_2 and θ on the $\mathcal{O}(\epsilon)$ scale is depicted in Fig. 3. The figure suggests that both I_1 and I_2 are practically constant on the ϵ scale. Indeed, it can be shown that the average of $x_1 y_2^2$ on any segment of length $\mathcal{O}(\epsilon)$ and larger is of order ϵ^2. Therefore, the averaged $\dot{I}_1$ is bounded independent of ϵ and I_1 evolves on the $\mathcal{O}(1)$ time scale, rather than the expected $\mathcal{O}(\epsilon)$. The time evolution of I_1 and I_2 on the slowest $\mathcal{O}(1)$ time scale is depicted in Fig. 4. In addition, the figure shows the results of the three-tier HMM integrator described in Sect. 4. The HMM algorithm approximates the slow $\mathcal{O}(1)$ dynamics using macroscopic steps which are independent of ϵ. The integration is done using a fourth order method (in the macroscopic step size) and its efficiency is essentially independent of ϵ.

5.2 An Example Motivated by the Fermi-Pasta-Ulam (FPU) Problem

The following example, which consists of three coupled oscillators, is motivated by a version of the FPUα model [15] with periodic boundary conditions. The system is described by the Hamiltonian

$$H = \frac{1}{2}\sum_{i=1}^{3} p_i^2 + \sum_{i=1}^{3}\left[\frac{1}{2}(q_{i+1} - q_i)^2 + \frac{\epsilon}{3}(q_{i+1} - q_i)^3\right], \tag{36}$$

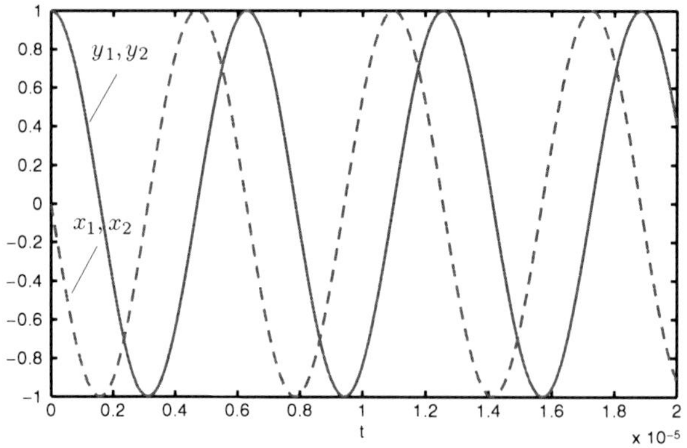

Fig. 2 The dynamics of (33) on the ϵ^2 time scale. $\epsilon = 10^{-3}$

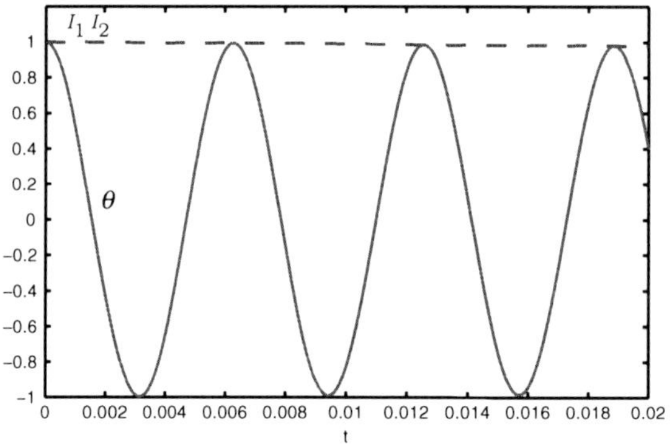

Fig. 3 The dynamics of (33) on the ϵ^1 time scale. $\epsilon = 10^{-3}$

where $q_0 = q_3$ and $q_4 = q_1$. The purpose of this example is to demonstrate the advantages of the HMM multiscale method for Hamiltonian systems compared to the standard Verlet method. From this case study, one may see the advantage of the proposed HMM over other standard reversible and symplectic integrators.

Rescaling time, $s = \epsilon^2 t$, and denoting $[\cdot]' = (d/ds)$, the dynamics is given by

$$\begin{cases} q_1' &= \frac{1}{\epsilon^2} p_1, \\ p_1' &= -\frac{1}{\epsilon^2}(2q_1 - q_3 - q_2) - \frac{1}{\epsilon}(q_2 - q_3)(2q_1 - q_3 - q_2), \\ q_2' &= \frac{1}{\epsilon^2} p_2, \\ p_2' &= -\frac{1}{\epsilon^2}(2q_2 - q_1 - q_3) - \frac{1}{\epsilon}(q_3 - q_1)(2q_2 - q_1 - q_3), \\ q_3' &= \frac{1}{\epsilon^2} p_3, \\ p_3' &= -\frac{1}{\epsilon^2}(2q_3 - q_2 - q_1) - \frac{1}{\epsilon}(q_1 - q_2)(2q_3 - q_2 - q_1). \end{cases} \tag{37}$$

Due to the periodic boundaries the total momentum is preserved. Hence, without loss of generality we pick initial conditions such that the center of mass is fixed

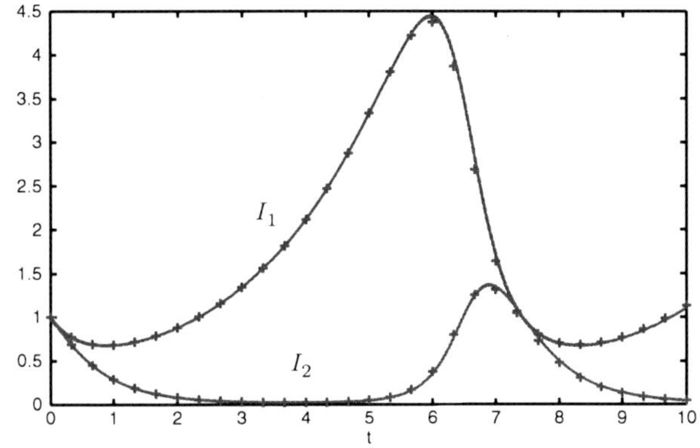

Fig. 4 The dynamics of (33) on the ϵ^0 time scale. $\epsilon = 10^{-3}$. Plus signs are results of a 3-tier HMM with fourth order RK on all scales. HMM parameters are H = 1/3, $\eta_j = 70.1\epsilon^j$, $h_j = \epsilon^j/10$, $j = 1,2$.

at the origin, $p_{\text{tot}} = p_1 + p_2 + p_3 = 0$ and $q_{\text{cm}} = q_1 + q_2 + q_3 = 0$. Using the algorithm detailed in Appendix, we identify the following five variables evolving on the ϵ^0 time scale:

$$\begin{aligned} p_{\text{tot}} &= p_1 + p_2 + p_3, \\ q_{\text{cm}} &= q_1 + q_2 + q_3, \\ I_1 &= 3q_2^2 + p_2^2, \\ I_2 &= 3q_3^2 + p_3^2, \\ \theta &= 3q_2q_3 + p_2p_3. \end{aligned} \tag{38}$$

Differentiating with respect to the rescaled time and using the fixed center of mass assumption yields

$$\begin{aligned} p'_{\text{tot}} &= 0, \\ q'_{\text{cm}} &= 0, \\ I'_1 &= -\frac{6}{\epsilon} p_2q_2(2q_3 + q_2), \\ I'_2 &= -\frac{6}{\epsilon} p_3q_3(2q_1 + q_3), \\ \theta' &= -\frac{3}{\epsilon}(q_3p_2(2q_1 + q_3) + p_3q_2(2q_3 + q_2)). \end{aligned} \tag{39}$$

We emphasize here that even I'_1, I'_2, and θ' are formally unbounded as $\epsilon \to 0$ for fixed values of p_1, p_2, p_3, q_1, q_2, and q_3, I_1, I_2, and θ all *evolve on the* ϵ^0 *time scale!*

Figure 5 compares the results computed by the proposed HMM with those by Verlet using the initial conditions $(q_1,q_2,q_3) = (-0.65, 0.35, 0.3)$ and $(p_1, p_2, p_3) = (0.3, -0.4, 0.1)$. Taking $\epsilon = 10^{-4}$, parameters are chosen to give an error of about 1%. HMM parameters are $H = 2$, $\eta_1 = 18.77\epsilon^1$, $h_1 = \epsilon^1/10$, $\eta_2 = 75.1\epsilon^2$,

$h_2 = \epsilon^2/20$. The tier 2 solver is Verlet. The tier 1 and 0 solvers are the midpoint rules and the kernel is exponential [13]. Solving the system using Verlet with $\epsilon = 10^{-4}$ is practically impossible. However, in order to achieve the desired accuracy the values of ϵ can be increased artificially [35]. Since we require a relative error of order 0.01, we take $\epsilon = 0.01$ and decrease step size until the energy drift is of the same order. This requires $h = \epsilon^2/100$. With these parameters HMM runs over 2,000 time faster. It is interesting to note that the efficiency of both methods is independent of ϵ. Hence, the gain in efficiency does not depend on ϵ as long as it is small enough. Lastly, the energy drift with HMM is small in this case, but it may increase in longer time intervals. The time reversible schemes, developed in [3], may be advantageous.

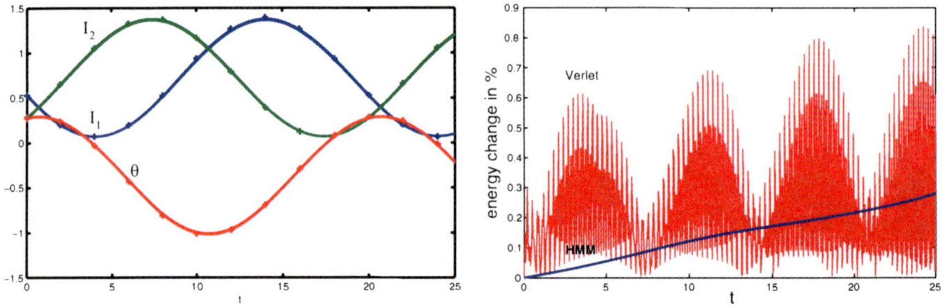

Fig. 5 *Left*: evolution of the slow variables. Solid curves are computed by Verlet with an adjusted value of ϵ. The values computed by the proposed HMM are shown by the plus signs. *Right*: energy drift. With simulation parameters tuned to give comparable errors in the total energy, HMM runs over 2,000 times faster

6 Summary

In this paper, we investigate several issues related to the design of multiscale algorithms for computing the effective behavior of highly oscillatory dynamical systems involving more than two separated scales. We discuss a type of effective behavior (slowly changing quantities) which do not have bounded derivatives. Homogenization technique based on multiscale expansions seem to be able to pick out such quantities which are one of the state variables in the given equations. We further demonstrate that this type of effective behavior cannot be ignored in our numerical examples.

Acknowledgements Research are performed under the support of NSF DMS-0714612. Tsai is partially supported by a Sloan Fellowship. We thank Seong Jun Kim for comments, corrections and a careful reading of the manuscript.

Appendix: Finding Polynomial Slow Variables

In many highly oscillatory physical systems the leading order oscillations are harmonic. It can then be shown [1,4] that slow variables can be polynomials. In [1] we suggest a variational method to automatically identify the polynomials making up maximally slow charts using the bounded derivative concept. As was demonstrated in this paper, with three or more well-separated scales, the bounded derivative concept is not sufficient to determine the time scale on which a variable evolves. Hence, it cannot be used to construct maximally slow charts. Accordingly, the purpose of this appendix is to modify the variational method of [1] to use the new concept of slow variable, Definition 3. The main idea is to compare two trajectories with different values of ϵ and find a polynomial that takes similar values on both trajectories.

Let $p(x)$ denote a polynomial in $\mathbb{R}^d$. Following Definition 3, $p(x)$ evolves on the ϵ^k time scale if, for all $m \geq i \geq k \geq 0$, the limit

$$\lim_{\epsilon \to 0} p(\mathbf{x}(\epsilon^i s; \epsilon, \mathbf{x}_0)) \tag{40}$$

exists for all $s \in [0, S]$ and $\mathbf{x}_0 \in \mathscr{D}_0$, a connected open set. Both $\mathscr{D}_0$ and S are independent of ϵ. Changing variables $\tau = \epsilon^{-k} t$, the general ODE (1) becomes

$$\frac{d}{d\tau}\mathbf{x} = \sum_{i=0}^{m-k} \epsilon^{-i} f_{i+k}(\mathbf{x}) + \mathscr{O}(\epsilon). \tag{41}$$

Let $\mathbf{x}_0 \in \mathscr{D}_0$ and consider two solutions of (41) with the same initial condition $\mathbf{x}_0$ but different ϵ. Using the notations of Sect. 1, the first solution, obtained with a small parameter ϵ is denoted by $\mathbf{x}(\tau; \epsilon, \mathbf{x}_0)$. The second solution, obtained with a small parameter 2ϵ, is denoted by $\mathbf{x}(\tau; 2\epsilon, \mathbf{x}_0)$. Furthermore, let $\mathbf{x}_1^\epsilon, \ldots, \mathbf{x}_j^\epsilon, \ldots, \mathbf{x}_N^\epsilon$ denote a numerical solution of $\mathbf{x}(\tau; \epsilon, \mathbf{x}_0)$ at times $\tau_j = Hj$ that are computed by some integrator with constant step size less than $\epsilon^{m-k} H$. Here, H is a constant which is independent of ϵ. Similarly, let $\mathbf{x}_1^{2\epsilon}, \ldots, \mathbf{x}_j^{2\epsilon}, \ldots, \mathbf{x}_N^{2\epsilon}$ denote a numerical solution that approximate $\mathbf{x}(\tau; 2\epsilon, \mathbf{x}_0)$ at times $s_j = Hj$. Then, if $p(\mathbf{x})$ is slow of order ϵ^k we have that

$$|p(\mathbf{x}_j^\epsilon) - p(\mathbf{x}_j^{2\epsilon})| = o(1) \tag{42}$$

for all $j = 1, \ldots, N$ with N independent of ϵ.

Consider,

$$Q(p) = \sum_{j=1}^{N} \left[p(\mathbf{x}_j^{\epsilon}) - p(\mathbf{x}_j^{2\epsilon}) \right]^2. \tag{43}$$

Then, a polynomial $p(\mathbf{x})$ that minimizes $Q(p)$ is a good candidate for a variable that evolves on the ϵ^k scale. Since $Q(p)$ is quadratic in the coefficients of p, the minimizer can be found using least squares.

Finally, the process described above can be repeated, starting with the slowest order ϵ^0 and gradually moving to faster time scales. Each time a slow variables is identified, one can add a penalty that forces subsequent minimizers to be orthogonal to it in the space of polynomial coefficients. The method is described in [1].

References

1. G. Ariel, B. Engquist, and R. Tsai. A multiscale method for highly oscillatory ordinary differential equations with resonance. *Math. Comp.*, 78:929–956, 2009.
2. G. Ariel, B. Engquist, and R. Tsai. Numerical multiscale methods for coupled oscillators. *Multi. Mod. Simul.*, 7:1387–1404, 2009.
3. G. Ariel, B. Engquist, and R. Tsai. A reversible multiscale integration method. *Comm. Math. Sci.*, 7:595–610, 2009.
4. V.I. Arnol'd. *Mathematical methods of classical mechanics*. Springer-Verlag, New York, 1989.
5. Z. Artstein, I. G. Kevrekidis, M. Slemrod, and E. S. Titi. Slow observables of singularly perturbed differential equations. *Nonlinearity*, 20(11):2463–2481, 2007.
6. Z Artstein, J. Linshiz, and E. S. Titi. Young measure approach to computing slowly advancing fast oscillations. *Multiscale Model. Simul.*, 6(4):1085–1097, 2007.
7. C. Beck. Brownian motion from deterministic dynamics. *Phys. A*, 169:324–336, 1990.
8. M. P. Calvo and J. M. Sanz-Serna. Instabilities and inaccuracies in the integration of highly oscillatory problems. *SIAM J. Sci. Comput.*, 31(3):1653–1677, 2009.
9. W. E. Analysis of the heterogeneous multiscale method for ordinary differential equations. *Commun. Math. Sci.*, 1(3):423–436, 2003.
10. W. E and B. Engquist. The heterogeneous multiscale methods. *Commun. Math. Sci.*, 1(1):87–132, 2003.
11. W. E, B. Engquist, X. Li, W. Ren, and E. Vanden-Eijnden. Heterogeneous multiscale methods: A review. *Comm. Comput. Phys.*, 2:367–450, 2007.
12. W. E, D. Liu, and E. Vanden-Eijnden. Analysis of multiscale methods for stochastic differential equations. *Commun. on Pure and Applied Math.*, 58:1544–1585, 2005.
13. B. Engquist and Y.-H. Tsai. Heterogeneous multiscale methods for stiff ordinary differential equations. *Math. Comp.*, 74(252):1707–1742, 2005.
14. I. Fatkullin and E. Vanden-Eijnden. A computational strategy for multiscale chaotic systems with applications to Lorenz 96 model. *J. Comp. Phys.*, 200:605–638, 2004.
15. E. Fermi, J. Pasta, and S. Ulam. Studies of the nonlinear problems, i. *Los Alamos Report LA-1940*, 1955. Later published in *Collected Papers of Enrico Fermi*, ed. E. Segre, Vol. II (University of Chicago Press, 1965) p.978.
16. B. García-Archilla, J. M. Sanz-Serna, and R. D. Skeel. Long-time-step methods for oscillatory differential equations. *SIAM J. Sci. Comput.*, 20(3):930–963, 1999.
17. C. W. Gear and I. G. Kevrekidis. Projective methods for stiff differential equations: problems with gaps in their eigenvalue spectrum. *SIAM J. Sci. Comput.*, 24(4):1091–1106 (electronic), 2003.
18. C. W. Gear and I. G. Kevrekidis. Constraint-defined manifolds: a legacy code approach to low-dimensional computation. *J. Sci. Comput.*, 25(1-2):17–28, 2005.

19. C.W. Gear and K.A. Gallivan. Automatic methods for highly oscillatory ordinary differential equations. In *Numerical analysis (Dundee, 1981)*, volume 912 of *Lecture Notes in Math.*, pages 115–124. Springer, 1982.
20. D. Givon and R. Kupferman. White noise limits for discrete dynamical systems driven by fast deterministic dynamics. *Phys. A*, 335(3-4):385–412, 2004.
21. D. Givon, R. Kupferman, and A.M. Stuart. Extracting macroscopic dynamics: model problems and algorithms. *Nonlinearity*, 17:R55–R127, 2004.
22. M. Holland and I. Melbourne. Central limit theorems and invariance principles for Lorenz attractors. *J. Lond. Math. Soc. (2)*, 76(2):345–364, 2007.
23. H.-O. Kreiss. Problems with different time scales for ordinary differential equations. *SIAM J. Numer. Anal.*, 16(6):980–998, 1979.
24. H.-O. Kreiss. Problems with different time scales. In *Acta numerica, 1992*, pages 101–139. Cambridge Univ. Press, 1992.
25. H.-O. Kreiss and J. Lorenz. Manifolds of slow solutions for highly oscillatory problems. *Indiana Univ. Math. J.*, 42(4):1169–1191, 1993.
26. A.M. Majda, I. Timofeyev, and E. Vanden-Eijnden. Stochastic models for selected slow variables in large deterministic systems. *Nonlinearity*, 19:769–794, 2006.
27. I. Melbourne and A. M. Stuart. A note on diffusion limits of chaotic skew product flows. *Nonlinearity*, 24:1361–1367, 2011.
28. G. Papanicolaou. Introduction to the asymptotic analysis of stochastic equations. In *Modern modeling of continuum phenomena*, volume 16 of *Lectures in Applied Mathematics*, pages 47–109. Amer. Math. Soc., Providence, RI, 1977.
29. G. A. Pavliotis and A. M. Stuart. *Multiscale Methods: Averaging and Homogenization*. Number 53 in Texts in Applied Mathematics. Springer-Verlag, New York, 2008.
30. R.L. Petzold, O.J. Laurent, and Y. Jeng. Numerical solution of highly oscillatory ordinary differential equations. *Acta Numerica*, 6:437–483, 1997.
31. J. M. Sanz-Serna. Modulated Fourier expansions and heterogeneous multiscale methods. *IMA J. Numer. Anal.*, 29(3):595–605, 2009.
32. R.E. Scheid. The accurate numerical solution of highly oscillatory ordinary differential equations. *Mathematics of Computation*, 41(164):487–509, 1983.
33. M. Tao, H. Owhadi, and J. Marsden. Non-intrusive and structure preserving multiscale integration of stiff odes, sdes and hamiltonian systems with hidden slow dynamics via flow averaging. *Multiscale Model. Simul.*, 8:1269-1324, 2010.
34. E. Vanden-Eijnden. Numerical techniques for multi-scale dynamical systems with stochastic effects. *Comm. Math. Sci.*, 1:385–391, 2003.
35. E. Vanden-Eijnden. On HMM-like integrators and projective integration methods for systems with multiple time scales. *Commun. Math. Sci.*, 5:495–505, 2007.

Variance Reduction in Stochastic Homogenization: The Technique of Antithetic Variables

Xavier Blanc, Ronan Costaouec, Claude Le Bris, and Frédéric Legoll

Abstract This work is a follow up to previous articles of the same authors (Costaouec, Le Bris, and Legoll, Boletin Soc. Esp. Mat. Apl. 50:9–27, 2010; Blanc, Costaouec, Le Bris, and Legoll, Markov Processes and Related Fields, in press). It has been shown there, both numerically and theoretically, that the technique of antithetic variables successfully applies to stochastic homogenization of divergence-form linear elliptic problems and allows to reduce variance in computations. In (Costaouec, Le Bris, and Legoll, Boletin Soc. Esp. Mat. Apl. 50:9–27, 2010), variance reduction was assessed numerically for the diagonal terms of the homogenized matrix, in the case when the random field, that models uncertainty on some physical property at microscale, has a simple form. The numerical experiments have been complemented in Blanc, Costaouec, Le Bris, and Legoll (Markov Processes and Related Fields, in press) by a theoretical study. The main objective of this work is to proceed with some numerical experiments in a broader set of cases. We show the efficiency of the approach in each of the settings considered.

1 Introduction

Several settings in homogenization require the solution of corrector problems posed on the entire space $\mathbb{R}^d$. In practice, truncations of these problems over bounded domains are considered and the homogenized coefficients are obtained in the

R. Costaouec · C. Le Bris (✉) · F. Legoll
École Nationale des Ponts et Chaussées, 6 & 8 avenue Blaise Pascal, 77455 Marne-La-Vallée Cedex 2 and INRIA Rocquencourt, MICMAC team-project, Domaine de Voluceau, B.P. 105, 78153 Le Chesnay Cedex, France
e-mail: costaour@cermics.enpc.fr; lebris@cermics.enpc.fr; legoll@lami.enpc.fr

X. Blanc
CEA, DAM, DIF, 91297 Arpajon, France
e-mail: blanc@ann.jussieu.fr; Xavier.Blanc@cea.fr

B. Engquist et al. (eds.), *Numerical Analysis of Multiscale Computations*, Lecture Notes in Computational Science and Engineering 82, DOI 10.1007/978-3-642-21943-6_3,

limit of large domains. The question arises as to accelerate such computations. In the random case, the main difficulty is related to the intrinsic noise present in the simulation. Although very well investigated in other application fields such as financial mathematics, variance reduction techniques seem to have not been applied to the context of stochastic homogenization. In a previous article (see [9]), we have presented a first attempt to reduce the variance in stochastic homogenization using antithetic random variables. For this purpose, we have considered a simple situation. In particular, the equation under consideration was an elliptic equation in divergence form, with a *scalar* coefficient. In addition, the coefficient was assumed to consist of independent, identically distributed random variables set on a simple mesh (see (8) below). Though a bit restrictive, this situation pointed out that using antithetic variables results practically in diminishing the variance for the diagonal terms of the approximated homogenized matrix. We thus obtained an effective gain in computational time at fixed accuracy. Beyond this practical validation, we have also demonstrated, on a theoretical level and for some sufficiently simple situations, that the technique does reduce variance. The theoretical arguments of [3] not only apply to the examples of scalar random fields that we previously considered in [9] but they extend to a wider range of random fields. We present here some numerical tests that show that the technique still efficiently reduces variance in the presence of correlations and for matrices more general than those considered in our previous contributions. We also investigate variance reduction for eigenproblems.

For convenience of the reader and consistency, we devote the rest of the present section to a brief introductory exposition of random homogenization, the related numerical challenges, and the technique of antithetic variables. We turn in Sect. 2 to homogenization problems for materials that have correlations or that are anisotropic. Section 3 discusses variance reduction for eigenproblems.

More comprehensive tests that are not included here will be presented in [8]. Likewise, other variance reduction techniques, such as techniques based on control variates, will be the subject of future investigations and will be reported on elsewhere.

1.1 Homogenization Theoretical Setting

To begin with, we introduce the basic setting of stochastic homogenization we will employ. We refer to [10] for a general, numerically oriented presentation, and to [2, 7, 11] for classical textbooks. We also refer to [4, 5] or [13] for a presentation of our particular setting. Throughout this article, $(\Omega, \mathcal{F}, \mathbb{P})$ is a probability space and we denote by $\mathbb{E}(X) = \int_\Omega X(\omega) d\mathbb{P}(\omega)$ the expectation value of any random variable $X \in L^1(\Omega, d\mathbb{P})$. We next fix $d \in \mathbb{N}^*$ (the ambient physical dimension), and assume that the group $(\mathbb{Z}^d, +)$ acts on Ω. We denote by $(\tau_k)_{k\in\mathbb{Z}^d}$ this action, and assume that it preserves the measure $\mathbb{P}$, that is, for all $k \in \mathbb{Z}^d$ and all $A \in \mathcal{F}$, $\mathbb{P}(\tau_k A) = \mathbb{P}(A)$. We assume that the action τ is *ergodic*, that is, if $A \in \mathcal{F}$ is such that $\tau_k A = A$ for any $k \in \mathbb{Z}^d$, then $\mathbb{P}(A) = 0$ or 1. In addition, we define the following notion of stationarity (see [5]): any $F \in L^1_{\text{loc}}\left(\mathbb{R}^d, L^1(\Omega)\right)$ is said to be *stationary* if, for all $k \in \mathbb{Z}^d$,

$$F(x+k,\omega) = F(x,\tau_k\omega), \tag{1}$$

almost everywhere in x and almost surely. In this setting, the ergodic theorem [12, 15] can be stated as follows: *Let $F \in L^\infty\left(\mathbb{R}^d, L^1(\Omega)\right)$ be a stationary random variable in the above sense. For $k = (k_1, k_2, \dots, k_d) \in \mathbb{Z}^d$, we set $|k|_\infty = \sup_{1 \le i \le d} |k_i|$. Then*

$$\frac{1}{(2N+1)^d} \sum_{|k|_\infty \le N} F(x,\tau_k\omega) \underset{N\to\infty}{\longrightarrow} \mathbb{E}\left(F(x,\cdot)\right) \quad \textit{in } L^\infty(\mathbb{R}^d), \textit{ almost surely.}$$

This implies that (denoting by Q the unit cube in $\mathbb{R}^d$)

$$F\left(\frac{x}{\epsilon},\omega\right) \underset{\epsilon\to 0}{\overset{*}{\longrightarrow}} \mathbb{E}\left(\int_Q F(x,\cdot)dx\right) \quad \text{in } L^\infty(\mathbb{R}^d), \text{ almost surely.}$$

Besides technicalities, the purpose of the above setting is simply to formalize that, even though realizations may vary, the function F at point $x \in \mathbb{R}^d$ and the function F at point $x+k$, $k \in \mathbb{Z}^d$, share the same law. In the homogenization context we now turn to, this means that the local, microscopic environment (encoded in the matrix A) is everywhere the same *on average*. From this, homogenized, macroscopic properties will follow. In addition, and this is evident reading the above setting, the microscopic environment considered has a relation to an underlying *periodic* structure (thus the integer shifts k in (1)).

We now consider the elliptic boundary value problem

$$\begin{cases} -\text{div}\left(A\left(\dfrac{x}{\epsilon},\omega\right)\nabla u^\epsilon\right) = f & \text{in } \mathcal{D}, \\ u^\epsilon = 0 & \text{on } \partial\mathcal{D}, \end{cases} \tag{2}$$

set on an open, regular, bounded domain $\mathcal{D} \subset \mathbb{R}^d$. The right-hand side is an L^2 function f on $\mathcal{D}$. The random symmetric matrix A is assumed stationary in the sense (1) defined above. We also assume that A is bounded and that, in the sense of quadratic forms, A is positive and almost surely bounded away from zero. The mathematical theory of homogenization states that when ϵ goes to zero, u^ϵ converges to a *deterministic* function $u^\star$ that is solution of the so-called homogenized problem

$$\begin{cases} -\text{div}\left(A^\star\nabla u^\star\right) = f & \text{in } \mathcal{D}, \\ u^\star = 0 & \text{on } \partial\mathcal{D}. \end{cases} \tag{3}$$

In contrast to problem (2), problem (3) is *deterministic* and *does not involve the small scale* ϵ. It is hence easier to solve. Yet, the practical computation of the homogenized matrix $A^\star$, necessary for solving (3), is challenging. In our specific setting, this matrix reads

$$A_{ij}^{\star} = \int_Q \mathbb{E}\Big[(\nabla w_{e_j}(x,\cdot) + e_j)^T \; A(x,\cdot) \; (\nabla w_{e_i}(x,\cdot) + e_i)\Big] \, dx,$$

where, for any vector $p \in \mathbb{R}^d$, the *corrector* w_p is the solution (unique up to the addition of a random constant) in $\{w \in L^2_{\mathrm{loc}}(\mathbb{R}^d, L^2(\Omega)), \nabla w \in L^2_{\mathrm{unif}}(\mathbb{R}^d, L^2(\Omega))\}$ to

$$\begin{cases} -\mathrm{div}\big[A(\nabla w_p + p)\big] = 0 \quad \text{in } \mathbb{R}^d \ \text{ a.s.}, \\ \nabla w_p \text{ is stationary in the sense of (1)}, \\ \displaystyle\int_Q \mathbb{E}(\nabla w_p) = 0. \end{cases} \tag{4}$$

We have used the notation L^2_{unif} for the *uniform* L^2 space, that is the space of functions for which, say, the L^2 norm on a ball of unit size is bounded above independently from the center of the ball.

The major practical difficulty of random homogenization lies in the fact that the above problem (4), necessary for determining the homogenized matrix $A^\star$, is posed on the entire space $\mathbb{R}^d$.

1.2 Numerical Approach

In practice, the corrector problem (4), posed on the whole space $\mathbb{R}^d$, is approximated by the *truncated* corrector problem

$$\begin{cases} -\mathrm{div}\left(A(\cdot,\omega)\left(p + \nabla w_p^N(\cdot,\omega)\right)\right) = 0 \quad \text{in} \quad \mathbb{R}^d, \\ w_p^N(\cdot,\omega) \text{ is } Q_N\text{-periodic}, \end{cases}$$

posed on the cube $Q_N = (-N - 1/2, N + 1/2)^d$, centered at the origin. Correspondingly, the matrix $A^\star$ is then approximated by the *random* matrix

$$\left[A_N^\star\right]_{ij}(\omega) = \frac{1}{|Q_N|} \int_{Q_N} \left(e_i + \nabla w_{e_i}^N(y,\omega)\right)^T A(y,\omega) \left(e_j + \nabla w_{e_j}^N(y,\omega)\right) dy.$$

Although $A^\star$ itself is a deterministic object, its practical approximation $A_N^\star$ is random. It is only in the limit of infinitely large domains Q_N that the deterministic value is attained [6].

Besides the homogenized matrix $A^\star$ itself, other related quantities, such as the eigenelements of the matrix $A^\star$, the solution $u^\star$ of the homogenized problem (3), and the eigenelements of the operator $L_{A^\star} = -\mathrm{div}(A^\star \nabla \cdot)$, are of major interest. They all reflect some property of the homogenized material. As is the case for $A^\star$, only *random* approximations of those quantities are accessible. We formalize this by saying that all these quantities, denoted by $\mathscr{F}(A^\star)$, are approximated by the corresponding random variables $\mathscr{F}\left(A_N^\star(\omega)\right)$ obtained by truncation and approximation (using a Monte Carlo method). For simplicity, we will suppose throughout this article that $\mathscr{F}$ is scalar valued. Let $(A^{\mathrm{m}}(x,\omega))_{1\le \mathrm{m}\le M}$ denote M independent and identically distributed underlying random fields. We define a family $\left(A_N^{\star,\mathrm{m}}\right)_{1\le \mathrm{m}\le M}$ of i.i.d. homogenized matrices by, for any $1\le i,j\le d$,

$$\left[A_N^{\star,\mathrm{m}}\right]_{ij}(\omega)=\frac{1}{|Q_N|}\int_{Q_N}\left(e_i+\nabla w_{e_i}^{N,\mathrm{m}}(\cdot,\omega)\right)^T A^{\mathrm{m}}(\cdot,\omega)\left(e_j+\nabla w_{e_j}^{N,\mathrm{m}}(\cdot,\omega)\right),$$

where $w_{e_j}^{N,\mathrm{m}}$ is the solution of the truncated corrector problem associated to A^{m}. Then we define for each quantity $\mathscr{F}\left(A_N^\star\right)$ the empirical mean and variance

$$\begin{aligned}\mu_M\left(\mathscr{F}\left(A_N^\star\right)\right)&=\frac{1}{M}\sum_{\mathrm{m}=1}^{M}\mathscr{F}\left(A_N^{\star,\mathrm{m}}\right),\\ \sigma_M\left(\mathscr{F}\left(A_N^\star\right)\right)&=\frac{1}{M-1}\sum_{\mathrm{m}=1}^{M}\left(\mathscr{F}\left(A_N^{\star,\mathrm{m}}\right)-\mu_M\left(\mathscr{F}\left(A_N^\star\right)\right)\right)^2.\end{aligned}\tag{5}$$

Since the matrices $A_N^{\star,\mathrm{m}}$ are i.i.d. the strong law of large numbers applies:

$$\mu_M\left(\mathscr{F}\left(A_N^\star\right)\right)(\omega)\underset{M\to+\infty}{\longrightarrow}\mathbb{E}\left(\mathscr{F}\left(A_N^\star\right)\right)\text{ almost surely.}$$

The central limit theorem then yields

$$\sqrt{M}\left(\mu_M\left(\mathscr{F}\left(A_N^\star\right)\right)-\mathbb{E}\left(\mathscr{F}\left(A_N^\star\right)\right)\right)\underset{M\to+\infty}{\overset{\mathscr{L}}{\longrightarrow}}\sqrt{\mathrm{Var}(\mathscr{F}(A_N^\star))}\,\mathscr{N}(0,1),$$

where the convergence holds in law, and $\mathscr{N}(0,1)$ denotes the standard Gaussian law. Introducing its 95 percent quantile, it is standard to consider that the exact mean $\mathbb{E}\left(\mathscr{F}\left(A_N^\star\right)\right)$ lies in the interval

$$\left[\mu_M\left(\mathscr{F}\left(A_N^\star\right)\right)-1.96\frac{\sqrt{\sigma_M\left(\mathscr{F}\left(A_N^\star\right)\right)}}{\sqrt{M}},\mu_M\left(\mathscr{F}\left(A_N^\star\right)\right)+1.96\frac{\sqrt{\sigma_M\left(\mathscr{F}\left(A_N^\star\right)\right)}}{\sqrt{M}}\right].\tag{6}$$

The value $\mu_M\left(\mathscr{F}\left(A_N^\star\right)\right)$ is thus, for both M and N sufficiently large, adopted as the approximation of the exact value $\mathscr{F}\left(A^\star\right)$.

Our aim is to design a numerical technique that, for finite N, allows to compute a better approximation of $\mathbb{E}\left(\mathscr{F}\left(A_N^\star\right)\right)$, *e.g.* an approximation with smaller variance.

1.3 The Technique of Antithetic Variables

The application of variance reduction using the antithetic variable technique, a classical variance reduction technique ubiquitous in many applied fields, to the specific framework of stochastic homogenization was first performed in [9]. For the sake of completeness we outline here the basic steps of the approach in our specific context. For an elementary introduction to the technique, we refer to [14].

Fix $M = 2\mathcal{M}$ and suppose that we give ourselves $\mathcal{M}$ i.i.d. copies $\big(A^{\mathbf{m}}(x,\omega)\big)_{1\le\mathbf{m}\le\mathcal{M}}$ of $A(x,\omega)$. Construct next $\mathcal{M}$ i.i.d. *antithetic* random fields

$$B^{\mathbf{m}}(x,\omega) = T\left(A^{\mathbf{m}}(x,\omega)\right), \quad 1 \le \mathbf{m} \le \mathcal{M},$$

from the $(A^{\mathbf{m}}(x,\omega))_{1\le\mathbf{m}\le\mathcal{M}}$. The map T transforms the random field $A^{\mathbf{m}}$ into another, so-called *antithetic*, field $B^{\mathbf{m}}$. Given the random field $A(x,\omega)$, there is a large variety of possible choices for the antithetic field B. However, the transformation is to be performed in such a way that, for each $\mathbf{m}$, $B^{\mathbf{m}}$ has the same law as $A^{\mathbf{m}}$. Besides this constraint, only practice dictates appropriate choices. We are providing examples below, see *e.g.* (22)–(23). Somewhat vaguely stated, if the coefficient was obtained in a coin tossing game (using a fair coin), then the antithetic coefficient would be head each time the original coefficient is tail and vice versa. Then, for each $1 \le \mathbf{m} \le \mathcal{M}$, we solve two corrector problems. One is associated to the original $A^{\mathbf{m}}$, the other one is associated to the antithetic field $B^{\mathbf{m}}$. Using its solution $v_p^{N,\mathbf{m}}$, we define the *antithetic homogenized matrix* $B_N^{\star,\mathbf{m}}$, the entries of which read, for $1 \le i,j \le d$,

$$\left[B_N^{\star,\mathbf{m}}\right]_{ij}(\omega) = \frac{1}{|Q_N|}\int_{Q_N}\Big(e_i + \nabla v_{e_i}^{N,\mathbf{m}}(\cdot,\omega)\Big)^T B^{\mathbf{m}}(\cdot,\omega)\,\Big(e_j + \nabla v_{e_j}^{N,\mathbf{m}}(\cdot,\omega)\Big).$$

And finally we set, for any $1 \le \mathbf{m} \le \mathcal{M}$,

$$\widetilde{A}_N^{\star,\mathbf{m}}(\omega) := \frac{1}{2}\left(A_N^{\star,\mathbf{m}}(\omega) + B_N^{\star,\mathbf{m}}(\omega)\right).$$

Since $A^{\mathbf{m}}$ and $B^{\mathbf{m}}$ are identically distributed, so are $A_N^{\star,\mathbf{m}}$ and $B_N^{\star,\mathbf{m}}$. Thus, $\widetilde{A}_N^{\star,\mathbf{m}}$ is unbiased (that is, $\mathbb{E}\left(\tilde{A}_N^{\star,\mathbf{m}}\right) = \mathbb{E}\left(A_N^{\star,\mathbf{m}}\right)$). In addition, it satisfies:

$$\widetilde{A}_N^{\star,\mathbf{m}} \underset{N\to+\infty}{\longrightarrow} A^\star \text{ almost surely},$$

because B is ergodic. The matrix $\widetilde{A}_N^\star$ is thus an alternative random variable that converges almost surely to $A^\star$ when $N \to \infty$. In addition, for any N, the mean of $\widetilde{A}_N^\star$ is equal to that of $A_N^\star$. Consequently, $\widetilde{A}_N^\star$ can be used to define new estimators.

In order to compute an approximation of $\mathbb{E}\left(\mathscr{F}(A_N^\star)\right)$, we use the antithetic variable defined above, and define

$$\mu_{\mathcal{M}}\left(\widetilde{\mathscr{F}}\left(A_N^\star\right)\right) = \frac{1}{\mathcal{M}}\sum_{\mathbf{m}=1}^{\mathcal{M}} \widetilde{\mathscr{F}}\left(A_N^{\star,\mathbf{m}}\right) = \frac{1}{\mathcal{M}}\sum_{\mathbf{m}=1}^{\mathcal{M}} \frac{\mathscr{F}\left(A_N^{\star,\mathbf{m}}\right) + \mathscr{F}\left(B_N^{\star,\mathbf{m}}\right)}{2},$$

$$\sigma_{\mathcal{M}}\left(\widetilde{\mathscr{F}}\left(A_N^\star\right)\right) = \frac{1}{\mathcal{M}-1}\sum_{\mathbf{m}=1}^{\mathcal{M}}\left(\widetilde{\mathscr{F}}\left(A_N^{\star,\mathbf{m}}\right) - \mu_{\mathcal{M}}\left(\widetilde{\mathscr{F}}\left(A_N^\star\right)\right)\right)^2,$$

which require $2\mathcal{M}$ resolutions of corrector problems, i.e. as many as the classical estimators (5). Our new random variable has variance

$$\mathrm{Var}\left(\widetilde{\mathcal{F}}\left(A_N^\star\right)\right) = \frac{1}{2}\mathrm{Var}\left(\mathcal{F}\left(A_N^\star\right)\right) + \frac{1}{2}\mathrm{Cov}\left(\mathcal{F}\left(A_N^\star\right), \mathcal{F}\left(B_N^\star\right)\right). \qquad (7)$$

Applying the central limit theorem to $\widetilde{\mathcal{F}}\left(A_N^\star\right)$, we obtain

$$\sqrt{\mathcal{M}}\left(\mu_{\mathcal{M}}\left(\widetilde{\mathcal{F}}\left(A_N^\star\right)\right) - \mathbb{E}\left(\widetilde{\mathcal{F}}\left(A_N^\star\right)\right)\right) \xrightarrow[\mathcal{M}\to+\infty]{\mathcal{L}} \sqrt{\mathrm{Var}\left(\widetilde{\mathcal{F}}\left(A_N^\star\right)\right)}\, \mathcal{N}(0,1).$$

Similarly to (6), we deduce a confidence interval from this convergence. The exact mean $\mathbb{E}\left(\widetilde{\mathcal{F}}\left(A_N^\star\right)\right)$ is equal to $\mu_{\mathcal{M}}\left(\widetilde{\mathcal{F}}\left(A_N^\star\right)\right)$ within a typical margin of error

$$1.96\frac{\sqrt{\mathrm{Var}\left(\widetilde{\mathcal{F}}\left(A_N^\star\right)\right)}}{\sqrt{\mathcal{M}}}.$$

We see on (7) that, when

$$\mathrm{Cov}\left(\mathcal{F}\left(A_N^\star\right), \mathcal{F}\left(B_N^\star\right)\right) \leq 0,$$

the width of the interval of confidence has been reduced by the approach, and, consequently, the quality of approximation at given computational cost has improved.

1.4 A Brief Summary of Our Former Results

We have considered in [9] the case of an isotropic random field

$$A(x,\omega) = \sum_{k\in\mathbb{Z}^d} \mathbf{1}_{Q+k}(x) a_k(\omega)\mathrm{Id} = \sum_{k\in\mathbb{Z}^d} \mathbf{1}_{Q+k}(x) f\left(X_k(\omega)\right)\mathrm{Id} \qquad (8)$$

where $(X_k(\omega))_{k\in\mathbb{Z}^d}$ is a family of independent uniform random variables and f is a monotone function. For well-posedness, we assume that there exist $\alpha > 0$ and $\beta < \infty$ such that, for all k, $0 < \alpha \leq a_k \leq \beta < +\infty$ almost surely. Consequently, A is uniformly coercive and bounded. The quantity we mainly considered in [9] is $\mathcal{F}\left(A_N^\star\right) = \left[A_N^\star\right]_{ii}$, an approximation of a diagonal entry of the matrix $A^\star$. We have demonstrated numerically the efficiency of the approach. We have also discussed in the same reference how the approach affects (and indeed reduces) the variance of other outputs, such as the solution $u^\star$ to (3).

Another purpose of [9] was to investigate the approach theoretically. The one-dimensional setting was addressed. The study has been complemented by a study in higher dimensions in [3]. A particularly useful ingredient, theoretically, is, somewhat vaguely stated, the monotonicity of the homogenized objects in function of the original random field. More precisely, we proved in [3] that variance is indeed reduced as long as the output $\mathscr{F}\left(A_N^\star\right)$ we consider is monotone with respect to each of the uniform random variables. The arguments given in [3] apply under the following *structure hypothesis* on A: for any N, there exists an integer n (possibly $n = |Q_N|$, but not necessarily) and a function $\mathscr{A}$, defined on $Q_N \times \mathbb{R}^n$, such that the tensor $A(x,\omega)$ writes

$$\forall x \in Q_N, \quad A(x,\omega) = \mathscr{A}(x, X_1(\omega), \dots, X_n(\omega)) \quad \text{a.s.}, \tag{9}$$

where $\{X_k(\omega)\}_{1 \le k \le n}$ are independent scalar random variables, which are all distributed according to the uniform law $\mathscr{U}[0,1]$. Then the global monotonicity of $\mathscr{F}\left(A_N^\star\right)$ is related to the following composition scheme

$$\{X_k(\omega)\}_{1 \le k \le n} \xrightarrow{\mathscr{A}} A(x,\omega) \xrightarrow{\mathscr{H}} A_N^\star(\omega) \xrightarrow{\mathscr{F}} \mathscr{F}\left(A_N^\star\right),$$

where $\mathscr{H}$ denotes the application associated to periodic homogenization. Since $\mathscr{H}$ is increasing in the sense of symmetric matrices, the global monotonicity only depends on our way to model randomness $\mathscr{A}$ and the output $\mathscr{F}$ we are interested in. In [3], we proved that variance is indeed reduced by the approach described in Sect. 1.3 when $\mathscr{A}$ is non-decreasing with respect to each of its argument, and $\mathscr{F}$ is monotone.

2 Variance Reduction for Problems Involving Correlations or Anisotropy

Our theoretical results encourage us to apply the technique to more general cases than the simple cases addressed in [9]. We will subsequently consider in this section two specific situations:

- **Correlated isotropic fields**, that is matrices A in (2) of the form

$$A(x,\omega) = \sum_{k \in \mathbb{Z}^d} \mathbf{1}_{Q+k}(x) a_k(\omega) \mathrm{Id} = \sum_{k \in \mathbb{Z}^d} \mathbf{1}_{Q+k}(x) F\left(\left\{X_{k+j}\right\}_{|j|_\infty \le p}(\omega)\right) \mathrm{Id}, \tag{10}$$

  where p is some fixed non-negative integer, $\{X_k(\omega)\}_{k \in \mathbb{Z}^d}$ is a family of independent real-valued random variables and F is defined on $\mathbb{R}^{2p+1}$ and real valued;
- **I.i.d. anisotropic fields**, that is matrices A in (2) of the form

$$A(x,\omega) = \sum_{k \in \mathbb{Z}^d} \mathbf{1}_{Q+k}(x) A_k(\omega) = \sum_{k \in \mathbb{Z}^d} \mathbf{1}_{Q+k}(x) F(X_k(\omega)), \tag{11}$$

  where $\{X_k(\omega)\}_{k \in \mathbb{Z}^d}$ is a family of independent $\mathbb{R}^{N_{rv}}$-valued random vectors, the components of which are independent and identically distributed (we choose the uniform law). The function F, defined on $\mathbb{R}^{N_{rv}}$, is valued in the set of symmetric matrices.

Of course, we could combine the structure assumptions (10) and (11) to form correlated anisotropic random fields, but we will not proceed in this direction here.

In the case of correlated fields, in line with the theoretical observations of [3] recalled in the previous section, we assume that the function F is non-decreasing with respect to each of its arguments. In the case of anisotropic fields, we will first consider functions F that are non-decreasing. To check the robustness of the approach, we will second consider functions F that are non monotone.

We will specifically investigate four questions.

First, considering the correlated isotropic case, we will try to understand how correlation affects the efficiency of our variance reduction technique (see Sect. 2.1). To this end, we consider variance reduction of the diagonal entries of the matrix $A^\star$, first on the correlated case (10), second on an uncorrelated case, as we previously did in [9]. Comparing the two cases will outline the influence of correlation. In this context, the monotonicity assumptions are satisfied and we are thus proceeding on a sound theoretical ground.

Second, we will use anisotropic fields generated using *monotone* functions $\mathscr{A}$ in (9) (that is, monotone functions F in (11)), and that have homogenized matrices with non trivial off-diagonal terms (see Sect. 2.2, Example 1). We will double-check that variance is reduced on diagonal terms as was the case in our previous study. As for off-diagonal terms, which are *not* monotone functions of the random fields, we cannot rely on any theoretical guideline. As our experiments will show, we still reduce variance, though.

Third, we will consider anisotropic fields that do not correspond to monotone functions $\mathscr{A}$ (they are of the form (11) with a non-monotone F). Absent any theoretical analysis, we investigate numerically variance reduction on both diagonal and off-diagonal terms (see Sect. 2.2, Examples 2 and 3).

Fourth, again using anisotropic fields, we will consider variance reduction of eigenelements (see Sect. 3).

2.1 Correlated Cases

We consider a two dimensional situation and proceed computationally as explained in Sect. 1. We restrict ourselves to considering the first diagonal entry $\left[A_N^\star\right]_{11}$. In order to investigate the role of correlation, we consider random fields of form (10)

$$A(x,\omega) = \sum_{k\in\mathbb{Z}^d} \mathbf{1}_{Q+k}(x) F\left(\left\{a_{k+j}\right\}_{|j|_\infty \le p}(\omega)\right) \mathrm{Id}, \qquad (12)$$

with correlation length p. We begin with the case $p = 1$ and next consider some larger values of p. In order to focus on the effect of correlation, we will not only monitor the variance reduction for the homogenized matrix $A^\star$ associated to the above matrix A. We also consider a similar matrix, where the correlation is set to zero, and apply the variance reduction technique for its homogenization. More precisely, we introduce

$$C(x,\omega) = \sum_{k\in\mathbb{Z}^d} \mathbf{1}_{Q+k}(x) F\left(\left\{c_{k,j}\right\}_{|j|_\infty\le 1}(\omega)\right) \mathrm{Id}, \tag{13}$$

where $\left\{(c_{k,j})_{|j|_\infty\le 1}\right\}_{k\in\mathbb{Z}^d}$ denotes a family of i.i.d. random *vectors*, the components of which are *independent from one another* and share the exact same law as the a_k (which we take here as the uniform law). The local behaviour (meaning, the behaviour on a single unit cell) of the field C is similar to that of the field A. However, when it comes to the global fields seen as functions on the entire space, the behaviours differ, because correlation is turned off in the case of C. In the very peculiar one-dimensional situation (where homogenization is a local, pointwise, process), the homogenized matrices $A^\star$ and $C^\star$ respectively obtained from A and C are identical. The variance of the approximate matrices $A^\star_N$ and $C^\star_N$ can be different, though. Some elementary arguments, not included here and for which we refer to [8], allow to prove that in both cases we reduce variance using the technique of antithetic variable. In dimensions higher than or equal to 2, $A^\star \neq C^\star$. The matrix $C^\star$ serves as a useful reference to evaluate how correlation affects the efficiency of our variance reduction technique.

In the numerical examples below, the random variables

$$\{a_k\}_{k\in\mathbb{Z}^d} \quad \text{and} \quad \{c_{k,j}\}_{k\in\mathbb{Z}^d,|j|_\infty\le 1}$$

are all uniformly distributed between $\alpha = 3$ and $\beta = 20$.

2.1.1 Influence of Correlation: Identical Local Behaviour

We define the function F in (12) as

$$F\left(\left\{a_{k+q}\right\}_{|q|\le 1}\right) = \frac{1}{9}\sum_{|q|\le 1} a_{k+q}. \tag{14}$$

For comparison purposes, the field C of (13) is, as announced above, defined by

$$F\left(\left\{c_{k,l}\right\}_{|l|\le 1}\right) = \frac{1}{9}\sum_{|l|\le 1} c_{k,l}. \tag{15}$$

Our results for the variance reduction on $\left[A^\star_N\right]_{11}$ and $\left[C^\star_N\right]_{11}$ are reported in Tables 1 and 2, respectively. As mentioned in [9], because of isotropy and invariance by rotation of angle $\pi/2$, the corresponding approximated homogenized matrix $A^\star_N$ and $C^\star_N$ are, like the exact homogenized matrices $A^\star$ and $C^\star$, isotropic. In each table, the ratio of variance

$$R\left(\left[A^\star_N\right]_{11}\right) = \frac{\sigma_{100}\left(\left[A^\star_N\right]_{11}\right)}{2\sigma_{50}\left(\left[\widetilde{A}^\star_N\right]_{11}\right)} \tag{16}$$

Table 1 Correlated equidistributed case (12)–(14), $p = 1$: mean and standard deviation of $[A_N^\star]_{11}$

N	$\mu_{100}([A_N^\star]_{11})$	$\sqrt{\sigma_{100}}([A_N^\star]_{11})$	$\mu_{50}\left(\left[\widetilde{A}_N^\star\right]_{11}\right)$	$\sqrt{\sigma_{50}}\left(\left[\widetilde{A}_N^\star\right]_{11}\right)$	$R([A_N^\star]_{11})$
40	11.3791	0.0637	11.3824	0.0054	69.25
60	11.3794	0.0438	11.3818	0.0028	121.07
80	11.3765	0.0310	11.3821	0.0029	58.42
100	11.3794	0.0259	11.3818	0.0018	97.21

Table 2 Uncorrelated equidistributed case (13)–(15), $p = 1$: mean and standard deviation of $[C_N^\star]_{11}$

N	$\mu_{100}([C_N^\star]_{11})$	$\sqrt{\sigma_{100}}([C_N^\star]_{11})$	$\mu_{50}\left(\left[\widetilde{C}_N^\star\right]_{11}\right)$	$\sqrt{\sigma_{50}}\left(\left[\widetilde{C}_N^\star\right]_{11}\right)$	$R([C_N^\star]_{11})$
40	11.3843	0.0226	11.3859	0.0021	55.64
60	11.3850	0.0153	11.3858	0.0017	40.95
80	11.3863	0.0111	11.3858	0.0012	40.53
100	11.3863	0.0091	11.3860	0.0009	51.00

measures the reduction of uncertainty on estimations of $\mathbb{E}\left(A_N^\star\right)$ at fixed computational cost, that is, the efficiency of the variance reduction technique. It corresponds to the ratio of the square of the widths of intervals of confidence. We will use a similar ratio (with obvious definition and notation) for all the tables presented throughout this article.

From the consideration of Tables 1 and 2 we conclude that correlation does not affect the efficiency of the technique.

Note that we observe here a ratio R of the order of 40, better than in [9]. It owes to the fact that we deliberately considered in [9] more challenging test cases in order to prove that variance can be reduced in generic situations, even demanding ones in terms of normalized variance. Here *we are focusing on the effect of correlation only*, and our purpose, different in nature from that of [9], is to compare the correlated and the uncorrelated situations. Indeed, denoting by $A_k(\omega) = F\left(\left\{a_{k+q}\right\}_{|q|\leq 1}\right)$ and $C_k(\omega) = F\left(\left\{c_{k,l}\right\}_{|l|\leq 1}\right)$ the local values of the correlated field (12)–(14) and of the uncorrelated field (13)–(15) respectively, we see here that the corresponding normalized variance reads

$$\frac{\text{Var } A_k}{(\mathbb{E}\ A_k)^2} = \frac{\text{Var } C_k}{(\mathbb{E}\ C_k)^2} = \frac{\text{Var } c_{0,0}}{9(\mathbb{E}\ c_{0,0})^2} = \frac{(\beta-\alpha)^2}{27(\beta+\alpha)^2} \approx 0.0202.$$

In contrast, in the test case (iii) of [9], the random field is

$$A(x,\omega) = \sum_{k\in\mathbb{Z}^d} \mathbf{1}_{Q+k}(x) a_k(\omega)\,\mathrm{Id}, \tag{17}$$

where $\{a_k\}_{k\in\mathbb{Z}^d}$ is a family of i.i.d. variables uniformly distributed between $\alpha_0 = 3$ and $\beta_0 = 20$. The normalized variance of the local value of $A(x,\omega)$ hence reads

$$\frac{\mathrm{Var}\, a_k}{(\mathbb{E}\, a_k)^2} = \frac{(\beta_0-\alpha_0)^2}{3(\beta_0+\alpha_0)^2} \approx 0.182, \tag{18}$$

and is indeed 9 times as large as the normalized local variance considered here. Our formal considerations above are confirmed by the numerical results shown in Table 3, where we consider the test case (17), this time with $\alpha_0 = 3$ and $\beta_0 = 5$, so that the normalized local variance (which is now equal to 0.0208, in view of (18)) is close to the one of the fields (12)–(14) and (13)–(15). We again obtain an efficiency ratio close to 50. So, the normalized local variance of the fields (12)–(14), (13)–(15) (with $\alpha = 3$ and $\beta = 20$) and (17) (with $\alpha_0 = 3$ and $\beta_0 = 5$) are of the same order, and we indeed observe an efficiency ratio R of the same order.

Table 3 Uncorrelated case (17), where $a_k \sim \mathscr{U}[\alpha_0,\beta_0]$, with $\alpha_0 = 3$ and $\beta_0 = 5$: mean and standard deviation of $\left[A_N^\star\right]_{11}$

N	$\mu_{100}\left([A_N^\star]_{11}\right)$	$\sqrt{\sigma_{100}}\left([A_N^\star]_{11}\right)$	$\mu_{50}\left(\left[\widetilde{A}_N^\star\right]_{11}\right)$	$\sqrt{\sigma_{50}}\left(\left[\widetilde{A}_N^\star\right]_{11}\right)$	$R\left([A_N^\star]_{11}\right)$
40	3.9597	0.0071	3.9595	0.00069	52.87
60	3.9591	0.0049	3.9595	0.00045	59.18
80	3.9589	0.0037	3.9594	0.00035	55.66
100	3.9590	0.0030	3.9594	0.00025	71.53

Remark 1. Consider, in the one-dimensional setting, the case

$$A(x,\omega) = \sum_{k\in\mathbb{Z}} \mathbf{1}_{[k,k+1)}(x) a_k(\omega),$$

where $\{a_k\}_{k\in\mathbb{Z}}$ is a family of i.i.d. variables uniformly distributed between α_0 and β_0. Then the efficiency ratio (16), which we write here as

$$R_N = \frac{\mathrm{Var}\left(A_N^\star\right)}{2\mathrm{Var}\left(\widetilde{A}_N^\star\right)},$$

is analytically computable. After tedious but straightforward computations (see [8] for details), we obtain

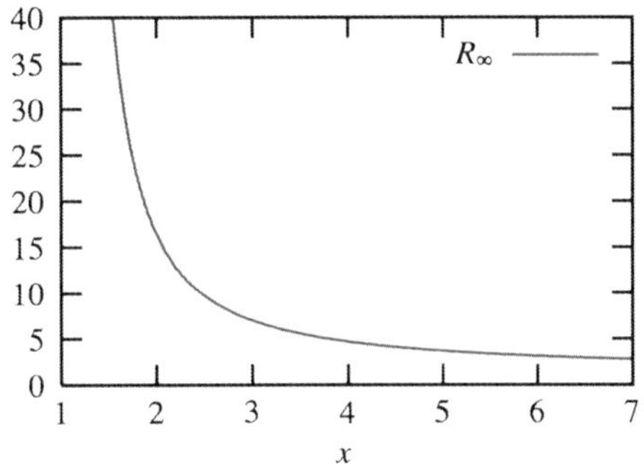

Fig. 1 Variance reduction efficiency R_∞ defined by (19), as a function of $x = \beta_0/\alpha_0$

$$\lim_{N\to\infty} R_N = R_\infty = \left[1 - \frac{g(x)\ \ln(x)}{(x-1)[1/x - (\ln(x)/(x-1))^2]}\right]^{-1} \tag{19}$$

where $x = \beta_0/\alpha_0 > 1$ and $g(x) = \ln(x)/(x-1) - 2/(1+x)$. On Fig. 1, we plot R_∞ as a function of x. For any x, we observe that $R_\infty > 1$, that is the variance reduction technique is indeed efficient, and provides a more accurate estimation of $\mathbb{E}\left(A_N^\star\right)$ for an equal computational cost. We also observe that R_∞ is a decreasing function of x, which tends to 1 as x tends to infinity. This one-dimensional study also confirms our considerations above: the technique always allows to reduce the variance, but is all the more efficient as the original normalized variance (here intuitively measured by the quotient x, and above measured by the ratio (18)) is small.

2.1.2 Centered vs Equidistributed Correlation Structure

We now compare two different correlation structures sharing the same correlation length $p = 1$. The first structure is the equidistributed case (14). As for the second structure, we consider

$$F\left(\{a_{k+q}\}_{|q|\le 1}\right) = \frac{1}{2} a_k + \frac{1}{16} \sum_{|q|\le 1; q\neq 0} a_{k+q}, \tag{20}$$

where, as in (14), $\{a_k\}_{k\in\mathbb{Z}^d}$ refers to a family of i.i.d. random variables uniformly distributed between $\alpha = 3$ and $\beta = 20$. From Tables 1 and 4, we see that the correlation structure affects the efficiency of the method, but the reduction remains significant.

Table 4 Correlated centered case (12)–(20), $p = 1$: mean and standard deviation of $[A_N^\star]_{11}$

N	$\mu_{100}([A_N^\star]_{11})$	$\sqrt{\sigma_{100}}([A_N^\star]_{11})$	$\mu_{50}\left(\left[\widetilde{A}_N^\star\right]_{11}\right)$	$\sqrt{\sigma_{50}}\left(\left[\widetilde{A}_N^\star\right]_{11}\right)$	$R([A_N^\star]_{11})$
40	11.1531	0.0655	11.1563	0.0083	31.17
60	11.1527	0.0448	11.1551	0.0050	39.89
80	11.1494	0.0319	11.1552	0.0045	25.43
100	11.1523	0.0267	11.1548	0.0028	45.52

2.1.3 Longer Correlation Lengths

We now let our parameter p modeling the correlation length increase, and consider (12), with F defined by

$$F\left(\{a_{k+q}\}_{|q|\leq p}\right) = \frac{1}{(2p+1)^2} \sum_{|q|\leq p} a_{k+q}, \tag{21}$$

with values $p = 2$, $p = 5$ and $p = 10$ (the case $p = 1$ corresponds to (14)). Results are reported in Tables 5–7. Comparing also with Table 1, we see that increasing the correlation length indeed affects, in fact advantageously, the efficiency of the variance reduction.

Table 5 Correlated equidistributed case (12)–(21), $p = 2$: mean and standard deviation of $[A_N^\star]_{11}$

N	$\mu_{100}([A_N^\star]_{11})$	$\sqrt{\sigma_{100}}([A_N^\star]_{11})$	$\mu_{50}\left(\left[\widetilde{A}_N^\star\right]_{11}\right)$	$\sqrt{\sigma_{50}}\left(\left[\widetilde{A}_N^\star\right]_{11}\right)$	$R([A_N^\star]_{11})$
40	11.4563	0.05697	11.4574	0.0033	145.75
60	11.4556	0.03641	11.4580	0.0023	121.15
80	11.4528	0.03053	11.4579	0.0017	169.72
100	11.4554	0.02641	11.4579	0.0013	214.58

Table 6 Correlated equidistributed case (12)–(21), $p = 5$: mean and standard deviation of $[A_N^\star]_{11}$

N	$\mu_{100}([A_N^\star]_{11})$	$\sqrt{\sigma_{100}}([A_N^\star]_{11})$	$\mu_{50}\left(\left[\widetilde{A}_N^\star\right]_{11}\right)$	$\sqrt{\sigma_{50}}\left(\left[\widetilde{A}_N^\star\right]_{11}\right)$	$R([A_N^\star]_{11})$
40	11.4871	0.0592	11.4912	0.0014	886.83
60	11.4853	0.0413	11.4914	0.0010	873.11
80	11.4882	0.0313	11.4915	0.0007	994.81
100	11.4888	0.0246	11.4914	0.0005	1015.36

Table 7 Correlated equidistributed case (12)–(21), $p = 10$: mean and standard deviation of $[A_N^\star]_{11}$

N	$\mu_{100}([A_N^\star]_{11})$	$\sqrt{\sigma_{100}}([A_N^\star]_{11})$	$\mu_{50}\left(\left[\widetilde{A}_N^\star\right]_{11}\right)$	$\sqrt{\sigma_{50}}\left(\left[\widetilde{A}_N^\star\right]_{11}\right)$	$R([A_N^\star]_{11})$
40	11.4970	0.0596	11.4978	0.0007	3360.50
60	11.4940	0.0392	11.4977	0.0006	2473.20
80	11.4963	0.0301	11.4977	0.0003	3719.40
100	11.4937	0.0251	11.4977	0.0002	4640.11

2.2 Anisotropic Cases

2.2.1 Test Cases

To begin with, we introduce three test cases we will focus on in the remainder of this section. They correspond to different *deterministic* functions $\mathscr{A}$, that is, different ways of constructing in (9) the field $A(x,\omega)$ from the random variables.

Example 1

We consider a random matrix

$$A_1(x,\omega) = P\left(\sum_{k\in\mathbb{Z}^2} \mathbf{1}_{Q+k}(x)\begin{pmatrix}\lambda_k^1(\omega) & 0\\ 0 & \lambda_k^2(\omega)\end{pmatrix}\right)P^T \quad \text{with} \quad P = \frac{1}{\sqrt{2}}\begin{pmatrix}1 & -1\\ 1 & 1\end{pmatrix}, \tag{22}$$

where $\{\lambda_k^1\}_{k\in\mathbb{Z}^2}$ and $\{\lambda_k^2\}_{k\in\mathbb{Z}^2}$ are two independent families of independent random variables uniformly distributed on $[\alpha,\beta]$ and $[\delta,\gamma]$ respectively. We assume that

$$\rho_1 = \min(\alpha,\delta) > 0,$$

so that, for all $k \in \mathbb{Z}^2$, $\lambda_k^1(\omega) \geq \rho_1 > 0$ and $\lambda_k^2(\omega) \geq \rho_1 > 0$ almost surely.

This case corresponds to the *deterministic* function

$$\mathscr{A}_1(x, \{(y_k, z_k)\}_{k\in\mathbb{Z}^2}) = P\left(\sum_{k\in\mathbb{Z}^2} \mathbf{1}_{Q+k}(x) \begin{pmatrix} \alpha + (\beta-\alpha)y_k & 0 \\ 0 & \delta + (\gamma-\delta)z_k \end{pmatrix}\right) P^T,$$

and to the choice

$$A_1(x,\omega) = \mathscr{A}_1(x, \{(Y_k(\omega), Z_k(\omega))\}_{k\in\mathbb{Z}^2}),$$

where Y_k and Z_k are i.i.d. random variables with uniform law on $[0,1]$. Note that $\mathscr{A}_1$ is indeed non-decreasing with respect to any y_k and z_k. The associated antithetic field is

$$B_1(x,\omega) = P\left(\sum_{k\in\mathbb{Z}^2} \mathbf{1}_{Q+k}(x) \begin{pmatrix} \alpha + \beta - \lambda_k^1(\omega) & 0 \\ 0 & \delta + \gamma - \lambda_k^2(\omega) \end{pmatrix}\right) P^T. \tag{23}$$

Parameters values are fixed at $\alpha = 5$, $\beta = 20$, $\delta = 25$ and $\gamma = 40$.

Example 2

We choose

$$A_2(x,\omega) = \sum_{k\in\mathbb{Z}^2} \mathbf{1}_{Q+k}(x) A_k(\omega) \text{ with } A_k(\omega) = \begin{pmatrix} a_k(\omega) & b_k(\omega) \\ b_k(\omega) & a_k(\omega) \end{pmatrix},$$

where $\{a_k\}_{k\in\mathbb{Z}^2}$ and $\{b_k\}_{k\in\mathbb{Z}^2}$ are two independent families of i.i.d. random variables uniformly distributed in $[\alpha,\beta]$ and $[\delta,\gamma]$ respectively, with

$$\alpha > 0.$$

The eigenvalues of A_k are $\lambda_k^1(\omega) = a_k(\omega) - b_k(\omega)$ and $\lambda_k^2(\omega) = a_k(\omega) + b_k(\omega)$. We thus assume that there exists $\rho_2 > 0$ such that

$$\text{for all } k \in \mathbb{Z}^2, \; a_k(\omega) - |b_k(\omega)| \geq \rho_2 \; \text{ almost surely},$$

so that A_2 is uniformly coercive. Note that the deterministic function $\mathscr{A}_2$ associated to A_2, which reads

$$\mathscr{A}_2(x, \{(y_k, z_k)\}_{k\in\mathbb{Z}^2}) = \sum_{k\in\mathbb{Z}^2} \mathbf{1}_{Q+k}(x) \begin{pmatrix} \alpha + (\beta-\alpha)y_k & \delta + (\gamma-\delta)z_k \\ \delta + (\gamma-\delta)z_k & \alpha + (\beta-\alpha)y_k \end{pmatrix},$$

is not monotone with respect to z_k. This case does not fall in the framework of [3]. The antithetic field we will consider is

$$B_2(x,\omega) = \sum_{k\in\mathbb{Z}^2} \mathbf{1}_{Q+k}(x) \begin{pmatrix} \alpha+\beta-a_k(\omega) & \gamma+\delta-b_k(\omega) \\ \gamma+\delta-b_k(\omega) & \alpha+\beta-a_k(\omega) \end{pmatrix}.$$

The numerical tests have been performed with the following parameters: $\alpha = 25$, $\beta = 40$, $\delta = 5$ and $\gamma = 20$.

Example 3

We define the random matrix

$$A_3(x,\omega) = \sum_{k\in\mathbb{Z}^2} \mathbf{1}_{Q+k}(x) A_k(\omega) \text{ with } A_k(\omega) = \begin{pmatrix} a_k(\omega) & c_k(\omega) \\ c_k(\omega) & b_k(\omega) \end{pmatrix},$$

where $\{a_k\}_{k\in\mathbb{Z}^2}$, $\{b_k\}_{k\in\mathbb{Z}^2}$ and $\{c_k\}_{k\in\mathbb{Z}^2}$ are three independent families of independent random variables, uniformly distributed in $[\alpha,\beta]$, $[\delta,\gamma]$ and $[\iota,\kappa]$ respectively, with

$$\alpha > 0, \quad \delta > 0 \quad \text{and} \quad \iota > 0. \tag{24}$$

Uniform coercivity holds if and only if the two eigenvalues of $A_k(\omega)$ are positive and uniformly bounded away from 0. A necessary condition is that the trace and the determinant of $A_k(\omega)$ are positive and uniformly bounded away from 0, which is guaranteed under the assumptions (24) and the existence of $\rho_3 > 0$ such that

$$\alpha\delta - \kappa^2 \geq \rho_3 > 0.$$

The lowest eigenvalue $\lambda_k^1(\omega)$ of $A_k(\omega)$ then reads

$$\lambda_k^1(\omega) = \frac{2\det A_k(\omega)}{\mathrm{Tr}\ A_k(\omega) + \sqrt{(\mathrm{Tr}\ A_k(\omega))^2 - 4\det A_k(\omega)}},$$

which is bounded from below as $\det A_k$ is bounded from below by $\rho_3 > 0$ and $\mathrm{Tr}\ A_k$ is bounded from above by $\beta + \gamma$.

The corresponding antithetic field reads

$$B_3(x,\omega) = \sum_{k\in\mathbb{Z}^d} \mathbf{1}_{Q+k}(x) \begin{pmatrix} \alpha+\beta-a_k(\omega) & \iota+\kappa-c_k(\omega) \\ \iota+\kappa-c_k(\omega) & \delta+\gamma-b_k(\omega) \end{pmatrix}.$$

The numerical tests have been performed with the following parameters: $\alpha = 15$, $\beta = 30$, $\delta = 20$, $\gamma = 40$, $\iota = 5$ and $\kappa = 15$.

2.2.2 Numerical Results

We consider both a diagonal term, namely $[A_N^\star]_{11}$ and an off-diagonal term, namely $[A_N^\star]_{12}$. For the former, in the case when monotonicity holds, we expect the results to be qualitatively good, since we have a theoretical result ensuring variance reduction. Our purpose is to evaluate the reduction *quantitatively*. When monotonicity does not hold, because of the particular structure considered, then we also test the reduction itself. We mention that the other diagonal entry $\left[A_N^\star\right]_{22}$ of the matrix would yield results qualitatively similar to those for $\left[A_N^\star\right]_{11}$.

Table 8 confirms that variance of the diagonal terms is reduced in our Example 1. The gain is rather significant. We also observe on Table 9 the same computational gain for the off-diagonal term, although our theoretical arguments in [3] do not cover this case. The other Tables (Tables 10–13) show that variance reduction is also obtained for our Examples 2 and 3, although no theoretical argument holds in these non-monotone settings.

Table 8 Example 1: mean and standard deviation of the diagonal term $\left[A_N^\star\right]_{11}$

N	$\mu_{100}([A_N^\star]_{11})$	$\sqrt{\sigma_{100}}([A_N^\star]_{11})$	$\mu_{50}\left(\left[\widetilde{A}_N^\star\right]_{11}\right)$	$\sqrt{\sigma_{50}}\left(\left[\widetilde{A}_N^\star\right]_{11}\right)$	$R([A_N^\star]_{11})$
40	22.0595	0.0362	22.0577	0.0066	14.88
60	22.0550	0.0240	22.0575	0.0036	21.80
80	22.0570	0.0161	22.0578	0.0020	32.96
100	22.0565	0.0166	22.0578	0.0025	21.74

Table 9 Example 1: mean and standard deviation of the off-diagonal term $\left[A_N^\star\right]_{12}$

N	$\mu_{100}([A_N^\star]_{12})$	$\sqrt{\sigma_{100}}([A_N^\star]_{12})$	$\mu_{50}\left(\left[\widetilde{A}_N^\star\right]_{12}\right)$	$\sqrt{\sigma_{50}}\left(\left[\widetilde{A}_N^\star\right]_{12}\right)$	$R([A_N^\star]_{12})$
40	−10.0897	0.0389	−10.0877	0.0043	41.32
60	−10.0902	0.0266	−10.0880	0.0024	63.93
80	−10.0899	0.0215	−10.0882	0.0022	48.82
100	−10.0892	0.0169	−10.0886	0.0019	39.10

3 Variance Reduction for Eigenproblems

As announced in the introduction, we now turn to the issue of variance reduction for eigenproblems.

Table 10 Example 2: mean and standard deviation of the diagonal term $[A_N^\star]_{11}$

N	$\mu_{100}([A_N^\star]_{11})$	$\sqrt{\sigma_{100}}([A_N^\star]_{11})$	$\mu_{50}\left(\left[\widetilde{A}_N^\star\right]_{11}\right)$	$\sqrt{\sigma_{50}}\left(\left[\widetilde{A}_N^\star\right]_{11}\right)$	$R([A_N^\star]_{11})$
40	31.8828	0.0498	31.8783	0.0098	12.85
60	31.8768	0.0337	31.8765	0.0049	23.48
80	31.8761	0.0284	31.8779	0.0036	31.50
100	31.8776	0.0242	31.8781	0.0032	29.30

Table 11 Example 2: mean and standard deviation of the off-diagonal term $[A_N^\star]_{12}$

N	$\mu_{100}([A_N^\star]_{12})$	$\sqrt{\sigma_{100}}([A_N^\star]_{12})$	$\mu_{50}\left(\left[\widetilde{A}_N^\star\right]_{12}\right)$	$\sqrt{\sigma_{50}}\left(\left[\widetilde{A}_N^\star\right]_{12}\right)$	$R([A_N^\star]_{12})$
40	12.6126	0.0561	12.6118	0.0085	21.69
60	12.6083	0.0342	12.6125	0.0051	16.94
80	12.6071	0.0270	12.6127	0.0042	20.86
100	12.6106	0.0226	12.6123	0.0038	18.18

Table 12 Example 3: mean and standard deviation of the diagonal term $[A_N^\star]_{11}$

N	$\mu_{100}([A_N^\star]_{11})$	$\sqrt{\sigma_{100}}([A_N^\star]_{11})$	$\mu_{50}\left(\left[\widetilde{A}_N^\star\right]_{11}\right)$	$\sqrt{\sigma_{50}}\left(\left[\widetilde{A}_N^\star\right]_{11}\right)$	$R([A_N^\star]_{11})$
20	22.0121	0.1239	22.0116	0.0125	49.44
40	22.0105	0.0571	22.0086	0.0073	30.54
60	22.0050	0.0387	22.0086	0.0046	34.39
80	22.0073	0.0282	22.0079	0.0037	28.55

We respectively denote by $\{\lambda_k^A(\omega)\}_{1\le k\le d}$ and $\{\lambda_k^B(\omega)\}_{1\le k\le d}$ the eigenvalues of the (approximate) homogenized matrix $A_N^\star(\omega)$ and the (approximate) homogenized matrix $B_N^\star(\omega)$ obtained using the antithetic field $B(x,\omega)$. We sort these eigenvalues in non-decreasing order.

Likewise, we denote by $\left(\Lambda_k^A(\omega), u_k^A(\omega)\right)_{k\in\mathbb{N}}$ the eigenelements of the operator $L_A = -\mathrm{div}\left[A_N^\star(\omega)\nabla\cdot\right]$ with homogeneous Dirichlet boundary conditions, i.e.

$$-\mathrm{div}\left[A_N^\star(\omega)\nabla u_k^A(\omega)\right] = \Lambda_k^A(\omega)\,u_k^A(\omega)$$

with $u_k^A(\omega) \in H_0^1(\mathscr{D})$ and $\|u_k^A(\omega)\|_{L^2(\mathscr{D})} = 1$. We proceed similarly for the matrix obtained using the antithetic field $B(x,\omega)$ and consider the eigenelements of

Table 13 Example 3: mean and standard deviation of the off-diagonal term $\left[A_N^\star\right]_{12}$

N	$\mu_{100}([A_N^\star]_{12})$	$\sqrt{\sigma_{100}}([A_N^\star]_{12})$	$\mu_{50}\left(\left[\widetilde{A}_N^\star\right]_{12}\right)$	$\sqrt{\sigma_{50}}\left(\left[\widetilde{A}_N^\star\right]_{12}\right)$	$R([A_N^\star]_{12})$
20	2.5031	0.0369	2.5024	0.0046	31.61
40	2.5017	0.0193	2.5021	0.0025	30.42
60	2.5012	0.0121	2.5028	0.0016	27.25
80	2.5012	0.0092	2.5012	0.0092	37.47

$L_B = -\mathrm{div}\left[B_N^\star(\omega)\nabla\cdot\right]$:

$$-\mathrm{div}\left[B_N^\star(\omega)\nabla u_k^B(\omega)\right] = \Lambda_k^B(\omega)\, u_k^B(\omega).$$

We also assume that, almost surely, $\Lambda_k^A(\omega)$ and $\Lambda_k^B(\omega)$ are sorted in non-decreasing order.

Our purpose here is to reduce the variance on $\mathscr{F}\left(A_N^\star\right) = \lambda_k^A$ or Λ_k^A for some $k \in \mathbb{N}$. Note that this is a monotone function of the random field $A(x,\omega)$ (see [3]). In the case when $\mathscr{A}$ is also monotone, the following result from [3] applies.

Proposition 1. *Define* $\widetilde{\lambda}_k(\omega) := \frac{1}{2}\left[\lambda_k^A(\omega) + \lambda_k^B(\omega)\right]$. *Then, for all* $1 \le k \le d$,

$$\mathbb{E}\left(\widetilde{\lambda}_k\right) = \mathbb{E}\left(\lambda_k^A\right) \quad \text{and} \quad \mathrm{Var}\left(\widetilde{\lambda}_k\right) \le \frac{1}{2}\mathrm{Var}\left(\lambda_k^A\right).$$

Define $\widetilde{\Lambda}_k(\omega) := \frac{1}{2}\left(\Lambda_k^A(\omega) + \Lambda_k^B(\omega)\right)$. *Then, for all* $k \in \mathbb{N}$,

$$\mathbb{E}\left(\widetilde{\Lambda}_k\right) = \mathbb{E}\left(\Lambda_k^A\right) \quad \text{and} \quad \mathrm{Var}\left(\widetilde{\Lambda}_k\right) \le \frac{1}{2}\mathrm{Var}\left(\Lambda_k^A\right).$$

This guarantees that the technique indeed reduces variance. We however need a *quantitative* evaluation of the efficiency of variance reduction.

To begin with, we mention that in the one-dimensional setting, or in the case of diagonal homogenized matrices, the question of variance reduction for eigenelements reduces to elementary questions already addressed. Indeed, in the one-dimensional setting, the approximate homogenized operator reads

$$-a_N^\star(\omega)\frac{d^2}{dx^2}$$

and thus its eigenfunctions are the deterministic eigenfunctions of the one-dimensional Laplacian, and its eigenvalues are likewise the deterministic eigenvalues of the one-dimensional Laplacian *multiplied* by the random quantity $a_N^\star(\omega)$. The variance reduction of the eigenelements comes down to that of $a_N^\star(\omega)$. Similarly, in the two-dimensional setting when the approximate homogenized matrix is diagonal, namely

$$A_N^\star(\omega) = \begin{pmatrix} a_N^\star(\omega) & 0 \\ 0 & b_N^\star(\omega) \end{pmatrix},$$

the eigenfunctions and eigenvalues may again be explicitly expressed in terms of those (deterministic) of the Laplacian. All is a matter of scaling, and again the question of variance reduction is elementary and already covered by that of reducing the variance on $A_N^\star(\omega)$.

Besides these oversimplified cases, additional numerical experiments are in order. We consider the three examples defined in Sect. 2.2.1. For each of them, and for the eigenvalues of the matrix $A_N^\star$ and the eigenvalues of the operator $L_{A_N^\star}$, we study an effectivity ratio R similar to that defined in (16).

Tables 14–16 illustrate the efficiency of the technique for the computation of the first eigenvalue for any structure of the random fields. Our results for the second eigenvalue are displayed on Tables 17–19. These results show the good efficiency of the approach, for all the test cases considered.

Tables 20–25 illustrate the variance reduction for the first two eigenvalues of L_A. Again, the approach performs very well. We omit to present here our results for higher eigenvalues of L_A. They lead to similar qualitative conclusions on the good efficiency of the approach.

Table 14 Example 1: mean and standard deviation of the first eigenvalue λ_1^A of the homogenized matrix

N	$\mu_{100}(\lambda_1^A)$	$\sqrt{\sigma_{100}}(\lambda_1^A)$	$\mu_{50}(\widetilde{\lambda}_1)$	$\sqrt{\sigma_{50}}(\widetilde{\lambda}_1)$	$R(\lambda_1^A)$
40	11.9703	0.0572	11.9702	0.0075	29.19
60	11.9650	0.0385	11.9696	0.0043	39.36
80	11.9670	0.0267	11.9696	0.0035	28.63
100	11.9672	0.0233	11.9692	0.0033	24.00

Table 15 Example 2: mean and standard deviation of the first eigenvalue λ_1^A of the homogenized matrix

N	$\mu_{100}(\lambda_1^A)$	$\sqrt{\sigma_{100}}(\lambda_1^A)$	$\mu_{50}(\widetilde{\lambda}_1)$	$\sqrt{\sigma_{50}}(\widetilde{\lambda}_1)$	$R(\lambda_1^A)$
40	19.2698	0.07870	19.2670	0.01353	16.91
60	19.2688	0.05414	19.2650	0.00799	22.96
80	19.2690	0.04196	19.2652	0.00593	25.02
100	19.2668	0.03411	19.2659	0.00557	18.70

Table 16 Example 3: mean and standard deviation of the first eigenvalue λ_1^A of the homogenized matrix

N	$\mu_{100}(\lambda_1^A)$	$\sqrt{\sigma_{100}}(\lambda_1^A)$	$\mu_{50}(\widetilde{\lambda}_1)$	$\sqrt{\sigma_{50}}(\widetilde{\lambda}_1)$	$R(\lambda_1^A)$
20	16.2994	0.0465	16.2998	0.0051	41.59
40	16.3016	0.0235	16.2995	0.0029	32.82
60	16.3005	0.0148	16.2991	0.0016	44.61
80	16.3001	0.0115	16.2989	0.0011	50.42

Table 17 Example 1: mean and standard deviation of the second eigenvalue λ_2^A of the homogenized matrix

N	$\mu_{100}(\lambda_2^A)$	$\sqrt{\sigma_{100}}(\lambda_2^A)$	$\mu_{50}(\widetilde{\lambda}_2)$	$\sqrt{\sigma_{50}}(\widetilde{\lambda}_2)$	$R(\lambda_2^A)$
40	32.1496	0.0480	32.1456	0.0049	47.62
60	32.1454	0.0327	32.1455	0.0031	56.60
80	32.1467	0.0270	32.1459	0.0022	78.10
100	32.1456	0.0241	32.1463	0.0019	79.77

Table 18 Example 2: mean and standard deviation of the second eigenvalue λ_2^A of the homogenized matrix

N	$\mu_{100}(\lambda_2^A)$	$\sqrt{\sigma_{100}}(\lambda_2^A)$	$\mu_{50}(\widetilde{\lambda}_2)$	$\sqrt{\sigma_{50}}(\widetilde{\lambda}_2)$	$R(\lambda_2^A)$
40	44.4951	0.0704	44.4905	0.0079	39.31
60	44.4854	0.0471	44.4900	0.0052	41.47
80	44.4832	0.0365	44.4905	0.0040	41.93
100	44.4880	0.0326	44.4905	0.0035	42.98

Table 19 Example 3: mean and standard deviation of the second eigenvalue λ_2^A of the homogenized matrix

N	$\mu_{100}(\lambda_2^A)$	$\sqrt{\sigma_{100}}(\lambda_2^A)$	$\mu_{50}(\widetilde{\lambda}_2)$	$\sqrt{\sigma_{50}}(\widetilde{\lambda}_2)$	$R(\lambda_2^A)$
20	23.1095	0.1053	23.1079	0.0099	56.70
40	23.1069	0.0472	23.1052	0.0060	31.18
60	23.1018	0.0337	23.1057	0.0043	31.24
80	23.1035	0.0249	23.1051	0.0032	29.77

Table 20 Example 1: mean and standard deviation of Λ_1^A

N	$\mu_{100}(\Lambda_1^A)$	$\sqrt{\sigma_{100}}(\Lambda_1^A)$	$\mu_{50}\left(\widetilde{\Lambda}_1\right)$	$\sqrt{\sigma_{50}}\left(\widetilde{\Lambda}_1\right)$	$R(\Lambda_1^A)$
40	842.9851	1.5576	842.9140	0.2084	27.94
60	842.7855	1.0364	842.8998	0.1284	32.60
80	842.8560	0.6882	842.9044	0.0900	29.21
100	842.8422	0.6977	842.9013	0.0891	30.67

Table 21 Example 1: mean and standard deviation of Λ_2^A

N	$\mu_{100}(\Lambda_2^A)$	$\sqrt{\sigma_{100}}(\Lambda_2^A)$	$\mu_{50}\left(\widetilde{\Lambda}_2\right)$	$\sqrt{\sigma_{50}}\left(\widetilde{\Lambda}_2\right)$	$R(\Lambda_2^A)$
40	1847.7161	4.4299	1847.5915	0.5976	27.47
60	1847.1983	2.9559	1847.5500	0.3593	33.86
80	1847.3861	1.9570	1847.5559	0.2678	29.45
100	1847.3705	1.8871	1847.5370	0.2614	26.05

Table 22 Example 2: mean and standard deviation of Λ_1^A

N	$\mu_{100}(\Lambda_1^A)$	$\sqrt{\sigma_{100}}(\Lambda_1^A)$	$\mu_{50}\left(\widetilde{\Lambda}_1\right)$	$\sqrt{\sigma_{50}}\left(\widetilde{\Lambda}_1\right)$	$R(\Lambda_1^A)$
40	611.2645	1.0533	611.1924	0.1631	27.87
60	611.1149	0.7222	611.1646	0.0985	26.86
80	611.1549	0.6078	611.1718	0.0657	42.82
100	611.1685	0.5140	611.1796	0.0639	32.33

Table 23 Example 2: mean and standard deviation of Λ_2^A

N	$\mu_{100}(\Lambda_2^A)$	$\sqrt{\sigma_{100}}(\Lambda_2^A)$	$\mu_{50}\left(\widetilde{\Lambda}_2\right)$	$\sqrt{\sigma_{50}}\left(\widetilde{\Lambda}_2\right)$	$R(\Lambda_2^A)$
40	1351.9939	2.9920	1351.8258	0.4960	19.95
60	1351.7016	2.0189	1351.7458	0.2958	23.29
80	1351.7995	1.6841	1351.7609	0.2063	33.31
100	1351.7827	1.3957	1351.7865	0.1986	24.69

Table 24 Example 3: mean and standard deviation of Λ_1^A

N	$\mu_{100}(\Lambda_1^A)$	$\sqrt{\sigma_{100}}(\Lambda_1^A)$	$\mu_{50}\left(\widetilde{\Lambda}_1\right)$	$\sqrt{\sigma_{50}}\left(\widetilde{\Lambda}_1\right)$	$R(\Lambda_1^A)$
20	389.7542	1.2979	389.7434	0.1258	53.26
40	389.7524	0.5981	389.7139	0.0774	29.86
60	389.6902	0.4072	389.7138	0.0467	38.06
80	389.7038	0.2967	389.7064	0.0375	31.37

Table 25 Example 3: mean and standard deviation of A_2^A

N	$\mu_{100}(A_2^A)$	$\sqrt{\sigma_{100}}(A_2^A)$	$\mu_{50}(\widetilde{A}_2)$	$\sqrt{\sigma_{50}}(\widetilde{A}_2)$	$R(A_2^A)$
20	901.8832	2.1182	901.8619	0.1934	60.01
40	901.9316	0.9934	901.8325	0.1221	33.09
60	901.8420	0.6740	901.8300	0.0650	53.95
80	901.8335	0.5028	901.8193	0.0514	53.95

Acknowledgements This work has been presented in the workshop *Numerical analysis of multiscale computations*, December 2009, at BIRS. CLB would like to thank the organizers of the workshop for their kind invitation. The work of RC, CLB and FL is partially supported by ONR under Grant 00014-09-1-0470.

References

1. A. Anantharaman, R. Costaouec, C. Le Bris, F. Legoll, and F. Thomines. Introduction to numerical stochastic homogenization and the related computational challenges: some recent developments. In W. Bao and Q. Du, editors, Multiscale Modeling and Analysis for Materials Simulations, volume 22 of Lecture Notes Series, Institute for Mathematical Sciences, National University of Singapore, 2011.
2. A. Bensoussan, J.-L. Lions, and G. Papanicolaou, Asymptotic analysis for periodic structures, Studies in Mathematics and its Applications, 5. North-Holland Publishing Co., Amsterdam-New York, 1978.
3. X. Blanc, R. Costaouec, C. Le Bris, and F. Legoll, Variance reduction in stochastic homogenization using antithetic variables, *Markov Processes and Related Fields*, in press.
4. X. Blanc, C. Le Bris, and P.-L. Lions, Une variante de la théorie de l'homogénéisation stochastique des opérateurs elliptiques [A variant of stochastic homogenization theory for elliptic operators], *C. R. Acad. Sci. Série I*, 343(11-12):717–724, 2006.
5. X. Blanc, C. Le Bris, and P.-L. Lions, Stochastic homogenization and random lattices, *J. Math. Pures Appl.*, 88(1):34–63, 2007.
6. A. Bourgeat and A. Piatnitski, Approximation of effective coefficients in stochastic homogenization, *Ann I. H. Poincaré - PR*, 40(2):153–165, 2004.
7. D. Cioranescu and P. Donato, An introduction to homogenization, Oxford Lecture Series in Mathematics and its Applications, 17. Oxford University Press, New York, 1999.
8. R. Costaouec, Thèse de l'Université Paris Est, in preparation.
9. R. Costaouec, C. Le Bris, and F. Legoll, Variance reduction in stochastic homogenization: proof of concept, using antithetic variables, *Boletin Soc. Esp. Mat. Apl.*, 50:9–27, 2010.
10. B. Engquist and P. E. Souganidis, Asymptotic and numerical homogenization, *Acta Numerica*, 17:147–190, 2008.
11. V. V. Jikov, S. M. Kozlov, and O. A. Oleinik, Homogenization of differential operators and integral functionals, Springer-Verlag, 1994.
12. U. Krengel, Ergodic theorems, de Gruyter Studies in Mathematics, vol. 6, de Gruyter, 1985.
13. C. Le Bris, Some numerical approaches for "weakly" random homogenization, in *Numerical mathematics and advanced applications*, Proceedings of ENUMATH 2009, Lect. Notes Comput. Sci. Eng., Springer, G. Kreiss, P. Lötstedt, A. Malqvist, M. Neytcheva Eds., 29–45, 2010.
14. J. S. Liu, Monte-Carlo strategies in scientific computing, Springer Series in Statistics, 2001.
15. A. N. Shiryaev, Probability, Graduate Texts in Mathematics, vol. 95, Springer, 1984.

A Stroboscopic Numerical Method for Highly Oscillatory Problems

Mari Paz Calvo, Philippe Chartier, Ander Murua, and Jesús María Sanz-Serna

Abstract We suggest a method for the integration of highly oscillatory systems with a single high frequency. The new method may be seen as a purely numerical way of implementing the analytical technique of stroboscopic averaging. The technique may be easily implemented in combination with standard software and may be applied with variable step sizes. Numerical experiments show that the suggested algorithms may be substantially more efficient than standard numerical integrators.

1 Introduction

We suggest a numerical method for the integration of highly oscillatory differential equations $dy/dt = f(y,t)$ with a single high frequency $2\pi/\epsilon$, $\epsilon \ll 1$. The new method may be seen as a purely numerical way of implementing the analytical technique of *stroboscopic averaging* [13] which constructs an averaged differential system $dY/dt = F(Y)$ whose solutions Y (approximately) interpolate the sought

J.M. Sanz-Serna (✉)
Departamento de Matemática Aplicada, Facultad de Ciencias, Universidad de Valladolid, Valladolid, Spain
e-mail: sanzsern@mac.uva.es

M.P. Calvo
Departamento de Matemática Aplicada, Facultad de Ciencias, Universidad de Valladolid, Valladolid, Spain
e-mail: maripaz@mac.uva.es

P. Chartier
INRIA Rennes, ENS Cachan Bretagne, Campus Ker-Lann, av. Robert Schumann, 35170 Bruz, France
e-mail: Philippe.Chartier@inria.fr

A. Murua
Konputazio Zientziak eta A. A. Saila, Informatika Fakultatea, UPV/EHU, E–20018 Donostia–San Sebastián, Spain e-mail: Ander.Murua@ehu.es

B. Engquist et al. (eds.), *Numerical Analysis of Multiscale Computations*, Lecture Notes in Computational Science and Engineering 82, DOI 10.1007/978-3-642-21943-6_4,

highly oscillatory solution y at times $t = t_0 + 2\pi\epsilon n$, (n integer). In the spirit of the heterogeneous multiscale methods (see [1, 5–8, 16], cf. [3, 14]), we integrate numerically the averaged system without using the analytic expression of F; all information on F required by the algorithm is gathered on the fly by numerically integrating the original system in small time windows. The technique may be easily implemented in combination with standard software and may be applied with *variable step sizes*.

Section 2, based on [4], presents the theoretical foundation of the algorithm. Section 3 contains a description of the new method along with a brief discussion of related literature. Examples of oscillatory systems that may be treated with our approach are provided in Sect. 4 and the final section presents numerical examples. It is found that the suggested algorithms may be substantially more efficient than standard numerical integrators.

2 A Modified Equation Approach to Averaging

We wish to integrate numerically initial value problems for differential systems of the form

$$\frac{d}{dt}y = f\left(y, \frac{t}{\epsilon}; \epsilon\right), \tag{1}$$

where y is a D-dimensional real vector, ϵ is a small parameter and the smooth function f is assumed to depend 2π-periodically on the variable t/ϵ. Our interest is in situations where, as $\epsilon \to 0$, the solutions or some of their derivatives with respect to t become *unbounded*; relevant examples will be presented in Sect. 4.

If we denote by $\varphi_{t_0,t;\epsilon} : \mathscr{R}^D \to \mathscr{R}^D$ the solution operator of (1), so that

$$y(t) = \varphi_{t_0,t;\epsilon}(y_0)$$

is the solution that satisfies the initial condition $y(t_0) = y_0$, then the *one-period map* $\Psi_{t_0;\epsilon} = \varphi_{t_0,t_0+2\pi\epsilon;\epsilon}$ depends on t_0 in a $2\pi\epsilon$-periodic manner; this is proved by noting that both $\varphi_{t_0,t;\epsilon}(y_0)$ and $\varphi_{t_0+2\pi\epsilon,t+2\pi\epsilon;\epsilon}(y_0)$ satisfy the same initial value problem

$$\frac{d}{dt}y(t) = f\left(y(t), \frac{t}{\epsilon}; \epsilon\right) = f\left(y(t), \frac{t+2\pi\epsilon}{\epsilon}; \epsilon\right), \qquad y(t_0) = y_0.$$

It follows that, at the *stroboscopic times* $t_n = t_0 + 2\pi\epsilon n$, $n = 0, \pm 1, \pm 2, \ldots$,

$$y(t_n) = \varphi_{t_0,t_n;\epsilon}(y_0) = \varphi_{t_{n-1},t_n;\epsilon}(\varphi_{t_0,t_{n-1};\epsilon}(y_0)) = \varphi_{t_0,t_0+2\pi\epsilon;\epsilon}(\varphi_{t_0,t_{n-1};\epsilon}(y_0))$$

and, hence, we arrive at the fundamental formula:

$$y(t_n) = (\Psi_{t_0;\epsilon})^n(y_0), \quad n = 0, \pm 1, \pm 2, \ldots \tag{2}$$

For the problems we are interested in (see Sect. 4) there is an expansion

$$\Psi_{t_0;\epsilon}(y_0) = y_0 + \sum_{j=1}^{\infty} \epsilon^j M_j(y_0), \tag{3}$$

with suitable smooth maps $M_j : \mathscr{R}^D \to \mathscr{R}^D$ independent of ϵ, and thus $\Psi_{t_0;\epsilon}$ is a smooth *near-to-identity map*. Standard results from the backward error analysis of numerical integrators [9, 15] show then the existence of an *autonomous* system (the modified system of $\Psi_{t_0;\epsilon}$)

$$\frac{d}{dt}Y = F(Y;\epsilon) = F_1(Y) + \epsilon F_2(Y) + \epsilon^2 F_3(Y) + \cdots \tag{4}$$

whose (formal) solutions satisfy that $Y(t_n) = \Psi_{t_0;\epsilon}(Y(t_{n-1}))$ for $n = 0, \pm 1, \pm 2, \ldots$ so that

$$Y(t_n) = (\Psi_{t_0;\epsilon})^n(Y_0), \quad n = 0, \pm 1, \pm 2, \ldots \tag{5}$$

(F and the F_j depend on t_0 – because $\Psi_{t_0;\epsilon}$ does–, but this dependence has not been incorporated to the notation.) We conclude from (2) and (5) that, if one chooses $Y(t_0) = y(t_0)$, then $Y(t)$ *exactly coincides with* $y(t)$ *at the stroboscopic times* $t_n = t_0 + 2\pi\epsilon n$. In this way it is possible in principle to find $y(t_n)$ by solving the system (4), *where all* t*-derivatives of* Y *remain bounded as* $\epsilon \to 0$. Furthermore y may be recovered from Y even at values of t that do not coincide with one of the stroboscopic times. In fact,

$$y(t) = (\varphi_{t_n,t;\epsilon} \circ \Phi_{t_n - t;\epsilon})(Y(t)), \tag{6}$$

where t_n is the largest stroboscopic time $\le t$ and $\Phi_{\cdot;\epsilon}$ denotes the flow of (4). In this way, y is 'enslaved' to Y through the mapping $\varphi_{t_n,t;\epsilon} \circ \Phi_{t_n - t;\epsilon}$ whose dependence on t is easily seen to be $2\pi\epsilon$-periodic.

For future reference we note that an alternative way of writing (5) is

$$\Psi^n_{t_0;\epsilon} \equiv \Phi_{2\pi\epsilon n;\epsilon}; \tag{7}$$

after a whole number n of periods the solution operator $\Psi^n_{t_0;\epsilon} = \varphi_{t_0,t_0+2\pi\epsilon n}$ of the non-autonomous system (1) coincides with the flow of the autonomous (4).

It is well known that the series (4) does not converge in general, and in order to get rigorous results one has to consider a truncated version ($J \ge 1$ is an arbitrarily large integer)

$$\frac{d}{dt}Y = F^{(J)}(Y;\epsilon) = F_1(Y) + \epsilon F_2(Y) + \epsilon^2 F_3(Y) + \cdots + \epsilon^{J-1} F_J(Y), \tag{8}$$

whose solutions satisfy that $Y(t_n) - \Psi_{t_0;\epsilon}(Y(t_{n-1})) = \mathscr{O}(\epsilon^{J+1})$. If Y solves (8) with $Y(t_0) = y(t_0)$ then $Y(t_n)$ and $y(t_n)$ differ by an $\mathscr{O}(\epsilon^J)$ amount, where the constant implied in the $\mathscr{O}$ notation is uniform as the stroboscopic time t_n ranges in a time interval $t_0 \le t_n \le t_0 + T$ with $T = \mathscr{O}(1)$ as $\epsilon \to 0$.

The process of obtaining the autonomous system (4) (or (8)) from the original system (1) is referred to in the averaging literature [13] as high-order stroboscopic averaging. As a rule, the amount of work required to find analytically the functions F_j is formidable, even when the interest is limited to lowest values of j.

3 A Numerical Method

In this section we propose a purely numerical method that bypasses the need for finding analytically the functions F_j. To simplify the exposition, we will ignore hereafter the $\mathcal{O}(\epsilon^J)$ remainder that arises from truncating (4), i.e. we will proceed as if the series (4) were convergent. Since J may be chosen arbitrarily large, the disregarded truncation errors are, as $\epsilon \to 0$, negligible when compared with other errors present in the method to be described.

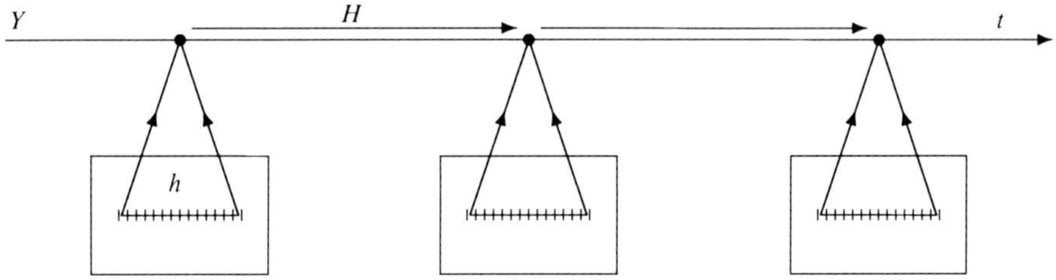

Fig. 1 Schematic view of the numerical integration. The t-axis above represents the macro-integration of the averaged system with (large) macro-steps H. Whenever the macro-solver requires information on the averaged system, the algorithm carries out a micro-integration of the original problem in a small time-window. The micro-step size h is small with respect to ϵ

In order to integrate the highly oscillatory system (1) with initial condition $y(t_0) = y_0$, we (approximately) compute the corresponding smooth interpolant $Y(t)$, i.e. the solution of the initial value problem specified by the *averaged system* (4) along with the initial condition $Y(t_0) = y_0$. We integrate (4) by a standard numerical method, the so-called *macro-solver*, with a macro-step H that ideally should be substantially larger than the small period $2\pi\epsilon$. In the spirit of heterogeneous multiscale methods, the information on F required by the macro-solver is gathered on the fly by integrating, with a micro-step h, the original system (1) in time-windows of length $\mathcal{O}(\epsilon)$. These auxiliary integrations are also performed by means of a standard numerical method, the *micro-solver*, see Fig. 1. (It is not necessary that the choices of macro and micro-solver coincide.)

If the macro-solver is a linear multistep or Runge-Kutta method, then the only information on the system (4) required by the solver are function values $F(Y^*;\epsilon)$ at given values of the argument Y^*. Since, by definition, $\Phi_{t;\epsilon}$ is the flow of (4) we may write

$$F(Y^*;\epsilon) = \frac{d}{dt}\Phi_{t;\epsilon}(Y^*)\bigg|_{t=0},$$

or, after approximating the time-derivative by central differences,

$$F(Y^*;\epsilon) = \frac{1}{2\delta}[\Phi_{\delta;\epsilon}(Y^*) - \Phi_{-\delta;\epsilon}(Y^*)] + \mathscr{O}(\delta^2).$$

We now set $\delta = 2\pi\epsilon$ and use the identity (7) to get

$$F(Y^*;\epsilon) = \frac{1}{4\pi\epsilon}[\Psi_{t_0;\epsilon}(Y^*) - \Psi_{t_0;\epsilon}^{-1}(Y^*)] + \mathscr{O}(\epsilon^2), \tag{9}$$

a formula that may be used to compute approximately $F(Y^*;\epsilon)$ since $\Psi_{t_0;\epsilon}(Y^*)$ and $\Psi_{t_0;\epsilon}^{-1}(Y^*)$ may be found numerically through micro-integrations. In fact one has to integrate (1) with initial condition $y(t_0) = Y^*$, first from $t = t_0$ to $t = t_0 + 2\pi\epsilon$ and then from $t = t_0$ to $t = t_0 - 2\pi\epsilon$.

Some important remarks are in order. The initial condition for each micro-integration is *always* prescribed at $t = t_0$, regardless of the point of the time axis the macro-solver may have reached when the micro-integration is performed. We have tried to make this fact apparent in Fig. 1 by enclosing different micro-integrations in boxes that are not connected by a common time-axis (cf. Fig. 1.1 in [8] or Fig. 2 in [16]). All micro-integrations find solutions of (1) in the interval $[t_0 - 2\pi\epsilon, t_0 + 2\pi\epsilon]$. With the terminology of [3] we may say that the algorithm suggested here is *asynchronous*. Figure 2 may be of assistance in understanding the situation. This figure should also make it clear that it is not at all necessary that the step-points used by the macro-integrator be stroboscopic times; this is a particularly valuable feature if the macro-solver employs variable steps. We also emphasize that if the macro-solver outputs (an approximation to) the averaged solution Y at a stroboscopic time t_n, then the output is an approximation to $y(t_n)$; if output occurs at a non-stroboscopic value of t it is still possible to recover an approximation to $y(t)$ by using (6).

Of course, other difference formulae may also be used instead of (9). For instance, we may approximate $F(Y^*;\epsilon)$ with an $\mathscr{O}(\epsilon^4)$ error by means of

$$\begin{aligned}&\frac{1}{24\pi\epsilon}\Big(-\Phi_{4\pi\epsilon;\epsilon}(Y^*) + 8\Phi_{2\pi\epsilon;\epsilon}(Y^*) - 8\Phi_{-2\pi\epsilon;\epsilon}(Y^*) + \Phi_{-4\pi\epsilon;\epsilon}(Y^*)\Big)\\ &\quad = \frac{1}{24\pi\epsilon}\Big(-\Psi_{t_0;\epsilon}^{2}(Y^*) + 8\Psi_{t_0;\epsilon}(Y^*) - 8\Psi_{t_0;\epsilon}^{-1}(Y^*) + \Psi_{t_0;\epsilon}^{-2}(Y^*)\Big).\end{aligned} \tag{10}$$

Now the integrations to be carried out to find $\Psi_{t_0;\epsilon}^{2}(Y^*) = \varphi_{t_0,t_0+4\pi\epsilon;\epsilon}(Y^*)$ and $\Psi_{t_0;\epsilon}^{-2}(Y^*) = \varphi_{t_0,t_0-4\pi\epsilon;\epsilon}(Y^*)$ work in the intervals $t_0 \le t \le t_0 + 4\pi\epsilon$ and $t_0 \ge t \ge t_0 - 4\pi\epsilon$ respectively. Difference formulae of arbitrarily high orders may also be employed, but higher order implies a wider stencil and costlier micro-integrations.

The approach suggested here is related to methods called envelop-following or multi-revolution (see [2, 12] and their references) that go back to the 1960s and have been successfully used in a number of application areas, including celestial

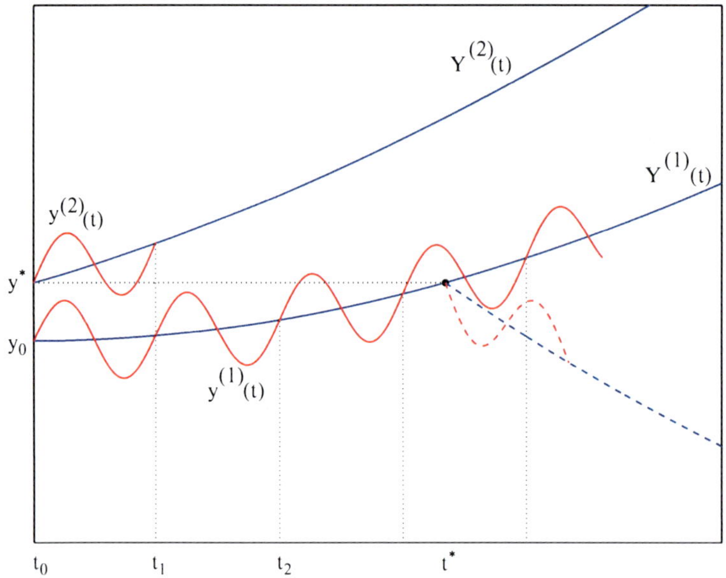

Fig. 2 The wiggly solid lines represent the solutions $y^{(1)}(t)$ and $y^{(2)}(t)$ of the oscillatory problem with initial conditions $y^{(1)}(t_0) = y_0$ and $y^{(2)}(t_0) = y^*$. We have also represented the solutions of the averaged system with $Y^{(1)}(t_0) = y_0$ and $Y^{(2)}(t_0) = y^*$; the graphs of $Y^{(1)}(t)$ and $Y^{(2)}(t)$ are translates along the time-axis of one another because the averaged system is autonomous. At stroboscopic times each oscillatory solution $y^{(i)}(t)$ coincides with the corresponding averaged solution $Y^{(i)}(t)$. Now assume that we are computing numerically $Y^{(1)}$, that the macro-solver has reached the point (t^*, y^*) (t^* is not a stroboscopic time) and that it requires the value of the slope $F(y^*;\epsilon)$. The correct procedure is based on the fact that the slope of $Y^{(1)}(t)$ at (t^*, y^*) coincides with the slope of $Y^{(2)}(t)$ at (t_0, y^*); micro-integrations on the intervals $t_0 \leq t \leq t_0 + 2\pi\epsilon$ and $t_0 \geq t \geq t_0 - 2\pi\epsilon$ (this is not shown in the figure) are performed to find $y^{(2)}(t_0 \pm 2\pi\epsilon) = Y^{(2)}(t_0 \pm 2\pi\epsilon)$ and the values $Y^{(2)}(t_0 \pm 2\pi\epsilon)$ are then used to find the slope by means of finite differences. Micro-integrating in the intervals $t^* \leq t \leq t^* + 2\pi\epsilon$ and $t^* \geq t \geq t^* - 2\pi\epsilon$ will not do: the averaged system depends on t_0 – see Sect. 2 – and such micro-integrations (discontinuous wiggly lines) would provide information on a solution (discontinuous line without wiggles) of the wrong averaged system

mechanics and circuit theory. Note that, while in this paper both the macro- and micro-integrators are standard ODE solvers, the multi-revolution technique requires the construction of new special formulae. The closest relative of the algorithm described above is perhaps the LIPS method of Kirchgraber [10] that, in lieu of the finite difference formulae employed here, retrieves values of $F(Y^*;\epsilon)$ through Runge-Kutta like formulae. Again those formulae have to be build on purpose and reference [10] provides coefficients for the orders $\mathcal{O}(\epsilon^2)$, $\mathcal{O}(\epsilon^3)$, $\mathcal{O}(\epsilon^4)$.[1]

[1] The possibility of using finite-difference formulae to approximate modified equations – this is essentially the problem solved by Kirchgraber's formulae – was already pointed out in reference [11], page 228.

4 Examples

In order that a highly-oscillatory problem (1) may be integrated by the procedure outlined above, it is necessary that the corresponding one-period map $\Psi_{t_0;\epsilon}$ be a smooth near-to-identity transformation as in (3). In this section we present families of systems that satisfy this condition.

(i) If f in (1) is of the form

$$f(y,\tau;\epsilon) = \sum_{j=1}^{\infty} \epsilon^{j-1} f_j(y,\tau). \tag{11}$$

where the $f_j(y,\tau)$ are smooth 2π-periodic functions of τ, then $f = \mathscr{O}(1)$ as $\epsilon \to 0$ and therefore $y(t) - y(t_0)$ undergoes $\mathscr{O}(\epsilon)$ changes in the interval $t_0 \le t \le t_0 + 2\pi\epsilon$ and (3) holds. Presented in [4] is a way of systematically constructing with the help of rooted trees the functions M_j that feature in (3). The format (11) is the standard starting point to perform analytically averaging so that any system to be averaged has first to be brought to that format via suitable changes of variables. We show next that those preliminary changes of variables are not needed to implement the numerical method of Sect. 3.

(ii) Consider second order systems of the form

$$\frac{d^2}{dt^2} q = G\left(q, \frac{t}{\epsilon}; \epsilon\right), \tag{12}$$

where $q \in \mathscr{R}^d$ and the force G has an expansion

$$G(q,\tau;\epsilon) = \sum_{j=0}^{\infty} \epsilon^{j-1} G_j(q,\tau)$$

(the G_j are 2π-periodic in τ).
To treat this case, we begin by rewriting (12) as a first order system

$$\frac{d}{dt} q = p, \qquad \frac{d}{dt} p = G\left(q, \frac{t}{\epsilon}; \epsilon\right) \tag{13}$$

for the vector $y = (q,p)$ in $\mathscr{R}^D$, $D = 2d$. Note that here $G = \mathscr{O}(1/\epsilon)$ and the solution y will undergo $\mathscr{O}(1)$ changes in the interval $t_0 \le t \le t_0 + 2\pi\epsilon$. However if the leading term $(1/\epsilon)G_0$ of G averages to zero over one period, i.e.

$$\int_0^{2\pi} G_0(q,\tau)\, d\tau = 0, \tag{14}$$

then (3) holds as proved in [4], a reference that presents a technique for explicitly constructing the functions M_j. An alternative proof will be given here. Consider the system

$$\frac{d}{dt}q = 0, \qquad \frac{d}{dt}p = \frac{1}{\epsilon}G_0\left(q, \frac{t}{\epsilon}\right), \tag{15}$$

denote by $\widehat{\varphi}_{t_0,t;\epsilon}(q_0, p_0)$ the corresponding solution operator and introduce the time-dependent change of variables

$$(q(t), p(t)) = \widehat{\varphi}_{t_0,t;\epsilon}\big(\widehat{q}(t), \widehat{p}(t)\big).$$

Of course, this change reduces the system (15) to the trivial form $(d/dt)\widehat{q}=0$ and $(d/dt)\widehat{p} = 0$. When applied to the full (13), the change reduces the system to the format (11) (i.e. the new right-hand side contains no $\mathcal{O}(1/\epsilon)$ term). From case (i) above we conclude that (3) holds *after changing variables*. However the solution operator is explicitly given by

$$\widehat{\varphi}_{t_0,t;\epsilon}(q_0, p_0) = \left(q_0, p_0 + \int_{t_0}^{t} \frac{1}{\epsilon}G_0\left(q_0, \frac{t'}{\epsilon}\right) dt'\right)$$

an expression that, in tandem with (14), shows that the associated one-period map $\widehat{\varphi}_{t_0,t_0+2\pi\epsilon;\epsilon}$ is the identity. Therefore at stroboscopic times t_n the values of the new $(\widehat{q}, \widehat{p})$ variables coincide with the values of the old variables (q, p) and (3) also holds *without changing variables*. As a consequence the numerical method works for the given system (13) without any need to previously perform any analytic manipulations.

Note that the expression of the change of variables reveals that in the interval $t_0 \le t \le t_0 + 2\pi\epsilon$, the variations of the variable $p(t)$ are $\mathcal{O}(1)$ and those in $q(t)$ are $\mathcal{O}(\epsilon)$. At *the end of the interval*, both $q(t_0 + 2\pi\epsilon)$ and $p(t_0 + 2\pi\epsilon)$ are $\mathcal{O}(\epsilon)$ away from their initial values $q(t_0)$ and $p(t_0)$ in view of (3).

A well known example of (12) is given by the vibrated inverted pendulum equation

$$\frac{d^2}{dt^2}q = G\left(q, \frac{t}{\epsilon}; \epsilon\right) = \left(\frac{1}{\epsilon}\frac{v_{max}}{\ell}\cos\left(\frac{t}{\epsilon} + \theta_0\right) + \frac{g}{\ell}\right)\sin q. \tag{16}$$

(iii) The reader is referred to [10] and [4] for further examples (including perturbed Kepler problems, perturbed harmonic oscillators, Fermi-Pasta-Ulam like problems) of systems for which (3) holds because they may be brought to the format (11) through a change of variables that coincides with the identity map at stroboscopic times.

5 Numerical Experiments

Our aim in this section is to illustrate by means of simple examples the use of the stroboscopic technique described in Sect. 3. For this reason we only report on experiments performed when the macro-integrator is either the 'classical' fourth-order, four stages Runge-Kutta (RK) method with constant step-sizes or the variable-step code ode45 from MATLAB. Extensive numerical experiments, including detailed comparisons with alternative techniques and wider choices of macro- and micro-solvers, will be presented elsewhere.

As a test problem, we integrate in the interval $t_0 = 0 \le t \le \pi$ the inverted (Kapitsa) pendulum equation (16) with parameter values $v_{max} = 4$, $\ell = 0.2$, $\theta_0 = 2$, $g = 9.8$, and initial conditions $q(0) = 0.25$, $p(0) = 0$. This equation has been used as a test example in [16] to illustrate the power of the heterogeneous multiscale approach (see also [3, 4, 14]). Unlike the algorithms described in this paper, those analyzed in [16] require some preliminary analytical work to derive formulae that relate macro- and micro-states.

5.1 Constant Step-Sizes

We first take the classical RK method with constant step-sizes as macro- and micro-integrator. This is run, for different values of ϵ, for combinations of macro- and micro-steps (H, h) of the form $(2\pi 2^{-\nu}/50, 2\pi\epsilon 2^{-\nu}/4)$, $\nu = 0, 1, 2, \ldots$ and with either second- or fourth-order differences (see (9) or (10) respectively).[2] The results are summarized in Tables 1 and 2 respectively. In the former, the symbol *** means that the corresponding run was not carried out: when H is smaller than $2\pi\epsilon$ the stroboscopic algorithm does not make any sense.

Table 1 Errors in stroboscopic algorithm: 2nd-order finite differences

H	Micro evaluations	$1/\epsilon$			
		3,200	6,400	12,800	25,600
$2\pi/50$	3,200	3.12(−1)	3.12(−1)	3.12(−1)	3.12(−1)
$2\pi/100$	12,800	2.14(−2)	2.16(−2)	2.17(−2)	2.17(−2)
$2\pi/200$	51,200	3.22(−3)	2.17(−3)	1.94(−3)	1.88(−3)
$2\pi/400$	204,800	1.59(−3)	5.31(−4)	2.67(−4)	2.02(−4)
$2\pi/800$	819,200	1.42(−3)	3.65(−4)	1.01(−4)	3.54(−5)
$2\pi/1{,}600$	3,276,800	1.41(−3)	3.53(−4)	8.88(−5)	2.29(−5)
$2\pi/3{,}200$	13,107,200	1.41(−3)	3.52(−4)	8.80(−5)	2.20(−5)
$2\pi/6{,}400$	52,428,800	***	3.52(−4)	8.79(−5)	2.20(−5)
$2\pi/12{,}800$	209,715,200	***	***	8.79(−5)	2.20(−5)
$2\pi/25{,}600$	838,860,800	***	***	***	2.20(−5)

[2] Our experience indicates that standard central differences of order 6 are not competitive in terms of efficiency with those of orders 2 or 4.

Table 2 Errors in stroboscopic algorithm: 4th-order finite differences

H	Micro evaluations	$1/\epsilon$			
		3,200	6,400	12,800	25,600
$2\pi/50$	6,400	3.12(−1)	3.12(−1)	3.12(−1)	3.12(−1)
$2\pi/100$	25,600	2.18(−2)	2.17(−2)	2.17(−2)	2.17(−2)
$2\pi/200$	102,400	1.87(−3)	1.86(−3)	1.86(−3)	1.86(−3)
$2\pi/400$	409,600	1.81(−4)	1.81(−4)	1.80(−4)	1.80(−4)
$2\pi/800$	1,638,400	1.36(−5)	1.35(−5)	1.34(−5)	1.34(−5)
$2\pi/1{,}600$	6,553,600	1.05(−6)	9.18(−7)	9.09(−7)	9.04(−7)
$2\pi/3{,}200$	26,214,400	2.01(−7)	6.74(−8)	5.89(−8)	5.45(−8)

Let us first discuss the computational cost. Since each micro-integration takes place in an interval of width $4\pi\epsilon$ (or $8\pi\epsilon$) and, for given H, the value of h is chosen to be proportional to ϵ, the cost of the algorithm is *independent of* ϵ. Furthermore when H is halved so is h and therefore the total number of micro-steps in a run is multiplied by four (see the second column of the tables that display the total number of function evaluations required by the micro-integrations).

We report errors measured as the maximum, over all macro-step-points, of the (absolute value of the) difference between the q component of a very accurate numerical approximation to the true solution of the oscillatory problem and the solution Q provided by the stroboscopic algorithm; errors in p behave in exactly the same way as those in q. There are three sources of error (cf. [14]): (i) the recovery of the right-hand side F of the averaged system by the finite-difference formula (9) (or (10)), (ii) the replacement in (9) (or (10)) of the exact values of $\Psi^k_{t_0;\epsilon}(Y^*)$ by numerical approximations based on micro-integrations, (iii) the discretization error introduced by the macro-integrator. We consider these sources in turn.

As H and h tend to 0, the errors arising from (ii) and (iii) vanish and only the source (i) remains. *At each evaluation of* F the error from this source is $\mathcal{O}(\epsilon^2)$ (or $\mathcal{O}(\epsilon^4)$) and, due to the stability of the macro-solver, these evaluation errors introduce $\mathcal{O}(\epsilon^2)$ (or $\mathcal{O}(\epsilon^4)$) errors in the values of Q. This is apparent in Table 1, where the errors at the bottom of the different columns, clearly behave as $\mathcal{O}(\epsilon^2)$. For fourth-order differences Table 2 does not report results for very small H and h due to the cost of obtaining a sufficiently accurate reference solution to measure errors.

To analyze the micro-integration errors, it is best to rewrite (16) in terms of the fast, non-dimensional time $\tau = t/\epsilon$, i.e.

$$\frac{d}{d\tau}q = \epsilon p, \qquad \frac{d}{d\tau}p = \epsilon G\left(q, \frac{t}{\epsilon}; \epsilon\right) = \left(\frac{v_{max}}{\ell}\cos(\tau + \theta_0) + \epsilon\frac{g}{\ell}\right)\sin q. \quad (17)$$

Now the force ϵG is bounded as $\epsilon \to 0$, the micro-integrations span intervals of fixed length 4π (or 8π) and (because the micro-step h in the variable t is chosen proportional to ϵ) the step-length h/ϵ in τ is also independent of ϵ. Therefore, standard results show that the error in finding each value $\Psi^k_{t_0;\epsilon}(Y^*)$ is $\mathcal{O}\big((h/\epsilon)^4\big)$. *Furthermore it can be shown that the constant C implied in the $\mathcal{O}$ notation is itself $\mathcal{O}(\epsilon)$*[3]; the extra factor in C makes up for the factor ϵ that features in the denominator of (9) (or (10)) and therefore, in each evaluation of F, the error due to the micro-integrator is $\mathcal{O}\big((h/\epsilon)^4\big)$, where the implied constant is ϵ-independent. Again the stability of the macro-solver entails that the corresponding effect in the macro-solution Q is itself $\mathcal{O}\big((h/\epsilon)^4\big)$, or, with our choice of H and h, $\mathcal{O}(H^4)$. Since the error due to discretizing the averaged equation is itself $\mathcal{O}(H^4)$, we conclude that the combined effect of sources (ii) and (iii) is $\mathcal{O}(H^4)$, *uniformly in ϵ*. In this way, the overall algorithm yields approximations to the true q and p of sizes $\mathcal{O}(\epsilon^\mu + H^4)$, where the implied constant is independent of ϵ and $\mu = 2$ or $\mu = 4$ for second and fourth-order differences respectively. Thus, unless H is chosen to be so small that the contribution of size ϵ^μ manifests itself, the algorithm yields errors that behave as $O(H^4)$ *uniformly in ϵ at a cost that is also independent of ϵ*. Once more this is borne out by the tables, where the errors in the top rows are independent of ϵ and of the finite-difference formula and show a reduction by a factor of ≈ 16 when H is halved.

Figures 3 and 4 are based on Tables 1 and 2 and compare the efficiency of the stroboscopic algorithm with second or fourth-order differences with that of a straightforward integration of the oscillatory problem with the classical RK method. For errors of size $\approx 10^{-2}$, Fig. 3 reveals that for $\epsilon = 1/3{,}200$ the second-difference algorithm needs an amount of work that is less than 1/5 of that required by the classical method. For $\epsilon = 1/25{,}600$, we see in Fig. 4 that the same ratio is less than 1/30. Also note that for the algorithm based on fourth-order differences, the lines in Figs. 3 and 4 are virtually identical, indicating an ϵ-independent behavior. The line corresponding to the classical RK method undergoes a marked translation to the right when ϵ is decreased, indicating an efficiency loss. For the algorithm with second-order differences, the lines in both figures coincide for larger values of the errors (larger values of H); however in Fig. 3 errors saturate at a larger value than that in Fig. 4 in agreement with earlier discussions. Finally we point out that the

[3] The proof of the estimate $C = \mathcal{O}(\epsilon)$ is easy after noting that for $\epsilon = 0$ the RK micro-integrator finds the solution of (17) at $\tau = 2\pi$ *without any error*. (In fact finding the solution at $\tau = 2\pi$ of (17) with $\epsilon = 0$ essentially requires the computation of the integral in (14); the RK numerical solution may be written down in closed form as a trigonometric sum whose value vanishes.) The key point here is that the micro-integrator is such that when applied to the system (15) it generates a one-period map that *exactly* coincides with the identity, thus mimicking a key property of the system being integrated. For micro-integrators that do not possess this property the error behavior is not so favorable as for those considered here because estimates suffer from the factor ϵ in the denominator of the finite-difference formulae (cf. our analysis with that in [10]). Similarly, when integrating perturbed Kepler problems, perturbed harmonic oscillators, etc. as in [10] or [4], it is important that the micro-integration be performed in such a way that for the unperturbed problem ($\epsilon = 0$) it results in a one-period map that coincides *exactly* with the identity. This may be achieved by using splitting methods.

lines of the stroboscopic algorithms possess a smaller slope than those of the RK method: while to divide the error by a factor of 16 the classical method has to work twice as hard, the new algorithms must toil four times as hard, as they require both more macro-steps and more accurate micro-integrations.

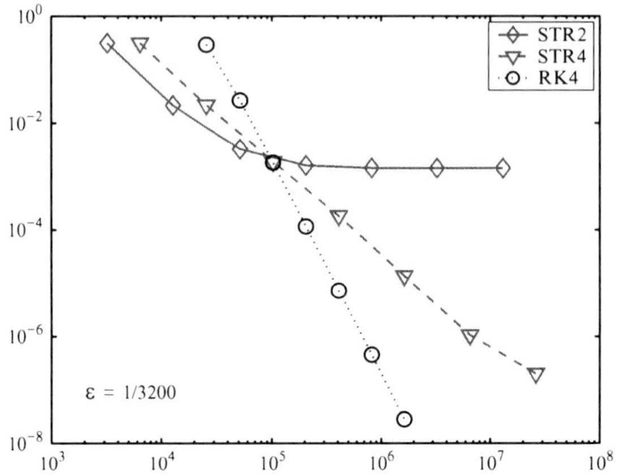

Fig. 3 Efficiency comparison: errors vs. number of evaluations of the micro-force. Constant step-sizes, 'larger' ϵ

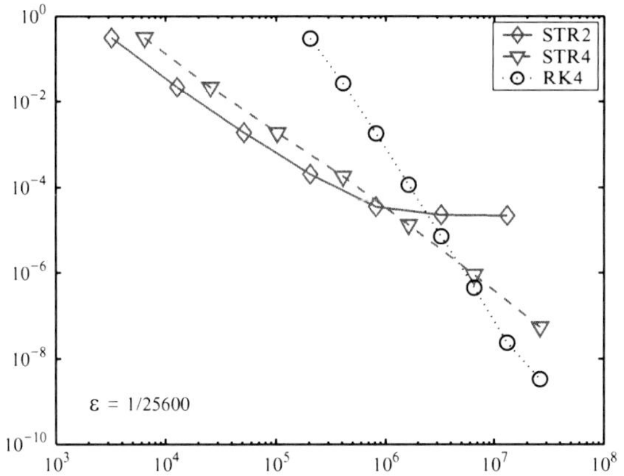

Fig. 4 Efficiency comparison: errors vs. number of evaluations of the micro-force. Constant step-sizes, smaller ϵ

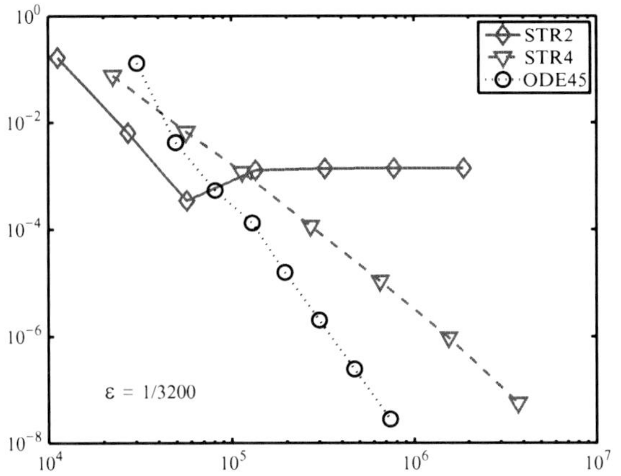

Fig. 5 Efficiency comparison: errors vs. number of evaluations of the micro-force. Variable-step macro-solver, 'larger' ϵ

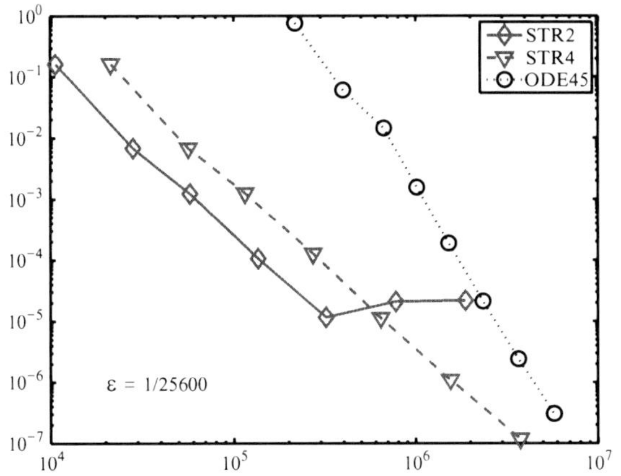

Fig. 6 Efficiency comparison: errors vs. number of evaluations of the micro-force. Variable-step macro-solver, smaller ϵ

5.2 Variable Step-Sizes

To illustrate the use of the stroboscopic algorithm with variable macro-step sizes we ran the ode45 MATLAB as macro-integrator with absolute error tolerances Tol from the sequence $10^{-2}, 10^{-3}, \dots, 10^{-8}$ (the relative error tolerance was taken to be equal to the absolute tolerance). For reasons discussed in the preceding subsection is important that the micro-integration is performed by a method that solves (17) exactly at $\tau = 2\pi$ for $\epsilon = 0$; we decided to micro-integrate, *with constant step-sizes,* by means of the fifth-order RK formula of the pair used by ode45.[4] We took $h = (2\pi\epsilon)/\nu$ where ν is the smallest integer for which $(2\pi/\nu)^5 \leq 1000 \times Tol$; this equilibrates the accuracy of the macro- and micro-integrations in a way similar to that analyzed in the preceding subsection. (The values of ν for the seven values of Tol turn out to be 4, 7, 10, 16, 26, 40, 63.) The variable-step macro-integrator chooses step-points that of course do not coincide with stroboscopic times but, as discussed in Sect. 3, this causes no problem to the stroboscopic algorithm. To measure errors we took advantage of the dense output capabilities of ode45 and generated output of the macro-integration at each stroboscopic time. Errors were then measured as the maximum, over all stroboscopic times, of the (absolute value of the) difference between the q component of the reference solution and the output Q provided by the algorithms.

Figures 5 and 6 compare the efficiency of the stroboscopic algorithms with that of a straightforward integration of the oscillatory problem with ode45. Again the stroboscopic algorithm exhibits a behavior that, unless Tol is so small that errors saturate, is ϵ-independent. Clearly, for small values of ϵ, this uniformity renders them more efficient than the conventional integrator, whose performance is degraded as $\epsilon \downarrow 0$.

Acknowledgements This research has been supported by 'Acción Integrada entre España y Francia' HF2008-0105. M.P. Calvo and J.M. Sanz-Serna are also supported by project MTM2007-63257 (Ministerio de Educación, España). A. Murua is also supported by projects MTM2007-61572 (Ministerio de Educación, España) and EHU08/43 (Universidad del País Vasco/Euskal Herriko Unibertsitatea).

References

1. Ariel, G., Engquist, B., Tsai, R.: A multiscale method for highly oscillatory ordinary differential equations with resonance. Math. Comput. **78**, 929–956 (2009)
2. Calvo, M., Jay, L.O., Montijano, J.I., Rández, L.: Approximate compositions of a near identity map by multi-revolution Runge-Kutta methods. Numer. Math. **97**, 635–666 (2004)
3. Calvo, M.P., Sanz-Serna, J.M.: Heterogeneous Multiscale Methods for mechanical systems with vibrations. SIAM J. Sci. Comput. 32, 2029–2046, (2010)
4. Chartier, Ph., Murua, A., Sanz-Serna, J.M.: Higher-order averaging, formal series and numerical integration I: B-series. Found. Comput. Math 10, 695–727 (2010)
5. E., W.: Analysis of the heterogeneous multiscale method for ordinary differential equations. Comm. Math. Sci. **1**, 423–436 (2003)
6. E., W., Engquist, B.: The heterogeneous multiscale methods. Comm. Math. Sci. **1**, 87–132 (2003)
7. E., W., Engquist, B., Li, X., Ren, W., Vanden-Eijnden, E.: Heterogeneous multiscale methods: A review. Commun. Comput. Phys. **2**, 367–450 (2007)
8. Engquist, B., Tsai, R.: Heterogeneous multiscale methods for stiff ordinary differential equations. Math. Comput. **74**, 1707–1742 (2005)

[4] The use of the variable-step code ode45 as micro-integrator for (17) with $\epsilon = 0$ yields errors that, after one period, are small but not exactly zero.

9. Hairer, E., Lubich, Ch., Wanner, G.: Geometric Numerical Integration, 2nd ed. Springer, Berlin (2006)
10. Kirchgraber, U.: An Ode-solver based on the method of averaging. Numer. Math. **53**, 621–652 (1988)
11. Murua, A.: Formal series and numerical integrators, Part I: Systems of ODEs and symplectic integrators. Appl. Numer. Math. **29**, 221–251 (1999)
12. Petzold, L.R., Jay, L.O., Yen, J.: Numerical solution of highly oscillatory ordinary differential equations. Acta Numerica **6**, 437–484 (1997)
13. Sanders, J.A., Verhulst, F., Murdock, J.: Averaging Methods in Nonlinear Dynamical Systems, 2nd ed. Springer, New York (2007)
14. Sanz-Serna, J.M.: Modulated Fourier expansions and heterogeneous multiscale methods. IMA J. Numer. Anal. **29**, 595–605 (2009)
15. Sanz-Serna, J.M., Calvo, M.P.: Numerical Hamiltonian Problems. Chapman and Hall, London (1994)
16. Sharp, R., Tsai, Y.-H., Engquist, B.: Multiple time scale numerical methods for the inverted pendulum problem. In: Engquist, B., Lötsdedt, P., Runborg, O. (eds) Multiscale Methods in Science and Engineering, Lect. Notes Comput. Sci. Eng. **44**, pp. 241–261. Springer, Berlin (2005)

The Microscopic Origin of the Macroscopic Dielectric Permittivity of Crystals: A Mathematical Viewpoint

Éric Cancès, Mathieu Lewin, and Gabriel Stoltz

Abstract The purpose of this paper is to provide a mathematical analysis of the Adler-Wiser formula relating the macroscopic relative permittivity tensor to the microscopic structure of the crystal at the atomic level. The technical level of the presentation is kept at its minimum to emphasize the mathematical structure of the results. We also briefly review some models describing the electronic structure of finite systems, focusing on density operator based formulations, as well as the Hartree model for perfect crystals or crystals with a defect.

1 Introduction

Insulating crystals are dielectric media. When an external electric field is applied, such an insulating material polarizes, and this induced polarization in turn affects the electric field. At the macroscopic level and in the time-independent setting, this phenomenon is modeled by the constitutive law

$$D = \epsilon_0 \epsilon_{\mathrm{M}} E \qquad (1)$$

specifying the relation between the macroscopic displacement field D and the macroscopic electric field E. The constant ϵ_0 is the dielectric permittivity of the vacuum, and ϵ_{M} the macroscopic relative permittivity of the crystal, a 3×3 symmetric tensor such that $\epsilon_{\mathrm{M}} \geq 1$ in the sense of symmetric matrices ($\mathbf{k}^T \epsilon_{\mathrm{M}} \mathbf{k} \geq |\mathbf{k}|^2$

É. Cancès (✉) · G. Stoltz
Université Paris-Est, CERMICS, Project-team Micmac, INRIA-Ecole des Ponts, 6 & 8 avenue Blaise Pascal, 77455 Marne-la-Vallée Cedex 2, France
e-mail: cances@cermics.enpc.fr; stoltz@cermics.enpc.fr

M. Lewin
CNRS & Laboratoire de Mathématiques UMR 8088, Université de Cergy-Pontoise, 95300 Cergy-Pontoise, France
e-mail: Mathieu.Lewin@math.cnrs.fr

B. Engquist et al. (eds.), *Numerical Analysis of Multiscale Computations*, Lecture Notes in Computational Science and Engineering 82, DOI 10.1007/978-3-642-21943-6_5,

for all $\mathbf{k} \in \mathbb{R}^3$). This tensor is proportional to the identity matrix for isotropic crystals. Recall that D is related to the so-called free charge ρ_{f} by the Gauss law $\mathrm{div}(D) = \rho_{\mathrm{f}}$ and that the macroscopic electric field E is related to the macroscopic potential V by $E = -\nabla V$, yielding the macroscopic Poisson equation

$$-\,\mathrm{div}(\epsilon_M \nabla V) = \rho_{\mathrm{f}}/\epsilon_0. \tag{2}$$

In the time-dependent setting, (1) becomes a time-convolution product:

$$D(\mathbf{r},t) = \epsilon_0 \int_{-\infty}^{+\infty} \epsilon_{\mathrm{M}}(t-t')E(\mathbf{r},t')\,dt'. \tag{3}$$

Fourier transforming in time, we obtain

$$\mathcal{F}D(\mathbf{r},\omega) = \mathcal{F}\epsilon_{\mathrm{M}}(\omega)\mathcal{F}E(\mathbf{r},\omega),$$

where, as usual in Physics, we have used the following normalization convention for the Fourier transform with respect to the time-variable:

$$\mathcal{F}f(\mathbf{r},\omega) = \int_{-\infty}^{+\infty} f(\mathbf{r},t)\,\mathrm{e}^{i\omega t}\,dt$$

(note that there is no minus sign in the phase factor). The time-dependent tensor ϵ_{M} in (1) can be seen as the zero-frequency limit of the frequency-dependent tensor $\mathcal{F}\epsilon_{\mathrm{M}}(\omega)$.

Of course, the constitutive laws (1) (time-independent case) and (3) (time-dependent case) are only valid in the *linear response regime*. When strong dielectric field are applied, the response can be strongly nonlinear.

The purpose of this paper is to provide a mathematical analysis of the Adler-Wiser formula [1, 36] relating the macroscopic relative permittivity tensor ϵ_{M} (as well as the frequency-dependent tensor $\mathcal{F}\epsilon_{\mathrm{M}}(\omega)$) to the microscopic structure of the crystal at the atomic level.

In Sect. 2, we discuss the modeling of the electronic structure of finite molecular systems. We introduce in particular the Hartree model (also called reduced Hartree-Fock model in the mathematical literature), which is the basis for our analysis of the electronic structure of crystals. This model is an approximation of the electronic N-body Schrödinger equation allowing to compute the ground state electronic density of a molecular system containing M nuclei considered as classical particles (Born-Oppenheimer approximation) and N quantum electrons, subjected to Coulomb interactions. The only empirical parameters in this model are a few fundamental constants of Physics (the reduced Planck constant $\hbar$, the mass of the electron m_{e}, the elementary charge e, and the dielectric permittivity of the vacuum ϵ_0) and the masses and charges of the nuclei. In this respect, this is an ab initio, or first-principle, model in the sense that it does not contain any empirical parameter specific to the molecular system under consideration.

We then show, in Sect. 3, how to extend the Hartree model for molecular systems (finite number of particles) to crystals (infinite number of particles). We first deal with perfect crystals (Sect. 3.2), then with crystals with local defects (Sect. 3.3). The mathematical theory of the electronic structure of crystals with local defects presented here (and originally published in [7]) has been strongly inspired by previous works on the mathematical foundations of quantum electrodynamics (QED) [18–20]. In some sense, a defect embedded in an insulating or semi-conducting crystal behaves similarly as a nucleus embedded in the polarizable vacuum of QED.

In Sect. 4, we study the dielectric response of a crystal. First, we focus on the response to an effective time-independent potential V, and expand it in powers of V (Sect. 4.1). The linear response term allows us to define the (microscopic) dielectric operator ϵ and its inverse ϵ^{-1}, the (microscopic) dielectric permittivity operator, and also to define a notion of renormalized charge for defects in crystals (Sect. 4.2). In Sect. 4.3, we derive the Adler-Wiser formula from the Hartree model, by means of homogenization arguments. Loosely speaking, a defect in a crystal generates an external field and thereby a dielectric response of the crystal. If a given local defect is properly rescaled, it produces a macroscopic charge (corresponding to the free charge $\rho_{\rm f}$ in (2)) and the total Coulomb potential converges to the macroscopic potential V solution to (2) where $\epsilon_{\rm M}$ is the tensor provided by the Adler-Wiser formula. A similar strategy can be used to obtain the frequency-dependent tensor $\mathcal{F}\epsilon_{\rm M}(\omega)$ (Sect. 4.4).

As trace-class and Hilbert-Schmidt operators play a central role in the mathematical theory of electronic structure, their definitions and some of their basic properties are recalled in the Appendix for the reader's convenience.

The mathematical results contained in this proceeding have been published [7–9], or will be published very soon [10]. The proofs are omitted. A pedagogical effort has been made to present this difficult material to non-specialists.

As usual in first-principle modeling, we adopt the system of atomic units, obtained by setting

$$\hbar = 1, \quad m_{\rm e} = 1, \quad e = 1, \quad \frac{1}{4\pi\epsilon_0} = 1,$$

so that (2) reads in this new system of units:

$$-\operatorname{div}(\epsilon_M \nabla V) = 4\pi\rho_{\rm f}. \tag{4}$$

For simplicity, we omit the spin variable, but taking the spin into account does not add any difficulty. It simply makes the mathematical formalism a little heavier.

2 Electronic Structure Models for Finite Systems

Let $\mathscr{H}$ be a Hilbert space and $\langle\cdot|\cdot\rangle$ its inner product (bra-ket Dirac's notation). Recall that if A is a self-adjoint operator on $\mathscr{H}$ and ϕ and ψ are in $D(A)$, the domain of A, then $\langle\phi|A|\psi\rangle := \langle\phi|A\psi\rangle = \langle A\phi|\psi\rangle$. If A is bounded from below, the bilinear form $(\phi,\psi) \mapsto \langle\phi|A|\psi\rangle$ can be extended in a unique way to the form domain of A. For instance, the operator $A = -\Delta$ with domain $D(A) = H^2(\mathbb{R}^d)$ is self-adjoint on $L^2(\mathbb{R}^d)$. Its form domain is $H^1(\mathbb{R}^d)$ and $\langle\phi|A|\psi\rangle = \int_{\mathbb{R}^d} \nabla\phi\cdot\nabla\psi$. In the sequel, we denote by $\mathscr{S}(\mathscr{H})$ the vector space of *bounded* self-adjoint operators on $\mathscr{H}$.

For $k = 0$, 1 and 2, and with the convention $H^0(\mathbb{R}^3) = L^2(\mathbb{R}^3)$, we denote by

$$\bigwedge_{i=1}^{N} H^k(\mathbb{R}^3) := \left\{ \Psi \in H^k(\mathbb{R}^{3N}) \middle| \Psi(\mathbf{r}_{p(1)},\ldots,\mathbf{r}_{p(N)}) = \epsilon(p)\Psi(\mathbf{r}_1,\ldots,\mathbf{r}_N), \forall p \in \mathbb{S}_N \right\}$$

(where $\mathbb{S}_N$ is the group of the permutations of $\{1,\ldots,N\}$ and $\epsilon(p)$ the parity of p) the antisymmetrized tensor product of N spaces $H^k(\mathbb{R}^3)$. These spaces are used to describe the electronic state of an N electron system. The antisymmetric constraint originates from the fact that electrons are fermions.

2.1 The N-Body Schrödinger Model

Consider a molecular system with M nuclei of charges $z_1,\ldots,z_M$. As we work in atomic units, z_k is a positive integer. Within the Born-Oppenheimer approximation, the nuclei are modeled as classical point-like particles. This approximation results from a combination of an adiabatic limit (the small parameter being the square root of the ratio between the mass of the electron and the mass of the lightest nucleus present in the system), and a semi-classical limit. We refer to [2, 3] and references therein for the mathematical aspects.

Usually, nuclei are represented by point-like particles. If the M nuclei are located at points $\mathbf{R}_1,\ldots,\mathbf{R}_M$ of $\mathbb{R}^3$, the nuclear charge distribution is modeled by

$$\rho^{\text{nuc}} = \sum_{k=1}^{M} z_k \delta_{\mathbf{R}_k},$$

where $\delta_{\mathbf{R}_k}$ is the Dirac measure at point $\mathbf{R}_k$. The Coulomb potential generated by the nuclei and seen by the electrons then reads

$$V^{\text{nuc}}(\mathbf{r}) := -\sum_{k=1}^{M} \frac{z_k}{|\mathbf{r}-\mathbf{R}_k|}$$

(the minus sign comes from the fact that the interaction between nuclei and electrons is attractive). In order to avoid some technical difficulties due to the singularity of the potential generated by point-like nuclei, the latter are sometimes replaced with smeared nuclei:

$$\rho^{\text{nuc}}(\mathbf{r}) = \sum_{k=1}^{M} z_k \chi(\mathbf{r} - \mathbf{R}_k),$$

where χ is a smooth approximation of the Dirac measure δ_0, or more precisely a non-negative smooth radial function such that $\int_{\mathbb{R}^3} \chi = 1$, supported in a small ball centered at 0. In this case,

$$V^{\text{nuc}}(\mathbf{r}) := -(\rho^{\text{nuc}} \star |\cdot|^{-1})(\mathbf{r}) = -\int_{\mathbb{R}^3} \frac{\rho^{\text{nuc}}(\mathbf{r}')}{|\mathbf{r} - \mathbf{r}'|}\, d\mathbf{r}'$$

is a smooth function. We will sometimes denote this smooth function by $V_{\rho^{\text{nuc}}}$ in order to emphasize that the potential is generated by a non-singular charge distribution.

The main quantity of interest in our study is the electrostatic potential generated by the total charge, which is by definition the sum of the nuclear charge ρ^{nuc} and the electronic charge ρ^{el}. According to the Born-Oppenheimer approximation, electrons are in their ground state, and ρ^{el} is the density associated with the ground state wavefunction Ψ_0. Let us make this definition more precise.

Any (pure) state of a system of N electrons is entirely described by a wavefunction $\Psi \in \bigwedge_{i=1}^{N} L^2(\mathbb{R}^3)$ satisfying the normalization condition $\|\Psi\|_{L^2(\mathbb{R}^{3N})} = 1$. The density associated with Ψ is the function ρ_Ψ defined by

$$\rho_\Psi(\mathbf{r}) = N \int_{\mathbb{R}^{3(N-1)}} |\Psi(\mathbf{r}, \mathbf{r}_2, \ldots, \mathbf{r}_N)|^2\, d\mathbf{r}_2 \cdots d\mathbf{r}_N. \tag{5}$$

Clearly,

$$\rho_\Psi \geq 0, \quad \rho_\Psi \in L^1(\mathbb{R}^3), \quad \text{and} \quad \int_{\mathbb{R}^3} \rho_\Psi = N.$$

It can be checked that if $\Psi \in \bigwedge_{i=1}^{N} H^1(\mathbb{R}^3)$, then $\sqrt{\rho} \in H^1(\mathbb{R}^3)$, which implies in particular that $\rho_\Psi \in L^1(\mathbb{R}^3) \cap L^3(\mathbb{R}^3)$.

The ground state wavefunction Ψ_0 is the lowest energy, normalized eigenfunction of the time-independent Schrödinger equation

$$H_N \Psi = E\Psi, \quad \Psi \in \bigwedge_{i=1}^{N} H^2(\mathbb{R}^3), \quad \|\Psi\|_{L^2(\mathbb{R}^{3N})} = 1, \tag{6}$$

where H_N is the electronic Hamiltonian. The latter operator is self-adjoint on $\bigwedge_{i=1}^{N} L^2(\mathbb{R}^3)$, with domain $\bigwedge_{i=1}^{N} H^2(\mathbb{R}^3)$ and form domain $\bigwedge_{i=1}^{N} H^1(\mathbb{R}^3)$, and is defined as

$$H_N = -\frac{1}{2}\sum_{i=1}^{N} \Delta_{\mathbf{r}_i} + \sum_{i=1}^{N} V^{\text{nuc}}(\mathbf{r}_i) + \sum_{1 \leq i < j \leq N} \frac{1}{|\mathbf{r}_i - \mathbf{r}_j|}. \tag{7}$$

The first term in the right-hand side of (7) models the kinetic energy of the electrons, the second term the Coulomb interaction between nuclei and electrons and the third term the Coulomb interaction between electrons. For later purposes, we write

$$H_N = T + V_{\rm ne} + V_{\rm ee},$$

where

$$T = -\frac{1}{2}\sum_{i=1}^{N}\Delta_{\mathbf{r}_i}, \quad V_{\rm ne} = \sum_{i=1}^{N} V^{\rm nuc}(\mathbf{r}_i), \quad V_{\rm ee} = \sum_{1\le i<j\le N}\frac{1}{|\mathbf{r}_i-\mathbf{r}_j|}.$$

It is proved in [37] that if the molecular system is neutral ($\sum_{k=1}^{M} z_k = N$) or positively charged ($\sum_{k=1}^{M} z_k \ge N$), then the essential spectrum of H_N is an interval of the form $[\Sigma_N, +\infty)$ with $\Sigma_N \le 0$ and $\Sigma_N < 0$ if $N \ge 2$, and its discrete spectrum is an increasing infinite sequence of negative eigenvalues converging to Σ_N. This guarantees the existence of Ψ_0. If E_0, the lowest eigenvalue of H_N is non-degenerate, Ψ_0 is unique up to a global phase, and $\rho^{\rm el} = \rho_{\Psi^0}$ is therefore uniquely defined by (5). If E_0 is degenerate, then the ground state electronic density is not unique. As the usual Born-Oppenheimer approximation is no longer valid when E_0 is degenerate, we will assume from now on that E_0 is a simple eigenvalue.

Note that Ψ_0 can also be defined variationally: It is the minimizer of

$$\inf\left\{\langle\Psi|H_N|\Psi\rangle,\ \Psi\in\bigwedge_{i=1}^{N} H^1(\mathbb{R}^3),\ \|\Psi\|_{L^2(\mathbb{R}^{3N})}=1\right\}. \tag{8}$$

Otherwise stated, it is obtained by minimizing the energy $\langle\Psi|H_N|\Psi\rangle$ over the set of all normalized, antisymmetric wavefunctions Ψ of finite energy.

Let us mention that, as in the absence of magnetic field, the N-body Hamiltonian is real (in the sense that it transforms a real-valued function into a real-valued function), there is no loss of generality in working in the space of real-valued N-body wavefunctions. Under the assumption that E_0 is non-degenerate, (8) has exactly two minimizers, Ψ_0 and $-\Psi_0$, both of them giving rise to the same electronic density.

2.2 The N-Body Schrödinger Model for Non-interacting Electrons

Neither the Schrödinger equation (6) nor the minimization problem (8) can be solved with standard numerical techniques when N exceeds two or three. On the other hand, these problems become pretty simple when the interaction between electrons is neglected. In this case, the N-body Hamiltonian is separable and reads

$$H_N^0 = T + V_{\rm ne} = \sum_{i=1}^{N} h_{\mathbf{r}_i} \quad \text{where} \quad h_{\mathbf{r}_i} = -\frac{1}{2}\Delta_{\mathbf{r}_i} + V^{\rm nuc}(\mathbf{r}_i)$$

is a self-adjoint operator on $L^2(\mathbb{R}^3)$ with domain $H^2(\mathbb{R}^3)$ and form domain $H^1(\mathbb{R}^3)$, acting on functions of the variable $\mathbf{r}_i$. It is known that the essential spectrum of h is $[0,+\infty)$ and that the discrete spectrum of h is an increasing infinite sequence of negative eigenvalues converging to 0. Let us denote by $\epsilon_1 < \epsilon_2 \leq \epsilon_3 \leq \cdots$ the eigenvalues of h counted with their multiplicities (it can be shown that ϵ_1 is simple) and let $(\phi_i)_{i\geq 0}$ be an orthonormal family of associated eigenvectors:

$$h\phi_i = \epsilon_i \phi_i, \quad \epsilon_1 < \epsilon_2 \leq \epsilon_3 \leq \cdots, \quad \phi_i \in H^2(\mathbb{R}^3), \quad \langle \phi_i | \phi_j \rangle_{L^2(\mathbb{R}^3)} = \delta_{ij}.$$

The eigenfunctions ϕ_i are called (molecular) orbitals and the eigenvalues ϵ_i are called (one-particle) energy levels.

It is easy to check that if $\epsilon_N < \epsilon_{N+1}$, then

$$\inf\left\{ \langle \Psi | H_N^0 | \Psi \rangle,\ \Psi \in \bigwedge_{i=1}^N H^1(\mathbb{R}^3),\ \|\Psi\|_{L^2(\mathbb{R}^{3N})} = 1 \right\} \tag{9}$$

has a unique solution (up to a global phase) given by the Slater determinant

$$\Psi_0(\mathbf{r}_1,\cdots,\mathbf{r}_N) = \frac{1}{\sqrt{N!}} \begin{vmatrix} \phi_1(\mathbf{r}_1) & \phi_1(\mathbf{r}_2) & \cdots & \phi_1(\mathbf{r}_N) \\ \phi_2(\mathbf{r}_1) & \phi_2(\mathbf{r}_2) & \cdots & \phi_2(\mathbf{r}_N) \\ \cdot & \cdot & \cdots & \cdot \\ \cdot & \cdot & \cdots & \cdot \\ \cdot & \cdot & \cdots & \cdot \\ \phi_N(\mathbf{r}_1) & \phi_N(\mathbf{r}_2) & \cdots & \phi_N(\mathbf{r}_N) \end{vmatrix}, \tag{10}$$

and that the ground state electronic density (5) takes the simple form

$$\rho^{\mathrm{el}}(\mathbf{r}) = \sum_{i=1}^N |\phi_i(\mathbf{r})|^2.$$

The above description of the electronic states of a set of N non-interacting electrons in terms of orbitals cannot be easily extended to infinite systems such as crystals (the number of orbitals becoming infinite). For this reason, we introduce a new formulation based on the concept of one-particle density operator, here abbreviated as density operator.

2.3 Density Operators

The (one-particle) density operator of a system of N electrons is an element of the convex set

$$\mathcal{D}_N = \left\{ \gamma \in \mathcal{S}(L^2(\mathbb{R}^3)) \mid 0 \leq \gamma \leq 1,\ \mathrm{Tr}(\gamma) = N \right\}.$$

Recall that if A and B are two bounded self-adjoint operators on a Hilbert space $\mathcal{H}$, the notation $A \leq B$ means that $\langle \psi | A | \psi \rangle \leq \langle \psi | B | \psi \rangle$ for all $\psi \in \mathcal{H}$.

Any density operator $\gamma \in \mathscr{D}_N$ is trace-class, hence compact (the basic properties of trace-class operators are recalled in the Appendix). It can therefore be diagonalized in an orthonormal basis:

$$\gamma = \sum_{i=1}^{+\infty} n_i |\phi_i\rangle\langle\phi_i| \quad \text{with} \quad \langle\phi_i|\phi_j\rangle = \delta_{ij}. \tag{11}$$

The eigenvalues n_i are called occupation numbers; the eigenfunctions ϕ_i are called natural orbitals. The conditions $0 \le \gamma \le 1$ and $\mathrm{Tr}(\gamma) = N$ are respectively equivalent to

$$0 \le n_i \le 1 \quad \text{and} \quad \sum_{i=1}^{+\infty} n_i = N.$$

The fact that $0 \le n_i \le 1$ is a mathematical translation of the Pauli exclusion principle, stipulating that each quantum state $|\phi_i\rangle$ is occupied by at most one electron. The sum of the occupation numbers is equal to N, the number of electrons in the system. The density associated with γ is defined by

$$\rho_\gamma(\mathbf{r}) = \sum_{i=1}^{+\infty} n_i |\phi_i(\mathbf{r})|^2, \tag{12}$$

this definition being independent of the choice of the orthonormal basis $(\phi_i)_{i\ge1}$ in (11) and satisfies

$$\rho_\gamma \ge 0, \quad \rho_\gamma \in L^1(\mathbb{R}^3), \quad \text{and} \quad \int_{\mathbb{R}^3} \rho_\gamma = N.$$

The kinetic energy of the density operator γ is defined as

$$T(\gamma) := \frac{1}{2}\mathrm{Tr}(|\nabla|\gamma|\nabla|),$$

and can be finite or infinite. Recall that the operator $|\nabla|$ is the unbounded self-adjoint operator on $L^2(\mathbb{R}^3)$ with domain $H^1(\mathbb{R}^3)$ defined by

$$\forall \phi \in H^1(\mathbb{R}^3), \quad (\mathscr{F}(|\nabla|\phi))(\mathbf{k}) = |\mathbf{k}|(\mathscr{F}(\phi))(\mathbf{k})$$

where $\mathscr{F}$ is the unitary Fourier transform

$$\mathscr{F}\phi(\mathbf{k}) = \widehat{\phi}(\mathbf{k}) = \frac{1}{(2\pi)^{3/2}} \int_{\mathbb{R}^3} \phi(\mathbf{r})\, \mathrm{e}^{-i\mathbf{k}\cdot\mathbf{r}}\, d\mathbf{r}.$$

The kinetic energy of a density operator γ decomposed as (11) is finite if and only if each ϕ_i is in $H^1(\mathbb{R}^3)$ and $\sum_{i=1}^{+\infty} n_i \|\nabla\phi_i\|_{L^2(\mathbb{R}^3)}^2 < \infty$, in which case

$$T(\gamma) = \frac{1}{2}\sum_{i=1}^{+\infty} n_i \|\nabla\phi_i\|_{L^2(\mathbb{R}^3)}^2.$$

As $|\nabla|$ is the square root of $-\Delta$ (*i.e.* $|\nabla|$ is self-adjoint, positive and $|\nabla|^2 = -\Delta$), the element $\mathrm{Tr}(|\nabla|\gamma|\nabla|)$ of $\mathbb{R}_+ \cup \{+\infty\}$ is often denoted by $\mathrm{Tr}(-\Delta\gamma)$. Using this notation, we can define the convex set $\mathscr{P}_N$ of the density operators of finite energy as

$$\mathscr{P}_N = \left\{\gamma \in \mathscr{S}(L^2(\mathbb{R}^3)) \mid 0 \le \gamma \le 1,\ \mathrm{Tr}(\gamma) = N,\ \mathrm{Tr}(-\Delta\gamma) < \infty\right\}.$$

Lastly, it is sometimes useful to introduce the integral kernel of a density operator $\gamma \in \mathscr{P}_N$, which is called a (one-particle) density matrix, and is usually also denoted by γ. It is by definition the function $\gamma \in L^2(\mathbb{R}^3 \times \mathbb{R}^3)$ such that

$$\forall \phi \in L^2(\mathbb{R}^3), \quad (\gamma\phi)(\mathbf{r}) = \int_{\mathbb{R}^3} \gamma(\mathbf{r},\mathbf{r}')\phi(\mathbf{r}')\,d\mathbf{r}'. \tag{13}$$

The expression of the density matrix γ in terms of natural orbitals and occupation numbers thus reads

$$\gamma(\mathbf{r},\mathbf{r}') = \sum_{i=1}^{+\infty} n_i \phi_i(\mathbf{r})\phi_i(\mathbf{r}').$$

Formally $\rho_\gamma(\mathbf{r}) = \gamma(\mathbf{r},\mathbf{r})$ and this relation makes sense rigorously as soon as the density matrix γ has a trace on the three-dimensional vector subspace $\left\{(\mathbf{r},\mathbf{r}),\ \mathbf{r} \in \mathbb{R}^3\right\}$ of $\mathbb{R}^3 \times \mathbb{R}^3$.

Let us now clarify the link between the description of electronic structures in terms of wavefunctions and the one in terms of density operators.

The density matrix associated with a wavefunction $\Psi \in \wedge_{i=1}^N L^2(\mathbb{R}^3)$ such that $\|\Psi\|_{L^2(\mathbb{R}^{3N})} = 1$ is the function of $L^2(\mathbb{R}^3 \times \mathbb{R}^3)$ defined as

$$\gamma_\Psi(\mathbf{r},\mathbf{r}') = N \int_{\mathbb{R}^{3(N-1)}} \Psi(\mathbf{r},\mathbf{r}_2,\ldots,\mathbf{r}_N)\Psi(\mathbf{r}',\mathbf{r}_2,\ldots,\mathbf{r}_N)\,d\mathbf{r}_2\cdots d\mathbf{r}_N \tag{14}$$

(recall that we are dealing with real-valued wavefunctions), and the corresponding density operator by

$$\forall \phi \in L^2(\mathbb{R}^3), \quad (\gamma_\Psi\phi)(\mathbf{r}) = \int_{\mathbb{R}^3} \gamma_\Psi(\mathbf{r},\mathbf{r}')\phi(\mathbf{r}')\,d\mathbf{r}'. \tag{15}$$

It is easy to see that the density operator γ_Ψ is in $\mathscr{D}_N$. Under the additional assumption that $\Psi \in \wedge_{i=1}^N H^1(\mathbb{R}^3)$, it is even in $\mathscr{P}_N$. Besides, the definition (5) of the density associated with Ψ agrees with the definition (12) of the density associated with γ_Ψ, *i.e.*

$$\rho_\Psi = \rho_{\gamma_\Psi},$$

and the same holds with the definition of the kinetic energy if $\Psi \in \wedge_{i=1}^N H^1(\mathbb{R}^3)$:

$$\langle\Psi|T|\Psi\rangle = T(\gamma_\Psi).$$

Remark 1. The maps $\left\{\Psi \in \bigwedge_{i=1}^N L^2(\mathbb{R}^3) \,\middle|\, \|\Psi\|_{L^2(\mathbb{R}^{3N})} = 1\right\} \ni \Psi \mapsto \gamma_\Psi \in \mathscr{D}_N$ and $\left\{\Psi \in \bigwedge_{i=1}^N H^2(\mathbb{R}^3) \,\middle|\, \|\Psi\|_{L^2(\mathbb{R}^{3N})} = 1\right\} \ni \Psi \mapsto \gamma_\Psi \in \mathscr{P}_N$ are not surjective. This means that an element of $\mathscr{D}_N$ (resp. of $\mathscr{P}_N$) is not necessarily the density operator associated with some *pure* state. However any $\gamma \in \mathscr{D}_N$ (resp. any $\gamma \in \mathscr{P}_N$) is the (one-particle) density operator associated with some *mixed* state (represented by some N-particle density operator). This property is referred to as the N-representability property of density operators.

We can now reformulate the electronic structure problem *for a system of N non-interacting electrons*, in terms of density operators:

1. The energy of a wavefunction $\Psi \in \wedge_{i=1}^N H^1(\mathbb{R}^3)$ is a linear form with respect to the density operator γ_Ψ:

$$\langle \Psi | H_N^0 | \Psi \rangle = E_{\rho^{\rm nuc}}^0(\gamma_\Psi) \quad \text{where} \quad E_{\rho^{\rm nuc}}^0(\gamma) = \mathrm{Tr}\left(-\frac{1}{2}\Delta\gamma\right) + \int_{\mathbb{R}^3} \rho_\gamma V^{\rm nuc}.$$

2. The ground state density matrix, that is the density operator associated with the ground state wavefunction Ψ^0 defined by (9), is the orthogonal projector (for the L^2 inner product) on the space $\mathrm{Span}(\phi_1, \ldots, \phi_N)$:

$$\gamma_{\Psi^0} = \sum_{i=1}^N |\phi_i\rangle\,\langle\phi_i|.$$

3. The ground state energy and the ground state density operators are obtained by solving the minimization problem

$$\inf\left\{E_{\rho^{\rm nuc}}^0(\gamma),\ \gamma \in \mathscr{S}(L^2(\mathbb{R}^3)),\ 0 \le \gamma \le 1,\ \mathrm{Tr}(\gamma) = N,\ \mathrm{Tr}(-\Delta\gamma) < \infty\right\}. \quad (16)$$

The advantages of the density operator formulation, which are not obvious for finite systems, will clearly appear in Sect. 3, where we deal with crystals.

2.4 The Hartree Model and Other Density Operator Models of Electronic Structures

Let us now reintroduce the Coulomb interaction between electrons, taking as a starting point the non-interacting system introduced in Sect. 2.2. The models presented in this section are density operator models in the sense that the ground state energy and density are obtained by minimizing some *explicit* functional $E_{\rho^{\rm nuc}}(\gamma)$ over the set of N-representable density operators $\mathscr{P}_N$.

All these models share the same mathematical structure. They read:

$$\inf\left\{E_{\rho^{\rm nuc}}(\gamma),\ \gamma \in \mathscr{S}(L^2(\mathbb{R}^3)),\ 0 \le \gamma \le 1,\ \mathrm{Tr}(\gamma) = N,\ \mathrm{Tr}(-\Delta\gamma) < \infty\right\}, \quad (17)$$

with

$$E_{\rho^{\mathrm{nuc}}}(\gamma) = \mathrm{Tr}\left(-\frac{1}{2}\Delta\gamma\right) + \int_{\mathbb{R}^3} \rho_\gamma V_{\rho^{\mathrm{nuc}}} + \frac{1}{2} D(\rho_\gamma, \rho_\gamma) + \widetilde{E}(\gamma),$$

where

$$D(f,g) = \int_{\mathbb{R}^3}\int_{\mathbb{R}^3} \frac{f(\mathbf{r})\, g(\mathbf{r}')}{|\mathbf{r}-\mathbf{r}'|}\, d\mathbf{r}\, d\mathbf{r}' \tag{18}$$

is the classical Coulomb interaction and $\widetilde{E}(\gamma)$ some correction term. Note that $D(f,g)$ is well defined for f and g in $L^{6/5}(\mathbb{R}^3)$, see for instance [30, Sect. IX.4]. Recall also that for each $\gamma \in \mathscr{P}_N$, $\rho_\gamma \in L^1(\mathbb{R}^3) \cap L^3(\mathbb{R}^3) \hookrightarrow L^{6/5}(\mathbb{R}^3)$.

The Hartree model, on which we will focus in this proceeding, corresponds to $\widetilde{E}(\gamma) = 0$:

$$E^{\mathrm{Hartree}}_{\rho^{\mathrm{nuc}}}(\gamma) = \mathrm{Tr}\left(-\frac{1}{2}\Delta\gamma\right) + \int_{\mathbb{R}^3} \rho_\gamma V_{\rho^{\mathrm{nuc}}} + \frac{1}{2} D(\rho_\gamma, \rho_\gamma).$$

The reason why we study this model is that it has much nicer mathematical properties than other models with $\widetilde{E}(\gamma) \neq 0$ (see below).

The Kohn-Sham models [24] originate from the Density Functional Theory (DFT) [13]. In this kind of models, $\widetilde{E}(\gamma)$ is an explicit functional of the density ρ_γ, called the exchange-correlation functional:

$$E^{\mathrm{KS}}_{\rho^{\mathrm{nuc}}}(\gamma) = \mathrm{Tr}\left(-\frac{1}{2}\Delta\gamma\right) + \int_{\mathbb{R}^3} V_{\rho^{\mathrm{nuc}}}\rho_\gamma + \frac{1}{2} D(\rho_\gamma, \rho_\gamma) + E^{\mathrm{xc}}(\rho_\gamma). \tag{19}$$

If follows from the Hohenberg-Kohn theorem [21] (see [27] for a more mathematical presentation of this result) that there exists some functional $E^{\mathrm{xc}}(\rho)$ depending only on the density ρ, such that minimizing (17) with $E_{\rho^{\mathrm{nuc}}} = E^{\mathrm{KS}}_{\rho^{\mathrm{nuc}}}$ provides the *exact* ground state energy and density, whatever the nuclear charge distribution ρ^{nuc}. Note however, that the Kohn-Sham ground state density operator obtained by minimizing (17) is not the ground state density operator corresponding to the ground state wavefunction Ψ^0. Unfortunately, the exact exchange-correlation functional is not known. Many approximate functionals have been proposed, and new ones come up on a regular basis. For the sake of illustration, the simplest approximate exchange-correlation functional (but clearly not the best one) is the so-called Xα functional

$$E^{\mathrm{xc}}_{\mathrm{X}\alpha}(\rho) = -C_{\mathrm{X}\alpha}\int_{\mathbb{R}^3} \rho^{4/3},$$

where $C_{\mathrm{X}\alpha}$ is a positive constant.

Lastly, the models issued from the Density-Matrix Functional Theory (DMFT) involve functionals $\widetilde{E}(\gamma)$ depending explicitly on the density operator γ, but not only on the density ρ_γ. Similar to DFT, there exists an *exact* (but unknown) functional $\widetilde{E}(\gamma)$ for which minimizing (17) gives the exact ground state energy and density, whatever the nuclear charge distribution ρ^{nuc}. However, unlike the exact DFT functional, the exact DMFT functional also provides the exact ground state density operator. Several approximate DMFT functionals have been proposed.

Note that the Hartree-Fock model, which is usually defined as the variational approximation of (8) obtained by restricting the minimization set to the set of finite energy Slater determinants, can also be seen as a DMFT functional

$$E^{\mathrm{HF}}_{\rho^{\mathrm{nuc}}}(\gamma) = \mathrm{Tr}\left(-\frac{1}{2}\Delta\gamma\right) + \int_{\mathbb{R}^3} \rho_\gamma V_{\rho^{\mathrm{nuc}}} + \frac{1}{2}D(\rho_\gamma,\rho_\gamma) - \frac{1}{2}\int_{\mathbb{R}^3}\int_{\mathbb{R}^3} \frac{|\gamma(\mathbf{r},\mathbf{r}')|^2}{|\mathbf{r}-\mathbf{r}'|}\,d\mathbf{r}\,d\mathbf{r}',$$

where, as above, $\gamma(\mathbf{r},\mathbf{r}')$ denotes the integral kernel of γ.

The existence of a solution to (17) for a neutral or positively charged system is established in [34] for the Hartree model ($E^{\mathrm{xc}} = 0$), in [26] for the Hartree-Fock model, in [4] for the Xα and the standard LDA model, and in [15] for the Müller DMFT functional.

The key-property allowing for a comprehensive mathematical analysis of the bulk limit for the Hartree model is that the ground state *density* is unique (which is not the case for the other models presented in this section). This means that in the Hartree framework, all the minimizers to (17) share the same density. This follows from the fact that the ground state Hartree density solves the variational problem

$$\inf\left\{\mathscr{E}(\rho),\ \rho \geq 0,\ \sqrt{\rho} \in H^1(\mathbb{R}^3),\ \int_{\mathbb{R}^3} \rho = N\right\}, \tag{20}$$

where

$$\mathscr{E}(\rho) = F(\rho) + \int_{\mathbb{R}^3} \rho V_{\rho^{\mathrm{nuc}}} + \frac{1}{2}D(\rho,\rho)$$

and

$$F(\rho) = \inf\left\{\mathrm{Tr}\left(-\frac{1}{2}\Delta\gamma\right),\ \gamma \in \mathscr{S}(L^2(\mathbb{R}^3)),\right.$$
$$\left. 0 \leq \gamma \leq 1,\ \mathrm{Tr}(\gamma) = N,\ \mathrm{Tr}(-\Delta\gamma) < \infty,\ \rho_\gamma = \rho\right\}.$$

As the functional $\mathscr{E}(\rho)$ is strictly convex on the convex set

$$\left\{\rho \geq 0,\ \sqrt{\rho} \in H^1(\mathbb{R}^3),\ \int_{\mathbb{R}^3} \rho = N\right\},$$

uniqueness follows.

The Euler equation for the Hartree model reads

$$\left\{\begin{array}{l} \gamma^0 = \sum_{i=1}^{+\infty} n_i |\phi_i\rangle\langle\phi_i|, \quad \rho^0(\mathbf{r}) = \rho_{\gamma^0}(\mathbf{r}) = \sum_{i=1}^{+\infty} n_i |\phi_i(\mathbf{r})|^2, \\ H^0\phi_i = \epsilon_i\phi_i, \quad \langle\phi_i|\phi_j\rangle = \delta_{ij}, \\ n_i = 1 \text{ if } \epsilon_i < \epsilon_{\mathrm{F}},\ 0 \le n_i \le 1 \text{ if } \epsilon_i = \epsilon_{\mathrm{F}},\ n_i = 0 \text{ if } \epsilon_i > \epsilon_{\mathrm{F}}, \quad \sum_{i=1}^{+\infty} n_i = N, \\ H^0 = -\frac{1}{2}\Delta + V^0, \\ -\Delta V^0 = 4\pi(\rho^0 - \rho^{\mathrm{nuc}}). \end{array}\right. \tag{21}$$

It can be proved that the essential spectrum of the self-adjoint operator H^0 is equal to $\mathbb{R}_+$ and that, for a neutral or positively charged system, H^0 has at least N negative eigenvalues. The scalar ϵ_{F}, called the Fermi level, can be interpreted as the Lagrange multiplier of the constraint $\mathrm{Tr}(\gamma^0) = N$.

Assuming that $\epsilon_N < \epsilon_{N+1}$, the ground state density operator γ^0 of the Hartree model is unique: It is the orthogonal projector

$$\gamma^0 = \sum_{i=1}^{N} |\phi_i\rangle\langle\phi_i|.$$

In this case, (21) can be rewritten under the more compact form

$$\left\{\begin{array}{l} \gamma^0 = 1_{(-\infty,\epsilon_{\mathrm{F}}]}(H^0), \quad \rho^0 = \rho_{\gamma^0}, \\ H^0 = -\frac{1}{2}\Delta + V^0, \\ -\Delta V^0 = 4\pi(\rho^0 - \rho^{\mathrm{nuc}}), \end{array}\right. \tag{22}$$

for any $\epsilon_N < \epsilon_{\mathrm{F}} < \epsilon_{N+1}$. In this equation, the notation $1_{(-\infty,\epsilon_F]}(H^0)$ is used for the spectral projector of H^0 corresponding to the spectrum in the interval $(-\infty, \epsilon_F]$.

Lastly, we remark that if smeared nuclei are used, then $D(\rho^{\mathrm{nuc}}_{\mathrm{per}}, \rho^{\mathrm{nuc}}_{\mathrm{per}})$ is well defined (and finite). This allows us to reformulate the Hartree ground state problem as

$$\inf\left\{\widetilde{E}^{\mathrm{Hartree}}_{\rho^{\mathrm{nuc}}}(\gamma),\ \gamma \in \mathscr{S}(L^2(\mathbb{R}^3)),\ 0 \le \gamma \le 1,\ \mathrm{Tr}(\gamma) = N,\ \mathrm{Tr}(-\Delta\gamma) < \infty\right\}, \tag{23}$$

where

$$\widetilde{E}^{\mathrm{Hartree}}_{\rho^{\mathrm{nuc}}}(\gamma) = \mathrm{Tr}\left(-\frac{1}{2}\Delta\gamma\right) + \frac{1}{2}D(\rho^{\mathrm{nuc}} - \rho_\gamma, \rho^{\mathrm{nuc}} - \rho_\gamma).$$

The main interest of this new formulation of the Hartree problem is that the functional $\widetilde{E}^{\mathrm{Hartree}}_{\rho^{\mathrm{nuc}}}$ is the sum of two non-negative contributions: the kinetic energy and the Coulomb energy of the total charge distribution $\rho^{\mathrm{nuc}} - \rho_\gamma$. The presence of the unphysical terms corresponding to the self-interaction of nuclei in $D(\rho^{\mathrm{nuc}}_{\mathrm{per}}, \rho^{\mathrm{nuc}}_{\mathrm{per}})$ is not a problem for our purpose.

The time-dependent version of the Hartree model formally reads

$$i\frac{d\gamma}{dt}(t) = \left[-\frac{1}{2}\Delta + (\rho_{\gamma(t)} - \rho^{\rm nuc}(t)) \star |\cdot|^{-1}, \gamma(t)\right],$$

where $[A, B] = AB - BA$ denotes the commutator of the operators A and B. We are not going to elaborate further on the precise mathematical meaning of this formal equation for finite systems, but refer the reader to [5] and references therein (see in particular [12, Sect. XVII.B.5]) for further precisions on the mathematical meaning of the above equation. We will define and study a mild version of it in the case of crystals with defects in Sect. 4.4.

3 The Hartree Model for Crystals

The Hartree model presented in the previous section describes a *finite* system of N electrons in the electrostatic potential created by a nuclear density of charge $\rho^{\rm nuc}$. Our goal is to describe an *infinite* crystalline material obtained in the bulk limit. In fact we shall consider two such systems. The first one is the periodic crystal obtained when, in the bulk limit, the nuclear density approaches the periodic nuclear distribution of the perfect crystal:

$$\rho^{\rm nuc} \to \rho^{\rm nuc}_{\rm per}, \tag{24}$$

$\rho^{\rm nuc}_{\rm per}$ being a $\mathscr{R}$-periodic distribution. The set $\mathscr{R}$ is a periodic lattice of $\mathbb{R}^3$:

$$\mathscr{R} = \mathbb{Z}\mathbf{a}_1 + \mathbb{Z}\mathbf{a}_2 + \mathbb{Z}\mathbf{a}_3, \tag{25}$$

where $(\mathbf{a}_1, \mathbf{a}_2, \mathbf{a}_3)$ is a given triplet of linearly independent vectors of $\mathbb{R}^3$. The second system is the previous crystal in the presence of a local defect:

$$\rho^{\rm nuc} \to \rho^{\rm nuc}_{\rm per} + m, \tag{26}$$

m representing the nuclear charge of the defect. The functional spaces in which $\rho^{\rm nuc}_{\rm per}$ and m are chosen are made precise below.

3.1 Basics of Fourier and Bloch-Floquet Theories

A perfect crystal is characterized by a lattice $\mathscr{R}$ of $\mathbb{R}^3$ and a $\mathscr{R}$-periodic nuclear charge distribution $\rho^{\rm nuc}_{\rm per}$. Not surprisingly, Fourier and Bloch-Floquet theories, which allow to conveniently exploit the periodicity of the problem, play essential roles in the mathematical description of the electronic structure of crystals.

Let $\mathscr{R}^*$ be the reciprocal lattice of the lattice $\mathscr{R}$ defined in (25) (also called dual lattice):

$$\mathscr{R}^* = \mathbb{Z}\mathbf{a}_1^* + \mathbb{Z}\mathbf{a}_2^* + \mathbb{Z}\mathbf{a}_3^*, \quad \text{where} \quad \mathbf{a}_i \cdot \mathbf{a}_j^* = 2\pi\delta_{ij}.$$

Denote by Γ a unit cell of $\mathcal{R}$. Recall that a unit cell is a semi-open bounded polytope of $\mathbb{R}^3$ such that the cells $\Gamma + \mathbf{R} = \{(\mathbf{r} + \mathbf{R}), \mathbf{r} \in \Gamma\}$ for $\mathbf{R} \in \mathcal{R}$ form a tessellation of the space $\mathbb{R}^3$ (*i.e.* $(\Gamma + \mathbf{R}) \cap (\Gamma + \mathbf{R}') = 0$ if $\mathbf{R} \neq \mathbf{R}'$ and $\cup_{\mathbf{R} \in \mathcal{R}} (\Gamma + \mathbf{R}) = \mathbb{R}^3$). A possible choice for Γ is $\{x_1 \mathbf{a}_1 + x_2 \mathbf{a}_2 + x_3 \mathbf{a}_3, \ -1/2 < x_i \leq 1/2\}$. Another choice is the Wigner-Seitz cell of $\mathcal{R}$, which is by definition the semi-open Voronoi cell of the origin for the lattice $\mathcal{R}$. Lastly, we denote by Γ^* the first Brillouin zone, that is the Wigner-Seitz cell of the dual lattice. Let us illustrate these concepts on the simplest example, the cubic lattice, for which $\mathcal{R} = a\mathbb{Z}^3$ (for some $a > 0$). In this particular case, $\mathcal{R}^* = \frac{2\pi}{a}\mathbb{Z}^3$, the Wigner-Seitz cell is $\Gamma = (-a/2, a/2]^3$ and $\Gamma^* = (-\pi/a, \pi/a]^3$.

For each $\mathbf{K} \in \mathcal{R}^*$, we denote by $e_{\mathbf{K}}(\mathbf{r}) = |\Gamma|^{-1/2} \mathrm{e}^{i\mathbf{K}\cdot\mathbf{r}}$ the Fourier mode with wavevector $\mathbf{K}$. According to the theory of Fourier series, each $\mathcal{R}$-periodic distribution v can be expanded in Fourier series as

$$v = \sum_{\mathbf{K} \in \mathcal{R}^*} c_{\mathbf{K}}(v)\, e_{\mathbf{K}}, \tag{27}$$

where $c_{\mathbf{K}}(v)$ is the $\mathbf{K}$th Fourier coefficient of v, the convergence of the series holding in the distributional sense. We introduce the usual $\mathcal{R}$-periodic L^p spaces defined by

$$L^p_{\mathrm{per}}(\Gamma) := \left\{ v \in L^p_{\mathrm{loc}}(\mathbb{R}^3) \ \middle| \ v \ \mathcal{R}\text{-periodic} \right\},$$

and endow them with the norms

$$\|v\|_{L^p_{\mathrm{per}}(\Gamma)} := \left(\int_\Gamma |v|^p \right)^{1/p} \quad \text{for } 1 \leq p < \infty \quad \text{and} \quad \|v\|_{L^\infty_{\mathrm{per}}(\Gamma)} := \text{ess-sup}\,|v|.$$

In particular,

$$\|v\|_{L^2_{\mathrm{per}}(\Gamma)} = (v, v)^{1/2}_{L^2_{\mathrm{per}}(\Gamma)} \quad \text{where} \quad (v, w)_{L^2_{\mathrm{per}}(\Gamma)} := \int_\Gamma \overline{v} w.$$

Any function $v \in L^2_{\mathrm{per}}(\Gamma)$ can be expanded in Fourier modes according to (27), the Fourier coefficients being given by the simple formula

$$c_{\mathbf{K}}(v) = \frac{1}{|\Gamma|^{1/2}} \int_\Gamma v(\mathbf{r})\, \mathrm{e}^{-i\mathbf{K}\cdot\mathbf{r}}\, d\mathbf{r},$$

and the convergence of the series (27) also holds in $L^2_{\mathrm{per}}(\Gamma)$. Besides,

$$\forall (v, w) \in L^2_{\mathrm{per}}(\Gamma) \times L^2_{\mathrm{per}}(\Gamma), \quad (v, w)_{L^2_{\mathrm{per}}(\Gamma)} = \sum_{\mathbf{K} \in \mathcal{R}^*} \overline{c_{\mathbf{K}}(v)}\, c_{\mathbf{K}}(w).$$

For each $s \in \mathbb{R}$, the $\mathcal{R}$-periodic Sobolev space of index s is defined as

$$H^s_{\mathrm{per}}(\Gamma) := \left\{ v = \sum_{\mathbf{K} \in \mathcal{R}^*} c_{\mathbf{K}}(v) e_{\mathbf{K}} \ \middle| \ \sum_{\mathbf{K} \in \mathcal{R}^*} (1 + |\mathbf{K}|^2)^s |c_{\mathbf{K}}(v)|^2 < \infty \right\},$$

and endowed with the inner product

$$(v,w)_{H^s_{\rm per}(\Gamma)} := \sum_{\mathbf{K}\in\mathscr{R}^*} (1+|\mathbf{K}|^2)^s \overline{c_{\mathbf{K}}(v)} c_{\mathbf{K}}(w).$$

The Bloch-Floquet theory was introduced by Floquet for periodic differential equations and generalized by Bloch to periodic partial differential equations. We just recall the basic results of this theory used in this proceeding and refer the reader to [31] for further precisions.

Any function $f \in L^2(\mathbb{R}^3)$ can be decomposed by the Bloch-Floquet transform as

$$f(\mathbf{r}) = \fint_{\Gamma^*} f_{\mathbf{q}}(\mathbf{r})\, \mathrm{e}^{i\mathbf{q}\cdot\mathbf{r}} d\mathbf{q},$$

where $\fint_{\Gamma^*}$ is a notation for $|\Gamma^*|^{-1}\int_{\Gamma^*}$ and where the functions $f_{\mathbf{q}}$ are defined by

$$f_{\mathbf{q}}(\mathbf{r}) = \sum_{\mathbf{R}\in\mathscr{R}} f(\mathbf{r}+\mathbf{R})\mathrm{e}^{-i\mathbf{q}\cdot(\mathbf{r}+\mathbf{R})} = \frac{(2\pi)^{3/2}}{|\Gamma|} \sum_{\mathbf{K}\in\mathscr{R}^*} \widehat{f}(\mathbf{q}+\mathbf{K})\mathrm{e}^{i\mathbf{K}\cdot\mathbf{r}}. \tag{28}$$

For almost all $\mathbf{q}\in\mathbb{R}^3$, $f_{\mathbf{q}} \in L^2_{\rm per}(\Gamma)$. Besides, $f_{\mathbf{q}+\mathbf{K}}(\mathbf{r}) = f_{\mathbf{q}}(\mathbf{r})\mathrm{e}^{-i\mathbf{K}\cdot\mathbf{r}}$ for all $\mathbf{K}\in\mathscr{R}^*$ and almost all $\mathbf{q}\in\mathbb{R}^3$. Lastly,

$$\|f\|^2_{L^2(\mathbb{R}^3)} = \fint_{\Gamma^*} \|f_{\mathbf{q}}\|^2_{L^2_{\rm per}(\Gamma)} d\mathbf{q}.$$

For $\mathbf{R}\in\mathbb{R}^3$, we denote by $\tau_{\mathbf{R}}$ the translation operator defined by

$$\forall v \in L^2(\mathbb{R}^3), \quad (\tau_{\mathbf{R}} v)(\mathbf{r}) = v(\mathbf{r}-\mathbf{R}).$$

The main interest of the Bloch-Floquet transform (28) is that it provides a "block diagonalization" of any $\mathscr{R}$-periodic operator, that is of any operator on $L^2(\mathbb{R}^3)$ which commutes with $\tau_{\mathbf{R}}$ for all $\mathbf{R}\in\mathscr{R}$. Consider first a bounded $\mathscr{R}$-periodic operator A on $L^2(\mathbb{R}^3)$. Then there exists a family $(A_{\mathbf{q}})_{\mathbf{q}\in\Gamma^*}$ of bounded operators on $L^2_{\rm per}(\Gamma)$ such that

$$\forall v \in L^2(\mathbb{R}^3), \quad (Av)_{\mathbf{q}} = A_{\mathbf{q}} v_{\mathbf{q}} \quad \text{for almost all } q \in \Gamma^*. \tag{29}$$

If, in addition, A is self-adjoint on $L^2(\mathbb{R}^3)$, then $A_{\mathbf{q}}$ is self-adjoint on $L^2_{\rm per}(\Gamma)$ for almost all $\mathbf{q}\in\Gamma^*$ and

$$\sigma(A) = \overline{\bigcup_{\mathbf{q}\in\Gamma^*} \sigma(A_{\mathbf{q}})}.$$

In particular, the translation operators $(\tau_{\mathbf{R}})_{\mathbf{R}\in\mathscr{R}}$, which obviously commute with each other, are homotheties in the Bloch-Floquet representation

$$\forall \mathbf{R}\in\mathscr{R}, \quad (\tau_{\mathbf{R}})_{\mathbf{q}} = \mathrm{e}^{i\mathbf{q}\cdot\mathbf{R}} 1_{L^2_{\rm per}(\Gamma)}.$$

As $(e_{\mathbf{K}})_{\mathbf{K}\in\mathscr{R}^*}$ form an orthonormal basis of $L^2_{\mathrm{per}}(\Gamma)$, it follows from (29) that any bounded $\mathscr{R}$-periodic operator on $L^2(\mathbb{R}^3)$ is completely characterized by the Bloch-Floquet matrices $(([A_{\mathbf{K},\mathbf{K}'}(\mathbf{q})])_{(\mathbf{K},\mathbf{K}')\in\mathscr{R}^*\times\mathscr{R}^*})_{\mathbf{q}\in\Gamma^*}$ defined for almost all $\mathbf{q}\in\Gamma^*$ by

$$A_{\mathbf{K},\mathbf{K}'}(\mathbf{q}) := \langle e_{\mathbf{K}}, A_{\mathbf{q}} e_{\mathbf{K}'}\rangle_{L^2_{\mathrm{per}}(\Gamma)}.$$

In particular, it holds

$$\forall v \in L^2(\mathbb{R}^3), \quad \widehat{(Av)}(\mathbf{q}+\mathbf{K}) = \sum_{\mathbf{K}'\in\mathscr{R}^*} A_{\mathbf{K},\mathbf{K}'}(\mathbf{q})\widehat{v}(\mathbf{q}+\mathbf{K}'),$$

for all $(\mathbf{K},\mathbf{K}')\in\mathscr{R}^*\times\mathscr{R}^*$ and almost all $\mathbf{q}\in\Gamma^*$.

For unbounded operators, the situation is a little bit more intricate. Let us limit ourselves to the case of $\mathscr{R}$-periodic Schrödinger operators of the form

$$H = -\frac{1}{2}\Delta + V_{\mathrm{per}}$$

with $V_{\mathrm{per}} \in L^2_{\mathrm{per}}(\Gamma)$. By the Kato-Rellich theorem and [31, Theorem XIII.96], the operator H is self-adjoint on $L^2(\mathbb{R}^3)$, with domain $H^2(\mathbb{R}^3)$. It can also be decomposed as follows:

$$\forall v \in H^2(\mathbb{R}^3), \quad v_{\mathbf{q}} \in H^2_{\mathrm{per}}(\Gamma) \quad \text{and} \quad (Hv)_{\mathbf{q}} = H_{\mathbf{q}} v_{\mathbf{q}} \quad \text{for almost all } \mathbf{q}\in\Gamma^*,$$

where $H_{\mathbf{q}}$ is the self-adjoint operator on $L^2_{\mathrm{per}}(\Gamma)$ with domain $H^2_{\mathrm{per}}(\Gamma)$, defined by

$$H_{\mathbf{q}} = -\frac{1}{2}\Delta - i\mathbf{q}\cdot\nabla + \frac{|\mathbf{q}|^2}{2} + V_{\mathrm{per}}.$$

It is easily seen that for each $\mathbf{q}\in\Gamma^*$, $H_{\mathbf{q}}$ is bounded below and has a compact resolvent. Consequently, there exists a sequence $(\epsilon_{n,\mathbf{q}})_{n\geq 1}$ of real numbers going to $+\infty$, and an orthonormal basis $(u_{n,\mathbf{q}})_{n\geq 1}$ of $L^2_{\mathrm{per}}(\Gamma)$ such that

$$H_{\mathbf{q}} = \sum_{n=1}^{+\infty} \epsilon_{n,\mathbf{q}} |u_{n,\mathbf{q}}\rangle\langle u_{n,\mathbf{q}}|.$$

As the mapping $\mathbf{q}\mapsto H_{\mathbf{q}}$ is polynomial on $\mathbb{R}^3$, it is possible to number the eigenvalues $\epsilon_{n,\mathbf{q}}$ in such a way that $(\epsilon_{n,0})_{n\geq 1}$ is non-decreasing and that for each $n\geq 1$, the mapping $\mathbf{q}\mapsto\epsilon_{n,\mathbf{q}}$ is analytic in each direction. Then (see Fig. 1)

$$\sigma(H) = \overline{\bigcup_{\mathbf{q}\in\Gamma^*}\sigma(H_{\mathbf{q}})} = \bigcup_{n\geq 1}\left[\Sigma_n^-, \Sigma_n^+\right],$$

with

$$\Sigma_n^- = \min_{\mathbf{q}\in\overline{\Gamma^*}} \epsilon_{n,\mathbf{q}}, \quad \Sigma_n^+ = \max_{\mathbf{q}\in\overline{\Gamma^*}} \epsilon_{n,\mathbf{q}}. \tag{30}$$

The interval $[\Sigma_n^-, \Sigma_n^+]$ is called the nth band of the spectrum of H. It is possible to prove that the spectrum of H is purely absolutely continuous [35]. In particular, H has no eigenvalues.

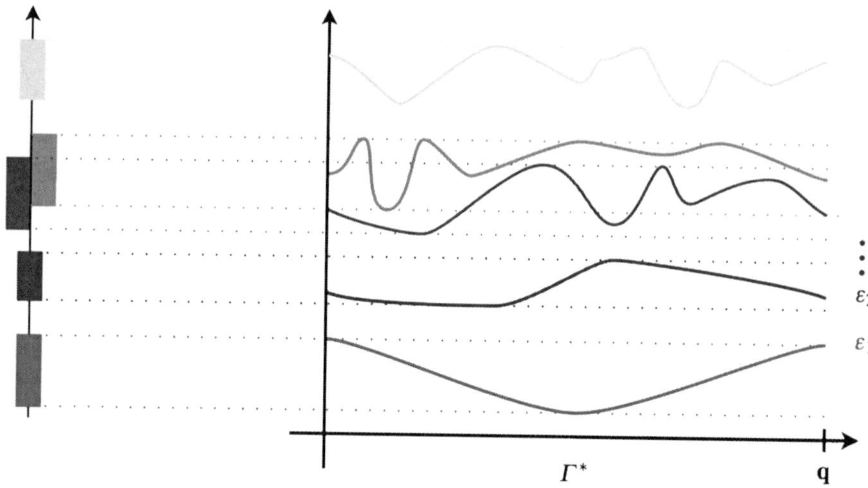

Fig. 1 The spectrum of a periodic Schrödinger operator is a union of bands, as a consequence of the Bloch-Floquet decomposition

3.2 Perfect Crystals

The purpose of this section is to formally construct, then justify with mathematical arguments, a Hartree model for the electronic structure of perfect crystals.

As announced, we begin with a formal argument and consider a sequence of finite nuclear distributions $(\rho_n^{\text{nuc}})_{n\in\mathbb{N}}$ converging to the periodic distribution $\rho_{\text{per}}^{\text{nuc}}$ of the perfect crystal when n goes to infinity. For instance, we can take

$$\rho_n^{\text{nuc}} = \rho_{\text{per}}^{\text{nuc}} \left(\sum_{\mathbf{R}\in\mathscr{R}\,|\,|\mathbf{R}|\leq n} 1_{\Gamma+\mathbf{R}} \right)$$

(we assume that the function describing the nuclear charge in the unit cell of the perfect crystal is supported in some compact set included in the interior of Γ). We solve the Hartree problem for each ρ_n^{nuc} with the constraint that the system remains neutral for each n. Assuming that when n goes to infinity:

- The Hartree ground state density converges to some $\mathscr{R}$-periodic density $\rho_{\text{per}}^0 \in L^1_{\text{per}}(\Gamma)$
- The Coulomb potential generated by the total charge converges to some $\mathscr{R}$-periodic potential V_{per}^0
- The Hartree ground state density operator converges to some operator γ_{per}^0

- The Fermi level converges to some $\epsilon_F^0 \in \mathbb{R}$

we obtain by *formally* passing to the limit in (22), the self-consistent equations

$$\begin{cases} \gamma_{\rm per}^0 = 1_{(-\infty,\epsilon_F^0]}(H_{\rm per}^0), \quad \rho_{\rm per}^0 = \rho_{\gamma_{\rm per}^0}, \\ H_{\rm per}^0 = -\frac{1}{2}\Delta + V_{\rm per}^0, \\ -\Delta V_{\rm per}^0 = 4\pi(\rho_{\rm per}^0 - \rho_{\rm per}^{\rm nuc}). \end{cases} \tag{31}$$

Let us comment on this system of equations. First, we notice that for the periodic Coulomb equation $-\Delta V_{\rm per}^0 = 4\pi(\rho_{\rm per}^0 - \rho_{\rm per}^{\rm nuc})$ to have a solution, each unit cell must be neutral:

$$\int_\Gamma \rho_{\rm per}^0 = \int_\Gamma \rho_{\rm per}^{\rm nuc} = Z, \tag{32}$$

where Z is the number of electrons, and also the number of protons, per unit cell. Second, as $V_{\rm per}^0$ is $\mathscr{R}$-periodic (and belongs to $L^2_{\rm per}(\Gamma)$ even for point-like nuclei), we can apply the result of the previous section and write down the Bloch-Floquet decomposition of $H_{\rm per}^0$:

$$(H_{\rm per}^0)_{\mathbf{q}} = -\frac{1}{2}\Delta - i\mathbf{q}\cdot\nabla + \frac{|\mathbf{q}|^2}{2} + V_{\rm per}^0 = \sum_{n=1}^{+\infty} \epsilon_{n,\mathbf{q}} |u_{n,\mathbf{q}}\rangle\langle u_{n,\mathbf{q}}|. \tag{33}$$

The operator $\gamma_{\rm per}^0 = 1_{(-\infty,\epsilon_F^0]}(H_{\rm per}^0)$ is then a bounded self-adjoint operator which commutes with the translations $(\tau_{\mathbf{R}})_{\mathbf{R}\in\mathscr{R}}$, and its Bloch-Floquet decomposition reads

$$(\gamma_{\rm per}^0)_{\mathbf{q}} = \sum_{n=1}^{+\infty} 1_{\epsilon_{n,\mathbf{q}} \le \epsilon_F^0} |u_{n,\mathbf{q}}\rangle\langle u_{n,\mathbf{q}}|.$$

Actually, the set $\{q \in \Gamma^* \mid \exists n \ge 1 \text{ s.t. } \epsilon_{n,\mathbf{q}} = \epsilon_F^0\}$ is of measure zero (the spectrum of $H_{\rm per}^0$ is purely continuous). It follows that $\gamma_{\rm per}^0$ is always an orthogonal projector, even if ϵ_F^0 belongs to the spectrum of $H_{\rm per}^0$.

Using the Bloch decomposition of $\gamma_{\rm per}^0$, we can write the density $\rho_{\rm per}^0$ as

$$\rho_{\rm per}^0(\mathbf{r}) = \fint_{\Gamma^*} \sum_{n=1}^{+\infty} 1_{\epsilon_{n,q} \le \epsilon_F^0} |u_{n,\mathbf{q}}(\mathbf{r})|^2 \, d\mathbf{q}.$$

Integrating on Γ, and using (32) and the orthonormality of the functions $(u_{n,\mathbf{q}})_{n\ge1}$ in $L^2_{\rm per}(\Gamma)$, we obtain

$$Z = \frac{1}{|\Gamma^*|} \sum_{n=1}^{+\infty} \left|\{\mathbf{q} \in \Gamma^* \mid \epsilon_{n,\mathbf{q}} \le \epsilon_F^0\}\right|. \tag{34}$$

It is easy to see that if the periodic Coulomb potential is shifted by a uniform constant C, and if ϵ_F^0 is replaced with $\epsilon_F^0 + C$, then $\gamma_{\rm per}^0$ and $\rho_{\rm per}^0$ remain unchanged.

The formal bulk limit argument presented above has been rigorously founded by Catto, Le Bris and Lions in [11], for $\rho_{\rm per}^{\rm nuc} = \sum_{\mathbf{R}\in\mathbb{Z}^3} \chi(\cdot - \mathbf{R})$ (smeared nuclei of unit charge disposed on the cubic lattice $\mathbb{Z}^3$). It is also possible to justify the periodic Hartree model by passing to the limit on the supercell model with artificial periodic boundary conditions (see [7]). The latter approach is less physical, but technically much easier, and its extension to arbitrary crystalline structures (including point-like nuclei) is straightforward. It results from these mathematical works that the Hartree model for perfect crystals is well-defined. More precisely:

1. The Hartree ground state density operator $\gamma_{\rm per}^0$ and density $\rho_{\rm per}^0$ of a crystal with periodic nuclear density $\rho_{\rm per}^{\rm nuc}$ (composed of point-like or smeared nuclei) are uniquely defined.
2. The ground state density $\rho_{\rm per}^0$ satisfies the neutrality charge constraint (32).
3. The periodic Coulomb potential $V_{\rm per}^0$ is uniquely defined up to an additive constant.
4. The ground state density operator $\gamma_{\rm per}^0$ is an infinite rank orthogonal projector satisfying the self-consistent equation (31).
5. $\gamma_{\rm per}^0$ can be obtained by minimizing some periodic model set on the unit cell Γ (see [11] for details).

In the remainder of the paper we assume that the system is an insulator (or a semi-conductor) in the sense that the $N^{\rm th}$ band is strictly below the $(N+1)^{\rm st}$ band:

$$\Sigma_N^+ < \Sigma_{N+1}^-,$$

where $\Sigma_n^\pm$ are defined in (30). In this case, one can choose for $\epsilon_{\rm F}^0$ any number in the range $(\Sigma_N^+, \Sigma_{N+1}^-)$. The electronic state of the perfect crystal is the same whatever the value of $\epsilon_{\rm F}^0$ in the gap $(\Sigma_N^+, \Sigma_{N+1}^-)$. On the other hand, as will be seen in the next section, fixing the value of $\epsilon_{\rm F}^0$ may change the electronic state of the crystal in the presence of a local defect.

In this paper however, we are only interested in the dielectric response of the crystal, which corresponds to the limit of small defects (in a sense that will be made precise later), and in this limit, the value of $\epsilon_{\rm F}^0$ does not play any role as long as it remains inside the gap $(\Sigma_N^+, \Sigma_{N+1}^-)$. For simplicity, we consider in the following

$$\epsilon_{\rm F}^0 = \frac{\Sigma_N^+ + \Sigma_{N+1}^-}{2}.$$

Lastly, we denote by

$$g = \Sigma_{N+1}^- - \Sigma_N^+ > 0 \tag{35}$$

the band gap.

3.3 Crystals with Local Defects

We now describe the results of [7] dealing with the modeling of local defects in crystals in the framework of the Hartree model. The main idea is to seek the ground state density operator of a crystal with a local defect characterized by the nuclear charge distribution (26) under the form

$$\gamma_{m,\epsilon_F^0} = \gamma_{\rm per}^0 + Q_{m,\epsilon_F^0}.$$

In this formalism, the defect is seen as a quasi-molecule with nuclear charge distribution m and electronic ground state density operator Q_{m,ϵ_F^0} (and ground state electronic density $\rho_{Q_{m,\epsilon_F^0}}$), embedded in the perfect crystal. Here, the charge of the defect is controlled by the Fermi level (the chemical potential). The dual approach, in which the charge of the defect is imposed, is also dealt with in [7]. It should be noticed that neither m nor $\rho_{Q_{m,\epsilon_F^0}}$ are a priori non-negative. For instance, the nuclear distribution of a defect corresponding to the replacement of a nucleus of charge z located at point $\mathbf{R} \in \mathbb{R}^3$ with a nucleus of charge z' is $m = (z'-z)\delta_{\mathbf{R}}$ and can therefore be positively or negatively charged depending on the value of $z'-z$. Regarding the electronic state, the constraints $(\gamma_{m,\epsilon_F^0})^* = \gamma_{m,\epsilon_F^0}$, $0 \le \gamma_{m,\epsilon_F^0} \le 1$ and $\rho_{\gamma_{m,\epsilon_F^0}} \ge 0$, respectively read $(Q_{m,\epsilon_F^0})^* = Q_{m,\epsilon_F^0}$, $-\gamma_{\rm per}^0 \le Q_{m,\epsilon_F^0} \le 1-\gamma_{\rm per}^0$ and $\rho_{Q_{m,\epsilon_F^0}} \ge -\rho_{\rm per}^0$.

The next step is to exhibit a variational model allowing to compute Q_{m,ϵ_F^0} from m, ϵ_F^0 and the ground state of the perfect crystal.

First, we perform the following formal calculation of the difference between the Hartree free energy of some trial density operator $\gamma = \gamma_{\rm per}^0 + Q$ subjected to the nuclear potential generated by $\rho_{\rm per}^{\rm nuc} + m$, and the Hartree free energy of the perfect crystal:

$$\begin{aligned}
&\left(\widetilde{E}^{\rm Hartree}_{\rho_{\rm per}^{\rm nuc}+m}(\gamma_{\rm per}^0 + Q) - \epsilon_F^0 {\rm Tr}(\gamma_{\rm per}^0 + Q)\right) - \left(\widetilde{E}^{\rm Hartree}_{\rho_{\rm per}^{\rm nuc}}(\gamma_{\rm per}^0) - \epsilon_F^0 {\rm Tr}(\gamma_{\rm per}^0)\right) \\
&\overset{\rm formal}{=} \quad {\rm Tr}\left(-\frac{1}{2}\Delta Q\right) + \int_{\mathbb{R}^3} \rho_Q V_{\rm per}^0 - \int_{\mathbb{R}^3} \rho_Q V_m + \frac{1}{2} D(\rho_Q, \rho_Q) - \epsilon_F^0 {\rm Tr}(Q) \\
&\qquad - \int_{\mathbb{R}^3} m V_{\rm per}^0 + \frac{1}{2} D(m,m). \qquad (36)
\end{aligned}$$

The last two terms are constants that we can discard. Of course, the left-hand side of (36) does not have any mathematical sense since it is the difference of two energies both equal to plus infinity. On the other hand, we are going to see that it is possible to give a mathematical meaning to the sum of the first five terms of the right-hand side when Q belongs to some functional space $\mathscr{Q}$ defined below, and to characterize the ground state density operator Q_{m,ϵ_F^0} of the quasi-molecule, by minimizing the so-defined energy functional on a closed convex subset $\mathscr{K}$ of $\mathscr{Q}$.

For this purpose, we first need to extend the definition (18) of the Coulomb interaction to the Coulomb space $\mathscr{C}$ defined as

$$\mathscr{C} := \left\{ f \in \mathscr{S}'(\mathbb{R}^3) \;\middle|\; \widehat{f} \in L^1_{\rm loc}(\mathbb{R}^3),\; D(f,f) := 4\pi \int_{\mathbb{R}^3} \frac{|\hat{f}(k)|^2}{|k|^2}\, dk \right\},$$

where $\mathscr{S}'(\mathbb{R}^3)$ is the space of tempered distributions on $\mathbb{R}^3$. Endowed with its natural inner product

$$\langle f, g\rangle_{\mathscr{C}} := D(f,g) := 4\pi \int_{\mathbb{R}^3} \frac{\overline{\hat{f}(k)}\,\hat{g}(k)}{|k|^2}\, dk, \tag{37}$$

$\mathscr{C}$ is a Hilbert space. It can be proved that $L^{6/5}(\mathbb{R}^3) \hookrightarrow \mathscr{C}$ and that for any $(f,g) \in L^{6/5}(\mathbb{R}^3) \times L^{6/5}(\mathbb{R}^3)$, it holds

$$4\pi \int_{\mathbb{R}^3} \frac{\overline{\hat{f}(k)}\,\hat{g}(k)}{|k|^2}\, dk = \int_{\mathbb{R}^3}\int_{\mathbb{R}^3} \frac{f(\mathbf{r})\,g(\mathbf{r}')}{|\mathbf{r}-\mathbf{r}'|}\, d\mathbf{r}\, d\mathbf{r}'.$$

Hence, the definition (37) of $D(\cdot,\cdot)$ on $\mathscr{C}$ is consistent with the usual definition (18) of the Coulomb interaction when the latter makes sense. The Coulomb space $\mathscr{C}$ therefore is the set of charge distributions of finite Coulomb energy.

Second, we introduce, for an operator A on $L^2(\mathbb{R}^3)$, the notation

$$A^{--} := \gamma^0_{\rm per} A \gamma^0_{\rm per}, \qquad A^{-+} := \gamma^0_{\rm per} A (1-\gamma^0_{\rm per}),$$
$$A^{+-} := (1-\gamma^0_{\rm per}) A \gamma^0_{\rm per}, \qquad A^{++} := (1-\gamma^0_{\rm per}) A (1-\gamma^0_{\rm per}),$$

and note that the constraints $Q = Q^*$ and $-\gamma^0_{\rm per} \leq Q \leq 1-\gamma^0_{\rm per}$ are equivalent to

$$Q^* = Q, \qquad Q^2 \leq Q^{++} - Q^{--}. \tag{38}$$

From the second inequality we deduce that it then holds $Q^{--} \leq 0$ and $Q^{++} \geq 0$. Using the fact that $\mathrm{Tr}(V^0_{\rm per} Q) = \int_{\mathbb{R}^3} \rho_Q V^0_{\rm per}$, we formally obtain

$$\begin{aligned}\mathrm{Tr}\left(-\frac{1}{2}\Delta Q\right) + \int_{\mathbb{R}^3} \rho_Q V^0_{\rm per} - \epsilon^0_{\rm F}\mathrm{Tr}(Q) &= \mathrm{Tr}((H^0_{\rm per} - \epsilon^0_{\rm F})Q)\\ &= \mathrm{Tr}((H^0_{\rm per} - \epsilon^0_{\rm F})^{++} Q^{++}) + \mathrm{Tr}((H^0_{\rm per} - \epsilon^0_{\rm F})^{--} Q^{--}).\end{aligned}$$

We now remark that, by definition of $\gamma^0_{\rm per}$, $(H^0_{\rm per} - \epsilon^0_{\rm F})^{++} \geq 0$ and $(H^0_{\rm per} - \epsilon^0_{\rm F})^{--} \leq 0$, so that the right-hand term of the above expression can be rewritten as

$$\mathrm{Tr}(|H^0_{\rm per} - \epsilon^0_{\rm F}|^{1/2}(Q^{++} - Q^{--})|H^0_{\rm per} - \epsilon^0_{\rm F}|^{1/2}). \tag{39}$$

The above expression is well defined in $\mathbb{R}_+ \cup \{+\infty\}$ for all Q satisfying the constraints (38). It takes a finite value if Q is chosen in the vector space

$$\begin{aligned}\mathscr{Q} = \{&Q \in \mathfrak{S}_2 \mid Q^* = Q,\; Q^{--} \in \mathfrak{S}_1,\; Q^{++} \in \mathfrak{S}_1,\\ &|\nabla| Q \in \mathfrak{S}_2,\; |\nabla| Q^{--} |\nabla| \in \mathfrak{S}_1,\; |\nabla| Q^{++} |\nabla| \in \mathfrak{S}_1\},\end{aligned} \tag{40}$$

where $\mathfrak{S}_1$ and $\mathfrak{S}_2$ respectively denote the spaces of trace-class and Hilbert-Schmidt operators on $L^2(\mathbb{R}^3)$ (see Appendix for details). Endowed with its natural norm, or with any equivalent norm such as

$$\|Q\|_{\mathcal{Q}} = \|(1+|\nabla|)Q\|_{\mathfrak{S}_2} + \|(1+|\nabla|)Q^{++}(1+|\nabla|)\|_{\mathfrak{S}_1} + \|(1+|\nabla|)Q^{--}(1+|\nabla|)\|_{\mathfrak{S}_1},$$

$\mathcal{Q}$ is a Banach space.

Before proceeding further, let us comment on the definition of $\mathcal{Q}$. As the trial density operators Q must satisfy the constraints (38), it is natural to impose $Q^* = Q$. Since $|H^0_{\rm per} - \epsilon^0_{\rm F}|^{1/2}(1+|\nabla|)^{-1}$ is a bounded operator with bounded inverse (see [7]), the four conditions $Q^{--} \in \mathfrak{S}_1$, $Q^{++} \in \mathfrak{S}_1$, $|\nabla|Q^{--}|\nabla| \in \mathfrak{S}_1$ and $|\nabla|Q^{++}|\nabla| \in \mathfrak{S}_1$ are necessary and sufficient conditions for the expression (39) with Q satisfying (38) being finite. The other constraints imposed to the elements of $\mathcal{Q}$ (that is, $Q \in \mathfrak{S}_2$ and $|\nabla|Q \in \mathfrak{S}_2$) follow from the fact that for any Q satisfying (38)

$$\begin{aligned} \left(Q^{--} \in \mathfrak{S}_1,\ Q^{++} \in \mathfrak{S}_1\right) \quad &\Rightarrow \quad \left(Q^2 \in \mathfrak{S}_1\right) \\ \left(|\nabla|Q^{--}|\nabla| \in \mathfrak{S}_1,\ |\nabla|Q^{++}|\nabla| \in \mathfrak{S}_1\right) \quad &\Rightarrow \quad \left(|\nabla|Q^2|\nabla| \in \mathfrak{S}_1\right). \end{aligned}$$

In order to simplify the notation, we set for $Q \in \mathcal{Q}$,

$$\begin{aligned} &\mathrm{Tr}_0(Q) := \mathrm{Tr}(Q^{++} + Q^{--}), \\ &\mathrm{Tr}_0((H^0_{\rm per} - \epsilon^0_{\rm F})Q) := \mathrm{Tr}(|H^0_{\rm per} - \epsilon^0_{\rm F}|^{1/2}(Q^{++} - Q^{--})|H^0_{\rm per} - \epsilon^0_{\rm F}|^{1/2}). \end{aligned}$$

An important result is that the linear mapping $Q \mapsto \rho_Q$ originally defined on the dense subset $\mathcal{Q} \cap \mathfrak{S}_1$ of $\mathcal{Q}$ can be extended in a unique way to a continuous linear mapping

$$\begin{aligned} \mathcal{Q} &\to L^2(\mathbb{R}^3) \cap \mathcal{C} \\ Q &\mapsto \rho_Q. \end{aligned}$$

Note that the density associated with a generic element of $\mathcal{Q}$ is not necessarily an integrable function. On the other hand, its Coulomb energy is always finite.

Let m be such that $V_m = (m \star |\cdot|^{-1}) \in \mathcal{C}'$. Here and in the sequel

$$\mathcal{C}' := \left\{ V \in L^6(\mathbb{R}^3) \,\middle|\, \nabla V \in (L^2(\mathbb{R}^3))^3 \right\}$$

denotes the dual space of $\mathcal{C}$, endowed with the inner product

$$\langle V_1, V_2 \rangle_{\mathcal{C}'} := \frac{1}{4\pi} \int_{\mathbb{R}^3} \nabla V_1 \cdot \nabla V_2 = \frac{1}{4\pi} \int_{\mathbb{R}^3} |k|^2 \overline{\hat{V}_1(k)}\, \hat{V}_2(k)\, dk.$$

It follows from the above arguments that the energy functional

$$E^{m,\epsilon_F^0}(Q) = \mathrm{Tr}_0((H_{\mathrm{per}}^0 - \epsilon_F^0)Q) - \int_{\mathbb{R}^3} \rho_Q V_m + \frac{1}{2} D(\rho_Q, \rho_Q)$$

is well defined on $\mathcal{Q}$ and that a good candidate for a variational model allowing to compute the ground state density operator Q_{m,ϵ_F^0} is

$$\inf \left\{ E^{m,\epsilon_F^0}(Q),\ Q \in \mathcal{K} \right\} \tag{41}$$

where

$$\mathcal{K} = \left\{ Q \in \mathcal{Q} \mid -\gamma_{\mathrm{per}}^0 \leq Q \leq 1 - \gamma_{\mathrm{per}}^0 \right\}. \tag{42}$$

Note that $\mathcal{K}$ is a closed convex subset of $\mathcal{Q}$.

The above formal construction of the model (41) is justified in [7] by means of rigorous bulk limit arguments. To summarize the situation, the Hartree ground state density operator of the crystal with nuclear charge density $\rho_{\mathrm{per}}^{\mathrm{nuc}} + m$ (the charge of the defect being controlled by the Fermi level) is given by

$$\gamma = \gamma_{\mathrm{per}}^0 + Q_{m,\epsilon_F^0}$$

where Q_{m,ϵ_F^0} is obtained by solving (41).

The existence of a Hartree ground state density operator for a crystal with a local defect, as well as the uniqueness of the corresponding density and some other important properties, are granted by the following theorem which gathers several results from [7] and [9].

Theorem 1. *Let m such that $(m \star |\cdot|^{-1}) \in L^2(\mathbb{R}^3) + \mathcal{C}'$. Then:*

1. *(41) has at least one minimizer Q_{m,ϵ_F^0}, and all the minimizers of (41) share the same density ρ_{m,ϵ_F^0}.*
2. *Q_{m,ϵ_F^0} is solution to the self-consistent equation*

$$Q_{m,\epsilon_F^0} = 1_{(-\infty,\epsilon_F^0)} \left(H_{\mathrm{per}}^0 + (\rho_{m,\epsilon_F^0} - m) \star |\cdot|^{-1} \right) - 1_{(-\infty,\epsilon_F^0]} \left(H_{\mathrm{per}}^0 \right) + \delta, \tag{43}$$

where δ is a finite-rank self-adjoint operator on $L^2(\mathbb{R}^3)$ such that $0 \leq \delta \leq 1$ and $Ran(\delta) \subset Ker\left(H_{\mathrm{per}}^0 + (\rho_{m,\epsilon_F^0} - m) \star |\cdot|^{-1} - \epsilon_F^0 \right)$.

The interpretation of the Euler equation (43), which also reads

$$\gamma_{\mathrm{per}}^0 + Q_{m,\epsilon_F^0} = 1_{(-\infty,\epsilon_F^0]}(H_{m,\epsilon_F^0}^0) + \delta$$

with

$$H_{m,\epsilon_F^0}^0 = H_{\mathrm{per}}^0 + (\rho_{m,\epsilon_F^0} - m) \star |\cdot|^{-1}, \quad 0 \leq \delta \leq 1, \quad \mathrm{Ran}(\delta) \subset \mathrm{Ker}(H_{m,\epsilon_F^0}^0 - \epsilon_F^0),$$

is the following. The mean-field Hamiltonian $H_{m,\epsilon_F^0}^0$ is uniquely defined, since all the minimizers of (41) share the same density ρ_{m,ϵ_F^0}. Besides, the operator $(\rho_{m,\epsilon_F^0} -$

$m) \star |\cdot|^{-1}$ being a relatively compact perturbation of $H^0_{\rm per}$, it results from the Weyl theorem (see [31, Sect. XIII.4]) that the Hamiltonians $H^0_{\rm per}$ and $H^0_{m,\epsilon^0_{\rm F}}$ have the same essential spectra. On the other hand, while $H^0_{\rm per}$ has no eigenvalues, $H^0_{m,\epsilon^0_{\rm F}}$ may have a countable number of isolated eigenvalues of finite multiplicities in the gaps as well as below the bottom of the essential spectrum. The only possible accumulation points of these eigenvalues are the edges of the bands.

If $\epsilon^0_{\rm F} \notin \sigma(H^0_{m,\epsilon^0_{\rm F}})$, then $\delta = 0$ and the ground state density operator of the crystal in the presence of the defect is the orthogonal projector $\gamma^0_{\rm per} + Q_{m,\epsilon^0_{\rm F}}$: All the energy levels lower than the Fermi level are fully occupied while the other ones are empty (see Fig. 2). In this case, $Q_{m,\epsilon^0_{\rm F}}$ is both a Hilbert-Schmidt operator and the difference of two projectors. It therefore follows from [18, Lemma 2] that

$$\mathrm{Tr}_0(Q_{m,\epsilon^0_{\rm F}}) \in \mathbb{N}. \tag{44}$$

Assuming that $m \in L^1(\mathbb{R}^3)$ and $\int_{\mathbb{R}^3} m \in \mathbb{N}$, the integer

$$\int_{\mathbb{R}^3} m - \mathrm{Tr}_0(Q_{m,\epsilon^0_{\rm F}})$$

can be interpreted as the *bare* charge of the defect (in contrast with the *screened* or *renormalized* charge to be defined later).

If $\epsilon^0_{\rm F} \in \sigma(H^0_{m,\epsilon^0_{\rm F}})$, then the energy levels with energy $\epsilon^0_{\rm F}$ may be fully or partially occupied, and it may a priori happen that (41) has several minimizers, differing from one another by a finite rank self-adjoint operator with range in $\mathrm{Ker}(H^0_{m,\epsilon^0_{\rm F}} - \epsilon^0_{\rm F})$.

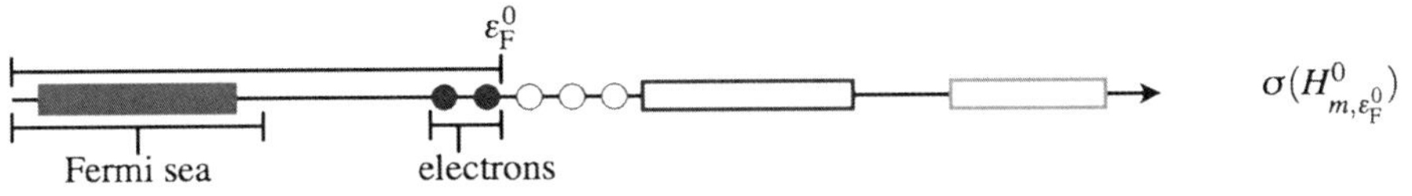

Fig. 2 General form of the spectrum of the self-consistent operator $H^0_{m,\epsilon^0_{\rm F}}$, in the presence of a defect and for a fixed chemical potential $\epsilon^0_{\rm F}$

4 Dielectric Response of a Crystal

In this section, we study the response of the electronic ground state of a crystal to a *small, effective* potential. In Sect. 4.1, we consider a time-independent perturbation $V \in L^2(\mathbb{R}^3) + \mathscr{C}'$, with $\|V\|_{L^2+\mathscr{C}'} < \alpha$ (for some $\alpha > 0$ small enough). It can be proved (see [9, Lemma 5]) that there exists $\beta > 0$ such that

$$\left(\|m \star |\cdot|^{-1}\|_{L^2+\mathscr{C}'} < \beta\right) \quad \Rightarrow \quad \left(\|(\rho_{m,\epsilon_F^0} - m) \star |\cdot|^{-1}\|_{L^2+\mathscr{C}'} < \alpha\right). \tag{45}$$

The results of Sect. 4.1 therefore directly apply to the case of a crystal with a local defect with nuclear charge distribution m, provided the defect is small enough (in the sense that $\|m \star |\cdot|^{-1}\|_{L^2+\mathscr{C}'} < \beta$).

In Sect. 4.4, we consider a time-dependent perturbation

$$v(t,\mathbf{r}) = (\rho(t,\cdot) \star |\cdot|^{-1})(\mathbf{r}) \qquad \text{with} \qquad \rho \in L^1_{\text{loc}}(\mathbb{R}, L^2(\mathbb{R}^3) \cap \mathscr{C}). \tag{46}$$

4.1 Series Expansion of the Time-Independent Response

For $V \in L^2(\mathbb{R}^3) + \mathscr{C}'$, the spectrum of $H^0_{\text{per}} + V$ depends continuously on V. In particular (see [9, Lemma 2]), there exists some $\alpha > 0$, such that if $\mathfrak{C}$ is a smooth curve in the complex plane enclosing the whole spectrum of H^0_{per} below ϵ_F^0, crossing the real line at ϵ_F^0 and at some $c < \inf \sigma(H^0_{\text{per}})$ and such that

$$d(\sigma(H^0_{\text{per}}), \Lambda) = \frac{g}{4} \quad \text{where} \quad \Lambda = \left\{ z \in \mathbb{C} \;\middle|\; d(z, \mathfrak{C}) \leq \frac{g}{4} \right\},$$

d denoting the Euclidean distance in the complex plane and g the band gap (35) (see Fig. 3), then $\sigma(H^0_{\text{per}} + V) \cap (-\infty, \epsilon_F^0]$ is contained in the interior of $\mathfrak{C}$ for all $V \in L^2(\mathbb{R}^3) + \mathscr{C}'$ such that $\|V\|_{L^2+\mathscr{C}'} < \alpha$.

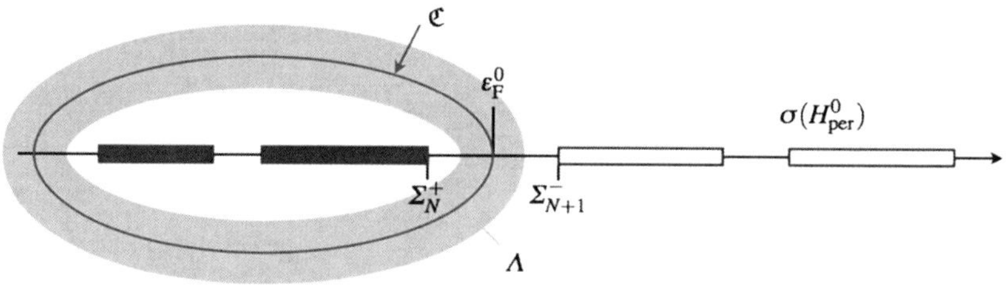

Fig. 3 Graphical representation of a contour $\mathfrak{C} \subset \mathbb{C}$ enclosing $\sigma(H^0_{\text{per}}) \cap (-\infty, \epsilon_F^0]$ and of the compact set Λ

As a consequence, we obtain that for all $V \in L^2(\mathbb{R}^3) + \mathscr{C}'$ such that $\|V\|_{L^2+\mathscr{C}'} < \alpha$,

$$\begin{aligned} Q_V &= 1_{(-\infty,\epsilon_F^0)}\left(H^0_{\text{per}} + V\right) - 1_{(-\infty,\epsilon_F^0]}\left(H^0_{\text{per}}\right) \\ &= \frac{1}{2i\pi} \oint_{\mathfrak{C}} \left(\left(z - H^0_{\text{per}} - V\right)^{-1} - \left(z - H^0_{\text{per}}\right)^{-1} \right) dz, \end{aligned} \tag{47}$$

where we have used the fact that $\epsilon_F^0 \notin \sigma(H^0_{\text{per}} + V)$ to establish the first equality, and the Cauchy formula to derive the second one.

Expanding (47) in powers of V, we obtain

$$Q_V = \sum_{n=1}^{N} Q_{n,V} + \widetilde{Q}_{N+1,V}, \tag{48}$$

where we have gathered the terms involving powers of V larger than N in a remainder $\widetilde{Q}_{N+1,V}$. The linear contribution is given by

$$Q_{1,V} = \frac{1}{2i\pi} \oint_{\mathfrak{C}} \left(z - H^0_{\rm per}\right)^{-1} V \left(z - H^0_{\rm per}\right)^{-1} dz. \tag{49}$$

The higher order contributions and the remainder are respectively given by

$$Q_{n,V} = \frac{1}{2i\pi} \oint_{\mathfrak{C}} \left(z - H^0_{\rm per}\right)^{-1} \left[V \left(z - H^0_{\rm per}\right)^{-1}\right]^n dz$$

and

$$\widetilde{Q}_{N+1,V} = \frac{1}{2i\pi} \oint_{\mathfrak{C}} \left(z - H^0_{\rm per} - V\right)^{-1} \left[V \left(z - H^0_{\rm per}\right)^{-1}\right]^{N+1} dz.$$

Proposition 1. *The terms of the perturbation expansion (48) enjoy the following properties.*

1. *The k-linear application*

$$(V_1, \ldots, V_n) \mapsto \frac{1}{2i\pi} \oint_{\mathfrak{C}} \left(z - H^0_{\rm per}\right)^{-1} V_1 \left(z - H^0_{\rm per}\right)^{-1} \cdots V_n \left(z - H^0_{\rm per}\right)^{-1} dz$$

is well-defined and continuous from $(L^2(\mathbb{R}^3) + \mathscr{C}')^n$ to $\mathscr{Q}$ for all $n \geq 1$, and from $(L^2(\mathbb{R}^3) + \mathscr{C}')^n$ to $\mathfrak{S}_1$ for all $n \geq 6$. In particular, for all $V \in L^2(\mathbb{R}^3) + \mathscr{C}'$, $Q_{n,V} \in \mathscr{Q}$ for all $n \geq 1$ and $Q_{n,V} \in \mathfrak{S}_1$ for all $n \geq 6$. Besides, for all $V \in L^2(\mathbb{R}^3) + \mathscr{C}'$, $\mathrm{Tr}_0(Q_{n,V}) = 0$ for all $n \geq 1$ and $\mathrm{Tr}(Q_{n,V}) = 0$ for all $n \geq 6$.
2. *If $V \in L^1(\mathbb{R}^3)$, $Q_{n,V}$ is in $\mathfrak{S}_1$ for each $n \geq 1$ and $\mathrm{Tr}(Q_{n,V}) = 0$.*
3. *For each $V \in L^2(\mathbb{R}^3) + \mathscr{C}'$ such that $\|V\|_{L^2+\mathscr{C}'} < \alpha$, the operator $\widetilde{Q}_{N+1,V}$ is in $\mathscr{Q}$ for all $N \geq 0$ with $\mathrm{Tr}_0(\widetilde{Q}_{N+1,V}) = 0$, and in $\mathfrak{S}_1$ for all $N \geq 5$, with $\mathrm{Tr}(\widetilde{Q}_{N+1,V}) = \mathrm{Tr}_0(\widetilde{Q}_{N+1,V}) = 0$.*

We are now in position to define some operators which play an important role in the sequel:

- The Coulomb operator v_c, which defines a bijective isometry between $\mathscr{C}$ and $\mathscr{C}'$:

$$v_c(\rho) := \rho \star |\cdot|^{-1}.$$

- The independent particle polarization operator χ_0 defined by

$$\chi_0(V) := \rho_{Q_{1,V}},$$

which provides the first order response of the electronic density of the crystal to a time-independent modification of the effective potential. The operator χ_0 is a

continuous linear application from $L^1(\mathbb{R}^3)$ to $L^1(\mathbb{R}^3)$ and from $L^2(\mathbb{R}^3)+\mathscr{C}'$ to $L^2(\mathbb{R}^3)\cap\mathscr{C}$.
- The linear operator $\mathscr{L}$ defined by

$$\mathscr{L} := -\chi_0 v_{\rm c},$$

which is a bounded nonnegative self-adjoint operator on $\mathscr{C}$. As a consequence, $(1+\mathscr{L})^{-1}$ is a well-defined bounded self-adjoint operator on $\mathscr{C}$.
- The dielectric operator $\epsilon = v_{\rm c}(1+\mathscr{L})v_{\rm c}^{-1}$, and its inverse, the dielectric permittivity operator

$$\epsilon^{-1} = v_{\rm c}(1+\mathscr{L})^{-1}v_{\rm c}^{-1},$$

both being continuous linear operators on $\mathscr{C}'$. Note that the hermitian dielectric operator, defined as $\widetilde{\epsilon} = v_{\rm c}^{-1/2}\epsilon v_{\rm c}^{1/2}$ is a self-adjoint, invertible, bounded operator on $L^2(\mathbb{R}^3)$ and is for this reason conveniently used in mathematical proofs.

We now focus our attention on the total Coulomb potential

$$V_m = (m-\rho_{m,\epsilon_{\rm F}^0}) \star |\cdot|^{-1} = v_{\rm c}(m-\rho_{m,\epsilon_{\rm F}^0}),$$

generated by some charge distribution m such that $\|m\star|\cdot|^{-1}\|_{L^2+\mathscr{C}'} < \beta$, and on the response $\rho_{m,\epsilon_{\rm F}^0}$ of the Fermi sea. In view of (45), we can apply the above results and deduce from (48) that

$$\begin{aligned}\rho_{m,\epsilon_{\rm F}^0} &= \rho_{Q_{-V_m}} = \rho_{Q_{1,-V_m}} + \rho_{\widetilde{Q}_{2,-V_m}} = -\chi_0 V_m + \rho_{\widetilde{Q}_{2,-V_m}} \\ &= \mathscr{L}(m-\rho_{m,\epsilon_{\rm F}^0}) + \rho_{\widetilde{Q}_{2,-V_m}}.\end{aligned} \tag{50}$$

The above relation, which also reads

$$(m-\rho_{m,\epsilon_{\rm F}^0}) = (1+\mathscr{L})^{-1}m - (1+\mathscr{L})^{-1}(\rho_{\widetilde{Q}_{2,-V_m}}) \tag{51}$$

or

$$V_m = v_{\rm c}(1+\mathscr{L})^{-1}m - v_{\rm c}(1+\mathscr{L})^{-1}(\rho_{\widetilde{Q}_{2,-V_m}}), \tag{52}$$

is fundamental since it allows to split the quantities of interest (the total charge $(m-\rho_{m,\epsilon_{\rm F}^0})$ or the total Coulomb potential V_m generated by the defect) into two components:

- A linear contribution in m, very singular, and responsible for charge renormalization at the microscopic level, and for the dielectric properties of the crystal at the macroscopic level.
- A nonlinear contribution which, in the regime under study ($\|m\star|\cdot|^{-1}\|_{L^2+\mathscr{C}'} < \beta$), is regular at the microscopic level and vanishes in the macroscopic limit.

4.2 Properties of Q_{m,ϵ_F^0} and ρ_{m,ϵ_F^0} for Small Amplitude Defects

The relation (50), combined with the properties of the operator $\mathscr{L}$ stated in Proposition 2 below, allows us to derive some interesting properties of Q_{m,ϵ_F^0} and ρ_{m,ϵ_F^0} and to propose a definition of the renormalized charge of the defect.

Proposition 2. *Let $\rho \in L^1(\mathbb{R}^3)$. Then, $\mathscr{L}(\rho) \in L^2(\mathbb{R}^3) \cap \mathscr{C}$, $\widehat{\mathscr{L}(\rho)}$ is continuous on $\mathbb{R}^3 \setminus \mathscr{R}^*$, and for all $\sigma \in S^2$ (the unit sphere of $\mathbb{R}^3$),*

$$\lim_{\eta \to 0+} \widehat{\mathscr{L}(\rho)}(\eta\sigma) = (\sigma^T L \sigma)\widehat{\rho}(0), \tag{53}$$

where $L \in \mathbb{R}^{3\times 3}$ is the non-negative symmetric matrix defined by

$$\forall \mathbf{k} \in \mathbb{R}^3, \quad \mathbf{k}^T L \mathbf{k} = \frac{8\pi}{|\Gamma|} \sum_{n=1}^{N} \sum_{n'=N+1}^{+\infty} \fint_{\Gamma^*} \frac{\left| \langle (\mathbf{k} \cdot \nabla_{\mathbf{r}}) u_{n,\mathbf{q}}, u_{n',\mathbf{q}} \rangle_{L^2_{\mathrm{per}}(\Gamma)} \right|^2}{\left(\epsilon_{n',\mathbf{q}} - \epsilon_{n,\mathbf{q}}\right)^3} \, d\mathbf{q}, \tag{54}$$

where the $\epsilon_{n,\mathbf{q}}$'s and the $u_{n,\mathbf{q}}$'s are the eigenvalues and eigenvectors arising in the spectral decomposition (33) *of $(H^0_{\mathrm{per}})_{\mathbf{q}}$. Additionally,*

$$L_0 = \frac{1}{3}\mathrm{Tr}(L) > 0. \tag{55}$$

Notice that the convergence of the series (54) is granted by the fact that $\epsilon_{n',\mathbf{q}} - \epsilon_{n,\mathbf{q}} \geq \Sigma^-_{n'} - \Sigma^+_n \geq g$ for all $n \leq N < n'$ and all $\mathbf{q} \in \Gamma^*$ (where $g > 0$ is the band gap), and the existence of $C \in \mathbb{R}_+$ such that $\|u_{n,\mathbf{q}}\|_{H^2_{\mathrm{per}}(\Gamma)} \leq C$ for all $1 \leq n \leq N$ and all $\mathbf{q} \in \Gamma^*$. Actually, the convergence of the series is rather fast since $\Sigma^-_{n'} \underset{n' \to \infty}{\sim} C n'^{2/3}$ (this estimate is obtained by comparing the eigenvalues of H^0_{per} with those of the Laplace operator on $L^2_{\mathrm{per}}(\Gamma)$).

We do not reproduce here the quite technical proof of Proposition 2. Let us however emphasize the essential role played by the long range character of the Coulomb potential. If $|\cdot|^{-1}$ is replaced by a potential $v_r \in L^1(\mathbb{R}^3)$, then for all $\rho \in L^1(\mathbb{R}^3)$, $\rho \star v_r \in L^1(\mathbb{R}^3)$, hence $\mathscr{L}(\rho) \in L^1(\mathbb{R}^3)$ and $L = 0$. More precisely, the Bloch-Floquet decomposition of the Coulomb kernel reads

$$(|\cdot|)_{\mathbf{q}}(\mathbf{r}) = \frac{4\pi}{|\Gamma|} \left(\frac{1}{|\mathbf{q}|^2} + \sum_{\mathbf{K} \in \mathscr{R}^* \setminus \{0\}} \frac{e^{i\mathbf{K}\cdot\mathbf{r}}}{|\mathbf{q}+\mathbf{K}|^2} \right),$$

and only the singular component $\frac{4\pi}{|\Gamma||\mathbf{q}|^2}$, which originates from the long-range of the Coulomb potential, gives a nonzero contribution to L.

We can deduce from (50) and Proposition 2 that, in general, the minimizer Q_{m,ϵ_F^0} to (41) is not trace-class and that the density ρ_{m,ϵ_F^0} is not an integrable function if the host crystal is anisotropic. Let us detail this point.

Consider some $m \in L^1(\mathbb{R}^3) \cap L^2(\mathbb{R}^3)$ such that $\int_{\mathbb{R}^3} m \neq 0$ and $\|m \star |\cdot|^{-1}\|_{L^2+\mathscr{C}'} < \beta$. In view of (45) and Proposition 1, it holds

$$\mathrm{Tr}_0(Q_{m,\epsilon_F^0}) = \mathrm{Tr}_0(Q_{1,-V_m} + \widetilde{Q}_{2,-V_m}) = 0. \tag{56}$$

Assume that ρ_{m,ϵ_F^0} is in $L^1(\mathbb{R}^3)$. Then a technical lemma (see [9, Lemma 4]) shows that the Fourier transform of the density $\rho_{\widetilde{Q}_{2,-V_m}}$, corresponding to the nonlinear response terms, is continuous and vanishes at zero. This means that, although it is not known whether $\rho_{\widetilde{Q}_{2,-V_m}}$ is in $L^1(\mathbb{R}^3)$, this density of charge behaves in the Fourier space as if it was integrable with an integral equal to zero. It follows from (50) and Proposition 1 that for each $\sigma \in S^2$,

$$\widehat{\rho}_{m,\epsilon_F^0}(0) = \lim_{\eta \to 0^+} \left(\mathscr{F}\mathscr{L}(\rho_{m,\epsilon_F^0} - m)\right)(\eta\sigma) = (\sigma^T L \sigma)(\widehat{\rho}_{m,\epsilon_F^0}(0) - \widehat{m}(0)). \tag{57}$$

As by assumption $\widehat{m}(0) \neq 0$ (since $\int_{\mathbb{R}^3} m \neq 0$), we reach a contradiction unless the matrix L is proportional to the identity matrix. Defining here an isotropic crystal as a crystal for which $L \neq L_0 1$, this proves that, in general, ρ_{m,ϵ_F^0} is not an integrable function for anisotropic crystals (and this *a fortiori* implies that Q_{m,ϵ_F^0} is not trace-class).

Let us now consider an isotropic crystal. If Q_{m,ϵ_F^0} were trace-class, then ρ_{m,ϵ_F^0} would be in $L^1(\mathbb{R}^3)$, and we would deduce from (56) that

$$(2\pi)^{3/2}\widehat{\rho}_{m,\epsilon_F^0}(0) = \int_{\mathbb{R}^3} \rho_{m,\epsilon_F^0} = \mathrm{Tr}(Q_{m,\epsilon_F^0}) = \mathrm{Tr}_0(Q_{m,\epsilon_F^0}) = 0.$$

Again, except in the very special case when $L = 1$, this contradicts (57) since $\widehat{m} \neq 0$ by assumption. Thus, in general, Q_{m,ϵ_F^0} is not trace-class, even for isotropic crystals. We do not know whether the electronic density ρ_{m,ϵ_F^0} generated by some $m \in L^1(\mathbb{R}^3) \cap L^2(\mathbb{R}^3)$ (this assumption implies $m \in L^{6/5}(\mathbb{R}^3) \hookrightarrow \mathscr{C}$) in an isotropic crystal is integrable or not. If it is, it follows from (57) that, still under the assumption that $\|m \star |\cdot|^{-1}\|_{L^2+\mathscr{C}'} < \beta$,

$$\int_{\mathbb{R}^3} m - \int_{\mathbb{R}^3} \rho_{m,\epsilon_F^0} = \frac{\int_{\mathbb{R}^3} m}{1 + L_0}.$$

This quantity can be interpreted as the renormalized charge of the defect, which differs from the bare charge $\int_{\mathbb{R}^3} m - \mathrm{Tr}_0(Q_{m,\epsilon_F^0}) = \int_{\mathbb{R}^3} m$ by a screening factor $\frac{1}{1+L_0}$. This is formally similar to the charge renormalization phenomenon observed in QED (see [17] for a mathematical analysis).

4.3 Dielectric Operator and Macroscopic Dielectric Permittivity

In this section, we focus again on the total potential

$$V_m = (m - \rho_{m,\epsilon_{\mathrm{F}}^0}) \star |\cdot|^{-1} \tag{58}$$

generated by the total charge of the defect, but we study it in a certain macroscopic limit.

For this purpose, we fix some $m \in L^1(\mathbb{R}^3) \cap L^2(\mathbb{R}^3)$ and introduce for all $\eta > 0$ the rescaled density

$$m_\eta(\mathbf{r}) := \eta^3 m(\eta \mathbf{r}).$$

We then denote by V_m^η the total potential generated by m_η and the corresponding electronic polarization, *i.e.*

$$V_m^\eta := (m_\eta - \rho_{m_\eta,\epsilon_{\mathrm{F}}^0}) \star |\cdot|^{-1}, \tag{59}$$

and define the rescaled potential

$$W_m^\eta(\mathbf{r}) := \eta^{-1}\, V_m^\eta\left(\eta^{-1}\mathbf{r}\right). \tag{60}$$

The scaling parameters have been chosen in a way such that in the absence of dielectric response (*i.e.* for $\mathscr{L} = 0$ and $\widetilde{\rho}_{Q_{2,-V_m^\eta}} = 0$), it holds $W_m^\eta = v_{\mathrm{c}}(m) = m \star |\cdot|^{-1}$ for all $\eta > 0$. To obtain a macroscopic limit, we let η go to zero.

As $\|(m_\eta \star |\cdot|^{-1})\|_{\mathscr{C}'} = \|m_\eta\|_{\mathscr{C}} = \eta^{1/2}\|m\|_{\mathscr{C}}$, we can apply the results of the previous sections as soon as η is small enough. Introducing the family of scaling operators $(U_\eta)_{\eta>0}$ defined by $(U_\eta f)(\mathbf{r}) = \eta^{3/2} f(\eta\mathbf{r})$ (each U_η is a bijective isometry of $L^2(\mathbb{R}^3)$), the equation linking the density of charge m to the rescaled potential W_m^η reads

$$W_m^\eta = v_{\mathrm{c}}^{1/2} U_\eta^* \tilde{\epsilon}^{-1} U_\eta v_{\mathrm{c}}^{1/2} m + \widetilde{w}_m^\eta, \tag{61}$$

where the nonlinear contribution $\widetilde{w}_m^\eta$ is such that there exists $C \in \mathbb{R}_+$ such that for η small enough, $\|\widetilde{w}_m^\eta\|_{\mathscr{C}'} \le C\eta$. The macroscopic limit of W_m^η therefore is governed by the linear response term, and is obtained as the limit when η goes to zero of the family $(U_\eta^* \widetilde{\epsilon}^{-1} U_\eta)_{\eta>0}$ of bounded self-adjoint operators on $L^2(\mathbb{R}^3)$.

If $\tilde{\epsilon}^{-1}$ was translation invariant, that is, if it was commuting with all the translations $\tau_{\mathbf{R}}$ for $\mathbf{R} \in \mathbb{R}^3$, it would be a multiplication operator in the Fourier space (*i.e.* such that for all $f \in L^2(\mathbb{R}^3)$, $\widehat{(\tilde{\epsilon}^{-1} f)}(\mathbf{k}) = \bar{\epsilon}^{-1}(\mathbf{k})\widehat{f}(\mathbf{k})$ for some function $\mathbb{R}^3 \ni \mathbf{k} \mapsto \bar{\epsilon}^{-1}(\mathbf{k}) \in \mathbb{C}$). Using the fact that the operator $v_{\mathrm{c}}^{1/2}$ is the multiplication operator by $(4\pi)^{1/2}/|\mathbf{k}|$ in the Fourier space, we would obtain in the limit

$$\lim_{\eta\to 0^+} \left(\frac{|\mathbf{k}|^2}{\bar{\epsilon}^{-1}(\eta\mathbf{k})}\right) \widehat{W}_m(\mathbf{k}) = 4\pi \widehat{m}(\mathbf{k}).$$

As the operator $\tilde{\epsilon}^{-1}$ actually commutes only with the translations of the lattice $\mathscr{R}$, the above argument cannot be applied. On the other hand, it can be proved, using Bloch-Floquet decomposition, that W_m^η has a limit W_m when η goes to zero, and that this limits satisfies

$$\lim_{\eta\to 0+}\left(\frac{|\mathbf{k}|^2}{[\tilde{\epsilon}^{-1}]_{00}(\eta\mathbf{k})}\right)\widehat{W}_m(\mathbf{k}) = 4\pi\widehat{m}(\mathbf{k}), \tag{62}$$

where $[\tilde{\epsilon}^{-1}]_{00}(\mathbf{q})$ is the entry of the Bloch matrix of the $\mathscr{R}$-periodic operator $\tilde{\epsilon}^{-1}$ corresponding to $\mathbf{K} = \mathbf{K}' = 0$. Besides,

$$\lim_{\eta\to 0+}\left(\frac{|\mathbf{k}|^2}{[\tilde{\epsilon}^{-1}]_{00}(\eta\mathbf{k})}\right) = \mathbf{k}^T\epsilon_{\mathrm{M}}\mathbf{k}, \tag{63}$$

where ϵ_{M} is a 3×3 symmetric, positive definite matrix. Transforming back (62) in the physical space, we obtain the macroscopic Poisson equation (4). Let us formalize this central result in a theorem.

Theorem 2. *There exists a 3×3 symmetric matrix $\epsilon_{\mathrm{M}} \geq 1$ such that for all $m \in L^1(\mathbb{R}^3)\cap L^2(\mathbb{R}^3)$, the rescaled potential W_m^η defined by (60) converges to W_m weakly in $\mathscr{C}'$ when η goes to zero, where W_m is the unique solution in $\mathscr{C}'$ to the elliptic equation*

$$-\mathrm{div}(\epsilon_{\mathrm{M}}\nabla W_m) = 4\pi m.$$

The matrix ϵ_{M} is proportional to the identity matrix if the host crystal has the symmetry of the cube.

From a physical viewpoint, the matrix ϵ_{M} is the electronic contribution to the macroscopic dielectric tensor of the host crystal. Note that the other contribution, originating from the displacements of the nuclei [29], is not taken into account in this study.

The matrix ϵ_{M} can be computed from the Bloch-Floquet decomposition of H^0_{per} as follows. The operator $\tilde{\epsilon} = v_{\mathrm{c}}^{-1/2}\epsilon v_{\mathrm{c}}^{1/2}$ being $\mathscr{R}$-periodic, it can be represented by the Bloch matrices $([\tilde{\epsilon}_{\mathbf{KK}'}(\mathbf{q})]_{\mathbf{K},\mathbf{K}'\in\mathscr{R}^*})_{\mathbf{q}\in\Gamma^*}$. It is proven in [9] that each entry of the Bloch matrix $\tilde{\epsilon}_{\mathbf{K},\mathbf{K}'}(\eta\sigma)$ has a limit when η goes to 0^+ for all fixed $\sigma \in S^2$. Indeed,

$$\lim_{\eta\to 0+}\tilde{\epsilon}_{0,0}(\eta\sigma) = 1+\sigma^T L\sigma$$

where L is the 3×3 non-negative symmetric matrix defined in (54). When $\mathbf{K},\mathbf{K}' \neq 0$, $\tilde{\epsilon}_{\mathbf{K},\mathbf{K}'}(\eta\sigma)$ has a limit at $\eta = 0$, which is independent of σ and which we simply denote by $\tilde{\epsilon}_{\mathbf{K},\mathbf{K}'}(0)$. When $\mathbf{K} = 0$ but $\mathbf{K}' \neq 0$, the limit is a linear function of σ: for all $\mathbf{K}' \in \mathscr{R}^*\setminus\{0\}$,

$$\lim_{\eta\to 0+}\tilde{\epsilon}_{0,\mathbf{K}'}(\eta\sigma) = \beta_{\mathbf{K}'}\cdot\sigma,$$

for some $\beta_{\mathbf{K}'} \in \mathbb{C}^3$. Both $\tilde{\epsilon}_{\mathbf{KK}'}(0)$ $(\mathbf{K},\mathbf{K}' \neq 0)$ and $\beta_{\mathbf{K}}$ can be computed from the eigenvalues $\epsilon_{n,\mathbf{q}}$ and eigenvectors $u_{n,\mathbf{q}}$ of the Bloch-Floquet decomposition of H^0_{per}

by formulae similar to (54). As already mentioned, the electronic contribution to the macroscopic dielectric permittivity is the 3×3 symmetric tensor defined as [6]

$$\forall \mathbf{k} \in \mathbb{R}^3, \quad \mathbf{k}^T \epsilon_{\mathrm{M}} \mathbf{k} = \lim_{\eta \to 0+} \frac{|\mathbf{k}|^2}{[\tilde{\epsilon}^{-1}]_{00}(\eta \mathbf{k})}. \tag{64}$$

By the Schur complement formula, it holds

$$\frac{1}{[\tilde{\epsilon}^{-1}]_{00}(\eta \mathbf{k})} = \tilde{\epsilon}_{00}(\eta \mathbf{k}) - \sum_{\mathbf{K},\mathbf{K}' \neq 0} \tilde{\epsilon}_{0,\mathbf{K}}(\eta \mathbf{k})[C(\eta \mathbf{k})^{-1}]_{\mathbf{K},\mathbf{K}'} \tilde{\epsilon}_{\mathbf{K}',0}(\eta \mathbf{k})$$

where $C(\eta \mathbf{k})^{-1}$ is the inverse of the matrix $C(\eta \mathbf{k}) = [\tilde{\epsilon}_{\mathbf{K}\mathbf{K}'}(\eta \mathbf{k})]_{\mathbf{K},\mathbf{K}' \in \mathscr{R}^* \setminus \{0\}}$. This leads to

$$\lim_{\eta \to 0+} \frac{|\mathbf{k}|^2}{[\tilde{\epsilon}^{-1}]_{00}(\eta \mathbf{k})} = |\mathbf{k}|^2 + \mathbf{k}^T L \mathbf{k} - \sum_{\mathbf{K},\mathbf{K}' \in \mathscr{R}^* \setminus \{0\}} (\beta_{\mathbf{K}} \cdot \mathbf{k})[C(0)^{-1}]_{\mathbf{K},\mathbf{K}'} \overline{(\beta_{\mathbf{K}'} \cdot \mathbf{k})}$$

where $C(0)^{-1}$ is the inverse of the matrix $C(0) = [\tilde{\epsilon}_{\mathbf{K}\mathbf{K}'}(0)]_{\mathbf{K},\mathbf{K}' \in \mathscr{R}^* \setminus \{0\}}$. Therefore,

$$\epsilon_{\mathrm{M}} = 1 + L - \sum_{\mathbf{K},\mathbf{K}' \in \mathscr{R}^* \setminus \{0\}} \beta_{\mathbf{K}} [C(0)^{-1}]_{\mathbf{K},\mathbf{K}'} \beta_{\mathbf{K}'}^*. \tag{65}$$

As already noticed in [6], it holds

$$1 \leq \epsilon_{\mathrm{M}} \leq 1 + L.$$

Formula (65) has been used in numerical simulations for estimating the macroscopic dielectric permittivity of real insulators and semiconductors [6, 14, 16, 22, 23]. Direct methods for evaluating ϵ_{M}, bypassing the inversion of the matrix $C(0)$, have also been proposed [25, 32].

4.4 Time-Dependent Response

We study in this section the variation of the electronic state of the crystal when the mean-field Hamiltonian H_{per}^0 of the perfect crystal is perturbed by a time-dependent effective potential $v(t, \mathbf{r})$ of the form (46). The mathematical proofs of the results announced in this section will be given in [10].

Let

$$H_v(t) = H_{\mathrm{per}}^0 + v(t, \cdot) = -\frac{1}{2}\Delta + V_{\mathrm{per}} + v(t, \cdot).$$

Under the assumption that $\rho_{\mathrm{per}}^{\mathrm{nuc}} \in L_{\mathrm{per}}^2(\Gamma)$ (smeared nuclei), the mean-field potential V_{per} is $\mathscr{R}$-periodic and in $C^0(\mathbb{R}^3) \cap L^\infty(\mathbb{R}^3)$. Besides, there exists a constant $C > 0$

such that $\|\rho \star |\cdot|^{-1}\|_{L^\infty} \leq C\|\rho\|_{L^2\cap\mathscr{C}}$ for all $\rho \in L^2(\mathbb{R}^3) \cap \mathscr{C}$, so that the time-dependent perturbation v belongs to $L^1_{\rm loc}(\mathbb{R}, L^\infty(\mathbb{R}^3))$.

Let us now define the propagator $(U_v(t,s))_{(s,t)\in\mathbb{R}\times\mathbb{R}}$ associated with the time-dependent Hamiltonian $H_v(t)$ following [30, Sect. X.12]. To this end, consider first the propagator $U_0(t) = e^{-itH^0_{\rm per}}$ associated with the time-independent Hamiltonian $H^0_{\rm per}$, and the perturbation in the so-called interaction picture:

$$v_{\rm int}(t) = U_0(t)^* v(t) U_0(t).$$

Standard techniques (see for instance [28, Sect. 5.1]) allow us to show the existence and uniqueness of the family of unitary propagators $(U_{\rm int}(t,s))_{(s,t)\in\mathbb{R}\times\mathbb{R}}$ associated with the bounded operators $(v_{\rm int}(t))_{t\in\mathbb{R}}$, with

$$U_{\rm int}(t,t_0) = 1 - i\int_{t_0}^t v_{\rm int}(s)U_{\rm int}(s,t_0)\,ds.$$

Therefore, $U_v(t,s) = U_0(t)U_{\rm int}(t,s)U_0(s)^*$ satisfies the integral equation

$$U_v(t,t_0) = U_0(t-t_0) - i\int_{t_0}^t U_0(t-s)v(s)U_v(s,t_0)\,ds. \tag{66}$$

Denoting by γ^0 the density operator of the crystal at time $t=0$, the dynamics of the system is governed by the evolution equation

$$\gamma(t) = U_v(t,0)\gamma^0 U_v(t,0)^*. \tag{67}$$

Note that the conditions $\gamma^0 \in \mathscr{S}(L^2(\mathbb{R}^3))$ and $0 \leq \gamma^0 \leq 1$ are automatically propagated by (67).

Considering $v(t)$ as a perturbation of the time-independent Hamiltonian $H^0_{\rm per}$, and $\gamma(t)$ as a perturbation of the ground state density operator $\gamma^0_{\rm per}$, it is natural to follow the same strategy as in the time-independent setting and introduce

$$Q(t) = \gamma(t) - \gamma^0_{\rm per}.$$

Using (66), (67), and the fact that $\gamma^0_{\rm per}$ is a steady state of the system in the absence of perturbation ($U_0(t)\gamma^0_{\rm per}U_0(t)^* = \gamma^0_{\rm per}$), an easy calculation shows that $Q(t)$ satisfies the integral equation

$$Q(t) = U_0(t)Q(0)U_0(t)^* - i\int_0^t U_0(t-s)[v(s), \gamma^0_{\rm per} + Q(s)]U_0(t-s)^*\,ds. \tag{68}$$

We now assume that $\gamma^0 = \gamma^0_{\rm per}$, *i.e.* $Q(0) = 0$, and write (formally for the moment) $Q(t)$ as the series expansion

$$Q(t) = \sum_{n=1}^{+\infty} Q_{n,v}(t), \tag{69}$$

where the operators $Q_{n,v}(t)$ are obtained, as in the time-independent case, by identifying terms involving n occurrences of the potential v. In particular, the linear response is given by

$$Q_{1,v}(t) = -i \int_0^t U_0(t-s) \left[v(s), \gamma_{\rm per}^0 \right] U_0(t-s)^* \, ds, \tag{70}$$

and the following recursion relation holds true

$$\forall n \geq 2, \quad Q_{n,v}(t) = -i \int_0^t U_0(t-s) \left[v(s), Q_{n-1,v}(s) \right] U_0(t-s)^* \, ds. \tag{71}$$

It is proved in [10] that for any $n \geq 1$ and any $t \geq 0$, the operator $Q_{n,v}(t)$ in (69) belongs to $\mathscr{Q}$ and satisfies

$$\forall \psi \in L^2(\mathbb{R}^3), \quad \langle \psi | Q_{n,v}(t) | \psi \rangle_{L^2} = 0.$$

In particular, $\mathrm{Tr}_0(Q_{n,v}(t)) = 0$. Besides, there exists $b \in \mathbb{R}_+$ such that for all $t \geq 0$

$$\| Q_{n,v}(t) \|_{\mathscr{Q}} \leq b^n \int_0^t \int_0^{t_1} \cdots \int_0^{t_{n-1}} \| \rho(t_1) \|_{L^2 \cap \mathscr{C}} \cdots \| \rho(t_n) \|_{L^2 \cap \mathscr{C}} \, dt_n \ldots dt_1,$$

and there exists $T > 0$ such that the series expansion (69) converges in $\mathscr{Q}$ uniformly on any compact subset of $[0, T)$. Lastly, $T = +\infty$ if $\rho \in L^\infty(\mathbb{R}_+, L^2(\mathbb{R}^3) \cap \mathscr{C})$.

As in the time-independent setting, the frequency-dependent dielectric properties of the crystal can be obtained from the linear response (70), by defining the time-dependent independent-particle polarization operator

$$\begin{aligned} \chi_0 \, : \, L^1(\mathbb{R}, v_{\rm c}(L^2(\mathbb{R}^3) \cap \mathscr{C})) &\rightarrow L^\infty(\mathbb{R}, L^2(\mathbb{R}^3) \cap \mathscr{C}) \\ v &\mapsto \rho_{Q_{1,v}} \end{aligned} \tag{72}$$

and the time-dependent operators $\mathscr{L} = -\chi_0 v_{\rm c}$, $\epsilon = v_{\rm c}(1 + \mathscr{L}) v_{\rm c}^{-1}$, $\epsilon^{-1} = v_{\rm c}(1 + \mathscr{L})^{-1} v_{\rm c}^{-1}$, and $\tilde{\epsilon} = v_{\rm c}^{-1/2} \epsilon v_{\rm c}^{1/2}$. Due to the invariance of the linear response with respect to translation in time, all these operators are convolutions in time. In addition they are $\mathscr{R}$-periodic in space. They can therefore be represented by frequency-dependent Bloch matrices $[T_{\mathbf{K},\mathbf{K}'}(\omega, \mathbf{q})]$, with $\mathbf{K}$, $\mathbf{K}'$ in $\mathscr{R}^*$, $q \in \Gamma^*$ and $\omega \in \mathbb{R}$. The Adler-Wiser formula states that the (electronic contribution of the) frequency-dependent macroscopic dielectric permittivity is given by the formula

$$\forall \mathbf{k} \in \mathbb{R}^3, \quad \mathbf{k}^T \mathcal{F} \epsilon_{\rm M}(\omega) \mathbf{k} = \lim_{\eta \to 0^+} \left(\frac{|\mathbf{k}|^2}{[\tilde{\epsilon}^{-1}]_{00}(\omega, \eta \mathbf{k})} \right).$$

The mathematical study of this formula and of its possible derivation from rigorous homogenization arguments, is work in progress.

We finally consider the self-consistent Hartree dynamics defined by

$$Q(t)=U_0(t)Q^0U_0(t)^*-i\int_0^t U_0(t-s)\Big[v(s)+v_c(\rho_{Q(s)}),\gamma_{\rm per}^0+Q(s)\Big]U_0(t-s)^*ds, \tag{73}$$

for an initial condition $Q^0 \in \mathscr{K}$, and for an external potential $v(t) = v_c(m(t))$, where $m(t) \in L^2(\mathbb{R}^3) \cap \mathscr{C}$ for all t. The solution $Q(t)$ of (73) is such that $\gamma(t) = \gamma_{\rm per}^0 + Q(t)$ satisfies, formally, the time-dependent Hartree equation

$$i\frac{d\gamma}{dt}(t) = \left[-\frac{1}{2}\Delta + (\rho_{\gamma(t)} - \rho_{\rm per}^{\rm nuc} - m(t)) \star |\cdot|^{-1}, \gamma(t)\right].$$

The following result [10] shows the well-posedness of the nonlinear Hartree dynamics.

Theorem 3. *Let $m \in L^1_{\rm loc}(\mathbb{R}_+, L^2(\mathbb{R}^3)) \cap W^{1,1}_{\rm loc}(\mathbb{R}_+, \mathscr{C})$. Then, for any $Q^0 \in \mathscr{K}$, the time-dependent Hartree equation* (73) *has a unique solution in $C^0(\mathbb{R}_+, \mathscr{Q})$. Besides, for all $t \geq 0$, $Q(t) \in \mathscr{K}$ and* $\mathrm{Tr}_0(Q(t)) = \mathrm{Tr}_0(Q^0)$.

Appendix: Trace-Class and Self-Adjoint Operators

It is well-known that any compact self-adjoint operator A on a separable Hilbert space $\mathscr{H}$ can be diagonalized in an orthonormal basis set:

$$A = \sum_{i=1}^{+\infty} \lambda_i \, |\phi_i\rangle \, \langle \phi_i|, \tag{74}$$

where $\langle \phi_i | \phi_j \rangle = \delta_{ij}$, and where the sequence $(\lambda_i)_{i \geq 1}$ of the (real) eigenvalues of A, counted with their multiplicities, converges to zero. We have formulated (74) using again Dirac's bra-ket notation. The conventional mathematical formulation for (74) reads

$$\forall \phi \in \mathscr{H}, \quad A\phi = \sum_{i=1}^{+\infty} \lambda_i \, \langle \phi_i | \phi \rangle \, \phi_i .$$

A compact self-adjoint operator A is called trace-class if

$$\sum_{i=1}^{+\infty} |\lambda_i| < \infty.$$

The trace of A is then defined as

$$\mathrm{Tr}(A) := \sum_{i=1}^{+\infty} \lambda_i = \sum_{i=1}^{+\infty} \langle e_i | A | e_i \rangle,$$

the right-hand side being independent of the choice of the orthonormal basis $(e_i)_{i \geq 1}$. Note that if A is a non-negative self-adjoint operator, the sum $\sum_{i=1}^{+\infty} \langle e_i | A | e_i \rangle$ makes sense in $\mathbb{R}_+ \cup \{+\infty\}$ and its values is independent of the choice of the orthonormal basis $(e_i)_{i\geq 1}$. We can therefore give a sense to $\mathrm{Tr}(A)$ for *any* non-negative self-adjoint operator A, and this number is finite if and only if A is trace-class.

The notion of trace-class operators can be extended to non-self-adjoint operators [31, 33], but we do not need to consider this generalization here.

By definition, a compact operator A is Hilbert-Schmidt if A^*A is trace-class. A compact self-adjoint operator A on $\mathscr{H}$ decomposed according to (74) is Hilbert-Schmidt if and only if

$$\sum_{i\geq 1} |\lambda_i|^2 < \infty.$$

Obviously any trace-class self-adjoint operator is Hilbert-Schmidt, but the converse is not true.

In this contribution, we respectively denote by $\mathfrak{S}_1$ and $\mathfrak{S}_2$ the spaces of trace-class and Hilbert-Schmidt operators acting on $L^2(\mathbb{R}^3)$. We also denote by $\mathscr{S}(L^2(\mathbb{R}^3))$ the vector space of the bounded self-adjoint operators on $L^2(\mathbb{R}^3)$.

A classical result states that if A is a Hilbert-Schmidt operator on $L^2(\mathbb{R}^3)$, then it is an integral operator with kernel in $L^2(\mathbb{R}^3 \times \mathbb{R}^3)$. This means that there exists a unique function in $L^2(\mathbb{R}^3 \times \mathbb{R}^3)$, also denoted by A for convenience, such that

$$\forall \phi \in L^2(\mathbb{R}^3), \quad (A\phi)(\mathbf{r}) = \int_{\mathbb{R}^3} A(\mathbf{r},\mathbf{r}')\,\phi(\mathbf{r}')\,d\mathbf{r}'. \tag{75}$$

Conversely, if A is an operator on $L^2(\mathbb{R}^3)$ for which there exists a function $A \in L^2(\mathbb{R}^3 \times \mathbb{R}^3)$ such that (75) holds, then A is Hilbert-Schmidt.

If A is a self-adjoint Hilbert-Schmidt operator on $L^2(\mathbb{R}^3)$ decomposed according to (74), then its kernel is given by

$$A(\mathbf{r},\mathbf{r}') = \sum_{i\geq 1} \lambda_i\, \phi_i(\mathbf{r})\phi_i(\mathbf{r}').$$

If, in addition A is trace-class, then the density ρ_A, defined as

$$\rho_A(\mathbf{r}) = \sum_{i=1}^{+\infty} \lambda_i |\phi_i(\mathbf{r})|^2,$$

is a function of $L^1(\mathbb{R}^3)$ and it holds

$$\mathrm{Tr}(A) = \sum_{i=1}^{+\infty} \lambda_i = \int_{\mathbb{R}^3} \rho_A(\mathbf{r})\,d\mathbf{r}.$$

For convenience, we use the abuse of notation which consists in writing $\rho_A(\mathbf{r}) = A(\mathbf{r},\mathbf{r})$ even when the kernel of A is not continuous on the diagonal $\{\mathbf{r} = \mathbf{r}'\} \subset \mathbb{R}^6$.

References

1. Adler, S.L.: Quantum theory of the dielectric constant in real solids. Phys. Rev. **126**, 413–420 (1962)
2. Ambrosio, L., Friesecke, G. Giannoulis, J.: Passage from quantum to classical molecular dynamics in the presence of Coulomb interactions. Commun. Part. Diff. Eq. **35**, 1490–1515 (2010)
3. Ambrosio, L., Figalli, A., Friesecke, G. Giannoulis, J., Paul, T.: Semiclassical limit of quantum dynamics with rough potentials and well posedness of transport equations with measure initial data. To appear in Comm. Pure Appl. Math., (2011)
4. Anantharaman, A., Cancès, É.: Existence of minimizers for Kohn-Sham models in quantum chemistry. Ann. I. H. Poincaré-An **26**, 2425–2455 (2009)
5. Arnold, A.: Self-consistent relaxation-time models in quantum mechanics, Commun. Part. Diff. Eq., **21**(3-4), 473–506 (1996)
6. Baroni, S., Resta, R.: Ab initio calculation of the macroscopic dielectric constant in silicon. Phys. Rev. B **33**, 7017–7021 (1986)
7. Cancès, É., Deleurence, A., Lewin, M.: A new approach to the modeling of local defects in crystals: the reduced Hartree-Fock case. Commun. Math. Phys. **281**, 129–177 (2008)
8. Cancès, É., Deleurence, A., Lewin, M.: Non-perturbative embedding of local defects in crystalline materials. J. Phys.: Condens. Mat. **20**, 294213 (2008)
9. Cancès, É., Lewin, M.: The dielectric permittivity of crystals in the reduced Hartree-Fock approximation. Arch. Ration. Mech. Anal. **197**, 139–177 (2010)
10. Cancès, É., Stoltz, G.: in preparation
11. Catto, I, Le Bris, C., Lions, P.-L.: On the thermodynamic limit for Hartree-Fock type models. Ann. I. H. Poincaré-An **18**, 687–760 (2001)
12. Dautray, R. and Lions, J.-L. Mathematical Analysis and Numerical Methods for Science and Technology. Vol. 5. Evolution Problems I. Springer-Verlag Berlin (1992)
13. Dreizler, R., Gross, E.K.U.: Density functional theory. Springer Verlag, Berlin (1990)
14. Engel, G. E., Farid, B.: Calculation of the dielectric properties of semiconductors. Phys. Rev. B **46**, 15812–15827 (1992)
15. Frank, R.L., Lieb, E.H., Seiringer, R., Siedentop, H.: Müllers exchange-correlation energy in density-matrix-functional theory. Phys. Rev. A **76**, 052517 (2007)
16. Gajdoš, M., Hummer, K., Kresse, G., Furthmüller, J., Bechstedt, F.: Linear optical properties in the projector-augmented wave methodology. Phys. Rev. B **73**, 045112 (2006)
17. Gravejat, P., Lewin, M, Séré, É.: Ground state and charge renormalization in a nonlinear model of relativistic atoms. Commun. Math. Phys. **286**, 179–215 (2009)
18. Hainzl, C., Lewin, M., Séré, É.: Existence of a stable polarized vacuum in the Bogoliubov-Dirac-Fock approximation. Commun. Math. Phys. **257**, 515–562 (2005)
19. Hainzl, C., Lewin, M., Séré, E., Solovej, J.P.: A minimization method for relativistic electrons in a mean-field approximation of quantum electrodynamics. Phys. Rev. A **76**, 052104 (2007)
20. Hainzl, C., Lewin, M., Solovej, J.P.: The mean-field approximation in Quantum Electrodynamics: the no-photon case. Commun. Pur. Appl. Math. **60**(4), 546–596 (2007)
21. Hohenberg, P., Kohn, W.: Inhomogeneous electron gas. Phys. Rev. **136**, B864-B871 (1964)
22. Hybertsen, M.S., Louie, S.G.: Ab initio static dielectric matrices from the density-functional approach. I. Formulation and application to semiconductors and insulators. Phys. Rev. B **35**, 5585–5601 (1987)
23. Hybertsen, M.S., Louie, S.G.:Ab initio static dielectric matrices from the density-functional approach. II. Calculation of the screening response in diamond, Si, Ge, and LiCl. Phys. Rev. B **35**, 5602–5610 (1987)
24. Kohn, W., Sham L. J.: Self-consistent equations including exchange and correlation effects. Phys. Rev. **140**, A1133 (1965)
25. Kunc, K., Tosatti, E.: Direct evaluation of the inverse dielectric matrix in semiconductors. Phys. Rev. B **29**, 7045–7047 (1984)

26. Lieb E.H.: Variational principle for many-fermion systems. Phys. Rev. Lett. **46**, 457–459 (1981)
27. Lieb, E.H.: Density Functional for Coulomb systems. Int. J. Quantum Chem. **24**, 243–277 (1983)
28. Pazy A.: Semigroups of Linear Operators and Applications to Partial Differential Equations, vol. 44 of Applied Mathematical Sciences. Springer, New York (1983)
29. Pick R. M., Cohen, M.H., Martin R. M.: Microscopic theory of force constants in the adiabatic approximation. Phys. Rev. B **1**, 910–920 (1970)
30. Reed, M., Simon, B.: Methods of Modern Mathematical Physics. II. Fourier Analysis, Self-Adjointness. Academic Press, New York (1975)
31. Reed, M., Simon, B.: Methods of Modern Mathematical Physics. IV. Analysis of Operators. Academic Press, New York (1978)
32. Resta, R., Baldereschi, A.: Dielectric matrices and local fields in polar semiconductors. Phys. Rev. B **23**, 6615–6624 (1981)
33. Simon, B.: Trace ideals and their applications, vol. 35 of London Mathematical Society Lecture Note Series. Cambridge University Press, Cambridge (1979)
34. Solovej, J.P.: Proof of the ionization conjecture in a reduced Hartree-Fock model. Invent. Math. **104**, 291–311 (1991)
35. Thomas, L.E.: Time dependent approach to scattering from impurities in a crystal. Commun. Math. Phys. **33**, 335–343 (1973)
36. Wiser, N.: Dielectric constant with local field effects included. Phys. Rev. **129**, 62–69 (1963)
37. Zhislin, G.M., Sigalov, A.G.: The spectrum of the energy operator for atoms with fixed nuclei on subspaces corresponding to irreducible representations of the group of permutations. Izv. Akad. Nauk SSSR Ser. Mat. **29**, 835–860 (1965)

Fast Multipole Method Using the Cauchy Integral Formula

Cristopher Cecka, Pierre-David Létourneau, and Eric Darve

Abstract The fast multipole method (FMM) is a technique allowing the fast calculation of long-range interactions between N points in $O(N)$ or $O(N \ln N)$ steps with some prescribed error tolerance. The FMM has found many applications in the field of integral equations and boundary element methods, in particular by accelerating the solution of dense linear systems arising from such formulations. Standard FMMs are derived from analytic expansions of the kernel, for example using spherical harmonics or Taylor expansions. In recent years, the range of applicability and the ease of use of FMMs has been extended by the introduction of black box (Fong and Darve, Journal of Computational Physics 228:8712–8725, 2009) or kernel independent techniques (Ying, Biros and Zorin, Journal of Computational Physics 196:591–626, 2004). In these approaches, the user only provides a subroutine to numerically calculate the interaction kernel. This allows changing the definition of the kernel with minimal change to the computer program. This paper presents a novel kernel independent FMM, which leads to diagonal multipole-to-local operators. This results in a significant reduction in the computational cost (Fong and Darve, Journal of Computational Physics 228:8712–8725, 2009), in particular when high accuracy is needed. The approach is based on Cauchy's integral formula and the Laplace transform. We will present a short numerical analysis of the convergence and some preliminary numerical results in the case of a single level one dimensional FMM.

E. Darve (✉)
Mechanical Engineering Department, Institute for Computational and Mathematical Engineering, Stanford University, CA, U.S.A.
e-mail: darve@stanford.edu

C. Cecka · P.-D. Létourneau
Institute for Computational and Mathematical Engineering, Stanford University, CA, U.S.A.
e-mail: ccecka@stanford.edu; pletou1@stanford.edu

B. Engquist et al. (eds.), *Numerical Analysis of Multiscale Computations*, Lecture Notes in Computational Science and Engineering 82, DOI 10.1007/978-3-642-21943-6_6,

1 Introduction

The fast multipole method (FMM) is a general class of methods to reduce the cost of computing:

$$\phi_i = \sum_{j=1}^{N} K(r_i, r_j)\sigma_j, \quad 1 \leq i \leq N, \tag{1}$$

when N is large. The basic approximation strategy in the FMM is a low-rank approximation of the kernel of the type:

$$K(r, r^*) = \sum_{m=1}^{p}\sum_{q=1}^{p} u_m(r)\; T_{mq}\; v_q(r^*) + \epsilon.$$

With a low-rank approximation of this type one can construct an $O(N)$ or $O(N \ln N)$ method to calculate the sum in (1).

Fast multipole methods have been derived for many different types of kernels, including the electrostatic kernel $1/r$ and Helmholtz kernel e^{ikr}/r. Efforts have been made to extend the method to more general kernels [9]. The authors for example have developed a fast technique applicable to any smooth kernels, i.e., non-oscillatory (see [7]).

In this paper, we consider the creation of an FMM with two goals in mind:

1. The method should be applicable to a wide class of kernels.
2. The multipole-to-local (M2L) operator T_{mq} should be diagonal.

Even though requirement (1) is satisfied by the method in [7], requirement (2) is not. The second requirement allows in principle reducing the computational cost of the method by reducing the number of operations involved in the multiplication of v_q by T_{mq}. This has the potential to significantly reduce the cost of these general FMM schemes to improve their applicability and efficiency.

Some of the applications we have in mind include the use of radial basis functions such as r, r^n (n odd), $r^n \log r$ (n even), $\exp(-cr^2)$, $\sqrt{r^2+c^2}$, $1/\sqrt{r^2+c^2}$, $1/(r^2+c^2), \ldots,$ for interpolation schemes. These are popular schemes for mesh deformation [5] and graphics applications [1]. Interpolation using these functions requires generalized FMMs capable of handling a wide class of functions.

The method we are proposing is based on Cauchy's integral formula for analytic functions. This is the class of kernels that is covered by this new scheme. In this paper, we are only presenting an outline of the method. Additional details will be presented in future publications. Contrary to some of the other approaches, this technique proposes a general framework to construct FMMs. Depending on the specific nature of the kernel, various optimizations and modifications can be made to this general scheme to improve its efficiency.

2 Cauchy's Integral Formula and Low-Rank Approximations

For simplicity, let us assume that $K(r, r^*)$ is translationally invariant so it can be expressed as $K(r - r^*)$. In addition we will start with the one dimensional (1D) case. In this paper we will not discuss in details the extensions to two and three dimensions (2D/3D). However, at the end of this section, we will briefly explain how this can be done.

We consider a kernel $K(x)$, $x \in \mathbb{R}$, and assume that, in some region $\Omega \subset \mathbb{C}$ around the point x, it is an analytic (holomorphic) function. That is, the function $K(z)$, $z \in \mathbb{C}$, is complex differentiable at every point $z \in \Omega$. Then, by Cauchy's integral formula:

$$K(x) = \frac{1}{2\pi i} \oint_{\Gamma} \frac{K(z)}{z - x}\, dz.$$

The curve $\Gamma = \partial\Omega$ is closed and contains x.

Assume that $\mathrm{Re}(z - x) > 0$ (Re is the real part of a complex number) then:

$$\frac{1}{z - x} = \int_0^{\infty} e^{-s(z-x)}\, ds.$$

The reason why we want to use this formula is that after approximating these integrals using a numerical quadrature, we will obtain a low rank approximation of the kernel, and later on a fast method.

Since Γ encloses x it is not always the case that $\mathrm{Re}(z - x) > 0$. However, by applying rotations in the complex plane we get:

$$\frac{1}{z - x} = -\int_0^{\infty} e^{s(z-x)}\, ds, \qquad \text{if } \mathrm{Re}(z - x) < 0, \tag{2}$$

$$\frac{1}{z - x} = -i \int_0^{\infty} e^{is(z-x)}\, ds, \qquad \text{if } \mathrm{Im}(z - x) > 0, \tag{3}$$

$$\frac{1}{z - x} = i \int_0^{\infty} e^{-is(z-x)}\, ds, \qquad \text{if } \mathrm{Im}(z - x) < 0. \tag{4}$$

Let us decompose Γ into four curves such that on Γ_1, $\mathrm{Re}(z - x) > 0$; on Γ_2, $\mathrm{Im}(z - x) > 0$; on Γ_3, $\mathrm{Re}(z - x) < 0$; and on Γ_4, $\mathrm{Im}(z - x) < 0$. Then:

$$\begin{aligned} K(x) = &+ \frac{1}{2\pi i} \int_{\Gamma_1} K(z) \int_0^{\infty} e^{-s(z-x)}\, ds\, dz \\ &- \frac{1}{2\pi} \int_{\Gamma_2} K(z) \int_0^{\infty} e^{is(z-x)}\, ds\, dz \\ &- \frac{1}{2\pi i} \int_{\Gamma_3} K(z) \int_0^{\infty} e^{s(z-x)}\, ds\, dz \\ &+ \frac{1}{2\pi} \int_{\Gamma_4} K(z) \int_0^{\infty} e^{-is(z-x)}\, ds\, dz. \end{aligned} \tag{5}$$

This is illustrated in Fig. 1.

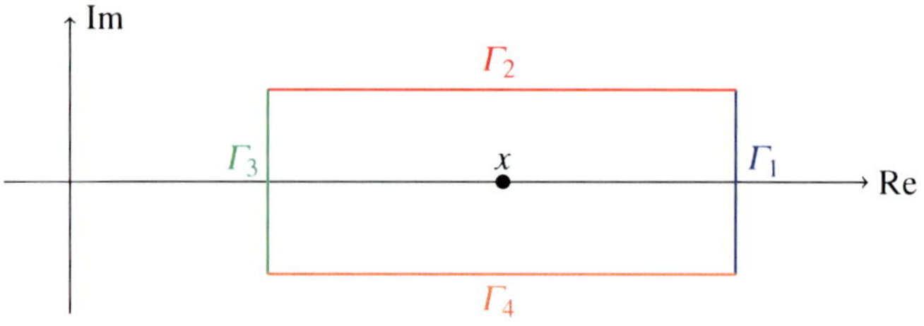

Fig. 1 Schematic of the four contour curves around x

How can we use this formula to construct a fast $O(N)$ method? We need to approximate $K(x-y)$ using a low rank decomposition. Let us consider the first integral along Γ_1. The variable x is now replaced by $x-y$. We assume we have obtained a quadrature along s, with weights w_q and points s_q, that approximates the integral along s. Then the contribution from Γ_1 to $K(x-y)$ can be approximated as:

$$\sum_q \left[\frac{w_q}{2\pi i}\int_{\Gamma_1} K(z)e^{-s_q z}\,dz\right] e^{s_q x}\, e^{-s_q y}.$$

Denote:

$$u_q(x) = e^{s_q x}, \qquad T_{qq} = \frac{w_q}{2\pi i}\int_{\Gamma_1} K(z)e^{-s_q z}\,dz, \qquad v_q(y) = e^{-s_q y},$$

and we see that this approximation is a low-rank approximation of the type (1). The M2L operator T_{qq} is diagonal. We immediately point out that this is not sufficient to construct an FMM scheme since we have not provided a method to gather multipole expansions from leaf nodes and propagate them up the tree, and a method to scatter local expansions from the root of the tree down to the leaves. However this formula provides the starting point for our analysis.

Extension to two and three dimensions. This extension relies on using a tensor product construction. For example, in 3D, the Cauchy formula reads:

$$K(u,v,w) = \frac{1}{(2\pi i)^3}\iiint_{\Gamma} \frac{K(z_1,z_2,z_3)}{(z_1-u)(z_2-v)(z_3-w)}\,dz_1\,dz_2\,dz_3,$$

where Γ is now a three dimensional domain in $\mathbb{C}^3$. Then each term $1/(z_1-u), \dots, 1/(z_3-w)$ is transformed into an integral with exponential functions. The final expression is rather long but the approximation of $K(\mathbf{r}-\mathbf{r}')$ with $\mathbf{r}=(u,v,w)$, $\mathbf{r}'=(u',v',w')$ involves terms like:

$$\sum_{q1,q_2,q_3} K_{q_1,q_2,q_3}\, e^{s_{q_1}u}\, e^{-s_{q_1}u'}\, e^{s_{q_2}v}\, e^{-s_{q_2}v'}\, e^{s_{q_3}w}\, e^{-s_{q_3}w'},$$

$$K_{q_1,q_2,q_3} = \frac{w_{q_1}w_{q_2}w_{q_3}}{(2\pi i)^3}\iiint_{\Gamma_1} K(z_1,z_2,z_3)\, e^{-s_{q_1}z_1}\, e^{-s_{q_2}z_2}\, e^{-s_{q_3}z_3}\,dz_1\,dz_2\,dz_3.$$

The 1D construction can then be extended to the 2D and 3D cases. In the rest of the paper, we will focus on the analysis of the one dimensional case.

3 Connection with Fourier and Laplace Transforms

The formula (5) can be viewed as an extension of the Fourier and Laplace transforms. Specifically, when the kernel satisfies some additional conditions at infinity, (5) can be related to Fourier and Laplace transforms. To understand the connection, let us assume that the paths Γ_i are straight segments:

$$\begin{aligned} K(x) = &+\frac{1}{2\pi}\int_{-a}^{a} K(c+it)\int_0^\infty e^{-ist}e^{-s(c-x)}\,ds\,dt \\ &+\frac{1}{2\pi}\int_{b}^{c} K(t+ia)\int_0^\infty e^{ist}e^{-isx-sa}\,ds\,dt \\ &+\frac{1}{2\pi}\int_{-a}^{a} K(b+it)\int_0^\infty e^{ist}e^{s(b-x)}\,ds\,dt \\ &+\frac{1}{2\pi}\int_{b}^{c} K(t-ia)\int_0^\infty e^{-ist}e^{isx-sa}\,ds\,dt. \end{aligned} \tag{6}$$

See Fig. 2 for the notations.

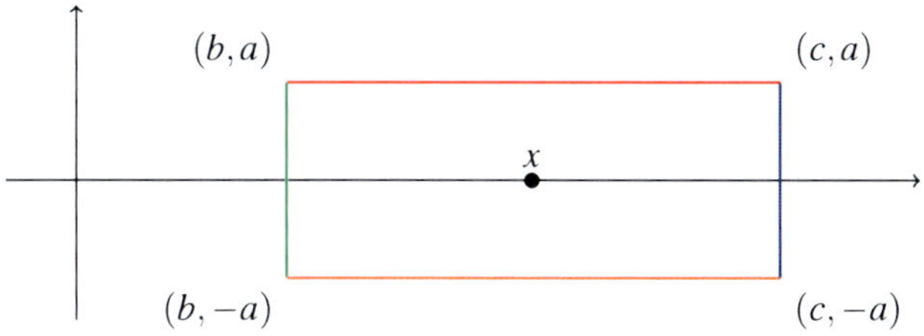

Fig. 2 Notations for straight segment contours

The Fourier or Laplace transforms can be recognized provided the kernel K satisfies some additional assumptions. If $K(t+ia)$ is in $\mathrm{L}^1(\mathbb{R})$ with respect to t, then we can choose $b=-\infty$ and $c=\infty$:

$$\begin{aligned} K(x) &= \frac{1}{2\pi}\Big(\int_{-\infty}^{\infty} K(t+ia)\int_0^\infty e^{ist}e^{-isx-sa}\,ds\,dt \\ &\quad+\int_{-\infty}^{\infty} K(t-ia)\int_0^\infty e^{-ist}e^{isx-sa}\,ds\,dt\Big) \\ &= \frac{1}{2\pi}\Big(\int_0^\infty e^{-isx-sa}\int_{-\infty}^{\infty} K(t+ia)e^{ist}\,dt\,ds \\ &\quad+\int_0^\infty e^{isx-sa}\int_{-\infty}^{\infty} K(t-ia)e^{-ist}\,dt\,ds\Big). \end{aligned}$$

After taking the limit $a \to 0$ and doing a change of variable $s \to 2\pi s$, we get:

$$K(x) = \int_{-\infty}^{0} e^{2\pi isx} \int_{-\infty}^{\infty} K(t)e^{-2\pi ist}\,dt\,ds + \int_{0}^{\infty} e^{2\pi isx} \int_{-\infty}^{\infty} K(t)e^{-2\pi ist}\,dt\,ds$$
$$= \int_{-\infty}^{\infty} e^{2\pi isx} \int_{-\infty}^{\infty} K(t)e^{-2\pi ist}\,dt\,ds.$$

This is the formula for the Fourier transform and its inverse transform. See an illustration on Fig. 3.

Similarly let us assume that $|K(z)| < A|z|^{-\epsilon}$ when $|z| \to \infty$, $\mathrm{Re}(z) \geq b$, $\epsilon > 0$. In that case we can choose $a \to \infty$ and $c \in \Theta(a)$ (bounded above and below by a, i.e., c and a go to infinity at the same rate). Then:

$$\lim_{\substack{a\to\infty \\ c\to\infty}} \frac{1}{2\pi} \int_{-a}^{a} K(c+it) \int_{0}^{\infty} e^{-ist} e^{-s(c-x)}\,ds\,dt = 0,$$

and similarly for the second and fourth integral in (6). We are only left with the third integral:

$$K(x) = \frac{1}{2\pi} \int_{-\infty}^{\infty} K(b+it) \int_{0}^{\infty} e^{ist} e^{s(b-x)}\,ds\,dt.$$

If we change the order of integration, assume that $K(b+it)$ is in $\mathrm{L}^1(\mathbb{R})$ with respect to t, and do the change of variable $z = b + it$, then:

$$K(x) = \frac{1}{2\pi i} \int_{0}^{\infty} e^{-sx} \int_{b-i\infty}^{b+i\infty} K(z)e^{sz}\,dz\,ds.$$

We recognize the Laplace transform and its inverse, expressed using the Bromwich integral. See Fig. 3.

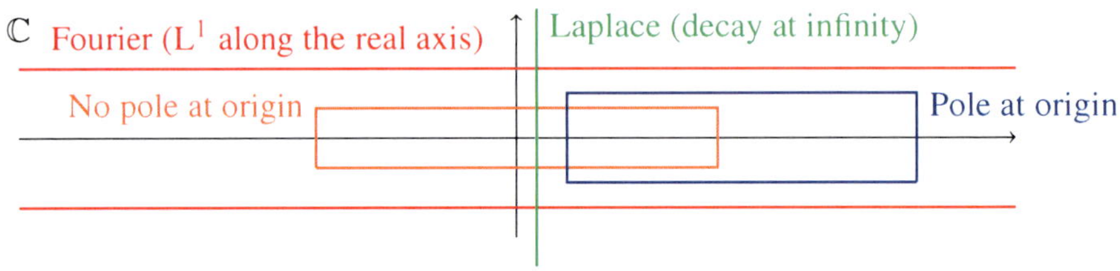

Fig. 3 This figure illustrates the different contours of integration that can be used depending on the properties of the kernel. We briefly recall for each contour the corresponding property of the kernel

4 Construction of Fast Methods

The problems of moving information up and down the tree in the fast multipole method and finding a suitable quadrature (s_k, w_k) are related. There are many ways to approach this question. We chose to use a spectral decomposition. As we will use Fourier analysis as our main tool, the functions e^{sx} are not suitable without making an appropriate change of variable. Assume for example that we have a cluster C of particles x_j with intensities σ_j. Then the multipole coefficients are of the form:

$$M_q(C) = \sum_{j,\, x_j \in C} e^{-s_q(C)x_j}\,\sigma_j,$$

where $s_q(C)$ denotes a quadrature adapted to cluster C. Let us assume that D is the parent cluster of C in the traditional FMM tree decomposition. Then we need to be able to calculate the contribution to $M_q(D)$ of particles in C that is:

$$\sum_{j,\, x_j \in C} e^{-s_q(D)x_j}\,\sigma_j.$$

However the quadrature points $s_q(D)$ are different from $s_q(C)$. A procedure to gather multipole expansions up the tree is therefore needed. Similarly we need a procedure to scatter down the tree local expansions.

These procedures can be developed if we use Gaussian functions instead of exponentials. Let us perform the change of variable $s \to s^2$ in Γ_1 for example

$$\int_0^{\infty} \left[\frac{s}{\pi i} \int_{\Gamma_1} K(z) e^{-s^2 z}\, dz\right] e^{s^2(x-y)}\, ds.$$

The functions $e^{s^2 x}$ are still not suitable. Depending on the sign of x they may be bounded or not. We need to enforce that the Gaussian functions decay. To reduce the number of symbols, we now assume that the clusters containing x and y have the same size. We add one more parameter l given by:

$$l = \min_{z \in \Gamma_1} \frac{\mathrm{Re}(z)}{2}. \tag{7}$$

With this parameter:

$$\int_0^{\infty} \left[\frac{s}{\pi i} \int_{\Gamma_1} K(z) e^{-s^2(z-2l)}\, dz\right] e^{-s^2(l-x)} e^{-s^2(l+y)}\, ds.$$

From now on (except in the next section) we assume that x is the displacement of a particle from the center of cluster C and similarly for y in cluster E, where C and E are two clusters that are well separated (according to the usual FMM prescription). Let us denote R the radius of each cluster. Since we must have, by definition of Γ_1, $\mathrm{Re}(z) > x - y$ for all x and y, we have $\mathrm{Re}(z) > 2R$. Consequently, it is possible to choose Γ_1 such that, from (7), $l > R$. This implies that $l - x > 0$ and $l + y > 0$. The two Gaussian functions $e^{-s^2(l-x)}$ and $e^{-s^2(l+y)}$ therefore decay as $s \to \infty$. In addition these Gaussian functions have a spectrum in Fourier space that decays rapidly. This will be essential to construct the interpolation and anterpolation, and in the error analysis.

5 Interpolation and Anterpolation

We now propose an approach for interpolating multipole data up the tree and anterpolating local data down the tree. Consider multipole coefficients of the form:

$$M_C(s_q) = \sum_{j,\, x_j \in C} e^{-s_q(C)^2(l_C - x_j + c_C)} \sigma_j.$$

We want to approximate the multipole coefficients for the parent cluster D:

$$M_D(s_q) = \sum_{j,\, x_j \in C} e^{-s_q(D)^2(l_D - x_j + c_D)} \sigma_j.$$

In this section only, we explicitly use the center of the clusters, c_C and c_D. In general since the radius of D is twice that of C, the Gaussian function decays much faster, and therefore the quadrature points $s_q(D)$ tend to cluster near 0 when compared to the quadrature points $s_q(C)$.

Let us assume that the quadrature points $s_q(C)$ have a spacing $\Delta s(C)$ while those for D have a spacing $\Delta s(D)$. In that case, we need to interpolate the function $M_C(s_q)$ and calculate its values at points with a spacing of $\Delta s(D)$. In general terms, this can be done by performing a fast Fourier transform of $M_C(s_q)$, padding with zeros, and performing an inverse transform. The coefficients $M_q(D)$ are obtained by keeping those samples in the desired interval: $[-L(D), L(D)]$ (using the notations of Sect. 6). The process is completed by a multiplication with

$$e^{-s_q(D)^2(l_D + c_D - l_C - c_C)}.$$

The coefficient $l_D + c_D - l_C - c_C$ is positive if we assume that $l_D - R_D > l_C - R_C$. This holds because:

$$l_D + c_D - l_C - c_C > l_D - l_C - R_C = l_D - R_D - l_C + R_C > 0,$$

since $R_D = 2R_C$. This procedure can be used to gather multipole coefficients going up the tree in the fast multipole method.

When going down, we need to scatter local coefficients from the root of the tree down to the leaves. The procedure is similar. However, the justification is more involved and requires a detailed numerical analysis. Additional justification is provided in the next two sections. We briefly outline the main argument. We have functions with high frequency components in s corresponding to a fast decay at infinity, for large clusters. However the final integration is against:

$$e^{-s^2(l+y)}$$

whose Fourier spectrum decays rapidly (since this is a Gaussian in s space that decays slowly). Therefore the high frequencies in the local multipole expansions can be removed when moving down the tree. This is done in a manner similar to the steps during the gathering (upward) phase of the FMM. First the local coefficients are multiplied by:

$$e^{-s_q(D)^2(l_D-c_D-l_C+c_C)}$$

if again D is the parent cluster of C. With the analysis above, we have that $l_D - c_D - l_C + c_C > 0$. Then the local expansion is padded with zeros to change its interval of definition from $[-L(D), L(D)]$ to $[-L(C), L(C)]$. Finally since we only need the low-frequency components, we Fourier transform the coefficients, remove the high frequencies that do not make significant contribution, and inverse Fourier transform the coefficients.

6 Error Analysis

We now discuss the error analysis and the construction of the scheme. Let us consider as an example the kernel $\sqrt{(x+x_0)^2+c^2}$ which is representative of a commonly found radial basis function. Let us assume that $c \in \mathbb{R}$. This kernel has two branch cuts starting at $-x_0 \pm ic$ on the imaginary axis. The contours of integration Γ_i must therefore avoid those branch cuts. This kernel does not have a Fourier or Laplace transform as described previously. Therefore we must resort to a contour curve formed by the four pieces Γ_1 through Γ_4.

The error analysis relies on a Fourier series approximation of the Gaussian functions $g(s) = e^{-s^2(l-x)}$ and $e^{-s^2(l+y)}$. In this section, we outline the main points of the analysis. The error analysis requires an integration from $-\infty$ to ∞ (instead of 0 to ∞) for s, otherwise $g(s)$ is effectively "discontinuous" and its Fourier spectrum decays slowly. We therefore modify the M2L operator and consider:

$$\int_{-\infty}^{\infty} \left[\frac{|s|}{2\pi i} \int_{\Gamma_1} K(z) e^{-s^2(z-2l)}\, dz\right] e^{-s^2(l-x)} e^{-s^2(l+y)}\, ds.$$

The Fourier spectrum of $g(s)$ decays rapidly. For example let us consider some tolerance ϵ and a bound L such that, when $|s| > L$, the function $g(s)$ is smaller than ϵ. On $[-L, L]$, we can expand $g(s)$ using a Fourier series. The Fourier series itself decays like a Gaussian function (up to ϵ terms). There is an integer P_0 such that all the Fourier coefficients of $g(s)$ beyond P_0 are smaller than ϵ: $|\hat{g}_k| < \epsilon$ for $|k| > P_0$. Let us now denote:

$$T(s) = \frac{|s|}{2\pi i} \int_{\Gamma_1} K(z) e^{-s^2(z-2l)}\, dz,$$

and $\hat{T}_k$ its kth Fourier coefficient. As a consequence of Parseval's theorem, the function

$$T^{\text{lb}}(s) = \sum_{k=-2P_0}^{2P_0} \hat{T}_k \, e^{2\pi i k s/2L},$$

is such that:

$$\int_{-\infty}^{\infty} T(s)\, e^{-s^2(l-x)} e^{-s^2(l+y)}\, ds = \int_{-\infty}^{\infty} T^{\text{lb}}(s)\, e^{-s^2(l-x)} e^{-s^2(l+y)}\, ds + O(\epsilon). \tag{8}$$

The key property of this change from T to T^{lb} is that the integral

$$\int_{-\infty}^{\infty} T^{\text{lb}}(s)\, e^{-s^2(l-x)} e^{-s^2(l+y)}\, ds$$

can be computed accurately, with error ϵ, using a trapezoidal rule with $4P_0 + 1$ points only. In addition, since the Gaussian functions are even, we can "fold" the M2L function and we really only need $2P_0 + 1$ points. This number is essential. This is the number of quadrature points s_q mentioned in Sect. 4. It drives the computational cost of the FMM.

Let us now try to understand the parameters that determine P_0. The function $g(s)$ is given again by:

$$g(s) = e^{-s^2(l-x)}.$$

The coefficient after s^2, $l - x$, is bounded by:

$$l - R \le l - x \le l + R.$$

When $s^2(l-x) \sim s^2(l-R)$, this corresponds to a slow decay and L needs to be large. When $s^2(l-x) \sim s^2(l+R)$, the Fourier spectrum decays slowly leading to a large bandwidth. This means that the spacing between quadrature points needs to be small. To achieve the desired accuracy, we need to satisfy both criteria: sufficiently large interval and sufficiently dense quadrature. After some analysis, the total number of quadrature points can be shown to be of order ($l > R$):

$$\ln(|\Gamma_1| \max_{\Gamma_1} |K(z)|/\epsilon) \sqrt{\frac{l+R}{l-R}}, \tag{9}$$

where $|\Gamma_1|$ is the length of the segment. Therefore we should try to make l as large as possible. This means the path Γ_1 should be moved to the right as much as possible. The path Γ_3 is similar. The Gaussian function is $e^{-s^2(l_3+x)}$ with

$$l_3 = -\max_z \frac{\text{Re}(z)}{2}.$$

(Note that along Γ_3, $\mathrm{Re}(z) < -2R$ and is therefore negative.) The number of quadrature points is of order:

$$\ln(|\Gamma_3| \max_{\Gamma_3} |K(z)|/\epsilon) \sqrt{\frac{l_3 + R}{l_3 - R}}, \tag{10}$$

so that the path Γ_3 should be moved to the left. Finally the path Γ_2 has $e^{-s^2(l_2+ix)}$ with

$$l_2 = \min_z \frac{\mathrm{Im}(z)}{2}.$$

The number of quadrature points is of order:

$$\ln(|\Gamma_2| \max_{\Gamma_2} |K(z)|/\epsilon) \sqrt{1 + \left(\frac{R}{l_2}\right)^2}, \tag{11}$$

so that the path Γ_2 should be moved up (large l_2). For kernel Ks that are real valued the path Γ_4 is the complex conjugate of Γ_2 so the same analysis applies. Note an interesting consequence. If K grows exponentially fast when $\mathrm{Im}(z)$ increases (to $+\infty$ or $-\infty$) we cannot increase l_2. This is true for oscillatory kernel. Take for example e^{iz}. In that case as R becomes large, the number of quadrature points must grow like $O(R)$ which is consistent with the behavior of known FMMs [2].

As a conclusion all paths must be essentially moved away from the origin. Note that we assumed that x and y are displacements from the center of a cluster so that we have already made the problem translationally "invariant." A constraint is that the poles $-x_0 \pm ic$ cannot be enclosed by the path. It is not essential to find the optimal path with great accuracy since the cost and accuracy of the scheme only weakly depend on picking the optimal path. Any reasonable choice away from the branch and from the origin is typically sufficient.

The choice of optimal contour does depend on the kernel. The method is very flexible and allows for many different types of optimizations. For example if we choose $1/\sqrt{(x+x_0)^2+c^2}$, the kernel has branch cuts starting at $-x_0 \pm ic$, and is unbounded at those points. However since $1/\sqrt{(z+x_0)^2+c^2}$ decays as $|z| \to \infty$, the contours Γ_1, Γ_2 and Γ_4 can be moved to infinity as explain in Sect. 3 for the Laplace transform. The optimal choice in this case is Γ_3 extending from $b - i\infty$ to $b + i\infty$, $b > -x_0$. The number of quadrature points is then of order:

$$\ln(|\Gamma_3| \max_{\Gamma_3} |K(z)|/\epsilon) \sqrt{\frac{x_0 + 2R}{x_0 - 2R}}.$$

This implies that $x_0 > 2R$. This condition is equivalent to saying that the two clusters need to be well separated (R is their radius).

In summary the basic procedure to choose the parameters in the method are as follows:

- Pick the optimal contour in the complex plane. This is done by moving the contour away from the $[-2R : 2R]$ interval on the real axis. The contour in general can be shaped like a rectangle. Examples are shown on Fig. 4.
- Poles in the kernel K cannot be included in the region. In addition, the extrema $\max_i \max_{\Gamma_i} |K(z)|$ of $|K|$ across all paths should not become large relative to the values in the $[-2R : 2R]$ interval. Some possible cases are illustrated in Fig. 4.
- Once the contour has been obtained, for each segment, we can obtain L by considering the decay in s space of the Gaussian functions associated with the segment.
- Finally given the interval $[-L : L]$ for a segment, we obtain the number of quadrature points p by considering the decay of the Gaussian functions in Fourier space [see (9), (10), and (11)].

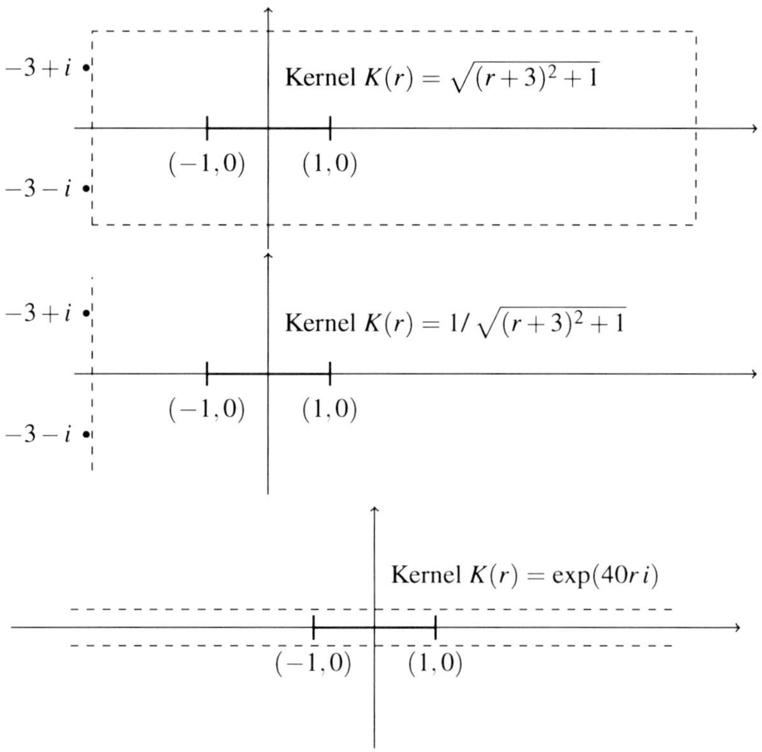

Fig. 4 Schematic of optimal contours for three different kernels in the complex plane. In all cases, we assume that $-1 \le r \le 1$. The poles of the kernel are shown using solid circles. The contour itself is shown with a dashed line – –. *Top figure*: this is the contour used for kernels that grow at infinity. *Middle figure*: this kernel decays sufficiently rapidly and a contour similar to the inverse Laplace transform can be used. *Bottom figure*: the kernel grows exponentially fast when moving away from the real axis. As a result the contour has to stay close to the real axis, resulting in a growth of the number of quadrature points with the size R of the clusters. This is consistent with the behavior of other FMMs

7 Detailed Error Analysis

A more detailed error analysis is useful to optimize the parameters in the method. Having sharp but easy to compute error bounds allows quickly varying the parameters to find optimal values. In additions, it guarantees a strict upper bound on the error. Although the final formula is complex we highlight the main points of the derivation and the final result.

The rotations in the complex plane in (2)–(4) can all be written as a multiplication by a complex number λ. Since we have in general up to four segments in our contour, we use a sum over an index k and express the kernel using:

$$K(x-y)=\sum_{k=1}^{n_{\text{seg}}} \int_{-\infty}^{\infty} e^{-s^2(\ell_k+\lambda_k y)}\Big[\frac{\lambda_k|s|}{2\pi i}\int_{\Gamma_k} K(z)e^{-s^2(\lambda_k z-2\ell_k)}dz\Big]e^{-s^2(\ell_k-\lambda_k x)}ds. \tag{12}$$

Two main approximations are required to obtain a low-rank approximation from the above formulation:

1. Reduce the infinite domain of the outer integral to a finite interval
2. Approximate the ensuing definite integral through an adequate quadrature

The symbol $:=$ will denote the definition of a new symbol.

In the first step, we introduce a parameter $L_k > 0$ for each path and approximate the outer integral through the expression by restricting the integration from $-L_k$ to L_k. Along each path, the error incurred in L^∞ norm is given by

$$\frac{1}{\pi}\left|\int_{L_k}^{\infty}\int_{\Gamma_k} K(z)\, s\, e^{-s^2(\lambda_k(z-x+y))}\, dz\, ds\right|. \tag{13}$$

The above expression can be bounded by

$$\leq \frac{e^{-L_k^2\eta_k}\,|\Gamma_k|}{2\pi\,\eta_k}\max_{z\in\Gamma_k}|K(z)| := B_0^k, \tag{14}$$

where $\eta_k = \text{Re}(\lambda_k(z-x+y)) > 0$.

The second major step involves an efficient quadrature of the integral over s. In this case, it is advantageous to apply Fourier analysis since we are using Gaussian functions in s. We will obtain the quadrature we are after once we can approximate the Gaussian functions by a truncated Fourier series:

$$e^{-s^2(\ell_k-\lambda_k x)} \approx \sum_{l=-P_k}^{P_k} b_l^k\, e^{i\pi l s/L_k}.$$

The process to find this approximation is to start by approximating the Gaussians by a *smooth* periodic function of period $2L_k$:

$$e^{-s^2(\ell_k-\lambda_k x)} \approx \sum_{n\in\mathbb{Z}} e^{-(s+2nL_k)^2(\ell_k-\lambda_k x)}, \qquad s\in[-L_k,L_k]. \tag{15}$$

The difference between these two functions can be bounded by:

$$\left| e^{-s^2(\ell_k - \lambda_k x)} - \sum_{n \in \mathbb{Z}} e^{-(s+2nL_k)^2(\ell_k - \lambda_k x)} \right| \leq A_k \, e^{-L_k^2 \operatorname{Re}(\gamma_k)} := B_1^k, \tag{16}$$

where $\gamma_k = \ell_k - \lambda_k x$, and A_k is a numerical constant close to 2. Since the Fourier transform of a Gaussian is analytically known (it is also a Gaussian), we can choose:

$$b_l^k = \frac{1}{2L_k} \sqrt{\frac{\pi}{\gamma_k}} \, e^{\frac{-\pi^2 l^2}{4\gamma_k L_k^2}}. \tag{17}$$

With this choice:

$$\left| \sum_{n \in \mathbb{Z}} e^{-(s+2nL_k)^2(\ell_k - \lambda_k x)} - \sum_{l=-P_k}^{P_k} b_l^k \, e^{\pi i s l / L_k} \right| = \left| \sum_{|l| > P_k} b_l^k \, e^{\pi i s l / L_k} \right|,$$

which can be bounded by

$$\leq \sqrt{\frac{|\gamma_k|}{\operatorname{Re}(\gamma_k)}} \operatorname{erfc}\left(\frac{P_k \pi}{2L_k |\gamma_k|} \sqrt{\operatorname{Re}(\gamma_k)} \right) := B_2^k.$$

As was explained earlier [see (8)], we should only use the low band part of

$$T_k(s) := \frac{\lambda_k}{2\pi i} |s| \int_{\Gamma_k} K(z) e^{-s^2(\lambda_k z - 2\ell_k)} \, dz. \tag{18}$$

This is required to reduce the number of quadrature points. Since the Gaussian functions can be accurately represented using only $2P_k + 1$ frequencies in Fourier space, the use of T^{lb} instead of T does not incur a large error [see (8)]. We denote a_n^k the nth Fourier series coefficient of the function of s in (18). The transfer function T^{lb} used in the FMM uses only those coefficients a_n^k with $|n| \leq 2P_k$.

Putting everything together, the total error accounting for all these approximations can be estimated using:

$$\text{FMM error} \sim \sum_{k=1}^{n_{\text{seg}}} \left[B_3^k (B_1^k + B_2^k) + B_0^k \right],$$

$$\text{with} \qquad B_3^k := 4 \sqrt{\frac{\pi}{\operatorname{Re}(\gamma_k)}} \max_{s \in [-L_k; L_k]} |T_k(s)|.$$

By further simplifying those expressions, we can recover the approximate formulas of Sect. 6.

8 Preliminary Numerical Results

We tested the method using different kernels that have different properties. Some of these results are preliminary as we do not yet have implemented an optimization scheme for the contours in the complex plane. We show only results corresponding to the one level method. Those numerical results serve to demonstrate the convergence of the method, and to establish that the Cauchy FMM approach is a rapidly converging scheme. Convergence to arbitrarily small errors (within the limits of the arithmetic precision) can be achieved. These results are consistent with the theoretical analysis of the previous sections.

We start with the inverse multiquadrics $1/\sqrt{z^2+1}$. In this case, the optimal choice is the inverse Laplace transform formula (see Fig. 3). In the figures below, R is the half-length of the cluster containing the points. We considered an interval of size $8R$ split into four clusters of size $2R$. The FMM expansion is applied to all clusters that are well separated. The other interactions are computed in a direct manner without the use of any accelerated scheme (Fig. 5).

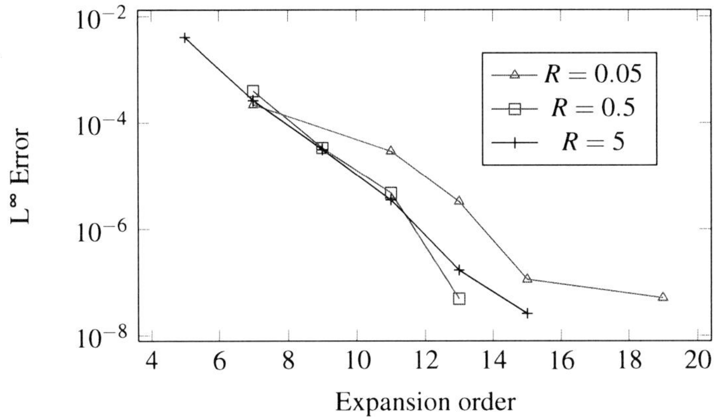

Fig. 5 Inverse multiquadrics $1/\sqrt{z^2+1}$ using the Laplace transform version of the Cauchy FMM

For comparison, we also used the contour associated with the Fourier transform case (horizontal infinite line) (Fig. 6). This case is sub-optimal since the poles at $\pm i$ prevent l_2 from growing. As as result, we observe a scaling of the order like R when R is large compared to 1. This is consistent with (11).

We then considered a kernel that requires four contours: the multiquadrics $\sqrt{z^2+1}$. This kernel grows as $|z|$ goes to ∞. These are preliminary results as the contour was not optimized. However we picked a contour that is "reasonable" so that these numerical results are meaningful. Results are shown on Fig. 7.

Finally we demonstrate the method with an oscillatory kernel, the Helmholtz kernel e^{2iz}/z using a closed contour with four segments. As was explained earlier, oscillatory kernels grow exponentially when $\text{Im}(z)$ increases (take for example e^{iz}). As a consequence, the contour needs to stay close to the x axis. As a result of (11),

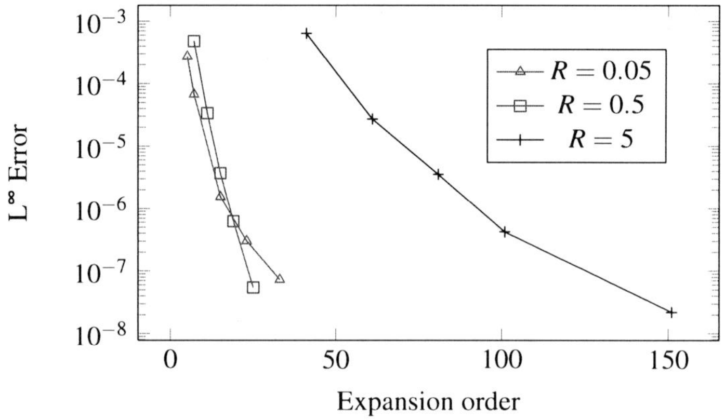

Fig. 6 Inverse multiquadrics $1/\sqrt{z^2+1}$ using the Fourier transform version of the Cauchy FMM. The scaling of the order with R is consistent with (11)

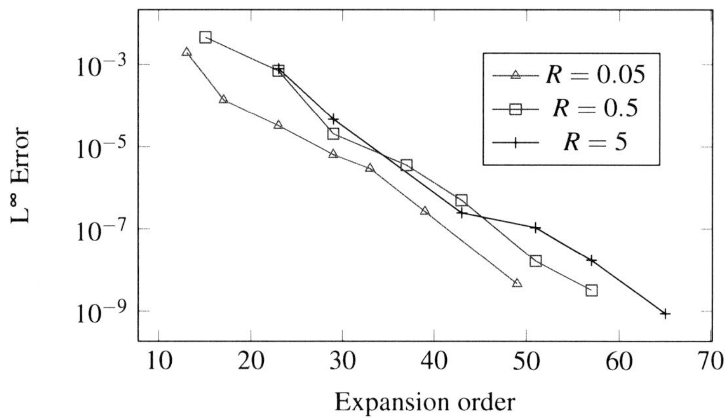

Fig. 7 Multiquadrics $\sqrt{z^2+1}$ using a full contour with four segments. The method is less efficient in this case since the four contours lead in general to more quadrature points compared to the Laplace version of the method. However we observe that the expansion order is largely independent of the cluster size which shows that the method is practical

the expansion order grows linearly with R in the high frequency regime. This is consistent with the results shown on Fig. 8. As a consequence, in 3D, the number of terms in the expansion grows like R^3. This is problematic when the points are distributed on a 2D manifold, which is typical for boundary element methods. In that case, the total number of points on the boundary surface, N, grows like the square of the object size and consequently the cost of the FMM grows like $N^{3/2}$, which is suboptimal. If the points are distributed in a volume the cost is constant at each level of the FMM leading to a complexity of $O(N \ln N)$. The cost of the FMM can be reduced to $N \ln N$ in both cases by using a variant of the Cauchy FMM presented in this paper. This variant uses the Cauchy FMM scheme along with the directional decomposition of Engquist et al. [6]. The details of this method will be described in a future publication.

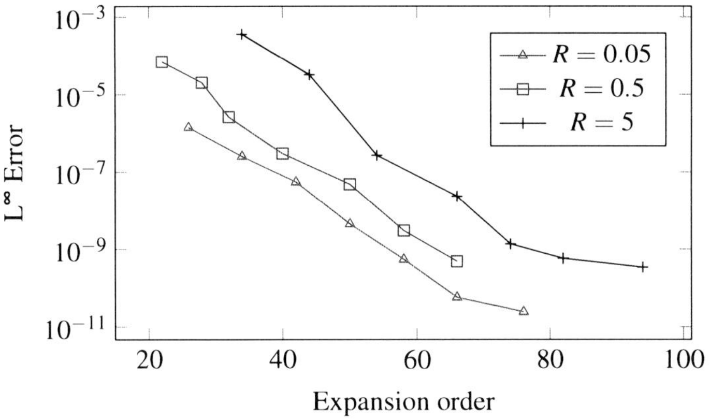

Fig. 8 Helmholtz kernel e^{2iz}/z. In that case, the number of expansion terms is roughly constant in the low frequency regime and grows linearly with the problem size in the high frequency regime

9 Conclusion

We have presented a framework to develop generalized fast multipole methods, in particular methods that are applicable to radial basis functions for interpolation schemes. This new approach has the potential to greatly speed up traditional FMMs since the multipole-to-local operators are diagonal. We have developed a numerical analysis of the error and schemes to select the optimal parameters in the method. The contour of integration in the complex plane needs to be optimized depending on the kernel and the size of the clusters. Numerical algorithms can be used for this optimization. This algorithm shares similarities with previously published methods [3, 4, 8]. The type of kernels that can be treated by the new approach has been extended significantly compared to [3, 4, 8]. It is probably the case that most kernels found in practical applications, including kernels known only numerically, can be treated by this method. An interesting aspect of this method is the fact that a diagonal multipole-to-local operator allows treating oscillatory kernels as well, for example of the type of $\exp(ikr)/r$. This is not the case for many black box or kernel independent FMMs.

References

1. Carr, J., Beatson, R., Cherrie, J., Mitchell, T., Fright, W., McCallum, B., Evans, T.: Reconstruction and representation of 3D objects with radial basis functions. In: Proceedings of the 28th annual conference on Computer graphics and interactive techniques, p. 76. ACM (2001)
2. Darve, E.: The fast multipole method: numerical implementation. Journal of Computational Physics **160**(1), 195–240 (2000)
3. Darve, E., Havé, P.: A fast multipole method for Maxwell equations stable at all frequencies. Philosophical Transactions: Mathematical, Physical and Engineering Sciences **362**(1816), 603–628 (2004)

4. Darve, E., Havé, P.: Efficient fast multipole method for low-frequency scattering. Journal of Computational Physics **197**(1), 341–363 (2004)
5. De Boer, A., Van der Schoot, M., Bijl, H.: Mesh deformation based on radial basis function interpolation. Computers and Structures **85**(11-14), 784–795 (2007)
6. Engquist, B., Ying, L.: Fast directional computation for the high frequency Helmholtz kernel in two dimensions. Arxiv preprint arXiv:0802.4115 (2008)
7. Fong, W., Darve, E.: The black-box fast multipole method. Journal of Computational Physics **228**(23), 8712–8725 (2009)
8. Greengard, L., Rokhlin, V.: A new version of the fast multipole method for the Laplace equation in three dimensions. Acta numerica **6**, 229–269 (2008)
9. Ying, L., Biros, G., Zorin, D.: A kernel-independent adaptive fast multipole algorithm in two and three dimensions. Journal of Computational Physics **196**(2), 591–626 (2004)

Tools for Multiscale Simulation of Liquids Using Open Molecular Dynamics

Rafael Delgado-Buscalioni

Abstract This work presents a review of recent tools for multiscale simulations of liquids, ranging from simple Newtonian fluids to polymer melts. Particular attention is given to the problem of imposing the desired macro state into *open* microscopic systems, allowing for mass, momentum and energy exchanges with the environmental state, usually provided by a continuum fluid dynamics (CFD) solver. This review intends to highlight that most of the different methods developed so far in the literature can be joined together in a general tool, which I call OPEN MD. The development of OPEN MD should be seen as an ongoing research program. A link between the micro and macro methods is the imposition of the external conditions prescribed by the macro-solver at or across the boundaries of a microscopic domain. The common methodology is the use of external particle forces within the so called particle buffer. Under this frame, OPEN MD requires minor modifications to perform state-coupling (i.e. imposing velocity and/or temperature) or flux exchange, or even any clever combination of both. This tool can be used either in molecular or mesoscopic-based research or in CFD based problems, which focus on mean flow effects arising from the underlying molecular nature. In this latter case an important goal is to allow for a general description of Non-Newtonian liquids, involving not only transfer of momentum in incompressible situations, but also mass and energy transfers between the micro and macro models.

1 Introduction

During this last decade the prefix *multi* has spread over many different disciplines, ranging from sociology to physics. In part, this is a consequence of dealing with new

R. Delgado-Buscalioni (✉)
Dpto. Física Teorica de la Materia Condensada, Universidad Autonóma de Madrid, Campus de Cantoblanco, Madrid E-28049, Spain
e-mail: rafael.delgado@uam.es

B. Engquist et al. (eds.), *Numerical Analysis of Multiscale Computations*, Lecture Notes in Computational Science and Engineering 82, DOI 10.1007/978-3-642-21943-6_7,

problems resulting from non-trivial interactions between entities of quite different nature. A natural approach to tackle these problems has been to design new methods from combinations of well-established theories. In this scenario *multiscale* has emerged as a new theoretical and computational paradigm in natural sciences. In particular, this work presents some tools for multiscale treatment of the liquid state. The general purpose is to connect the (classical) dynamics of an atomistic description of the liquid state (microscale) with other less involved descriptions, like the so called coarse-grained level, based on effective molecules (mesoscale) and with hydrodynamic and thermodynamic descriptions (macroscale). One can understand the different types of *multiscale* methods for fluids and soft condensed matter by dissecting the very term multiscale. First (abusing the latin root "multus") multi means at least two models which, might be solved concurrently (at the same time) or in a sequential (hierarchical) fashion (i.e. solve the fine description to extract information for the coarser level). The hierarchical strategy is in fact part of the coarse-graining methodology, which has usually been based on reproducing the essential microscopic structural (static) information. A recent challenge of coarse-graining is to incorporate dynamical information from the microscopic level [30]. On the other hand, concurrent coupling schemes deserve to qualify as *hybrids*. These hybrids can be divided in two types depending on how the space is decomposed. One can let the coupled models evolve in the same spatial domain or within different sub-domains. The first option is usually designed to treat solute-solvent flows: the solute (polymer, colloid, etc.) is solved with a particle approach while the solvent is treated using any preferred scheme (lattice Boltzmann [54], finite volume [9, 27], multiparticle collision dynamics [33], etc.). Depending on the flow regime, the solute-solvent coupling might be based on the Stokes friction force (low Reynolds number) or using more involved boundary conditions. By contrast, hybrids based on domain decomposition are required for many other types of problems which could depend on the interaction between a microscopic region and the outside hydrodynamic (or thermodynamic) state, or on how the stress is released by a microscopic model in a macroscopic flow (examples will be given later). Several types of domain decomposition strategies can be designed depending on the aspects of the multiscale research under study. In brief, there are two important issues (or categories) to be considered: first, the research might be focused either on the micro-dynamics or on the macroscopic level and second, the ratio of time scales for the *evolution* of the relevant micro and macroscopic process might be large $\tau_{mic}/\tau_{mac} \geq O(1)$ or small $\tau_{mic}/\tau_{mac} \ll 1$. I highlight "evolution" to warn about the fact that in a steady state $\tau_{mac} \to \infty$, so in practical terms, steady states can be grouped in the category of problems with *separation of time scales*. In liquids, the first category mentioned above separates continuum fluid dynamics (CFD) problems (such as polymeric fluid flow [5, 20, 38, 41, 56] or mean flow effects of singularities or defects in boundaries of micro or nano-fluidics [20, 29, 43]) from molecular based research (external flow effects on single molecules [2, 57], on membranes, melting or condensation [53] processes, wetting or sound-soft matter interaction [7, 11]). Logically, any macro-scale based research should be concerned about any possible gain in computational time from the separation of time and length scales, while if the molecular dynamics are the main concern this separation is irrelevant and the time-gap can only be reduced by molecular coarse-graining [30].

Most of the multiscale tools in this article originally come from molecular research but they can be quite useful in multiscale CFD programs because both share an important problem: how to impose the desired macro state on the micro domain. To frame this statement, let me briefly review some recent advances in CFD based multiscale research. An important class of domain decomposition hybrids is based on performing non-equilibrium microscopic simulations (using stochastic models, Brownian dynamics, molecular dynamics) at every (or some selected) nodes of a continuum solver grid. The local velocity gradient is imposed at each micro-solver box in order to measure the local stress used then to update the velocity field in the macro-solver. In the field of polymeric fluids this idea was probably introduced by Laso and Öttinger's CONFESSIT approach [38] and in recent years it has been continuously reappearing in the literature under many different flavours [5,41,52,56]. Two groups have set this multiscale approach in general mathematical frameworks, leading to the heterogeneous multiscale modeling (HMM) [21] or equation free models [34]. The HMM or equation-free formalism exploits time scale separation between micro and macro processes and gain computational time by sampling the micro-solver boxes over short temporal windows. However, the micro and macro clocks are the same so, after the macro solver is updated in time, the new dynamical state has to be imposed in the microscopic box. This operation (usually called lifting or reconstruction) might be a challenge in molecular simulation of liquids (see [39]). A clever asynchronous (multi-time) alternative, which avoids lifting, was recently proposed by E et al. [20] but indeed it also exploits time scale separation. Unfortunately, the interesting phenomena in complex liquids arise when the ratio τ_{mic}/τ_{mac} (i.e., Weissemberg or Deborah number, etc.) exceeds one. So in practise, gain in time can only be expected for flows of Newtonian liquids or probably to reach the steady state of a complex liquid flow at a faster pace. Spatio-temporal memory effects are essential in complex fluid dynamics, a relevant example being flow of polymer melts. Recent works have shown that the macro-solver is able to transmit spatial correlations between (otherwise independent) MD boxes [5,56] (i.e. gain in length) but, indeed, one needs to synchronize the micro and macro time updates (i.e. no gain in time). In polymers, local ordering effects induced by the trajectory of fluid packages can be important and difficult to implement. A possible solution is to use a Lagrangian-CFD solver [41] to feed the local state (velocity gradient) at each MD node.

A common feature of all these methods is that the microscopic domain is an open system which receives/send information from/to the outside world. Most of the CFD multiscale research have dealt with incompressible flows and thus the MD domains only need to receive momentum through their boundaries. However density and energy variations might be important in many kind of problems (e.g. thermal effects in sheared polymers) and as mentioned in recent works on the subject, some general and flexible formalism for "grand-canonical" molecular dynamics would be of great value [41]. In the same way, slip-stick motion at physical boundaries can only be described at molecular level and the tools described hereby could be deployed in a multiscale CFD scheme to solve this task (for instance, in unsteady polymeric flow under oscillatory walls [56]).

In what follows I will first describe a general formalism called OPEN MD, which enables to *open up* a molecular dynamic (MD) box so that it might exchange mass, momentum and energy with the outside world. Section 3 describes an adaptive resolution scheme acting as a coarse-grained particle interface model around the (atomistic) MD domain. This mesoscale interface makes feasible mass and momentum transfer in simulations of liquids made of large molecules. Section 4 discusses how to connect the molecular box with a continuum dynamics solver, including thermal fluctuations. Conclusions and perspectives are given in Sect. 5.

2 OPEN MD: Molecular Dynamics for Open Systems

The most delicate part of any hybrid scheme is the transmission of the state of the coarser description into the fine resolution model. The reason being that, in doing so, one needs to reconstruct microscopic degrees of freedom which should be consistent with the prescribed macroscopic state imposed. This task is sometimes called *one-way coupling*, lifting or reconstructing in the HMM and Equation-Free communities. The state of a solid is essentially given by the imposed stresses (forces) because the average velocity of a solid molecule is zero. By contrast, the thermo-hydrodynamic state of a gas require control over molecules velocity, as interaction forces are absent. An inherent complication in liquids is that their energy contains equal amount of kinetic and potential contributions and thus, control over both stresses and velocities is required. Of course, the stress and velocity fields are not independent and two strategies are possible: one can either choose to impose the average state variables (mean velocity, temperature) [40] or the fluxes of conserved variables (pressure tensor, energy flux) [7, 24]. In any case one is restricted to play with the set of microscopic mechanical quantities, namely, velocities and forces of the individual molecules of the system. As an aside, there are several popular methods to impose shear in *closed* MD boxes with periodic boundary conditions (BC), such as Lee-Edwards type BC or SLLOD dynamics used by many CFD multiscale works [20, 56]. Although they shall be not be reviewed here, a recent comparison between SLLOD and the type of boundary-driven imposition described hereby [31] showed some problems in SLLOD temperature homogenisation under shear. Also, alternatives based on Monte Carlo steps might also be possible [42], although the Metropolis algorithm does not preserves the system's dynamics and these will be not considered here either. Another relevant feature of liquids (and gases) is that they can be compressed, or when working with mixtures, vary in concentration. Compression effects may indeed arise from many different sources, such as sound transmission, or even shear rate pressure dependence in Non-Newtonian liquids. This means that, in general, one needs to devise some way to work with open molecular systems, i.e., with a variable number of molecules. A solution to this computational problem was proposed by Flekkøy *el al.* [24] some years ago and the idea, which I call OPEN MD, is still being generalised [17, 18]. In the original formulation of the OPEN MD scheme it is possible to impose the exact amount of work and heat into an open (variable density) molecular domain. OPEN MD has been used as the core of several particle-continuum hydrodynamic hybrids [7,18,37] but its range of applications is wider. In fact, by controlling the amount of work and heat introduced into an open system, one can study processes with different kinds of thermodynamic constraints. This makes OPEN MD a flexible method for many different fields ranging from confined systems [23] (where thermodynamic forces are driven by chemical potential gradients) to the rheology of Non-Newtonian liquids (where the normal pressure is known to depend on the imposed shear and constant volume measurements are not equivalent to constant pressure measures).

2.1 Open MD Setup

Figure 1 depicts a simple set-up of OPEN MD. A molecular system resolved by MD is extended with a buffer domain B, where the state of the outer domain is imposed. Particles are free to cross the hybrid interface but once inside B they will feel a certain external force $\mathbf{F}_i$ which should be prescribed to carry the desired information of the outside domain into the MD system. The objective of the original OPEN MD formulation [24] is to impose the desired momentum and heat flux ($\mathbf{P}$ and $\mathbf{J}_e$) across the so called *hybrid* interface H, i.e., it is based on flux-exchange. However the computational setup can be easily modified so as to impose the desired (external) velocity field $\mathbf{V}$ via constraint dynamics [43, 44] and in this paper I will unify both (flux and state coupling) approaches in the same framework.

The OPEN MD scheme can be divided in two main tasks:

1. Imposition of the desired macro-state via external forces and,
2. Control of mass and density profile at the buffer region.

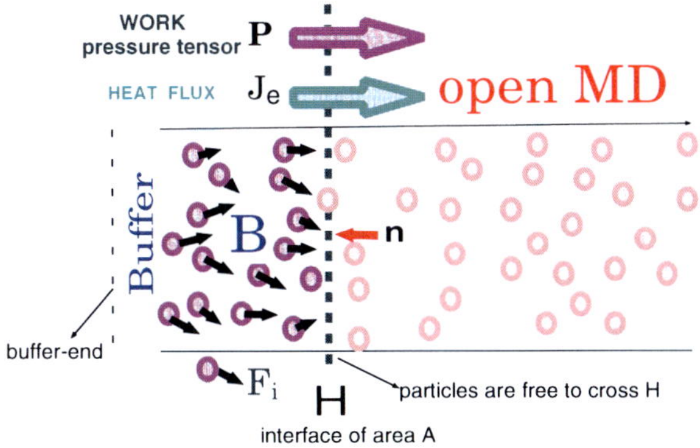

Fig. 1 Open molecular dynamics (OPEN MD) setup. A molecular dynamics box is extended using a buffer domain B where the state of the outer domain is imposed. In the flux-based scheme the external forces $\mathbf{F}_i^{ext}$ imposed to the buffer particles are prescribed so as to yield the desired momentum flux (stress tensor) $\mathbf{P}$ and heat flux $\mathbf{J}_e$ across the system interface H

2.2 Imposition of the Macro-State

As stated above the information prescribed at the buffer might either be a set of state variables (i.e. velocity, temperature) or the fluxes of conserved variables across H. A type of scheme based on variable coupling introduced by Patera and Hadjiconstantinou [28] use Maxwell daemons to modify the velocities of the buffer particles according to the local equilibrium distribution. This method is usually implemented in hybrids based on the Schwartz method, which alternatively imposes the local velocity of the adjacent domain at the overlapping layer until the steady state is reached. This is a good way to drive the (total particle+continuum) system towards a steady state (probably faster than its natural convergence rate) but it significantly alters the local dynamics (molecule diffusion, velocity time correlations) and it is restricted to closed systems (constant number of particles). Starting from the pioneer work of Thompson and O'Connell [44] several alternatives based on the imposition of external forces [12, 24, 43] were then developed. It is important to stress that external force imposition at the buffer allows for the implementation of either state and flux coupling. We shall now briefly discuss both approaches.

2.2.1 State-Coupling Based on Constrained Dynamics

The state-coupling approach comes from the continuum fluid dynamics community whose priority is to ensure that the external flow and convective forces are imposed into the molecular region. In this sense, the philosophy behind state-coupling is to treat the coarse (hydrodynamic) exterior domain as the *master* model and the microscopic dynamics as *slaved* one. Let us begin with the momentum transfer, which is carried out by imposing the desired (external) average velocity $\mathbf{V}$ at the particle buffer, i.e.

$$\frac{1}{N_B}\sum_{i\in B}\mathbf{v}_i = \mathbf{V}. \tag{1}$$

This might be seen as a constraint in the particle equations of motion $\ddot{\mathbf{r}}_i = \mathbf{f}_i/m$,[1] which can be written in terms of an external force $\mathbf{F}_i$ added to $\mathbf{f}_i$. An example is the Langevin type force used by O'Connell and Thompson [44] $\mathbf{F}_i = -\gamma\,(\mathbf{v}_i - \mathbf{V}) + \tilde{\mathbf{F}}$, with $\langle\tilde{\mathbf{F}}(t)\tilde{\mathbf{F}}(0)\rangle = 2k_B T\gamma\delta(t)$. Nie et al. modified this approach and proposed the following constrained dynamics at the buffer,

$$\ddot{\mathbf{r}}_i = \frac{1}{m}\left(\mathbf{f}_i - \langle\mathbf{f}\rangle\right) - \xi\left(\langle\mathbf{v}\rangle - \mathbf{V}\right), \tag{2}$$

where I have introduced the local microscopic average $\langle\mathbf{f}\rangle = \sum_i^{N_B}\mathbf{f}_i/N_B$ and the relaxation parameter ξ which, in principle, can be freely tuned. The idea underlying this approach is to substitute (at each time step) the average microscopic force $\langle\mathbf{f}\rangle$ at

---

[1] In what follows upper case letters indicate externally imposed quantities ($\mathbf{V}, \mathbf{F}$) while lower case ($\mathbf{v}_i, \mathbf{f}_i$) stands for microscopic variables.

the coupling domain for its macroscopic counterpart $MD\mathbf{V}/Dt$. In other words, by summing (2) over $i \in B$ one gets the total external force at the buffer $\xi M\,(\mathbf{V} - \langle \mathbf{v} \rangle)$, where $M = mN_B$ is the buffer mass. Thus, by choosing $\xi = \frac{1}{\Delta t_{MD}}$, as Nie *et al* [43] did, it is easy to show that the average microscopic velocity instantaneous relaxes to the imposed value, i.e., $\langle \mathbf{v} \rangle(t + \Delta t_{MD}) = \mathbf{V}(t + \Delta t_{MD})$. Instantaneous relaxation destroys the microscopic dynamics (altering the velocity time correlation) and this can be alleviated by increasing ξ (as originally proposed in [44]). If the imposed velocity changes (fast) in time, the price to pay is some lag (time delay) between the input $\mathbf{V}(t)$ and output $\langle \mathbf{v} \rangle(t)$ values.

Mass and energy transfer

In state-coupling methods the mass flux arising from the microscopic dynamics is in fact destroyed once the average local velocity at the system boundaries is *imposed*. In other to ensure mass continuity at the interface one thus needs to take the information from the coarser level (usually the Navier-Stokes equation), and modify the number of particles on the molecular system to an amount given by the continuum expression for the mass flow across H, $A\rho\mathbf{V}\cdot\mathbf{n}\Delta t/m$. Energy transfer might also be introduced in a state-coupling fashion by imposing the local temperature gradient at the buffer domain [12]. Particle-continuum hybrids can also impose heat transfer via temperature coupling by using a larger buffer (overlapping domain) with two parts: the local "continuum" temperature is imposed at the microscopic buffer, while the local microscopic temperature is imposed at the (adjacent) boundary of the continuum macro-solver [40]. In general, the use of unphysical artifacts (such as pure velocity rescaling [40] to impose a local temperature) introduces several drawbacks: for instance, transport coefficients (viscosity, thermal conductivity) need to be finely calibrated to control the amount of heat transferred via velocity and temperature gradients. Also, the length of the buffer (or the overlapping domain) will need to be increased so as to avoid error propagation into the MD domain, thus paying a larger computational price for the hybrid coupling.

2.2.2 The Flux-Based Scheme

The flux-coupling approach tries to reduce the number of unphysical artifacts at the buffer by retaining all possible information from the microscopic domain (e.g. fluctuations). In fact, hybrid models using flux exchange consider the coarse (hydrodynamic exterior) domain as the *slave* model while the microscopic dynamics stand the *master* model (see e.g. [1]). This approach permits, for instance, coupling of molecular dynamics and fluctuating hydrodynamics (FH) [7]. It should be the preferred one when dealing with problems where thermal fluctuations or molecular transport are relevant and they usually are at these nanoscopic scales and low or moderate Reynold numbers. The flux boundary conditions imposed at the buffer domain are specified by the normal component of the energy flux $j_\epsilon = \mathbf{J}_e \cdot \mathbf{n}$ and the normal component of the momentum flux $\mathbf{j}_p = \mathbf{P}\cdot\mathbf{n}$, where $\mathbf{P}$ is the pressure tensor and $\mathbf{n}$ the unit normal shown in Fig. 1. Both fluxes will in general include advective terms. In an open system, energy and momentum enters into the particle system both through the force $\mathbf{F}_i$ and via particle addition/removal. The prescribed momentum and energy fluxes need to take into account both effects i.e.

$$\mathbf{j}_p A dt = \sum_i \mathbf{F}_i dt + \sum_{i'} \Delta(m\mathbf{v}_{i'}), \tag{3}$$

$$j_\epsilon A dt = \sum_i \mathbf{F}_i \cdot \mathbf{v}_i dt + \sum_{i'} \Delta\epsilon_{i'}, \tag{4}$$

where i' runs only over the particles that have been added or removed during the last time step dt, and A is the buffer–bulk interface area. The momentum change is $\Delta(m\mathbf{v}_{i'}) = \pm m\mathbf{v}_{i'}$, where $(+)$ corresponds to inserted particles and (-) to removed ones (similarly for the energy change $\Delta\epsilon_{i'}$). The sums $\sum_i \mathbf{F}_i dt$ and $\sum_i \mathbf{F}_i \cdot \mathbf{v}_i dt$ are the momentum and energy inputs due to $\mathbf{F}_i$ during the time dt. In order to simplify (3) and (4) it is useful to define $\tilde{\mathbf{j}}_p$ and $\tilde{j}_\epsilon$ through the relations

$$A dt \tilde{\mathbf{j}}_p = A dt \mathbf{j}_p - \sum_{i'} \Delta(m\mathbf{v}_{i'}) = \sum_i \mathbf{F}_i dt\ , \tag{5}$$

$$A dt\, \tilde{j}_\epsilon = A dt\, j_\epsilon - \sum_{i'} \Delta\epsilon_{i'} = \sum_i \mathbf{F}_i \cdot \mathbf{v}_i dt\ . \tag{6}$$

Provided that the force $\mathbf{F}_i$ satisfies these conditions the correct energy and momentum fluxes into the particle system will result. To solve these set of equations the external force is decomposed into its average and fluctuating parts, $\mathbf{F}_i = \langle\mathbf{F}\rangle + \mathbf{F}'_i$. Momentum is introduced by the average component of $\mathbf{F}_i$ and thus,

$$\langle\mathbf{F}\rangle = \frac{A}{N_B}\tilde{\mathbf{j}}_p\ , \tag{7}$$

where $N_B(t)$ is the total number of particles receiving the external force at a given time t and A is the area of the interface H. On the other hand, the fluctuating part of the external force ($\sum_i \mathbf{F}'_i = 0$) introduces the desired heat via dissipation, i.e.,

$$\mathbf{F}'_i = \frac{A\mathbf{v}'_i}{\sum_{i=1}^{N_B} \mathbf{v}'^2_i}\left[\tilde{j}_\epsilon - \tilde{\mathbf{j}}_p \cdot \langle\mathbf{v}\rangle\right], \tag{8}$$

where we have used the fact that the total energy input by external forces is $\sum_{i=1}^{N_B} \mathbf{F}_i \cdot \mathbf{v}_i = N_B \mathbf{F}\cdot\langle\mathbf{v}\rangle + \sum_{i=1}^{N_B} \mathbf{F}'_i \cdot \mathbf{v}'_i$.

As shown in [24], by *exactly* controlling the amount of work and heat introduced into the particle system, one can implement several sorts of thermodynamic constraints. For instance, constant chemical potential (grand canonical ensemble), constant enthalpy, constant pressure. Also, steady (or unsteady) non-equilibrium states, such as constant heat flux or shear stress can be imposed. An interesting aspect of this OPEN MD method is that all these constraints occur at the particle boundaries (in fact, as it happens in any real system), so the dynamics of the molecular core are not unphysically modified *whatsoever*.

Mass flux

In the OPEN MD flux-based scheme, the mass flux across H naturally arises as a consequence of the pressure (or chemical potential) differences between the inner and outer domains. In other words, mass flux is not imposed. In a hybrid configuration, the microscopic domain will dictate the mass flux across H, which can be simply measured by counting the number of particles crossing the interface. Indeed, many problems crucially depends on the molecular transport (such as confined systems driven by the chemical potential difference between the interior and exterior) and the natural approach of the flux-based scheme permits one to retain this sort of microscopic information (e.g. fluctuations).

2.3 *Mass and Density Profile at the Buffer*

In a molecular simulation of an open fluid system one necessarily needs to decide what to do at the edges of the simulation box. The essential problem is to control the density profile normal to the interface H. Thus, two variables needs to be monitored: *i)* the shape of the density profile and *ii)* the total number of particles at the buffer.

2.3.1 Distribution of the External Force

The density profile depends on how the external force **F** is distributed at the buffer. For an interface of area A, pointing in the negative x direction, $\mathbf{n} = -\mathbf{i}$ (see Fig. 1), the force along the α direction on a buffer particle can, in general, be set as,

$$F_{i,\alpha} = \frac{g^{\alpha}(x_i)}{\sum_{i \in B} g_{\alpha}(x_i)} A P_{x\alpha} \quad \alpha = \{x, y, z\}, \tag{9}$$

where $P_{x\alpha}$ is the $x\alpha$ component of the pressure tensor (or any other total external force such as the one used in state-coupling). Although, most of the works done so far use a single distribution $g(x)$ for all directions, one is free to use different distributions $g_{\alpha}(x)$. In fact, depending on the problem, it might be useful to choose different shapes of g for tangent (shear) and normal forces (pressure).

Most of the concern in the literature on this subject logically corresponds to the shape of g_x (normal to the interface) because it directly determines the shape of the density profile $\rho(x)$. Figure 2 shows a qualitative picture on the relationship between $g(x)(= g_x(x))$ and $\rho(x)$.

Several values of g_i have been proposed in the literature. For instance, Flekkøy *et al.* [25] initially proposed a distribution $g(x)$ which tends to zero at H and diverges at the end of the particle buffer thus preventing particle to leave the system. Werder and Koutmoutsakos [55], showed that an evaluation of $g(x)$ from a previous calculation of the particle distribution function in the normal x coordinate, enables to maintain a constant density profile across the whole buffer (a comparison between several choices of g_i was also provided). Recently the group of Koutmousakos [36] introduced a feedback (relaxation) procedure to self-adapt a binned distribution according to the local density gradients, in such way that the fixed solution were the constant density profile. In some situations, such as the state-coupling approach, it is important to have a constant density profile at the buffer [36]. However, in flux-based schemes the most important aspect is to have a flat profile locally *across the interface* H [14] to avoid any spurious current.

Energy transfer and $g(x)$

It is important to note that flux based schemes implement the energy transfer via the power dissipated by the external force $\sum_{i\in B}\mathbf{F}_i\cdot\mathbf{v}_i$. Therefore, in this case, one is not free to choose $g(x)$ because heat will be produced in an uncontrolled way, at a rate $\sum_\alpha\sum_i g_\alpha(x_i)P_{x\alpha}v_{i,\alpha}/\sum_\alpha\sum_i g_\alpha(x_i)$. In fact when using any $g(x)$ with sharp gradients, strong thermostatting will be required to remove all this spurious heat. This is probably be the case of the state-coupling schemes based on temperature

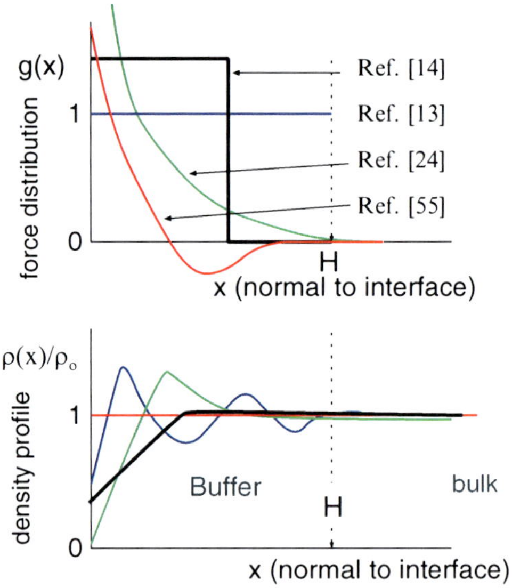

Fig. 2 Qualitative diagrams illustrating some types of external force distribution $g(x)$ used in open molecular dynamics (*top*) and the resulting density profile at the buffer (*bottom*)

imposition [40]. In order to keep control of the average energy dissipated by the external force, a pioneer work on energy transfer in hybrids [12] used $g(x) = 1$ and placed a couple of adjacent thermostats to transfer heat. The need of thermostats was finally avoided by the flux boundary condition method of [24], leading to (7) and (8) above. Note that (7) uses $g(x) = 1$ (or at least a step function, see Fig. 2) to distribute the mean external force. In this way the energy contribution of the mean external force is precisely the rate of reversible work done by the external forces $P_{xx}\langle v_x\rangle$, plus the rate of heat dissipation by shear forces $\sum_\alpha P_{x\alpha}\langle v_\alpha\rangle$, with $\alpha \neq x$. The entropic heat production is separately furnished by the fluctuating part of the external force.

2.3.2 The Buffer Mass: Particle Insertion and Deletion

The buffer domain can be understood as a *reservoir* which represents the outside world. This means that the number of particles at the buffer should be large enough to avoid important momentum and temperature fluctuations, which will certainly lead to numerical instabilities. In a typical flux-based method this means that, in average, the buffer should contain at least $\langle N_B\rangle \sim O(10^2)$ particles; which is not much considering that a typical MD simulation may contain $O(10^{[4-5]})$ or more. A simple way to keep the average $\langle N_B\rangle$ under control is to use a relaxation equation

$$\dot{N}_B = \frac{1}{\tau_B}\left(\langle N_B\rangle - N_B\right), \tag{10}$$

where $\tau_B \sim 10^2 \Delta t_{MD}$. As times goes on a number $\mathcal{N} = \mathrm{INT}[\Delta N_B]$ of particle insertions (or deletions) should be performed as soon as $\Delta N_B = \sum_i \dot{N}_B(t_i)\Delta t_{MD}$ becomes a positive (or negative) number with absolute value larger than one. Basic bookkeeping should then be performed to update N_B accordingly. Several possible tricks might be done with those particles reaching the buffer-end: one can just delete them randomize or reverse their velocity. In a conservative (flux-based) scheme the resulting momentum exchange should be accounted for in (3) (e.g. the later case would yield $-2\Delta m\mathbf{v}'_i \cdot \mathbf{n}$ per reversed particle). To minimise perturbations, particle deletions and insertions are usually done at the dilute region of the buffer (see Fig. 2), whenever it exists. A particularly delicate issue when dealing with open MD simulations of dense liquids is to avoid overlapping upon insertion (which results in extremely high energy jumps). In a pioneer work on open MD simulations we introduced a fast and efficient particle insertion method called USHER, now used in many hybrid particle-continuum simulations [35, 37]. It was initially designed for spherical (Lennard-Jones) particles [13] and then extended to polar molecules, such as water [8]. USHER is based on a Newton-Raphson algorithm with adaptable length step, which search some location within the complex multiparticle energy landscape where potential energy of the inserted particle equals the desired value. USHER solves this problem in a very efficient way (partly because it reboots any search once some increase in potential energy is performed) and can also explore low energy domains in the search for (however biased) chemical potential evaluation [49]. In open MD simulations it usually represents less than 5% of the computational task. The main limitation of USHER is that it is not suitable to insert big (or I should rather say complex) molecules. Typical examples could be star polymers in a melt. This limitation was sorted out recently [18] by introducing an adaptive coarse-grained layer at the buffer, whereby a coarse-grained version of the molecule is gradually and nicely decorated with its atomistic complexity as it enters into the MD core from the buffer corner, and vice versa. Implemented in a particle-continuum hybrid, this suggestive idea permits a macro-meso-micro zoom along the spatial coordinates, and has been called *triple-scale* method. Let us now comment on this approach.

3 Using Adaptive Resolution: The Mesoscopic Interface

The adaptive resolution scheme (AdResS) proposed by Praprotnik et al. [47, 48] is a type of domain decomposition based on coupling particle sub-domains with different resolution: from coarse-grained cg to explicit (i.e. atomistic) ex description. Figure 3 illustrates this idea in an open MD setup. The number of degrees of freedom of a molecules is modified (reduced/increased) as it crosses the "transition" layer, where a hybrid model (*hyb*) is deployed. In particular, the force $\mathbf{f}_{\alpha\beta}$ acting between centres of mass of molecules α and β is expressed as,

$$\mathbf{f}_{\alpha\beta} = w(x_\alpha)w(x_\beta)\mathbf{f}^{ex}_{\alpha\beta} + [1 - w(x_\alpha)w(x_\beta)]\mathbf{f}^{cg}_{\alpha\beta}, \tag{11}$$

where x_α and x_β are the molecule's position along he coupling coordinate. Pairwise atomic forces are $\mathbf{f}_{ij}$ and $\mathbf{f}^{ex}_{\alpha\beta} = \sum_{i\in\alpha, j\in\beta} \mathbf{f}_{ij}$ is the sum of all atomic interactions between molecules α and β. Finally $\mathbf{f}^{cg}_{\alpha\beta} = -\nabla_{\alpha\beta} U^{cg}$ results from the coarse-grained intermolecular potential. These interactions are weighted by a function $w(x)$ which switches from $w = 1$ at the ex region to $w = 0$ at the cg layer. Intermediate values $0 < w < 1$ might be understood as hybrid (hyb) model. With a few restrictions, one is rather free to choose the explicit form of $w(x)$, see e.g. [17, 47]. The great benefit of this sort of *on-the-fly* transition from coarse-grained to atomistic models is that molecule insertions can be quite easily performed at the cg end of the buffer because there, intermolecular interactions are soft. The whole set of hard-core atomic potentials is thus avoided in the open MD setup.

The key ingredient of AdResS is that (11) conserves momentum. Thus, it can be used in combination with any momentum conserving (flux-based) scheme [17, 18], and of course, it could be also combined in any state-coupling method. However, energy is not conserved by (11) and in fact, as a molecule moves towards the coarse-grained layer it looses all the kinetic energy associated with its internal degrees of freedom (sum of squared velocities w.r.t. centre-of-mass). This energy is lost forever and to maintain an equilibrium state (a flat free energy profile [46, 48]) it needs to be furnished by a thermostat, which usually is set to act along the whole simulation box (or at least within the buffer). A modification of AdResS solving this problem would certainly be an important contribution. Meanwhile, it might be still possible to allow for (averaged) energy exchange with the MD core, using some thermostatting tricks, although this idea is not still published.

Other issues which deserved some attention in the literature [17, 32] are related to how the change in resolution introduces differences in mass diffusion coefficient and viscosity in the *cg*, *hyb* and *ex* fluid models. This is a problem in a "pure" (i.e. closed) AdResS simulation and also if, for some reason, one is interested in placing the hybrid interface H of the open MD setup within the *cg* layer (note that Fig. 3 places H within the *ex* domain). In these cases one needs to take care to calibrate all the *cg*, *hyb* and *ex* viscosities and diffusion coefficients. This is, in general, not possible: in fact, either diffusivities or viscosities can possibly be matched at once [17] using either position dependent Langevin thermostats or DPD thermostat with variable tangential friction [32]. Liquid equilibrium structures (radial distribution functions $g(r)$) of the *cg* model can also be tuned to fit the *ex* one, using the standard tools [50]. This adds an extra pre-computational price to pay. However, as shown in [18] all this calibration burden (which needs to be repeated each time the thermodynamic state is changed) can be greatly alleviated by using the sort of setup illustrated in Fig. 3. In summary, variations in transport coefficient and fluid structure of the different fluid models within the buffer do not affect the proper transfer of momentum across H (which is guaranteed by fulfillment of the third Newton Law across all layers).

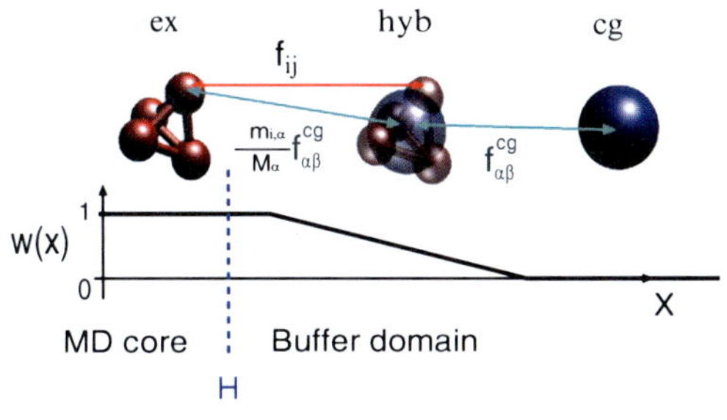

Fig. 3 Schematic setup of the adaptive resolution scheme (AdResS) being used within the buffer domain of an open MD simulation of a tetrahedral liquid. Molecules gradually transform from a coarse-grained cg to an explicit ex (atomistic) representation, as they cross the transition layer, hyb. The pairwise forces between atoms $\mathbf{f}_{ij}$ and molecules centre-of-mass $\mathbf{f}^{cg}_{\alpha\beta}$ are weighted by $w(x)$

4 HybridMD: Particle-Continuum Hybrid Scheme

This section discusses the most important details about the implementation of a particle-continuum hybrid based on domain decomposition. For a more complete view of each different method the reader is referred to the original papers cited hereby and references therein. There are (at least) three types of approaches to this problem: state coupling [43,44], flux coupling [14,25] and velocity-stress coupling [52, 56]. How to couple time marching algorithms of macro and micro solvers is a common problem for all kind of hybrids. The following is a brief discussion on this subject (see [20] for recent developments).

4.1 Time Coupling

In general, the time steps of the micro and macro solver (respectively δt and Δt) satisfy $\delta t \leq \Delta t$. However, the ratio of both quantities depends on the type of models to be coupled. For instance in hybrids of deterministic CFD and MD one can choose $\Delta t \gg \delta t$ but, solving fluctuating hydrodynamics (FH) requires much smaller time steps and Δt is only few time larger than δt [7]. On the other hand, communications between models occur after a certain time coupling interval Δt_c, which in general $\Delta t_c \geq \Delta t$ (an example of $\Delta t_c > \Delta t$ is discussed in [14]). Figure 4 illustrates some time coupling protocols. Concurrent algorithms (Fig. 4a) permit parallelization (tasks 1a and 1b) and might be quite useful if the computational load of micro and macro solvers is balanced by the implemented architecture (for instance use a fast GPU solver [6] for the MD domain and a slow CPU solver for a vast CFD region). Indeed, parallelization is most easily achieved if the need for performing averages in the micro domain are completely or substantially avoided. Examples are FH-MD hybrids [7] where the exchanged quantities are the actual MD and FH fluctuating variables at each coupling time (see Fig. 4a). Another example is the asynchronous time coupling devised by Weinman et al. which (if the signal-to-noise ratio is large enough) can work without explicit averaging because it is indirectly implemented in the deterministic macro-solver updates. Fluctuations are usually considered a nuisance in mean flow CFD problems [20,39] and microscopic averages might become necessary. Deterministic CFD-MD hybrids thus need to consider time synchronisation errors arising from performing MD averages lagging behind the coupling time (see Fig. 4b). These are $O(\Delta t_c)$ errors which might be significant if Δt_c is large compared with some relevant microscopic relaxation time. A possible solution, shown in Fig. 4c, consists on shifting the discretised time mesh of both models. The scheme of Fig.4d is an example of the synchronous coupling used in HMM-type schemes where the micro-solver is sampled during small time intervals, compared with the coupling time, and then lifted or reconstructed towards the updated state. This lifting operation might be a problem in MD of liquids so the new recent seamless multiscale asynchronous scheme, which avoids the MD reconstruction step [20], is particularly suited for solving flows where time separation applies, $\tau_{mic} \ll \tau_{mac}$.

4.2 Hybrids Based on State Coupling

State coupling relies on Dirichlet boundary conditions. This statement is valid either for hybrids based on the Schwartz scheme [28, 37] or for constraint dynamics [43]. I will focus on the constraint dynamics approach, whose typical setup is shown in Fig. 5a. The MD and CFD (finite differences in Fig. 5a) domains are connected in an overlapping region, sometimes called *handshaking* domain. The state variables of each model are mutually imposed at two different cells $P - C$ and $C - P$. At the $P - C$ cells the local average microscopic velocity $\langle \mathbf{v}_{PC} \rangle$ is imposed to the continuum as a Dirichlet BC, while at the $C - P$ cell the continuum velocity $\mathbf{V}_{CP}$ is imposed to the particle system, using the scheme explained in Sect. 2.2.1. The same strategy is for the imposition of local temperatures, so as to simulate an energy exchange between both models (see [40] for details). It is important to note that the $P - C$ and $C - P$ cells are some cells apart in order to let the particle system relax from all the dynamic constraints imposed at the $C - P$ cell (instantaneous velocity jumps of (2) and rescaling of peculiar velocity towards the desired temperature [40]). As an example in [40] the (linear) length of the overlapping domain is 4 cells of $\Delta x = 6.25\sigma$, thus a total of 25σ (where σ is the particle radius). When dealing with two or three dimensional flows, this is a relatively large computational load for the handshaking region. Molecular dynamics

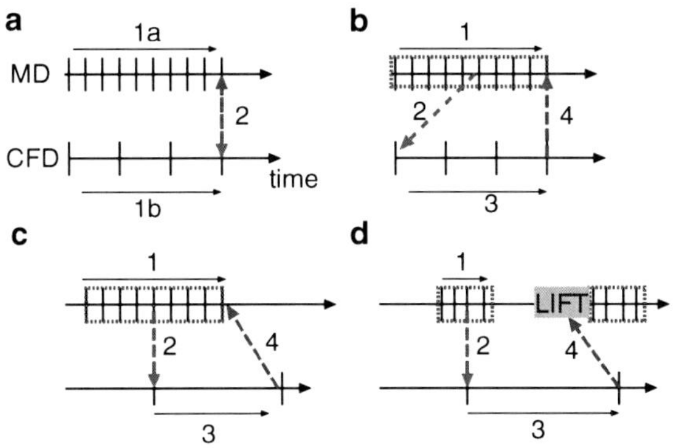

Fig. 4 Some possible (synchronous) time coupling schemes in hybrids. Horizontal arrows indicate time axis of each model (here molecular and continuum fluid dynamics, MD and CFD) and vertical lines their time steps. A dashed square means a time average operation and dashed arrows communications between models. Tasks are numbered in chronological order. (**a**) Concurrent coupling allowing parallelization. (**b**) Sequential coupling. (**c**) Sequential scheme avoiding time lag in MD averages. (d) HMM-type sequential coupling with a lifting step (4) to set the advanced state into the MD system

is by far the most expensive part of the hybrid algorithm and being able to reduce the length of the overlapping domain is a benefit one should take into account.

4.3 Hybrids Based on Flux Exchange

If our hybrid is going to be based on flux exchange, the most natural choice for continuum solver is one based on an integral conservative scheme, such as the finite volume method (see Patankar for an excellent textbook [45]). From the continuum solver perspective communications between continuum and molecular cells are essentially the same as those among continuum cells; i.e. there are only little modifications to do. Following the standard procedure of the finite volume method, the *whole* domain is divided into computational cells (see Fig 5c) which could be either w =boundary cells (walls) or f =fluid cells. The hybrid scheme introduces two more cell types, the m = molecular cells and the C cells. A conservative scheme simply calculates and sum up the amount of any conserved variable crossing the interface between every pair of adjacent cells. In particular $\Delta\Phi_H$ is the amount of Φ crossing the hybrid interface H over the coupling time Δt_c. The interface H separates cells C and the border molecular cells, sometimes called P (see Fig. 5b). If the flux across H is $J_H = \mathbf{J}_H \cdot \mathbf{n}$ then $\Delta\Phi_H = A J_H \Delta t_c$. Thus the central quantity is J_H: it will be imposed at the particle buffer following Sect. 2.2.2 and used to update the C cell, in the standard finite volume fashion. The interface flux J_H might be evaluated in different ways: one can perform a linear piecewise interpolation $J_H = (J_C + J_P)/2$ and evaluate J_P from microscopic expressions (Irving-Kirwood) or pass via the fluid constitutive relations using the hydrodynamic variables at the surrounding (fluid and molecular) cells into the selected type of discretised gradient.

Molecular-continuum hybrids are explicit in time and the time marching protocol is usually simple. Typically (Fig. 4a) the macro-solver updates all types of cells during a number $n_{FH}\,\Delta t_c/\Delta t$ of time steps[2] and at each coupling time it receives the hydrodynamic variables at the molecular m cells. The only modifications required on a standard code are set to ensure the mass conservation and momentum continuity at the C cells . The continuum solver structure is,

$$\Delta\Phi_i = \Delta t \mathtt{NS}\left[\{\Phi_j\}\right] + \delta_{fC}\Delta\phi^{MD}, \tag{12}$$

where Φ are the set of conserved variables (mass, momenta and energy), NS is a discretised Navier-Stokes operator (which may include hydrodynamic fluctuations [9, 19]) and the delta Kronecker δ_{fC} symbol is used to input the molecular flux corrections $\Delta\phi^{MD}$ into the C cells, as explained below.

[2] This number should not be large $1 \leq n_{FH} < O(10)$ [14].

4.4 Mass Transfer and Continuity in Flux Based Schemes

As stated above, the flux scheme permits one to minimise the number of unphysical artifacts imposed on the particle system by using the molecular system as the fundamental (or master) model which determines the mass flux and velocity at the interface. This philosophy was first proposed by Garcia *et al.* in an elegant hybrid model for gases [26]. Mass conservation is ensured by evaluating the molecular mass flux across H, ΔM_H^{MD} and releasing this mass to the C cell. A relaxation equation can be used for this sake, providing the following mass correction which needs to be added at C, according to (12),

$$\Delta M^{MD} = \frac{\Delta t_c}{\tau_M} \left(\Delta M_H^{MD} - \Delta M_H^{NS} \right). \tag{13}$$

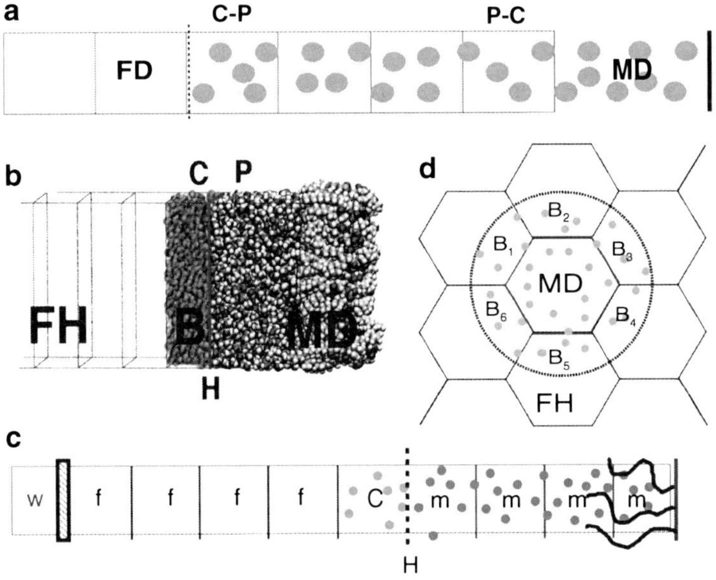

Fig. 5 (**a**) Typical setup used in state-coupling hybrids with a molecular dynamics (MD) and a deterministic Navier-Stokes finite difference (FD) solver domain. The overlapping domain contains $C - P$ cells where the local FD velocity is imposed to MD and $P - C$ cells where the average particle velocity is used as Dirichlet B.C. for the FD scheme. (**b**) Set-up of a flux based coupling MD (water wetting a lipid monolayer) and finite volume fluctuating hydrodynamics (FH). Exchange of fluxes between cells P∈MD and C∈FH are made through their interface H (no overlapping domain). (**c**) Arrangement of finite volume cells in a hybrid flux scheme (see text) (**d**) A possible 2D MD-FH flux-based coupling in an hexagonal lattice. Local pressure tensors and heat fluxes at each neighbour FH cell are imposed to MD from each sub-buffer B_i, flux exchange take place across each H_i interface

Here, τ_M is a relaxation time which usually can be set equal to the C-solver time step (instantaneous mass transfer and exact conservation) and $\Delta M_H^{NS} = -A\rho_H \mathbf{V}_H \cdot \mathbf{n}\, \Delta t$ is the mass crossing towards C according to the local hydrodynamic prediction (the NS solver).

On the other hand, a pure flux scheme does not impose velocity continuity and it has been shown to suffer from numerical instability, leading to velocity jumps at the interface [16, 51]. A simple solution to this problem, proposed in [16], is to add an extra relaxation term $M_C \Delta \mathbf{V}_C$ into the C cell momentum equation, where

$$\Delta \mathbf{V}^{MD} = \frac{\Delta t_c}{\tau_v} \left(\langle \mathbf{v}_C^{MD} \rangle_{[\delta t, \tau]} - \langle \mathbf{V}_C \rangle_{[\Delta t, \tau]} \right). \tag{14}$$

The relaxation time can usually be set to be rather fast $\tau_v \sim O(100)$ fs and $\mathbf{v}_C^{MD}$ is obtained from linear extrapolation of adjacent m cells (i.e. not from the buffer). Note that (14) is essentially the same idea used in the constrained molecular dynamics of the state-coupling hybrid, but here, it is the continuum velocity at the boundary C cells which is "constrained" to follow the molecular counterpart. A comparative study of continuity in several hybrids performed by Ren [51] confirmed the robustness of this approach. The averages used in (14) take into account the possibility of coupling two models with intrinsic fluctuations (such as FH and MD). I have defined $\langle \cdot \rangle_{[\delta t, \tau]}$ as a time average over τ sampling each δt steps. In a FH-MD hybrid τ is usually the coupling time Δt_c while Δt and δt are FH and MD time steps.

Exact momentum conservation

From the standpoint of momentum conservation, it is important to note that the particle buffer B is not part of the system. According to the OPEN MD procedure of Sect. 2.2.2, imposing the pressure tensor $\mathbf{P}_H$ at the buffer B will inject $A\mathbf{P} \cdot \mathbf{n} \Delta t_c$ momentum into MD+B, but one does not know how much of it will be transferred into MD across H. This source of momentum error is bounded by the mass of B (and was shown to be quite small in 1D coupling geometries [7]); however a slight modification of the scheme allows for exact momentum conservation. This might be necessary when dealing with two and three dimensional geometries (see e.g. Fig. 4d). The idea is to adjust the transfer towards each C cell so as to ensure global conservation along the interface contour. Consider Fig. 4b, the MD model is the first to move so $\Delta \phi^{MD} = \phi_{MD}(t_1) - \phi_{MD}(t_0)$ is known before the FH field is updated (in the concurrent scheme of Fig. 4a, the MD correction will be just transferred at the next time step). Local conservation means that $-\Delta \phi^{MD}$ crosses towards the C cell. In a general setup (see e.g. Fig. 5d), the interface H is divided in $h = \{1, \ldots, \mathcal{N}_H\}$ surface portions, each one h, facing a different C_h cell. We require conservation over the whole contour of the hybrid interface,

$$\sum_{h \in H} \Delta \Phi_h = -\Delta \phi^{MD}, \tag{15}$$

where $\Delta\Phi_h$ is the amount crossing the portion h of the interface H towards the corresponding C_h cell. The corresponding hydrodynamic prediction is $\Delta\Phi_h^{\text{pred}} = -A_h \mathbf{J}_h \cdot \mathbf{n} \Delta t_c$ where $\mathbf{J}_h$ the local flux, $\mathbf{n}$ points outwards C and A_h is the area of the h portion of H. The overall disagreement with respect to the molecular value is just,

$$E[\Phi^{\text{pred}}] = \left[(-\Delta\phi^{MD}) - \sum_{h \in H} \Delta\Phi_h^{\text{pred}} \right], \tag{16}$$

and in order to fulfill the conservation constraint (15) the transfer across each portion h of the H interface is corrected with,

$$\Delta\Phi_h = \Delta\Phi_h^{\text{pred}} + \frac{1}{\mathcal{N}_{\mathcal{H}}} E[\Phi^{\text{pred}}]. \tag{17}$$

5 Conclusions and Perspectives

In writing this review I realised that the number of papers including the prefix *multi* has boomed in recent years. It might well be that like in many other disciplines (art is an example), the stamp "multi" is able to promote some works with no real significance. For instance, in many processes continuum liquid hydrodynamics are known to remain valid up to quite small length scales [10, 12], thus making useless an hybrid particle-continuum, CFD based, approach (this is not the case for rarefied gases [1]). However, after the initial phase of "topic heating" the main relevant ideas and application fields will soon settle down. In my opinion, multiscale techniques for molecular liquid modeling will become a standard tool in commercial or open source packages (see [3] for recent work in this direction). The state of the art will soon benefit from modern faster parallel computing in cheaper architectures, which may also be grid-distributed [4]. To this end, the multiscale algorithms should allow for maximum flexibility with minimum computing modifications. It is easy to imagine that a farm of parallel MD simulations talking with a single macro-solver in a velocity-stress coupling scheme will soon permit to solve unsteady flow of non-Newtonian liquids with the desired molecular structure and (molecular) boundaries. For certain applications these MD simulations will necessarily need to describe open systems of nanoscopic size, evolving with the minimum amount of unphysical artifacts. The present review intends to highlight that a single computing framework should be able to allow for a flexible formulation of this sort of open molecular dynamics, which for instance, can combine state and flux coupling in the same hybrid scheme (see e.g. [19]).

This review is clearly not complete and some issues have been omitted for the sake of space. Other recent reviews can be found in [35,37]. Some comments should have been made on the tests required at each level of description [14, 18] (molecular structure, radial distributions transport coefficients, fluctuations, hydrodynamics and thermodynamics) or how to couple fluctuations in hybrids of fluctuating hydrodynamic and MD [7,14] or variants of Direct Simulation Monte Carlo [19]. As stated in the abstract, OPEN MD is an ongoing research program. Some interesting research lines are the molecular implementation of the open boundary conditions for fluctuating hydrodynamics [15], mass transfer involving multiple species [46] or polymer melts flow under constant external pressure (i.e. in open domains) using an OPEN MD-AdResS combined strategy [18]. Finally important challenges remain to be solved, such as a first-principle generalisation of the adaptive resolution scheme to allow for energy conservation (maybe based on the Mori-Zwanzig formalism [22]) or the extension of OPEN MD to nematic or ionic liquids.

References

1. Alexander, F.J., Garcia, A.L., Tartakovsky, D.M.: Algorithm refinement for stochastic partial differential equations. J. Comp. Phys **182**, 47 (2002)
2. Barsky, S., Delgado-Buscalioni, R., Coveney, P.V.: Comparison of molecular dynamics with hybrid continuum-molecular dynamics computational fluid dynamics for a single tethered polymer in a solvent. J. Chem. Phys. **121**, 2403 (2004)
3. Borg, M., Reese, J.: A hybrid particle-continuum framework. Proceedings of the 6th International Conference on Nanochannels, Microchannels, and Minichannels, ICNMM2008 A p. 995 (2008)
4. Buch, I., Harvey, M.J., Giorgino, T., Anderson, D.P., Fabritiis, G.D.: High-throughput all-atom molecular dynamics simulations using distributed computing. Journal of Chemical Information and Modeling **50**, 397 (2010)
5. De, S., Fish, J., Shephard, M.S., Keblinski, P., Kumar, S.K.: Multiscale modeling of polymer rheology. Phys. Rev. E **74**(3), 030,801 (2006)
6. De Fabritiis, G.: Performance of the cell processor for biomolecular simulations. Comp. Phys. Commun. **176**, 660 (2007)
7. De Fabritiis, G., Delgado-Buscalioni, R., Coveney, P.: Modelling the mesoscale with molecular specificity. Phys. Rev. Lett **97**, 134,501 (2006)
8. De Fabritiis, G., Delgado-Buscalioni, R., Coveney, P.V.: Energy controlled insertion of polar molecules in dense fluids. J. Chem. Phys. **121**, 12,139 (2004)
9. De Fabritiis, G., Serrano, M., Delgado-Buscalioni, R., Coveney, P.V.: Fluctuating hydrodynamic modelling of fluids at the nanoscale. Phys. Rev. E **75**, 026,307 (2007)
10. Delgado-Buscalioni, R., Chacon, E., Tarazona, P.: Hydrodynamics of nanoscopic capillary waves. Phys. Rev. Lett. **101**, 106,102 (2008)
11. Delgado-Buscalioni, R., Coveney, P.V., Fabritiis, G.D.: Towards multiscale modelling of complex liquids using hybrid particle-continuum schemes. Proceedings of the Institution of Mechanical Engineers, Part C: Journal of Mechanical Engineering Science, **222**(5), pp 769–776 (2008)
12. Delgado-Buscalioni, R., Coveney, P.V.: Continuum-particle hybrid coupling for mass, momentum and energy transfers in unsteady fluid flow. Phys. Rev. E **67**, 046,704 (2003)
13. Delgado-Buscalioni, R., Coveney, P.V.: USHER: an algorithm for particle insertion in dense fluids. J. Chem. Phys. **119**, 978 (2003)
14. Delgado-Buscalioni, R., De Fabritiis, G.: Embedding molecular dynamics within fluctuating hydrodynamics in multiscale simulations of liquids. Phys. Rev. E **76**, 036,709 (2007)
15. Delgado-Buscalioni, R., Dejoan, A.: Nonreflecting boundaries for ultrasound in fluctuating hydrodynamics of open systems. Phys. Rev. E **78**, 046,708 (2008)
16. Delgado-Buscalioni, R., Flekkøy, E., Coveney, P.V.: Fluctuations and continuity in particle-continuum hybrid simulations of unsteady flows based on flux-exchange. Europhys. Lett. **69**, 959 (2005)

17. Delgado-Buscalioni, R., Kremer, K., Praprotnik, M.: Concurrent triple-scale simulations of molecular liquids. J. Chem. Phys. **128**, 114,110 (2008)
18. Delgado-Buscalioni, R., Kremer, K., Praprotnik, M.: Coupling atomistic and continuum hydrodynamics through a mesoscopic model: Application to liquid water. J. Chem. Phys. **131**, 244,107 (2009)
19. Donev, A., Bell, J.B., Garcia, A.L., Alder, B.J.: A hybrid particle-continuum method for hydrodynamics of complex fluids. Multiscale Model. Simul. **8**, 871 (2010)
20. E, W., Ren, W., Vanden-Eijden, E.: A general strategy for designing seamless multiscale methods. J. Comp. Phys. **228**, 5437 (2009)
21. Engquist, W.E.B., Li, X., Ren, W., Vanden-Eijden, E.: Heterogeneous multiscale methods: A review. Commun. Comput. Phys. **2**, 367 (2007)
22. Español, P.: Hybrid description of complex molecules. EuroPhys. Lett. **88** (2009)
23. Faraudo, J., Bresme, F.: Anomaluos dielectric behaviour of water in ionic newton black films. Phys. Rev. Lett. **92**, 236,102 (2004)
24. Flekkoy, E.G., Delgado-Buscalioni, R., Coveney, P.V.: Flux boundary condition on particle systems. Phys. Rev. E **72**, 026,703 (2005)
25. Flekkøy, E.G., Wagner, G., Feder, J.: Hybrid model for combined particle and continuum dynamics. Europhys. Lett. **52**, 271 (2000)
26. Garcia, A., Bell, J., Crutchfield, W.Y., Alder, B.: Adaptive mesh and algorithm refinement using direct simulation monte carlo. J. Comp. Phys. **154**, 134 (1999)
27. Giupponi, G., De Fabritiis, G., Coveney, P.: An hybrid model for the simulation of macromolecular dynamics. J. Chem. Phys **126**, 154,903 (2007)
28. Hadjiconstantinou, N., Patera, A.: Int. J. Mod. Phys. C **8**, 967 (1997)
29. Hadjiconstantinou, N.G.: Hybrid atomistic-continuum formulations and the moving contact-line problem. J. Comput. Phys. **154**(2), 245–265 (1999)
30. Hijón, C., Español, P., Vanden-Eijnden, E., Delgado-Buscalioni, R.: Morizwanzig formalism as a practical computational tool. Faraday Discuss. **144**, 301 (2010)
31. Hoover, W.G., Hoover, C.G., Petravic, J.: Simulation of two- and three-dimensional dense-fluid shear flows via nonequilibrium molecular dynamics: Comparison of time-and-space-averaged stresses from homogeneous doll's and sllod shear algorithms with those from boundary-driven shear. Phys. Rev. E **78**, 046,701 (2008)
32. Junghans, C., Praprotnik, M., Kremer, K.: Transport properties controlled by a thermostat: An extended dissipative particle dynamics thermostat. Soft Matter **4**, 156–161 (2008)
33. Kapral, R.: Multiparticle collision dynamics: Simulation of complex systems on mesoscales. Advances in Chemical Physics **140**, 89 (2008)
34. Kevrekidis, I.G., Gear, C.W., Hummer, G.: Equation free: The computer-aided analysis of complex multiscale systems. AIChE J. **50**, 1346 (2004)
35. K.M.Mohamed, Mohamad, A.: A review of the development of hybrid atomistic-continuum methods for dense fluids. Microfluid Nanofluidic **8**, 283 (2010)
36. Kotsalis, E.M., Walther, J.H., Koumoutsakos, P.: Control of density fluctuations in atomistic-continuum simulations of dense liquids. Phys. Rev. E **76**(1), 016709 (2007)
37. Koumoutsakos, P.: Multiscale flow simulations using particles. Ann. Rev. Fluid Mech. **37**, 457 (2005)
38. Laso, M., Öttinger, H.: Calculation of viscoelastic flow using molecular models: the connffessit approach. Journal of Non-Newtonian Fluid Mechanics **47**, 1–20 (1993)
39. Liu, J., Chen, S., Nie, X., Robbins, M.O.: A continuum-atomistic multi-timescale algorithm form micro/nano flows. Comm. Comp. Phys. **4**, 1279 (2007)
40. Liu, J., Chen, S., Nie, X., Robbins, M.O.: A continuum-atomistic simulation of heat transfer in micro- and nano-flows. J. Comput. Phys. **227**, 279 (2007)
41. Murashima, T., Taniguchi, T.: Multiscale lagrangian fluid dynamics simulations for polymeric fluid. J. Polymer Sci. B Polymer Phys. **48**, 886 (2010)
42. Nedea, S.V., Frijns, A.J.H., van Steenhoven, A.A., Markvoort, A.J., Hilbers, P.A.J.: Novel hybrid simulations for heat transfer at atomistic level. In: Proceedings of the 4th International Conference on Nanochannels, Microchannnels, and Minichannels, Pts A and B, pp. 1315–1322 (2006). 4th International Conference on Nanochannels, Microchannels, and Minichannels, Limerick, IRELAND, JUN 19-21, 2006

43. Nie, X.B., Chen, S.Y., E, W.N., Robbins, M.O.: A continuum and molecular dynamics hybrid method for micro- and nano-fluid flow. J. Fluid Mech **500**, 55 (2004)
44. O'Connel, S.T., Thompson, P.A.: Phys. Rev. E **52**, R5792 (1995)
45. Patankar, S.V.: Numerical Heat Transfer and Fluid Flow. Hemisphere, New York (1980)
46. Poblete, S., Praprotnik, M., Kremer, K., Site, L.D.: Coupling different levels of resolution in molecular simulations. J. Chem. Phys. **132**, 114,101 (2010)
47. Praprotnik, M., Delle Site, L., Kremer, K.: Adaptive resolution molecular dynamics simulation: Changing the degrees of freedom on the fly. J. Chem. Phys. **123**, 224,106 (2005)
48. Praprotnik, M., Delle Site, L., Kremer, K.: Multiscale simulation of soft matter: From scale bridging to adaptive resolution. Annu. Rev. Phys. Chem. **59**, 545–571 (2008)
49. R. Delgado-Buscalioni, De Fabritiis, G., Coveney, P.V.: Determination of the chemical potential using energy-biased sampling. J. Chem. Phys. **123**, 054,105 (2005)
50. Reith, D., Pütz, M., Müller-Plathe, F.: Deriving effective mesoscale potentials from atomistic simulations. J. Comput. Chem. **24**, 1624–1636 (2003)
51. Ren, W.: Analytical and numerical study of coupled atomistic-continuum methods for fluids. J. Chem. Phys. **227**, 1353–1371 (2007)
52. Ren, W., E, W.: Heterogeneous multiscale method for the modeling of complex fluids and micro-fluidics. J. Comp. Phys. **204**, 1 (2005)
53. Sun, J., He, Y.L., Tao, W.Q.: Molecular dynamics-continuum hybrid simulation for condensation of gas flow in a microchannel. Microfluid NanoFluid **7**(3), 407–422 (2009)
54. Usta, O.B., Ladd, A.J.C., Butler, J.E.: Lattice-boltzmann simulations of the dynamics of polymer solutions in periodic and confined geometries. J. Chem. Phys. **122**, 094,902 (2005)
55. Werder, T., Walther, J.H., Koumoutsakos, P.: Hybrid atomistic continuum method for the simulation of dense fluid flows. J. Comput. Phys. **205**, 373 (2005)
56. Yasuda, S., Yamamoto, R.: Multiscale modeling and simulation for polymer melt flows between parallel plates. Phys. Rev. E **81**, 036,308 (2010)
57. Zhang, Y., Donev, A., Weisgraber, T., Alder, B.J., Graham, M.D., de Pablo, J.J.: Tethered dna dynamics in shear flow. J. Chem. Phys **130** (2009)

Multiscale Methods for Wave Propagation in Heterogeneous Media Over Long Time

Björn Engquist, Henrik Holst, and Olof Runborg

Abstract Multiscale wave propagation problems are computationally costly to solve by traditional techniques because the smallest scales must be represented over a domain determined by the largest scales of the problem. We have developed and analyzed new numerical methods for multiscale wave propagation in the framework of the heterogeneous multiscale method (HMM). The numerical methods couple simulations on macro- and microscales for problems with rapidly oscillating coefficients. The complexity of the new method is significantly lower than that of traditional techniques with a computational cost that is essentially independent of the smallest scale, when computing solutions at a fixed time and accuracy. We show numerical examples of the HMM applied to long time integration of wave propagation problems in both periodic and non-periodic medium. In both cases our HMM accurately captures the dispersive effects that occur. We also give a stability proof for the HMM, when it is applied to long time wave propagation problems.

1 Introduction

We consider the initial boundary value problem for the scalar wave equation,

O. Runborg (✉)
Department of Numerical Analysis, CSC and Swedish e-Science Research Center (SeRC) KTH, 100 44 Stockholm, Sweden
e-mail: olofr@nada.kth.se

B. Engquist
Department of Mathematics and Institute for Computational Engineering and Sciences, The University of Texas at Austin, 1 University Station C1200, Austin TX 78712, U.S.A e-mail: engquist@ices.utexas.edu

H. Holst
Department of Numerical Analysis, CSC, KTH, 100 44 Stockholm, Sweden e-mail: holst@kth.se

B. Engquist et al. (eds.), *Numerical Analysis of Multiscale Computations*, Lecture Notes in Computational Science and Engineering 82, DOI 10.1007/978-3-642-21943-6_8,

$$\begin{cases} u^{\epsilon}_{tt} - \nabla \cdot A^{\epsilon} \nabla u^{\epsilon} = 0, & \Omega \times [0,T], \\ u^{\epsilon}(x,0) = f(x), \quad u^{\epsilon}_{t}(x,0) = g(x), & \forall x \in \Omega, \end{cases} \tag{1}$$

on a smooth domain $\Omega \subset \mathbb{R}^N$ where $A^{\epsilon}(x)$ is a symmetric, uniformly positive definite matrix. The expression $\nabla \cdot A^{\epsilon} \nabla u^{\epsilon}$ should be interpreted as $\nabla \cdot (A^{\epsilon} \nabla u^{\epsilon})$. For simplicity we assume that Ω is a hypercube in $\mathbb{R}^N$ with periodic boundary conditions. We assume that A^{ϵ} has oscillations on a scale proportional to $\epsilon \ll 1$. The solution of (1) will then be a sum of two parts: one coarse scale (macroscale) part, which is independent of ϵ, and an oscillatory (microscale) part which is highly oscillating in both time and spatial directions on the scale ϵ. These kinds of multiscale problems are typically very computationally costly to solve by traditional numerical techniques. The smallest scale must be well represented over a domain which is determined by the largest scale of interest. However most often one is only interested in the coarse scale part of the solution. The goal of our research here is to find an efficient way to compute it.

Recently, new frameworks for numerical multiscale methods have been proposed. These include the heterogeneous multiscale method (HMM) [5] and the equation free approach [14]. They couple simulations on macro- and microscales to compute the coarse scale solution efficiently. The HMM framework has been applied to a number of multiscale problems, for example, ODEs with multiple time scales [12], elliptic and parabolic equations with multiscale coefficients [1, 7, 17], kinetic schemes [6] and large scale MD simulations of gas dynamics [15]. In this paper we use HMM for the wave equation. Our method builds on [10] where we described a HMM multiscale method which captured the coarse scale behavior of (1) for finite time. See also [2]. The main aim here is to show that the HMM methodology in [10] works also for long time, where new macroscale phenomena occurs.

As an inspiration for designing our HMM we first consider the classical homogenization theory, in which the coarse scale properties of partial differential equations with rapidly oscillating coefficients, like (1), can be analyzed. For example, in the setting of composite materials consisting of two or more mixed constituents (i.e., thin laminated layers that are ϵ-periodic), homogenization theory gives the effective properties of the composite. It is an interesting remark that the effective properties often are different than the average of the individual constituents that makes up the composite [4]. The main homogenization result is that, under certain conditions, when the period of the coefficients in the PDE goes to zero, the solution approaches the solution to another PDE which has no oscillatory (microscale) part. This *homogenized* PDE is very useful from a numerical perspective. It gives a coarse scale solution and the coefficients in the PDE have no ϵ-dependency. That means that the homogenized PDE is inexpensive to solve with standard numerical methods. At the same time the solution is a good approximation of the coarse scale (macroscopic) part of the original equation. For our multiscale problem (1) with $A^{\epsilon}(x) = A(x, x/\epsilon)$ and where $A(x, y)$ is periodic in y, the homogenized PDE is of the form,

$$\begin{cases} \bar{u}_{tt} - \nabla \cdot \bar{A} \nabla \bar{u} = 0, & \Omega \times [0,T], \\ \bar{u}(x,0) = f(x), \quad \bar{u}_t(x,0) = g(x), & \forall x \in \Omega, \end{cases} \tag{2}$$

where $\bar{A}(x)$ is the homogenized or effective coefficient. We refer to [3, 4, 11, 13, 16, 18] for more details about homogenization.

Homogenization gives the limit PDE as $\epsilon \to 0$ for a constant T (independent of ϵ). The use of classical homogenization is limited by the fact that it does not describe the dispersive effects in (1) that occur when T becomes very large. Santosa and Symes [20] developed effective medium equations for wave propagation problems with $T = \mathcal{O}(\epsilon^{-2})$. In the one-dimensional case, when $A^\epsilon(x) = A(x/\epsilon)$ and A periodic, this equation will be of the form

$$\tilde{u}_{tt} - \bar{A}\tilde{u}_{xx} - \beta\epsilon^2\tilde{u}_{xxxx} = 0,$$

where $\bar{A}$ is the same coefficient as in (2) and β is a functional of A. The effective medium solution $\tilde{u}$ can be used as an approximation for longer time than the homogenized solution $\bar{u}$ with an error of the form $\mathcal{O}(\epsilon) + \mathcal{O}(\epsilon^3 t)$. See [20] for further details about this model.

We will now briefly describe the typical setting of HMM for multiscale problems and how it can be applied to (1). We assume that there exists two models, a micro model $h(u^\epsilon, d^\epsilon) = 0$ describing the full problem, where u^ϵ is the quantity of interest and d^ϵ is the problem data (i.e. initial conditions, boundary conditions, ...), and a coarse macro model $H(u,d) = 0$, with solution u and data d. The micro model is accurate but is expensive to compute by traditional methods. In our case this model is (1). The macro model gives a coarse scale solution u, assumed to be a good approximation of the microscale solution u^ϵ and is less expensive to compute. The model is however incomplete in some sense and requires additional data. In our case we use

$$u_{tt} - \nabla \cdot F = 0,$$

with the flux F unknown. This is inspired by the form of the homogenized equation (2). A key idea in the HMM is to provide the missing data in the macro model using local solutions of the micro model. Here (1) is solved locally on a small domain with size proportional to ϵ and F is given as an average of the resulting microscopic flux $A^\epsilon \nabla u^\epsilon$. The initial data and boundary conditions (d^ϵ) for this computation is constrained by the macroscale solution u.

It should be noted that even if our numerical methods use ideas from homogenization theory they do not solve any effective (e.g. homogenization or effective medium) equation directly. The goal is to develop computational techniques that can be used when there is no fully known macroscopic PDE.

The article is organized as follows: In Sect. 2 we describe our HMM for the wave equation for finite time. In Sect. 3 we describe the modifications made to our HMM for the long time problem and in Sect. 3.4 we describe the theory behind the long time problem. We also treat problems which do not fit the theory. In Sect. 3.3 where we solve a non-periodic problem for long time. We end this paper with our conclusions in the closing Sect. 4.

2 HMM for the Wave Equation and Finite Time

We continue here with the description of our HMM method for the wave equation (1) over finite time. By finite time we mean that the final time T is independent of ϵ. In the next section we will consider cases where $T = T(\epsilon) \to \infty$ as $\epsilon \to 0$.

The HMM method we suggest here is described in three separate steps. We follow the same strategy as in [1] for parabolic equations and in [19] for the one-dimensional advection equation. See [8], [10] and [2] for additional details and proofs. In step one we give the macroscopic PDE (i.e. the form $H(u,d) = 0$) and a corresponding numerical discretization. In step two we describe the microscale problem (microproblem). The initial data for the microproblem is based on local macroscopic data. Finally, in step three we describe how we approximate F from the computed microproblem by taking a weighted average of its solution.

We will assume that the domain $\Omega = Y \subset \mathbb{R}^d$ is a hypercube such that our microscopic PDE is of the form,

$$\begin{cases} u^\epsilon_{tt} - \nabla \cdot A^\epsilon \nabla u^\epsilon = 0, & Y \times [0,T], \\ u^\epsilon(x,0) = f(x), \quad u^\epsilon_t(x,0) = g(x), & \forall x \in Y. \end{cases} \tag{3}$$

and $u^\epsilon(x,t)$ is Y-periodic in x.

Step 1: Macro model and discretization

We suppose there exists a macroscale PDE of the form,

$$\begin{cases} u_{tt} - \nabla \cdot F(x,u,\nabla u,\dots) = 0, & Y \times [0,T], \\ u(x,0) = f(x), \quad u_t(x,0) = g(x), & \forall x \in Y, \\ u \text{ is } Y\text{-periodic}, \end{cases} \tag{4}$$

where F is a function of x, u and higher derivatives of u. We will use this assumption throughout the whole paper. Another assumption is that $u \approx u^\epsilon$ when ϵ is small. In the method we suppose that $F = F(x,\nabla u)$. In the clean homogenization case we would have $F = \bar{A}\nabla u$, but we will not assume knowledge of a homogenized equation.

We discretize (4) using central differences with time step K and spatial grid size H in all directions,

$$\begin{cases} U_m^{n+1} = 2U_m^n - U_m^{n-1} + \dfrac{K^2}{H} \displaystyle\sum_{i=1}^{d} \left(e_i^T F^n_{m+\frac{1}{2}e_i} - e_i^T F^n_{m-\frac{1}{2}e_i} \right), \\ F^n_{m-\frac{1}{2}e_k} = F(x_{m-\frac{1}{2}e_k}, P^n_{m-\frac{1}{2}e_k}), \quad k = 1,\dots,d, \\ \text{(Note that } F^n_{m-\frac{1}{2}e_k} \text{ is a vector.)} \end{cases} \tag{5}$$

where $F^n_{m-\frac{1}{2}e_k}$ is F evaluated at point $x_{m-\frac{1}{2}e_k}$. The quantity $P^n_{m-\frac{1}{2}e_k}$ approximates ∇u in the point $x_{m-\frac{1}{2}e_k}$.

Step 2: Micro problem

The evaluation of $F^n_{m-\frac{1}{2}e_k}$ in each grid point is done by solving a micro problem to evaluate the flux values in (5). Given the parameters $x_{m-\frac{1}{2}e_k}$ and $P^n_{m-\frac{1}{2}e_k}$, we solve

$$\begin{cases} u^\epsilon_{tt} - \nabla \cdot A^\epsilon \nabla u^\epsilon = 0, & Y^\epsilon \times [-\tau, \tau], \\ u^\epsilon(x,0) = (P^n_{m-\frac{1}{2}e_k}) \cdot x, \quad u^\epsilon_t(x,0) = 0, & \forall x \in Y^\epsilon, \\ u^\epsilon - u^\epsilon(x,0) \text{ is } Y^\epsilon\text{-periodic}, \end{cases} \tag{6}$$

where $x - x_{m-\frac{1}{2}e_k} \mapsto x$ and $t - t_n \mapsto t$. The initial data $u^\epsilon(x,0)$ is a linear polynomial approximating the macroscopic solution locally, modulo a constant term; since we only consider the derivative of u^ϵ when computing F below, the constant term does not affect the result.

We keep the sides of the micro box Y^ϵ of order ϵ. We note that the solution u^ϵ is an even function with respect to t (i.e. $u^\epsilon(x,-t) = u^\epsilon(x,t)$) due to the initial condition $u^\epsilon_t(x,0) = 0$.

Step 3: Reconstruction step

After we have solved for u^ϵ for all $Y^\epsilon \times [-\tau, \tau]$ we approximate $F^n_{m-\frac{1}{2}e_k}$ by a weighted average of $f^\epsilon = A^\epsilon \nabla u^\epsilon$ over $[-\eta, \eta]^d \times [-\tau, \tau]$ where $[-\eta, \eta]^d \subset Y^\epsilon$. We choose η, τ sufficiently small so that information will not propagate into the region $[-\eta, \eta]^d$ from the boundary of the micro box Y^ϵ in $[-\tau, \tau]$ time. More precisely, we consider averaging kernels K described in [12]: We let $\mathbb{K}^{p,q}$ denote the kernel space of functions K such that $K \in C^q_c(\mathbb{R})$ with *supp* $K = [-1, 1]$ and

$$\int K(t) t^r \mathrm{d}t = \begin{cases} 1, & r = 0; \\ 0, & 1 \leq r \leq p. \end{cases}$$

Furthermore we will denote K_η as a scaling $K_\eta(x) := \eta^{-1} K(x/\eta)$ with compact support in $[-\eta, \eta]$. We then approximate

$$F^n_{m-\frac{1}{2}e_k} \approx \tilde{F}(x_{m-\frac{1}{2}e_k}, P^n_{m-\frac{1}{2}e_k}) = \iint K_\tau(t) K_\eta(x_1) \cdots K_\eta(x_d) f^\epsilon_k(x,t) \mathrm{d}x \mathrm{d}t, \tag{7}$$

where $f^\epsilon(x,t) = A^\epsilon(x + x_{m-\frac{1}{2}e_k}) \nabla u^\epsilon(x,t)$.

We proved in [10] that if we apply the HMM to the problem (1) with $A^\epsilon(x) = A(x/\epsilon)$ where A is a Y-periodic symmetric positive matrix the HMM generates results close to a direct discretization of the homogenized equation (2). In particular, we showed that

$$\tilde{F}(x,y) = F(x,y) + \mathcal{O}\left(\left(\frac{\epsilon}{\eta}\right)^{q+2}\right).$$

The function $\tilde{F}$ and F are defined in (7) and (4) respectively and we note that here $F(x,y) = \bar{A}y$. The integer q depends on the smoothness of the kernel used to compute the weighted average of f^ϵ in (7).

Theorem 1. *Let $\tilde{F}(x_0,y)$ be defined by* (7) *where u^ϵ solves the micro problem* (6) *exactly, $A^\epsilon(x) = A(x/\epsilon)$ and A is Y-periodic and C^∞. Moreover suppose $K \in \mathbb{K}^{p,q}$, f and g are C^∞ and $\tau = \eta$. Then for $y \neq 0$ and any dimension,*

$$\frac{1}{\|y\|}\left|\tilde{F}(x_0,y) - F(x_0,y)\right| \leq C\left(\frac{\epsilon}{\eta}\right)^{q+2},$$

where C is independent of ϵ and η. Furthermore, for the numerical approximation given in (5) *in one dimension, with $H = n\epsilon$ for some integer n and smooth initial data, we have the error estimate*

$$|U_m^n - \bar{u}(x_m,t_n)| \leq C(T)\left(H^2 + (\epsilon/\eta)^{q+2}\right), \qquad 0 \leq t_n \leq T,$$

where $\bar{u}$ is the homogenized solution to (2).

Remark 1. The weighted integrals above are computed numerically with a simple trapezoidal rule in time and a midpoint rule in space.

Remark 2. In our implementation, the micro problem (6) is solved by the same numerical scheme as the macro problem (5).

Remark 3. We assume that our scheme for the microproblem can have a constant number of grid points per ϵ to maintain a fixed accuracy. This implies that a direct solver for (3) on the full domain has a cost of order $\epsilon^{-(d+1)}$. The total cost for on-the-fly HMM is of the form (cost of micro problem) $\times\, M_d$ where

$$M_d \sim \frac{1}{K} \cdot \frac{1}{H^d}$$

is the number of micro problems needed to be solved. The macro PDE can be discretized independently of ϵ therefore M_d does not depend on ϵ. If we choose η and τ proportional to ϵ the cost of a single micro problem $(\tau/\epsilon) \times (\eta/\epsilon)^d$ is also independent of ϵ. In conclusion our HMM method has a computational cost independent of ϵ.

Remark 4. We can to reduce the computational cost of the HMM process even further if the function $\tilde{F}$ in (7) is linear in some of its arguments. We can then apply the HMM process to a smaller number of micro problems and form linear combinations of those for any given $\tilde{F}$ computation. If $\tilde{F}$ depends on u or t it might not be beneficial to precompute $\tilde{F}$ this way. See [10] for further details.

Remark 5. The macro scheme suggested here is embarrassingly parallel in space. This fact has been exploited by the authors in a Fortran 90 code with MPI parallelization. We think that it would be possible to implement the same algorithm in a general purpose GPU environment and see a good utilization of the hardware.

2.1 One Numerical Example

We consider the one-dimensional problem of the form (1),

$$\begin{cases} u_{tt}^{\epsilon} - \partial_x A^{\epsilon} u_x = 0, & [0,1] \times [0,1], \\ u^{\epsilon}(x,0) = \exp(-100x^2) + \exp(-100(1-x)^)), \quad u_t(x,0) = 0, & x \in [0,1), \\ u^{\epsilon} \text{ is 1-periodic}, \end{cases} \tag{8}$$

for $A^{\epsilon}(x) = A(x/\epsilon)$ where $A(y) = 1.1 + \sin 2\pi y$. The homogenized equation will then have the form (2) with $\bar{A} = \left(\int_0^1 \frac{1}{A(s)} \mathrm{d}s\right)^{-1} = \sqrt{0.21}$. Since we have periodic boundary conditions, the solution to the homogenized equation will be periodic in time with period $1/\sqrt{\bar{A}} \approx 1.47722$. We show the solution after one full period. The numerical parameters are $H = 1.0 \cdot 10^{-2}, K = 1.0 \cdot 10^{-3}, \eta = \tau = 0.05, h = 1.5 \cdot 10^{-4}$ and $k = 7.8 \cdot 10^{-5}$. We take $\epsilon = 0.01$. See Fig. 1 for a plot of the result. We refer to [10] for further examples where HMM is applied to other finite time problems.

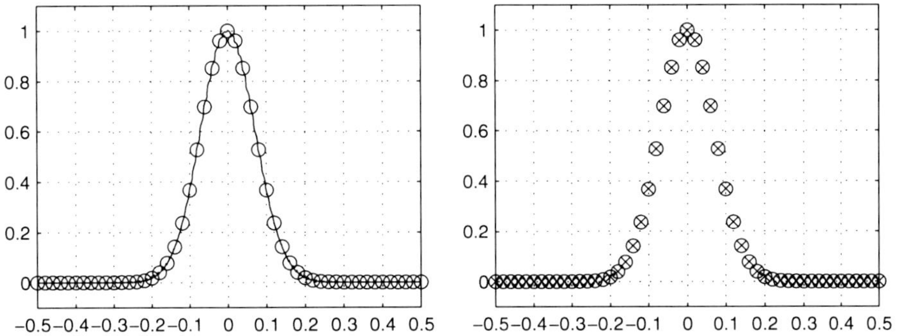

Fig. 1 Results from solving (8) with a finite difference method, DNS (direct numerical simulation), and the corresponding homogenized equation with highly accurate spectral method (*circles*), compared to our HMM method (*crosses*). The fast $\mathcal{O}(\epsilon)$ oscillations are visible as small fluctuations in the DNS computation

3 HMM for the Wave Equation over Long Time

Classical homogenization deals with constant T (i.e. independent of ϵ) and finds the limiting PDE as $\epsilon \to 0$. We demonstrated in the previous section that our HMM captures the same solution as homogenization (when applicable). In this section we will investigate how our HMM method, after some minor changes, handles the case when $T = \mathcal{O}(\epsilon^{-2})$. The microscopic variations in the medium introduces dispersive effects in the macroscopic behavior of the solution, after long time. Our goal is to show that our HMM method can capture the dispersion with less computational cost than just resolving the full equation.

Let us illustrate the dispersive effects by a numerical example. We consider the same one-dimensional example (8) as above, but solve it for a long time $T = \mathcal{O}(\epsilon^{-2})$. We compute the solutions after 1, 10 and 100 periods (≈ 1.47722) of the homogenized equation. We see in Fig. 2 that after 100 periods there is an $\mathcal{O}(1)$ error between the true solution u^{ϵ} and the homogenized solution $\bar{u}$ which thus fails to capture the dispersive behavior of the solution after long time.

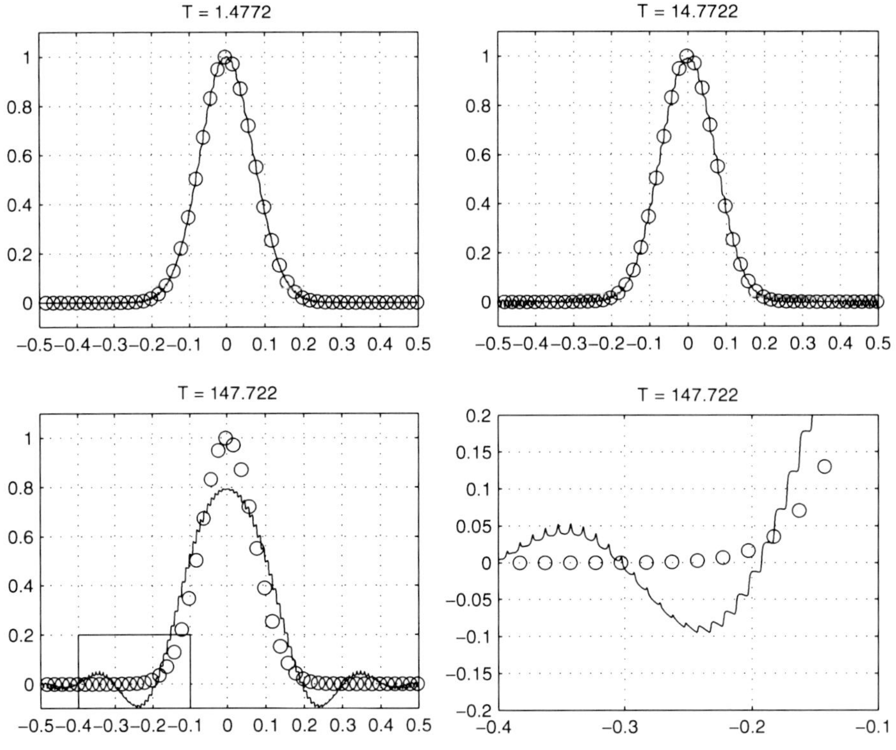

Fig. 2 Finite difference computation of (8) at $T = 1.47722$, $T = 14.7722$ and $T = 147.722$ (1, 10 and 100 periods of the homogenized solution) and the corresponding homogenized solution (*circles*). As we can see the homogenized solution does not capture the dispersive effects that occur

3.1 The HMM Algorithm for Long Time

We must make a few minor modifications to our original HMM algorithm in Sect. 2 in order to capture the dispersive effects seen in Fig. 2. We will now describe those changes. They can all be seen as modifications to increase accuracy.

Step 1: Macro model and discretization

We assume the macroscopic PDE still is of the form $u_{tt} - \nabla \cdot F = 0$ where F depends on u and its derivatives but we will use a higher order scheme instead of (5). The scheme below has better dispersive properties and hence allow us to better avoid some of the numerical dispersion,

$$U_m^{n+1} = 2U_m^n - U_m^{n-1} + \frac{K^2}{24H}\left(-F_{m+3/2}^n + 27F_{m+1/2}^n - 27F_{m-1/2}^n + F_{m-3/2}^n\right),$$

where $F_{m-\frac{1}{2}}^n$ is computed in the same fashion as before in step 2 and 3, defined below.

Step 2: Micro problem

The initial data for the micro problem for finite time (6) is modified to a cubic polynomial $Q(x)$,

$$\begin{cases} u_{tt}^\epsilon - \partial_x A^\epsilon u_x^\epsilon = 0, & Y^\epsilon \times [-\tau, \tau], \\ u^\epsilon(x,0) = Q(x), \quad u_t^\epsilon(x,0) = 0, & \forall x \in Y^\epsilon, \\ u^\epsilon \text{ is } Y^\epsilon\text{-periodic}, & \end{cases}$$

The state of the macroscopic solution is then more accurately represented by the initial data. The cubic polynomial is chosen as follows when computing the flux $F_{m+1/2}$. Let $\tilde{Q}(x)$ interpolate the macroscopic solution in the four grid points surrounding $x_{m+1/2}$. Then use $Q(x) = \tilde{Q}(x) - \gamma\epsilon^2\tilde{Q}''(x)$. The small correction is needed to get an initialization that is consistent with the macroscopic data $\tilde{Q}(x)$ to high order in ϵ. The factor γ can be determined numerically, see Sect. 3.4.

Step 3: Reconstruction step

The average is computed as before but we need to use a sufficiently accurate kernel K and take the average over a bit larger box, i.e. larger τ and η with respect to ϵ such that $\tau, \eta \sim \epsilon^{1-\alpha}$ with $\alpha > 0$. We will delay the discussion about α until Sect. 3.4.

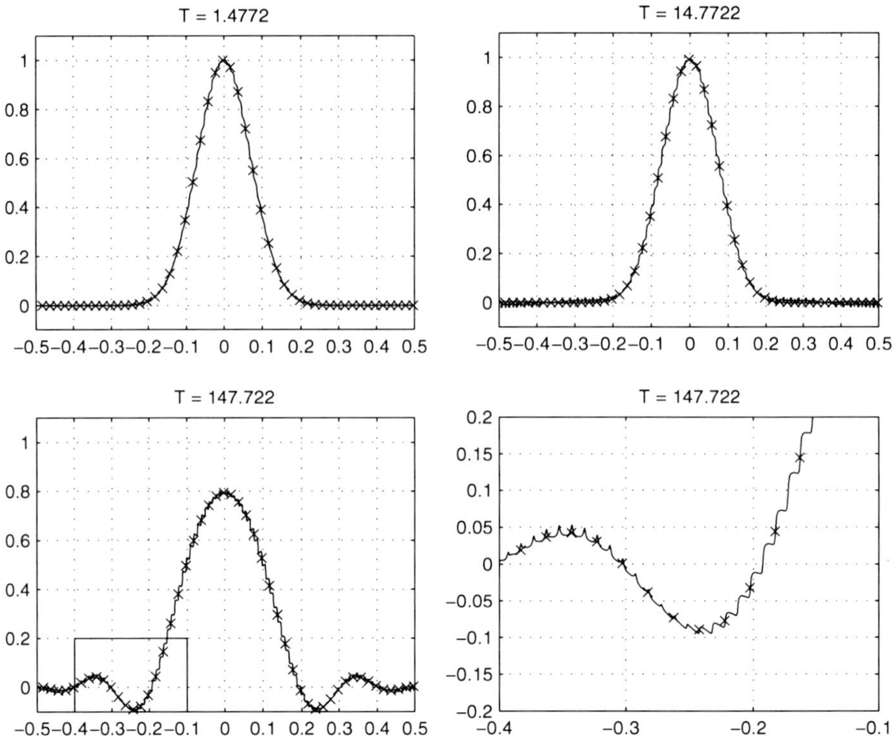

Fig. 3 A longtime DNS simulation (thin line) compared to an HMM solution (*crosses*) at $T = 1.47722$, $T = 14.7722$ and $T = 147.722$ (1, 10 and 100 periods of the homogenized solution) for the example in Sect. 3.2. As we can see, the HMM method gives good agreement with the true solution also after long time

3.2 A Long Time Computation with HMM

We solved the problem (8) using the HMM algorithm, with the improvements described above. As before we computed the solution after 1, 10 and 100 periods of the homogenized equation. In Fig. 3 we see that the HMM algorithm is able to accurately approximate the solution also after long time, and thus captures the correct dispersive effects.

The HMM solver uses $H = 5.7 \cdot 10^{-3}$, $K = 5.7 \cdot 10^{-4}$ and a kernel with $\tau = \eta = 0.5$ from $\mathbb{K}^{9,9}$ which is a 9 times continuously differentiable compact function with support $[-1, 1]$. The micro solver and the DNS solver uses $h = 7.8 \cdot 10^{-5}$ and $k = 3.9 \cdot 10^{-5}$. We take $\epsilon = 0.01$. Note that since the integration time T is very long we need to take H rather small to avoid dispersion errors in the macroscopic integration.

3.3 Non-periodic Material

We consider the problem (1) with a function A^ϵ which is not periodic on the microscale,

$$A^\epsilon(x) = A(rx/\epsilon, x/\epsilon), \quad \text{and} \quad A(y_1, y_2) = 1.1 + \frac{1}{2}(\sin 2\pi y_1 + \sin 2\pi y_2),$$

where r is an irrational number. We take $r = \sqrt{2}$. To be precise we take $r = 1.41$ and $\epsilon = 0.01$ to ensure A^ϵ is periodic on the macroscopic scale. There is no cell problem for this A^ϵ but it is well known that there is a homogenized equation of the form (2) with $\bar{A} = (\int_0^1 \frac{1}{A^\epsilon(x)} \mathrm{d}x)^{-1} = 0.744157$ and thus the period length is $1/\sqrt{\bar{A}} = 1.15922$. The initial data is $u(x,0) = \exp(-100x^2) + \exp(-100(1-x)^2)$ and $u_t(x,0) = 0$.

We compare our HMM-results with an accurate DNS computation after 10 and 100 periods. We use $\eta = \tau = 0.5$ and a kernel $K \in \mathbb{K}^{9,9}$. The numerical parameters are $H = 5.7 \cdot 10^{-3}$, $K = 5.7 \cdot 10^{-4}$, $h = 7.8 \cdot 10^{-5}$ and $k = 3.9 \cdot 10^{-5}$. See Fig. 4.

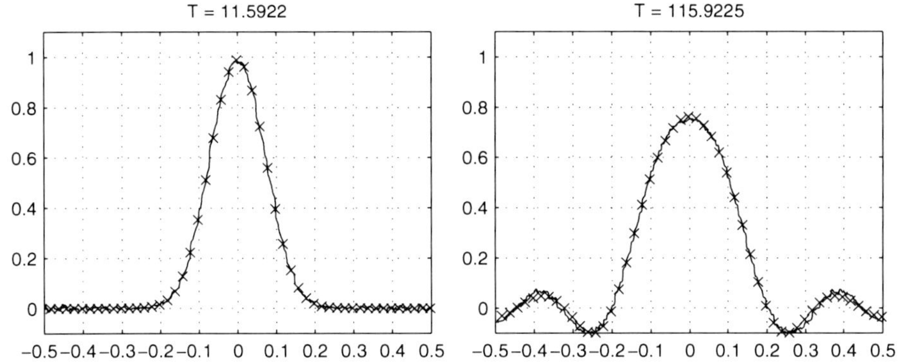

Fig. 4 Numerical result from the example in Sect. 3.3. A longtime DNS simulation (thin line) compared to an HMM solution (*crosses*) at $T = 11.5922$ and $T = 115.922$ (10 and 100 periods of the homogenized solution). The dispersion effects appearing after long time is captured by our HMM method

3.4 Theory

We will now give a motivation to why our HMM method works also for long time. In classical homogenization theory the homogenized solution $\bar{u}$ satisfies a homogenized PDE. The solution u is a good approximation to the true solution u^ϵ such that $||u^\epsilon(t,\cdot) - \bar{u}(t,\cdot)||_{L^2} = \mathcal{O}(\epsilon)$, upto a fixed time T independent of ϵ. Santosa and Symes derived an equation for a similar quantity $\tilde{u}$ which approximates

u^ϵ with an error of the form $\mathcal{O}(\epsilon) + \mathcal{O}(\epsilon^3 t)$ for $T = \mathcal{O}(\epsilon^{-2})$. We will now describe some of the theory presented in [20]. The theory thus extends the effective model (2) with additional terms, from $T = \mathcal{O}(1)$ up to time $T = \mathcal{O}(\epsilon^{-2})$.

Let us first give some definitions. Let ω_m^2 and ψ_m be the eigenvalues and eigenfunctions of the shifted cell (eigenvalue) problem [3, pp. 614],

$$\begin{cases} -\left(\partial_y + ik\right) A(y) \left(\partial_y + ik\right) \psi(y,k) = \omega^2(k)\psi(y,k), & Y \times Y, \\ \psi(y,k) \text{ is } Y\text{-periodic in } y, \end{cases}$$

where $Y = [0,1]^d$ and $k \in \mathbb{R}^d$. Let $v_m(x,k)$ be the scaled Bloch-waves,

$$v_m(x,k) = \psi_m(x/\epsilon, \epsilon k) \exp(ik \cdot x),$$

which satisfies

$$-\partial_x \left(a\left(\frac{x}{\epsilon}\right) \partial_x v_m \right) = \frac{1}{\epsilon^2} \omega_m^2(\epsilon k) v_m.$$

The functions U_m and $\hat{f}_m$ are defined as the projection of u^ϵ and f on v_m,

$$U_m(k,t) = \int u^\epsilon(x,t) v_m^*(x,k) \mathrm{d}x, \quad \hat{f}_m(k) = \int f(x) v_m^*(x,k) \mathrm{d}x.$$

Throughout this section we assume that the initial data $f(x)$ is a bandlimited function. The following theorem from [20] then states that if we expand the solution to the wave equation in the basis given by $\{v_m\}$, the terms with $m \geq 1$ are bounded by $\mathcal{O}(\epsilon)$ uniformly in time.

Theorem 2. *Suppose u^ϵ solves* (1) *with $g = 0$ and expand*

$$u^\epsilon(x,t) = \int_{Y/\epsilon} U_0(k,t) v_0(x,k) \mathrm{d}k + \sum_{m=1}^{\infty} \int_{Y/\epsilon} U_m(k,t) v_m(x,k) \mathrm{d}k. \tag{9}$$

Then

$$\int_{\mathbb{R}^3} \left| \sum_{m=1}^{\infty} \int_{Y/\epsilon} U_m(k,t) v_m(x,k) \mathrm{d}k \right|^2 \mathrm{d}x \leq C\epsilon^2.$$

Here C is independent of ϵ and t but depends on the H^2-norm of the initial data f and the L^∞-norm of a and ∇a.

See [20] for proof.

We denote the first term in (9) by u_0 and note that $\hat{f}_0(k)$ has compact support if $f(x)$ is band limited, see [20]. Then, for some fixed L,

$$\begin{aligned} u_0(x,t) &= \frac{1}{2} \int_{Y/\epsilon} \hat{f}_0(k) v_0(x,k) \exp\left(\pm i\omega_0(\epsilon k)t/\epsilon\right) \mathrm{d}k \\ &= \frac{1}{2} \int_{-L}^{L} \hat{f}_0(k) \psi_0(x/\epsilon, \epsilon k) \exp\left(ikx + i\omega_0(\epsilon k)t/\epsilon\right) \mathrm{d}k. \end{aligned}$$

We now Taylor expand ψ_0 in the second argument and use the fact that $\psi_0(x,0) \equiv 1$. This gives

$$
\begin{aligned}
u_0(x,t) &= \frac{1}{2}\int_{-L}^{L} \hat{f}_0(k)(\psi_0(x/\epsilon,0) + \mathscr{O}(\epsilon k))\exp(ikx + i\omega_0(\epsilon k)t/\epsilon)\mathrm{d}k \\
&= \frac{1}{2}\int_{-L}^{L} \hat{f}_0(k)\exp(ikx + i\omega_0(\epsilon k)t/\epsilon)\mathrm{d}k + \mathscr{O}(\epsilon).
\end{aligned}
$$

Next we Taylor expand $\omega_0(\epsilon k)$ around $k = 0$,

$$
\begin{aligned}
\omega_0(\epsilon k) &= \omega_0(0) + \epsilon k\omega_0'(0) + \frac{\epsilon^2 k^2}{2!}\omega_0''(0) + \frac{\epsilon^3 k^3}{3!}\omega_0^{(3)}(0) + \mathscr{O}(\epsilon^4 k^4) \\
&=: \tilde{\omega}_0(\epsilon k) + \mathscr{O}(\epsilon^4 k^4),
\end{aligned}
$$

and plug this expansion into the expression for u_0,

$$
\begin{aligned}
u_0(x,t) &= \frac{1}{2}\int_{-L}^{L} \hat{f}_0(k)\exp(ikx + i[\tilde{\omega}_0(\epsilon k) + \mathscr{O}(\epsilon^4 k^4)]t/\epsilon)\mathrm{d}k + \mathscr{O}(\epsilon) \\
&= \frac{1}{2}\int_{-L}^{L} \hat{f}_0(k)\exp(ikx + i\tilde{\omega}_0(\epsilon k)t/\epsilon)\mathrm{d}k + \mathscr{O}(\epsilon^3 t) + \mathscr{O}(\epsilon) \\
&=: \tilde{u}_0(x,t) + \mathscr{O}(\epsilon^3 t) + \mathscr{O}(\epsilon).
\end{aligned}
$$

Let us now differentiate the leading term $\tilde{u}_0(x,t)$ twice with respect to t,

$$
\partial_{tt}\tilde{u}_0(x,t) = \frac{1}{2}\int_{-L}^{L} -\frac{1}{\epsilon^2}(\tilde{\omega}_0(\epsilon k))^2\,\hat{f}_0(k)\exp(ikx + i\tilde{\omega}_0(\epsilon k)t/\epsilon)\mathrm{d}k
$$

and upon expanding the square of $\tilde{\omega}_0$ under the integral we obtain

$$
\begin{aligned}
\partial_{tt}\tilde{u}_0(x,t) = -\frac{1}{2}\int_{-L}^{L} \Big[&\epsilon^{-2}\omega_0(0)^2 2\epsilon^{-1}k\omega_0(0)\omega_0'(0) \\
&+ \frac{1}{2}k^2\left(2\omega_0(0)\omega_0''(0) + 2(\omega_0'(0))^2\right) \\
&+ \frac{1}{6}\epsilon k^3\left(2\omega_0(0)\omega_0^{(3)}(0) + 6\omega_0'(0)\omega_0''(0)\right) \\
&+ \frac{1}{24}\epsilon^2 k^4\left(8\omega_0'(0)\omega_0^{(3)}(0) + 6(\omega_0''(0))^2\right) \\
&+ \frac{1}{6}\epsilon^3 k^5\left(\omega_0''(0)\omega_0^{(3)}(0)\right) + \frac{1}{36}\epsilon^4 k^6(\omega_0^{(3)}(0))^2\Big]\times \\
&\hat{f}_0(k)\exp(ikx + i\tilde{\omega}_0(\epsilon k)t/\epsilon)\mathrm{d}k.
\end{aligned}
$$

We now use the facts that $\omega_0(0) = 0$ and that by symmetry all odd derivatives of $\omega_0^2(k)$ are zero when evaluated at $k = 0$. Then the expression for $\partial_{tt}\tilde{u}_0$ simplifies to

$$\begin{aligned}\partial_{tt}\tilde{u}_0(x,t) &= -\frac{1}{2!}\int_{-L}^{L}\left[\frac{1}{2}k^2 \left.\frac{\partial^2\omega_0^2(k)}{\partial k^2}\right|_{k=0} + \frac{1}{4!}\epsilon^2 k^4 \left.\frac{\partial^4\omega_0^2(k)}{\partial k^4}\right|_{k=0}\right. \\ &\quad \left. + \epsilon^3 k^5 R_1 + \epsilon^4 k^6 R_2\right]\hat{f}_0(k)\exp(ikx + i\tilde{\omega}_0(\epsilon k)t/\epsilon)\mathrm{d}k \\ &= \frac{1}{2!}\left.\frac{\partial^2\omega_0^2(k)}{\partial k^2}\right|_{k=0}\partial_{xx}\tilde{u}_0(x,t) - \epsilon^2\frac{1}{4!}\left.\frac{\partial^4\omega_0^2(k)}{\partial k^4}\right|_{k=0}\partial_{xxxx}\tilde{u}_0(x,t) \\ &\quad - i\epsilon^3 R_1\partial_{xxxxx}\tilde{u}_0(x,t) - \epsilon^4 R_2\partial_{xxxxxx}\tilde{u}_0(x,t), \end{aligned} \tag{10}$$

where R_1 and R_2 are some real numbers. This is approximated in [20] with the PDE

$$\tilde{u}_{tt} = \bar{A}\tilde{u}_{xx} + \beta\epsilon^2\tilde{u}_{xxxx}, \tag{11}$$

where

$$\bar{A} = \frac{1}{2!}\left.\frac{\partial^2\omega_0^2}{\partial k^2}\right|_{k=0}, \qquad \beta = -\frac{1}{4!}\left.\frac{\partial^4\omega_0^2}{\partial k^4}\right|_{k=0}.$$

The remaining $m \geq 1$ terms in (9) are as we said uniformly bounded by $\mathcal{O}(\epsilon)$ in L_2-norm, so that we can use $\tilde{u} \approx \tilde{u}_0$ as an $\mathcal{O}(\epsilon)$ approximation up to the time $t = \mathcal{O}(\epsilon^{-2})$. We present a final comparison based on the example (8) in Fig. 5

We arrive at three conclusions from the analysis above:

1. The long time effective equation (11) is of the form

$$\tilde{u}_{tt} - \partial_x F = 0, \qquad F = \bar{A}\tilde{u}_x + \beta\epsilon^2\tilde{u}_{xxx}.$$

   This fits into the assumed form of our macroscale PDE in (4) and we do not need to change the HMM algorithm to reflect a different macro model.
2. The flux F contains a third derivative of the macroscopic solution. In order to pass this information on to the micro simulation, the initial data must be at least a third order polynomial. This explains why the linear initial data used in the finite time HMM is not enough.
3. Since we need to accurately represent also the second term in the flux F, the error in the flux computation must be smaller than $O(\epsilon^2)$. The error term for F in Theorem 1 is of the form $(\epsilon/\eta)^{q+2}$. We thus need to chose q and η such that $(\epsilon/\eta)^{q+2} < \epsilon^2$, or $\eta > \epsilon^{1-\alpha}$ with $\alpha = \frac{2}{q+2}$. Recalling that in the finite time case we always take $\eta \sim \epsilon$, this hence explains why we need to have more accurate kernels or bigger micro boxes in the long time case. We note that in order to maintain a low computational cost we should have α small, which can be obtained by taking a large q, i.e. a very regular kernel.

We have left to discuss the correction to the initial data mentioned in Step 2 in Sect. 2. It is well established in HMM that initial data for the microscopic simulation should be consistent with the macroscopic data, in the sense that the reconstruction of the coarse variables from the microscopic simulation, evaluated at its initial point, should agree with actual macroscopic data at this point. In our setting we consider the macroscopic variables as the local average of the microscopic solution,

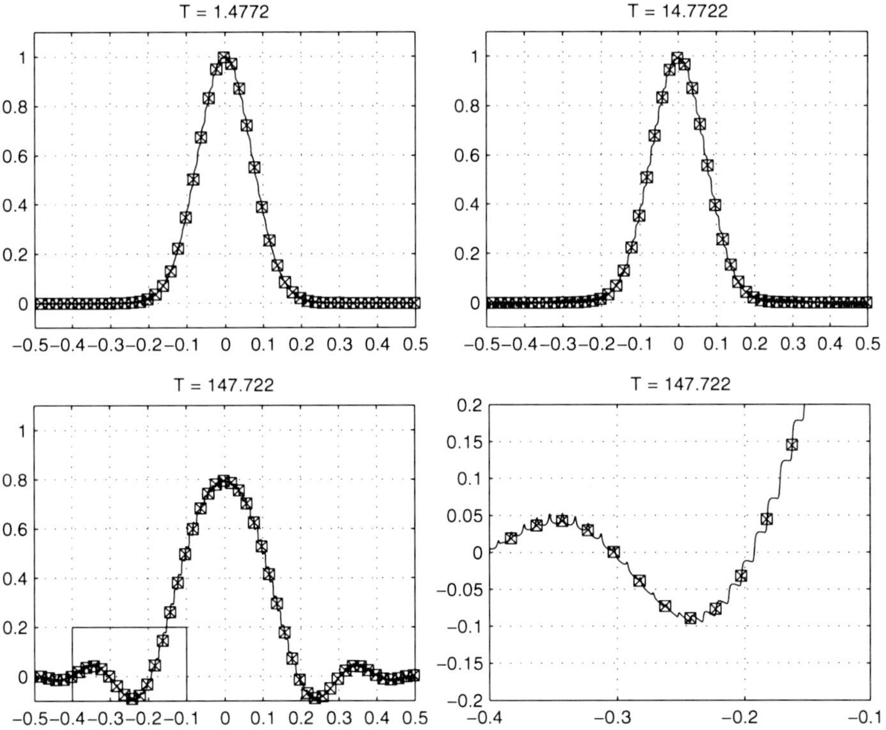

Fig. 5 Numerical result from example in Sect. 3.2: a long time DNS computation (*thin line*) compared to a direct discretization of the long time effective equation (11) with a coarse grid (*squares*) and our HMM method (*crosses*)

$$\tilde{u}(t,x) \sim \iint K_\eta(t')K_\eta(x')u^\epsilon(t+t',x+x')dt'dx'.$$

The given macroscopic data is the polynomial $\tilde{Q}(x)$, which interpolates the macroscopic solution at the initial point. The initial data $Q(x)$ for the microscopic simulation is therefore consistent if it generates a microscopic solution $u^\epsilon(t,x)$ such that

$$\tilde{Q}(x) = \iint K_\eta(t')K_\eta(x')u^\epsilon(t',x+x')dt'dx'.$$

Using the tools from the Bloch wave analysis above one can show [9] that

$$\iint K_\eta(t')K_\eta(x')u^\epsilon(t',x+x')dt'dx' = Q(x) + \epsilon^2\gamma Q''(x) + \text{h.o.t.}$$

if a sufficiently high order kernel is used. The coefficient γ can be computed analytically in some cases, but in general one needs to find it numerically by probing the dynamics once with the initial data $Q_{\text{probe}}(x) = x^2$ and taking

$$\gamma = 2(\tilde{u}_{\text{probe}}(0,x) - x^2)/\epsilon^2.$$

For the finite time case it is sufficient to take $Q(x) = \tilde{Q}(x)$, but in the long time case the first correction term of size $O(\epsilon^2)$ is important to include; recall that the flux must be computed with an accuracy that is better than $O(\epsilon^2)$. Using $\tilde{Q} - \epsilon^2\gamma\tilde{Q}''$ rather than $\tilde{Q}$ gives a higher order consistency.

3.5 Stability Analysis of the Macro Scheme for the Long Time Effective Equation

A complicating factor in Sect. 3.4 is the stability of the long time effective equation (11). In fact, (11) is ill-posed since $\beta > 0$. Perturbations in initial data grow without bounds as wave numbers become large. Since our HMM algorithm effectively discretizes (11) one must be concerned about the method's stability. In this section we show that as long as the macroscopic discretization is coarse enough, it is indeed stable.

Even though (11) is ill-posed, it can be used as an effective equation after regularization. Since we are interested in low frequency solutions it should be possible to use a regularized version of (11) where high frequencies are suppressed. The equation could for instance be regularized with a low-pass filter P_{low} applied at the macro level,

$$\tilde{u}_{tt} = P_{\text{low}}\left(\bar{A}\tilde{u}_{xx} + \beta\epsilon^2\tilde{u}_{xxxx}\right),$$

or by adding a small 6th order term,

$$\tilde{u}_{tt} = \bar{A}\tilde{u}_{xx} + \beta\epsilon^2\tilde{u}_{xxxx} + c\epsilon^4\tilde{u}_{xxxxxx},$$

cf. (10) above. Another regularization technique is to use large time and space grid sizes, which can be seen as a type of low-pass filtering. This is what we do in our HMM. We show here that this approach is stable when the coarse grid size H satisfies a standard CFL condition and in addition $H \geq C\epsilon$, for some constant C. This explains why our HMM is stable. Moreover, even with such a coarse grid the macroscopic solution can be computed accurately; In Fig. 5 we show an example of a solution obtained through a direct discretization of (11) on a coarse grid. The solution agrees very well with a direct numerical simulation of the full wave equation.

We now apply standard von Neumann stability analysis [21] to show stability of the macro scheme for periodic solutions,

$$\begin{cases} u_m^{n+1} = 2u_m^n - u_m^{n-1} + \dfrac{K^2}{24H}\left(-f_{m+3/2}^n + 27f_{m+1/2}^n - 27f_{m-1/2}^n + f_{m-3/2}^n\right), \\ f_m^n = (\bar{A}\partial_x + \beta\epsilon^2\partial_{xxx})p_m^n(x)\big|_{x=x_m}, \end{cases} \tag{12}$$

used in the HMM algorithm for the 1D problem and long time. Here we denote u_m^n as the numerical approximation of $u(x_m, t_n) = u(mH, nK)$ and K is the time step and H is the grid size. The scheme (12) is fourth order accurate with respect to K and second order with respect to H. We define the interpolation polynomial $p_{m-1/2}^n$ of degree three over four grid points $u_{m-2}^n, u_{m-1}^n, u_m^n$ and u_{m+1}^n. We assume a uniform grid and write down the polynomial $p_{m-1/2}$,

$$p_{m-1/2}^n(x) = c_1 + c_2(x - x_{m-2}) + c_3(x - x_{m-2})(x - x_{m-1}) \\ + c_4(x - x_{m-2})(x - x_{m-1})(x - x_m), \quad (13)$$

where the coefficients c_i are given by

$$\begin{cases} c_1 = u_{m-2}^n, \\ c_2 = \dfrac{u_{m-1}^n - u_{m-2}^n}{H}, \\ c_3 = \dfrac{u_m^n - 2u_{m-1}^n + u_{m-2}^n}{2H^2}, \\ c_4 = \dfrac{u_{m+1}^n - 3u_m^n + 3u_{m-1}^n - u_{m-2}^n}{6H^3}. \end{cases} \quad (14)$$

A numerical scheme is said to be stable if

$$\sum_j (u_j^n)^2 \le C(T) \sum_j (u_j^0)^2 \qquad n = 1, 2, \ldots, N, \quad Nk = T,$$

for some constant $C(T)$ independent of n. For the discretization (12) we can show stability if the ratio H/ϵ is large enough.

Theorem 3. *The finite difference scheme* (12) *applied on the effective equation* (11) *with 1-periodic boundary conditions, is stable for K and H satisfying*

$$\frac{\epsilon}{H} \le \sqrt{\frac{7\bar{A}}{24\beta}}, \quad (15)$$

and

$$\frac{K}{H} \le \frac{24}{\sqrt{\bar{A}}} \sqrt{h\left(\frac{24\epsilon^2\beta}{H^2\bar{A}}\right)}, \quad (16)$$

where

$$h(x) = \begin{cases} \dfrac{1}{784 - 112x}, & 0 \le x < \dfrac{21}{5}, \\ \dfrac{x^2 - 2x + 1}{128\left(2(x^2 - x + 1)^{3/2} - 2x^3 + 3x^2 + 3x - 2\right)}, & \dfrac{21}{5} \le x \le 7. \end{cases}$$

Proof. We plug in the value of interpolation polynomials (13) as replacements for the numerical fluxes $f^n_{m-1/2}$ and $f^n_{m+1/2}$ which depend on $u^n_{m-2}, u^n_{m-1}, u^n_m$ and u^n_{m+1}. By doing so, we see that the finite difference scheme (12) will be of the form

$$\begin{aligned} u^{n+1}_m &= 2u^n_m - u^{n-1}_m \\ &+ c\left(u^n_{m+3} - 54u^n_{m+2} + 783u^n_{m+1} - 1460u^n_m + 783u^n_{m-1} - 54u^n_{m-2} + u^n_{m-3}\right) \\ &+ cd\left(-u^n_{m+3} + 30u^n_{m+2} - 111u^n_{m+1} + 164u^n_m - 111u^n_{m-1} + 30u^n_{m-2} - u^n_{m-3}\right), \end{aligned} \tag{17}$$

where $c = K^2\bar{A}/(24^2H^2)$ and $d = 24\epsilon^2\beta/(H^2\bar{A})$. We perform the standard von Neumann stability analysis [21, Sect. 2.2] and replace $u^n_m = g^n \exp(imh\xi)$ in the scheme (17). After dividing with $\exp(imh\xi)$, we get a recurrence relation for g^n of the form,

$$g^{n+1} = (2 + cp(v))g^n - g^{n-1}, \tag{18}$$

where $p(v) = Av^3 + Bv^2 + Cv + D$ is a polynomial in $v = \cos\theta$ ($\theta = h\xi$) and where the coefficients A, B, C and D are affine functions in d,

$$A = -8d + 8,\ B = 120d - 216,\ C = -216d + 1560,\ D = 104d - 1352.$$

The difference equation (18) is stable if the roots r_1, r_2 of its characteristic polynomial $r^2 - (2 + cp(v))r + 1$ satisfy $|r_j| \leq 1$. It is well known that this happens precisely when $|2 + cp(v)| \leq 2$. Hence, the scheme (17) is stable if and only if $-4 \leq cp(v) \leq 0$. The domain of $p(v)$ is $[-1, 1]$ since $v = \cos\theta$. We now continue the proof to find conditions on c and d such that $-4 \leq cp(v) \leq 0$ is fulfilled for $|v| \leq 1$. We start by observing that,

$$p(v) = 8(v-1)(v-13)\left(v(1-d) - 13 + d\right),$$

$$p'(v) = 24\left[(1-d)v^2 + 2(5d-9)v + 65 - 9d\right], \quad p''(v) = 48(v(1-d) + 5d - 9), \tag{19}$$

and first consider the condition $p(v) \leq 0$ for $|v| \leq 1$. Since $p(1) = 0$ and $p'(1) = 1352 > 0$ independent of d, we just need to make sure that the root $(13-d)/(1-d) \notin [-1, 1]$. This happens when $0 \leq d \leq 7$, which gives (15). Next, we need to check that $p(v) \geq -4/c$ for $|v| \leq 1$. For this we use the derivatives of $p(v)$ in (19). We start with the interval $0 \leq d \leq 21/5$. This is chosen such that $p'(-1) = 96(21 - 5d) \geq 0$. Moreover, $p''(-1) = 48(6d - 10) \geq 0$ for $d \geq 5/3$ and $p''(1) = 48(4d - 8) \leq 0$ for $d \leq 2$. Therefore, recalling that $p'(1) = 1352$, the derivative $p'(v)$ must be positive when $|v| \leq 1$ for the d values considered. A necessary and sufficient condition is then that

$$-\frac{4}{c} \leq p(-1) = -448(7-d) \quad \Rightarrow \quad c \leq \frac{1}{112(7-d)}.$$

This gives the $0 \leq x < 21/5$ part of (16). By the same argument there is a point $v^* \in [-1, 1]$ where $p'(v^*) = 0$ when $21/5 \leq d \leq 7$. By solving $p'(v) = 0$ we obtain

$$v^* = \frac{9-5d}{1-d} - \sqrt{\left(\frac{9-5d}{1-d}\right)^2 + \frac{65-9d}{d-1}}. \tag{20}$$

As we showed above $p(v^*) < 0 = p(1)$. Therefore $p(v^*)$ is a minimum and it suffices to make sure that $p(v^*) \geq -4/c$ for $d \in [21/5, 7]$. Plugging (20) into this inequality gives the $21/5 \leq x \leq 7$ part of (16).

4 Conclusions

We have developed and analyzed numerical methods for multiscale wave equations with oscillatory coefficients. The methods are based on the framework of the heterogeneous multiscale method (HMM) and have substantially lower computational complexity than standard discretization algorithms. Convergence is proven in [10] for finite time approximation in the case of periodic coefficients and for multiple dimensions. The effective equation for long time is different from the finite time homogenized equation. After long time, dispersive effects enter and the method has to capture additional effects on the order of $\mathcal{O}(\epsilon^2)$ [20]. Numerical experiments show that the new techniques accurately and efficiently captures the macroscopic behavior for both finite and long time. It is emphasized that the HMM approach with just minor modifications accurately captures these dispersive phenomena. We prove that our method is stable if the spatial grid in the macro solver is sufficiently coarse.

References

1. Abdulle, A., E, W.: Finite difference heterogeneous multi-scale method for homogenization problems. J. Comput. Phys. **191**(1), 18–39 (2003)
2. Abdulle, A., Grote, M.J.: Finite element heterogeneous multiscale method for the wave equation. Preprint (2010)
3. Bensoussan, A., Lions, J.L., Papanicolaou, G.: Asymptotic Analysis in Periodic Structures. North-Holland Pub. Co. (1978)
4. Cioranescu, D., Donato, P.: An Introduction to Homogenization. No. 17 in Oxford Lecture Series in Mathematics and its Applications. Oxford University Press Inc. (1999)
5. E, W., Engquist, B.: The heterogeneous multiscale methods. Commun. Math. Sci. pp. 87–133 (2003)
6. E, W., Engquist, B., Li, X., Ren, W., Vanden-Eijnden, E.: Heterogeneous multiscale methods: A review. Commun. Comput. Phys. **2**(3), 367–450 (2007)
7. E, W., Ming, P., Zhang, P.: Analysis of the heterogeneous multiscale method for elliptic homogenization problems. J. Amer. Math. Soc. **18**(1), 121–156 (2004)
8. Engquist, B., Holst, H., Runborg, O.: Multiscale methods for the wave equation. In: Sixth International Congress on Industrial Applied Mathematics (ICIAM07) and GAMM Annual Meeting, vol. 7. Wiley (2007)
9. Engquist, B., Holst, H., Runborg, O.: Analysis of HMM for wave propagation problems over long time. Work in progress (2010)
10. Engquist, B., Holst, H., Runborg, O.: Multiscale methods for the wave equation. Comm. Math. Sci. **9**(1), 33–56 (2011)

11. Engquist, B., Souganidis, P.E.: Asymptotic and numerical homogenization. Acta Numer. **17**, 147–190 (2008)
12. Engquist, B., Tsai, Y.H.: Heterogeneous multiscale methods for stiff ordinary differential equations. Math. Comp. **74**(252), 1707–1742 (2005). DOI 10.1017/S0962492902000119
13. Jikov, V.V., Kozlov, S.M., Oleinik, O.A.: Homogenization of Differential Operators and Integral Functions. Springer (1991)
14. Kevrekidis, I.G., Gear, C.W., Hyman, J., Kevekidis, P.G., Runborg, O.: Equation-free, coarse-grained multiscale computation: Enabling microscopic simulators to perform system-level tasks. Comm. Math. Sci. pp. 715–762 (2003)
15. Li, X., E, W.: Multiscale modelling of the dynamics of solids at finite temperature. J. Mech. Phys. Solids **53**, 1650–1685 (2005)
16. Marchenko, V.A., Khruslov, E.Y.: Homogenization of partial differential equations. Progress in Mathematical Physics **46** (2006)
17. Ming, P., Yuen, X.: Numerical methods for multiscale elliptic problems. J. Comput. Phys. **214**(1), 421–445 (2005). DOI doi:10.1016/j.jcp.2005.09.024
18. Nguetseng, G.: A general convergence result for a functional related to the theory of homogenization. SIAM J. Math. Anal. **20**(3), 608–623 (1989). DOI http://dx.doi.org/10.1137/0520043
19. Samaey, G.: Patch dynamics: Macroscopic simulation of multiscale systems. Ph.D. thesis, Katholieke Universiteit Leuven (2006)
20. Santosa, F., Symes, W.W.: A dispersive effective medium for wave propagation in periodic composites. SIAM J. Appl. Math. **51**(4), 984–1005 (1991). DOI http://dx.doi.org/10.1137/0151049
21. Strikwerda, J.C.: Finite Difference Schemes and Partial Differential Equations (2nd ed.). SIAM (2004)

Numerical Homogenization via Approximation of the Solution Operator

Adrianna Gillman, Patrick Young, and Per-Gunnar Martinsson

Abstract The paper describes techniques for constructing simplified models for problems governed by elliptic partial differential equations involving heterogeneous media. Examples of problems under consideration include electro-statics and linear elasticity in composite materials, and flows in porous media. A common approach to such problems is to either up-scale the governing differential equation and then discretize the up-scaled equation, or to construct a discrete problem whose solution approximates the solution to the original problem under some constraints on the permissible loads. In contrast, the current paper suggests that it is in many situations advantageous to directly approximate the *solution operator* to the original differential equation. Such an approach has become feasible due to recent advances in numerical analysis, and can in a natural way handle situations that are challenging to existing techniques, such as those involving, *e.g.* concentrated loads, boundary effects, and irregular micro-structures. The capabilities of the proposed methodology are illustrated by numerical examples involving domains that are loaded on the boundary only, in which case the solution operator is a boundary integral operator such as, *e.g.*, a Neumann-to–Dirichlet operator.

1 Introduction

1.1 Background

The purpose of this report is to draw attention to a number of recent developments in computational harmonic analysis that may prove helpful to the construction of simplified models for heterogeneous media. We consider problems modeled by

A. Gillman · P. Young · P.-G. Martinsson (✉)
Department of Applied Mathematics, University of Colorado at Boulder, USA
e-mail: adrianna.gillman@colorado.edu; Patrick.Young@colorado.edu; martinss@colorado.edu

B. Engquist et al. (eds.), *Numerical Analysis of Multiscale Computations*, Lecture Notes in Computational Science and Engineering 82, DOI 10.1007/978-3-642-21943-6_9,

elliptic PDEs such as electrostatics and linear elasticity in composite materials, and Stokes' flow in porous media.

Many different solution approaches have been proposed for the type of problems under consideration. A classical technique that works relatively well in situations where there is a clear separation of length-scales is to derive so called *homogenized equations* which accurately model the macro-scale behavior of the constitutive equations without fully resolving the micro-structure. The homogenized equations can sometimes be derived analytically, but they are typically obtained from numerically solving a set of equations defined on a *Representative Volume Element* (RVE). An unfortunate aspect of this approach is that its accuracy is held hostage to many factors that are outside of the control of the modeler. Phenomena that tend to lead to less accurate solutions include:

1. Concentrated loads.
2. Boundaries, in particular non-smooth boundaries.
3. Irregular micro-structures.

The accuracy cannot readily be improved using generic techniques, but a number of strategies for developing coarse-grained models for specific situations have been developed. A popular class of such methods consists of variations of finite element methods in which a discretization on the macro-scale is constructed by solving a set of local problems defined on a representative collection of patches of fully resolved micro-structure [17, 20, 35].

We contend that it is in many situations advantageous to approximate the *solution operator*, rather than the *differential operator*. For the elliptic problems under consideration in this paper, the solution operator takes the form of an integral operator with the Green's function of the problem as its kernel. That such operators should in principle allow compressed representations has been known for some time (at least since [4]), but efficient techniques for actually computing them have become available only recently.

To illustrate the viability of the proposed techniques, we demonstrate how they apply to a couple of archetypical model problems. We first consider situations in which the micro-structure needs to be fully resolved and a coarse-grained model be constructed computationally. We show that this computation can be executed efficiently, and that once it has been, the reduced model allows for very fast solves, and is highly accurate even in situations that are challenging to existing coarse-graining methods. We then show that the proposed methods can fully exploit the simplifications possible when an accurate model of the material can be derived from computations on an RVE.

1.2 Mathematical Problem Formulation

While the ideas described are applicable in a broad range of environments, we will for expositional clarity focus on scalar elliptic boundary value problems defined on some regular domain $\Omega \subset \mathbb{R}^2$ with boundary Γ. Specifically, we consider Neumann problems of the form

$$\begin{cases} -\nabla \cdot \big(a(x) \cdot \nabla\, u(x)\big) = 0, & x \in \Omega, \\ u_n(x) = f(x), & x \in \Gamma, \end{cases} \tag{1}$$

where $a : \Omega \to \mathbb{R}^{2\times 2}$ is a matrix-valued function that varies "rapidly" (on the length-scale of the micro-structure), and where $u_n(x)$ denotes the normal derivative of u at $x \in \Gamma$. Our objective is to rapidly construct $u|_\Gamma$, from a given boundary function f. We are interested both in the situation where we are allowed a pre-computation involving some given function a, and in the situation in which a is specified probabilistically.

Some of our numerical work will focus on the special case where (1) represents a two-phase material. To be precise, we suppose that Ω can be partitioned into two disjoint "phases," $\bar{\Omega} = \bar{\Omega}_1 \cup \bar{\Omega}_2$, and that there exist constants a_1 and a_2 such that

$$a(x) = \begin{cases} a_1\, I, & x \in \Omega_1, \\ a_2\, I, & x \in \Omega_2, \end{cases}$$

where I is the identity matrix. We further suppose that $\bar{\Omega}_2$ is wholly contained inside Ω, and let Γ_{int} denote the boundary between Ω_1 and Ω_2, see Fig. 1. Then (1) can more clearly be written

$$\begin{cases} -a_1 \Delta u(x) = 0, & x \in \Omega_1, \\ -a_2 \Delta u(x) = 0, & x \in \Omega_2, \\ [u](x) = 0, & x \in \Gamma_{\text{int}}, \\ [a\, u_n](x) = 0, & x \in \Gamma_{\text{int}}, \\ u_n(x) = f(x), & x \in \Gamma, \end{cases} \tag{2}$$

where for $x \in \Gamma$, $[u](x)$ and $[a\, u_n](x)$ denote the jumps in the potential and in the *flow* $-a(x)\nabla u(x)$ in the normal direction, respectively.

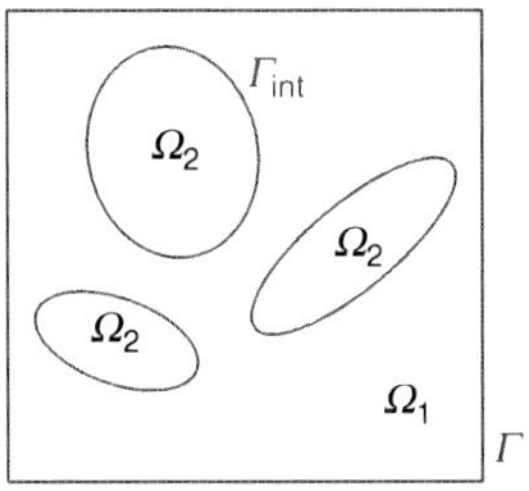

Fig. 1 A two phase domain

While the current paper concerns only situations modeled by equations of the types (1) and (2), the methodology extends to more general elliptic differential equations, see Sect. 5.

1.3 Coarse-Graining of the Differential Operator (Homogenization)

A classical technique [2, 14] for handling a problem such as (1) with a rapidly varying coefficient function a is to construct a function a_{hom} that varies on the macroscale only (or may even be constant) such that the solution u is in some sense approximated by the solution u_{hom} to

$$\begin{cases} -\nabla \cdot \big(a_{\text{hom}}(x)\cdot \nabla u_{\text{hom}}(x)\big) = 0, & x \in \Omega, \\ \partial_n\, u_{\text{hom}}(x) = f(x), & x \in \Gamma. \end{cases} \tag{3}$$

The derivation of an equation such as (3) typically relies on fairly strong assumptions on separation of length-scales, rendering this technique problematic in situations involving boundary effects, concentrated loads, multiple or undifferentiated length-scales, *etc*. A common technique for ameliorating these difficulties is to preserve a piece of the fully resolved micro-structure near the boundary, or the concentrated load, and then to "glue" the two models together.

Another common approach is to forego the construction of a coarse-grained continuum model and construct an equation involving a discretized differential operator whose solution in some sense captures the macro-scale behavior of the solution of (3), see *e.g.* [17]. The elements of the discretized matrix are typically constructed via local computations on patches of micro-structure.

1.4 Coarse-Graining of the Solution Operator

The premise of our work is that it is possible, and often advantageous, to approximate the *solution operator* of (1), rather than the differential operator itself. We will demonstrate that with this approach, many of the difficulties encountered in common coarse-graining strategies can be side-stepped entirely. To be precise, we note that mathematically, the solution to (1) takes the form

$$u(x) = [K\, f](x) = \int_\Gamma G(x,y)\, f(y)\, ds(y), \qquad x \in \Gamma, \tag{4}$$

where G is a kernel function that depends both on the function a, and on the domain Ω. It is known analytically only in the most trivial cases (such as a being constant, and Ω being a square or a circle). However, it turns out that the solution operator can be constructed numerically relatively cheaply, and that it admits very data-sparse representations.

Roughly speaking, our proposal is that instead of seeking an approximation of the form (3) of (1), it is often advantageous to seek an approximation of the form

$$u_{\text{hom}}(x) = [K_{\text{hom}}\, f](x) = \int_{\Gamma} G_{\text{hom}}(x, y)\, f(y)\, ds(y), \qquad x \in \Gamma.$$

of (4). The purpose of the manuscript is to demonstrate the basic viability and desirability of this approach. Specifically, we seek to:

1. Demonstrate via numerical examples that the solution operators can to high precision be approximated by "data-sparse" representations.
2. Illustrate a framework in which highly accurate reduced models can be constructed even for situations involving boundary effects, and concentrated loads.
3. Demonstrate that the reduced models can in many instances be computed inexpensively from statistical experiments on RVEs.
4. Demonstrate that in situations where the full micro-structure needs to be resolved, there exist highly efficient techniques for doing so, and that the resulting reduced models form natural building blocks in computational models.

Remark 1. In this paper, we focus on problems with no body load, such as (1). However, the ideas set out can equally well be applied to problems such as

$$\begin{cases} -\nabla \cdot \big(a(x) \cdot \nabla u(x)\big) = h(x), & x \in \Omega, \\ u_n(x) = f(x), & x \in \Gamma. \end{cases} \tag{5}$$

The mathematical solution operator then contains two terms, one corresponding to each of the two data functions f and h,

$$u(x) = \int_{\Gamma} G(x, y)\, f(y)\, ds(y) + \int_{\Omega} K(x, y)\, h(y)\, dA(y), \qquad x \in \Omega. \tag{6}$$

The second term in (6) is compressible in a manner very similar to that of the first.

Remark 2. A reason why approximation of the solution operator may prove advantageous compared to approximating the differential operator is hinted at by the spectral properties of the problem. For a bounded domain, an elliptic operator A such as the one defined by (1) or (2) typically has a discrete spectrum $(\lambda_n)_{n=1}^{\infty}$, where $\lambda_n \to \infty$, and where eigenfunctions get more oscillatory the larger λ_n is. In up-scaling A, we seek to construct an operator A_{hom} whose low eigenvalues and eigenfunctions approximate those of A. Measuring success is tricky, however, since the operator $A - A_{\text{hom}}$ is in many ways dominated by the high eigenvalues. One way of handling this is to consider multi-scale representations of the operators, see, *e.g.*, [1, 9, 16, 18, 19]. Another way is to try to approximate the *inverse* of the operator. We observe that A^{-1} is typically compact, and its dominant eigenmodes are precisely those that we seek to capture. Roughly speaking, we advocate the numerical construction of a finite dimensional operator T such that $||A^{-1} - T||$ is small.

Remark 3. Our goal with this paper is not to set up a mathematical analysis of the properties of kernels such as the function G in (4). However, to give a sense of the type of questions that arise, let us consider a situation where the function a in (1) represents a micro-structure with a characteristic length-scale λ. We then let d denote a cut-off parameter that separates the *near-field* from the *far-field*, say $d = 5\lambda$, and set

$$G_{\text{near}}(x,y) = \begin{cases} G(x,y), & |x-y| \leq d, \\ 0, & |x-y| > d, \end{cases} \qquad G_{\text{far}}(x,y) = \begin{cases} 0, & |x-y| \leq d, \\ G(x,y), & |x-y| > d, \end{cases}$$

and

$$u_{\text{near}}(x) = \int_{\Gamma} G_{\text{near}}(x,y)\, f(y)\, ds(y), \qquad u_{\text{far}}(x) = \int_{\Gamma} G_{\text{far}}(x,y)\, f(y)\, ds(y).$$

The function $y \mapsto G_{\text{near}}(x,y)$ depends strongly on the local micro-structure near x, and cannot easily be compressed. This part of the operator must be resolved sufficiently finely to fully represent the micro-structure. However, this is a local interaction, and u_{near} can be evaluated cheaply once G_{near} has been determined. In contrast, G_{far} is compressible. If Γ_1 and Γ_2 are two non-touching pieces of the boundary, then the integral operator

$$[T_{\Gamma_1 \leftarrow \Gamma_2} \sigma](x) = \int_{\Gamma_2} G_{\text{far}}(x,y)\, \sigma(y)\, ds(y), \qquad x \in \Gamma_1,$$

is not only compact, but its singular values typically decay exponentially fast, with the rate of decay depending on the sizes of Γ_1 and Γ_2, and on the distance between them. More careful analysis of these issues in an appropriate multi-scale framework can be found in [23].

2 Data-Sparse Matrices

A ubiquitous task in computational science is to rapidly perform linear algebraic operations involving very large matrices. Such operations typically exploit special *structure* in the matrix since the costs for methods capable of handling general matrices tend to scale prohibitively fast with matrix size: For a general $N \times N$ matrix, it costs $O(N^2)$ operations to perform a matrix-vector multiplication, $O(N^3)$ operations to perform Gaussian elimination or to invert the matrix, *etc.* A well-known form of structure in a matrix is sparsity. When at most a few entries in each row of the matrix are non-zero (as is the case, *e.g.*, for matrices arising upon the discretization of differential equations, or representing the link structure of the World Wide Web) matrix-vector multiplications can be performed in $O(N)$ operations instead of $O(N^2)$. The description *data-sparse* applies to a matrix that may be dense, but that shares the key characteristic of a sparse matrix that some linear algebraic operations, typically the matrix-vector multiplication, can to high precision be executed in fewer than $O(N^2)$ operations (often in close to linear time).

There are many different types of data-sparse representations of a matrix. In this paper, we will utilize techniques for so called *Hierarchically Semi-Separable* (HSS) matrices [11, 13, 33], which arise upon the discretization of many of the integral operators of mathematical physics, in signal processing, in algorithms for inverting certain finite element matrices, and in many other applications, see *e.g.* [12,29,33]. An HSS matrix is a dense matrix whose off-diagonal blocks are rank-deficient in a certain sense. Without going into details, we for now simply note that an HSS matrix A can be expressed via a recursive formula in L levels,

$$\mathsf{A}^{(\ell)} = \mathsf{U}^{(\ell)}\,\mathsf{A}^{(\ell-1)}\,\mathsf{V}^{(\ell)} + \mathsf{B}^{(\ell)}, \qquad \ell = 2, 3, \dots, L, \tag{7}$$

where $\mathsf{A} = \mathsf{A}^{(L)}$, and the sequence $\mathsf{A}^{(L)}, \mathsf{A}^{(L-1)}, \dots, \mathsf{A}^{(1)}$ consists of matrices that are successively smaller (typically, $\mathsf{A}^{(\ell-1)}$ is roughly half the size of $\mathsf{A}^{(\ell)}$). In (7), the matrices $\mathsf{U}^{(\ell)}$, $\mathsf{V}^{(\ell)}$ and $\mathsf{B}^{(\ell)}$ are all block-diagonal, so the formula directly leads to a fast technique for evaluating a matrix-vector product. The HSS property is similar to many other data-sparse representations in that it exploits rank-deficiencies in off-diagonal blocks to allow matrix-vector products to be evaluated rapidly; the Fast Multipole Method [24,25], Barnes-Hut [3], and panel clustering [26] are all similar in this regard. The HSS property is different from these other formats in that it also allows the rapid computation of a matrix inverse, of an LU factorization, *etc*, [10,11, 15,30,34]. The ability to perform algebraic operations other than the matrix-vector multiplication is also characteristic of the $\mathscr{H}$-matrix format of Hackbusch [28].

Remark 4. There currently is little consistency in terminology when it comes to "data-sparse" matrices. The property that we refer to as the "HSS" property has appeared under different names in, *e.g.*, [30–32, 34]. It is closely related to the "$\mathscr{H}^2$-matrix" format [5–7,27] which is more restrictive than the $\mathscr{H}$-matrix format, and often admits $O(N)$ algorithms.

Remark 5. This remark describes in which sense the off-diagonal blocks of a matrix that is compressible in the HSS-sense have low rank; it can safely be by-passed as the material here is referenced only briefly in Sect. 3.3. Let A denote an $N \times N$ HSS matrix A. Let I denote an index vector

$$I = [n+1, n+2, \dots, n+m],$$

where n and m are positive integers such that $n + m \le N$. Then we define the *HSS row block* R_I as the $m \times N$ matrix

$$\mathsf{R}_I = \left[\begin{array}{cccc|cccc|cccc} a_{n+1,1} & a_{n+1,2} & \cdots & a_{n+1,n} & 0 & 0 & \cdots & 0 & a_{n+1,n+m+1} & a_{n+1,n+m+2} & \cdots & a_{n+1,N} \\ a_{n+2,1} & a_{n+2,2} & \cdots & a_{n+2,n} & 0 & 0 & \cdots & 0 & a_{n+2,n+m+1} & a_{n+2,n+m+2} & \cdots & a_{n+2,N} \\ \vdots & \vdots & & \vdots & \vdots & \vdots & & \vdots & \vdots & \vdots & & \vdots \\ a_{n+m,1} & a_{n+m,2} & \cdots & a_{n+m,n} & 0 & 0 & \cdots & 0 & a_{n+m,n+m+1} & a_{n+m,n+m+2} & \cdots & a_{n+m,N} \end{array}\right]$$

In other words, R_I is an $m \times N$ sub-matrix of A corresponding to the rows marked by the index vector I, but with the diagonal block corresponding to I replaced by a zero matrix. The *HSS column block* C_I is analogously defined as the $N \times m$ matrix consisting of m columns of A with the diagonal block excised. The principal criterion for a matrix A to be compressible in the HSS sense is that its HSS blocks should have numerically low rank.

3 Case Study: A Discrete Laplace Equation on a Square

In this section, we illustrate how the coarse-graining techniques outlined in Sect. 1.4 can be applied to a discrete equation closely related to (1). This discrete equation can be viewed either as the result of discretizing (1) via a finite difference method, or as an equation that in its own right models, for instance, electro-statics on a discrete grid.

3.1 Problem Formulation

Given a positive integer N_{side}, we let Ω denote the $N_{\text{side}} \times N_{\text{side}}$ square subset of $\mathbb{Z}^2$ given by

$$\Omega = \{m = (m_1, m_2) \in \mathbb{Z}^2 : 1 \leq m_1 \leq N_{\text{side}} \text{ and } 1 \leq m_2 \leq N_{\text{side}}\}. \tag{8}$$

Figure 2a illustrates the definition. For a node $m \in \Omega$, we let $\mathbb{B}_m$ denote a list of of all nodes in Ω that directly connect to m. For instance, an interior node such as the node m shown in Fig. 2b would have the neighbor list

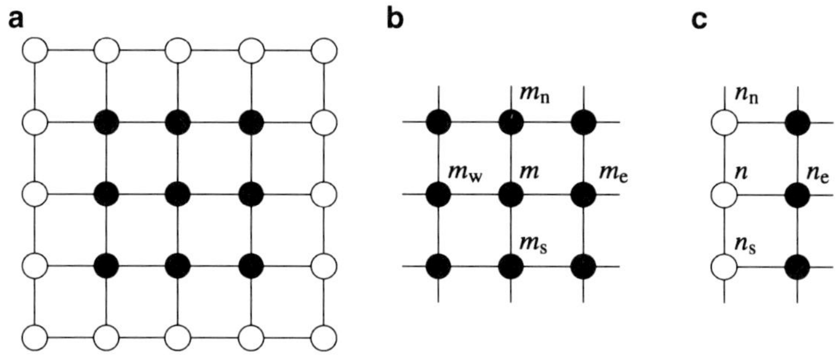

Fig. 2 Geometry of the lattice problem in Sect. 3.1. (**a**) The full lattice for $N_{\text{side}} = 5$. The boundary nodes in Ω_b are white and the interior nodes in Ω_i are black. (**b**) The four neighbors of an interior node m. (**c**) The three neighbors of a boundary node n

$$\mathbb{B}_m = \{m_s, m_e, m_n, m_w\},$$

while a node on a "western" boundary like n in Fig. 2c would have the neighbor list

$$\mathbb{B}_n = \{n_s, n_e, n_n\}.$$

For each pair $\{m, n\}$ of connected nodes, we let $\alpha_{m,n}$ denote a parameter indicating the *conductivity* of the link. For a function $u = u(m)$ where $m \in \Omega$, the *discrete Laplace operator* is then defined via

$$[\mathsf{A}\, u](m) = \sum_{n \in \mathbb{B}_m} \alpha_{m,n} \big[u(m) - u(n)\big]. \tag{9}$$

Example: For the case where $\alpha_{m,n} = 1$ for all connected nodes, we retrieve the standard five-point stencil associated with discretization of the Laplace operator. For instance, with column-wise ordering of the nodes in the lattice shown in Fig. 2a, we obtain the 25×25 matrix

$$\mathsf{A} = \begin{bmatrix} \mathsf{C} & -\mathsf{I} & 0 & 0 & 0 \\ -\mathsf{I} & \mathsf{D} & -\mathsf{I} & 0 & 0 \\ 0 & -\mathsf{I} & \mathsf{D} & -\mathsf{I} & 0 \\ 0 & 0 & -\mathsf{I} & \mathsf{D} & -\mathsf{I} \\ 0 & 0 & 0 & -\mathsf{I} & \mathsf{C} \end{bmatrix}, \tag{10}$$

where I is the 5×5 identity matrix and

$$\mathsf{C} = \begin{bmatrix} 2 & -1 & 0 & 0 & 0 \\ -1 & 3 & -1 & 0 & 0 \\ 0 & -1 & 3 & -1 & 0 \\ 0 & 0 & -1 & 3 & -1 \\ 0 & 0 & 0 & -1 & 2 \end{bmatrix}, \qquad \mathsf{D} = \begin{bmatrix} 3 & -1 & 0 & 0 & 0 \\ -1 & 4 & -1 & 0 & 0 \\ 0 & -1 & 4 & -1 & 0 \\ 0 & 0 & -1 & 4 & -1 \\ 0 & 0 & 0 & -1 & 3 \end{bmatrix}.$$

We let Ω_b denote the *boundary nodes* and we let Ω_i denote the *interior nodes* (*cf.* Fig. 2a). Partitioning the matrix A accordingly, the discrete analog of (1) becomes

$$\begin{bmatrix} \mathsf{A}_{b,b} & \mathsf{A}_{b,i} \\ \mathsf{A}_{i,b} & \mathsf{A}_{i,i} \end{bmatrix} \begin{bmatrix} u_b \\ u_i \end{bmatrix} = \begin{bmatrix} f_b \\ 0 \end{bmatrix}. \tag{11}$$

Solving for the boundary values of the potential, u_b, we find that[1]

$$u_b = \left(\mathsf{A}_{b,b} - \mathsf{A}_{b,i}\, \mathsf{A}_{i,i}^{-1}\, \mathsf{A}_{i,b}\right)^{-1} f_b.$$

In consequence, the discrete analog of the solution operator (in this case a discrete analog of the Neumann-to-Dirichlet operator) is

[1] Strictly speaking, the matrix $\mathsf{A}_{b,b} - \mathsf{A}_{b,i}\, \mathsf{A}_{i,i}^{-1}\, \mathsf{A}_{i,b}$ has a one-dimensional null-space formed by the constant functions and is not invertible. This is easily dealt with by a regularization that restricts attention to functions summing to zero. In what follows, such regularization will be employed where appropriate without further mention.

$$\mathsf{T} = \left(\mathsf{A}_{\mathrm{b,b}} - \mathsf{A}_{\mathrm{b,i}}\,\mathsf{A}_{\mathrm{i,i}}^{-1}\,\mathsf{A}_{\mathrm{i,b}}\right)^{-1}. \tag{12}$$

The operator T defined by (12) is dense, but turns out to be *data-sparse* in the sense described in Sect. 2. We will in this section substantiate this claim via numerical examples, and also outline strategies for rapidly constructing such an operator in different environments.

3.2 Model Problems

The compressibility of the solution operator T defined by (12) was investigated in the following five model environments:

Case A: Constant conductivities. In this model, all conductivities are identically one,

$$\alpha_{m,n} = 1 \qquad \text{for each connected pair } \{m, n\}. \tag{13}$$

For $N_{\mathrm{side}} = 5$, the resulting matrix A is the one given as an example in (10). Since in this case the matrix A can be viewed as a discretization of the Laplace operator $-\Delta$ on a square, the solution operator T can be viewed as a discrete analog of the standard Neumann-to-Dirichlet operator associated with Laplace's equation.

Case B: Smooth periodic conductivities. This case is a discrete analog of the equation

$$-\nabla\cdot\big(b(x)\nabla u(x)\big) = f(x), \qquad x \in [0,1]^2, \tag{14}$$

where b is a periodic function defined by

$$b(x) = 1 - 0.9\big(\cos(\pi\, N_{\mathrm{cells}}\, x_1)\big)^2\big(\cos(\pi\, N_{\mathrm{cells}}\, x_2)\big)^2, \qquad x = (x_1, x_2) \in [0,1]^2. \tag{15}$$

In other words, (14) models a medium whose conductivity repeats periodically across $N_{\mathrm{cells}} \times N_{\mathrm{cells}}$ cells in the square $[0,1]^2$. Figure 3a illustrates the function b for $N_{\mathrm{cells}} = 4$. A discrete analog of (14) is now obtained by setting

$$\alpha_{m,n} = b\left(\frac{m+n-2}{2\,(N_{\mathrm{side}}-1)}\right) \qquad \text{for each connected pair } \{m, n\}.$$

In our experiments, we chose N_{cell} so that 25 nodes were used to resolve each period, $N_{\mathrm{cell}} = (N_{\mathrm{side}} - 1)/25$ (for clarity, Fig. 3 shows a solution with only 15 nodes per period). In this case, the solutions are typically oscillatory on the boundary, *cf.* Fig. 3b. This is a basic two-scale problem that should be amenable to traditional homogenization techniques provided there is a sufficient separation of length-scales.

Case C: Random conductivities. The conductivities $\alpha_{m,n}$ are for each connected pair of nodes $\{m, n\}$ drawn independently from a uniform probability distribution on $[1, 2]$. In this case, there is no local regularity, but we would expect traditional

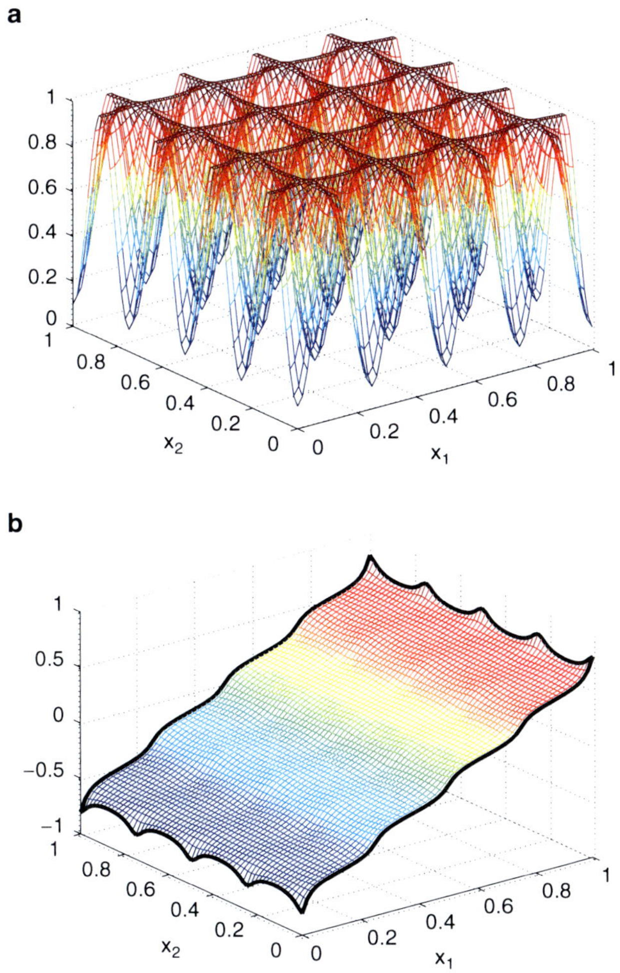

Fig. 3 The periodic problem described as *Case B* in Sect. 3.2 with $N_{\text{cells}} = 4$ and $N_{\text{side}} = 61$. (**a**) The function $b = b(x)$ defined by (15). (**b**) A solution to the Neumann problem (11) with a constant inflow at $x_1 = 1$ and a constant outflow at $x_1 = 0$

homogenization to give accurate results whenever the length-scales are sufficiently separated.

Case D: Sparsely distributed missing bars. In this model, all bars are assigned conductivity 1 (as in Case A), but then a small percentage p of bars are completely removed (in the examples reported, $p = 4\%$). In other words,

$$\alpha_{m,n} = \begin{cases} 1, \text{ with probability } 1-p & \text{if } \{m,n\} \text{ is a connected pair,} \\ 0, \text{ with probability } p & \text{if } \{m,n\} \text{ is a connected pair,} \\ 0, & \text{if } \{m,n\} \text{ is not a connected pair.} \end{cases}$$

As in Case C, there is no local regularity, but we would expect traditional homogenization to give accurate results whenever the length-scales are sufficiently separated.

Case E: A lattice with two long cracks. This model is similar to Case D in that a small number of links have been cut, and all the remaining ones have unit conductivity. However, we organized the cut links into two long cracks running through the lattice. Figure 4a illustrates for a case where $N_{\text{side}} = 50$. In larger lattices, the cracks have the same proportions, but the gap between the two cracks is kept constant at four links. In this case, solutions may exhibit major discontinuities. Figure 4b illustrate the electric field resulting from placing oppositely signed unit sources at the locations marked *source* and *sink* in Fig. 4a. We would expect analytic derivation of a simplified model to be very hard work in a situation such as this.

3.3 Compressibility of the Solution Operator

While the operator T defined by (12) is dense, it is in many situations of interest *data-sparse* in the sense described in Sect. 2. To illustrate this point, we computed the matrix T by brute force for several different lattices, compressed it into the HSS format to ten digits of accuracy (we enforced that local truncation errors be less than 10^{-10}), and looked at how much memory was required to store the result. Tables 1 and 2 show our findings for each of the five different models described in Sect. 3.2, and for differently sized lattices. To provide more detail, Table 3 reports the average ranks of the so called "HSS blocks" (as defined in Remark 5) of a $6\,396 \times 6\,396$ matrix T associated with a $1\,600 \times 1\,600$ square domain for each of the five examples.

An interesting aspect of the reported data is that the matrix T associated with the classical five-point stencil (represented by Case A) is highly compressible. To store it to ten digits of accuracy, less than 100 floating point numbers are required for each degree of freedom (see Table 2). This fact has been exploited in a series of recent papers, including [12, 23, 29]. What is perhaps more remarkable is that the compressibility property is extremely robust to small changes in the micro-structure. As Tables 1–3 show, there is almost no discernible difference in compressibility between the five models considered.

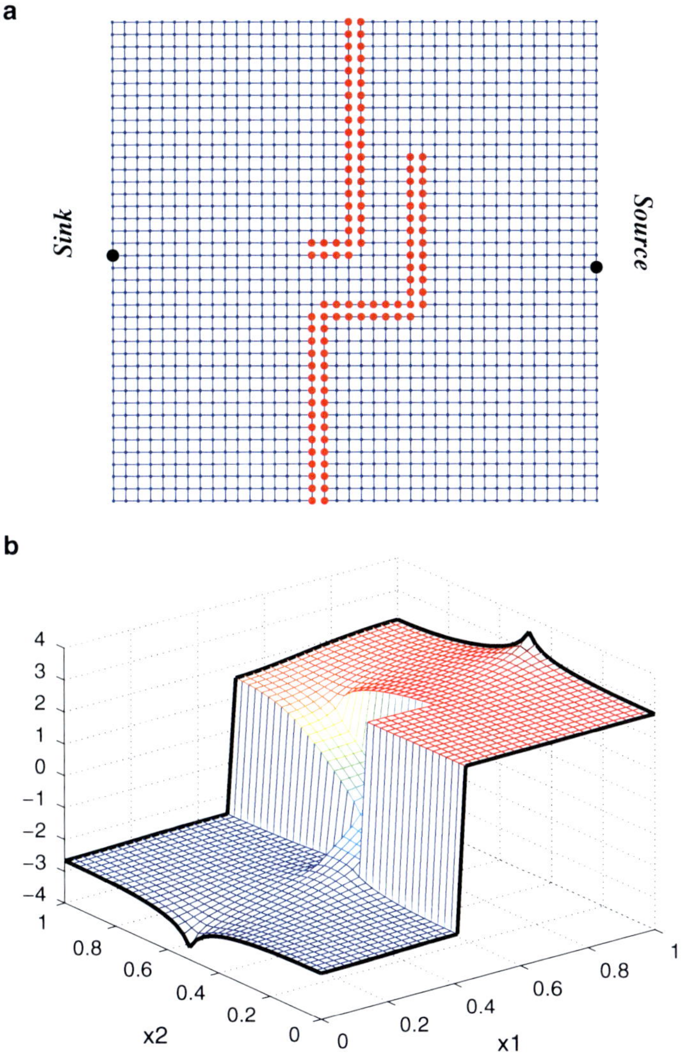

Fig. 4 (**a**) The lattice with cracks described as *Case E* in Sect. 3.2 for $N_{\text{side}} = 40$. (**b**) A solution to the Neumann problem (11) with a unit inflow at the location marked *source* in (**a**), and a unit outflow at the location marked *sink*

Table 1 Memory requirements in KB. The table shows the amount of memory (in KB) required for storing the matrix T defined by (12) for different problem sizes N_{size}. The first line gives the memory required for storing a general dense matrix of size $4(N_{side} - 1) \times 4(N_{side} - 1)$. The following lines give the amount of memory required to store T in the "HSS" data-sparse format described in Sect. 2 for each each of the five cases described in Sect. 3.2, to within precision 10^{-10}

	$N_{side} = 100$	200	400	800	1,600
General matrix	1.23e3	4.95e3	1.99e4	7.98e4	3.20e5
Case A (constant)	3.02e2	6.13e2	1.22e3	2.42e3	4.78e3
Case B (periodic)	2.97e2	6.06e2	1.21e3	2.38e3	4.69e3
Case C (random 1)	3.03e2	6.20e2	1.23e3	2.43e3	4.80e3
Case D (random 2)	2.96e2	6.06e2	1.20e3	2.38e3	4.70e3
Case E (cracks)	2.96e2	6.10e2	1.22e3	2.42e3	4.77e3

Table 2 Memory requirements in words per degree of freedom. The table shows the same data given in Table 1, but now scaled to demonstrate that the memory requirement scales linearly with problem size. To be precise, the entries given are the number of "words" (the memory required to store a floating point number to double precision accuracy) required per node on the boundary

	$N_{side} = 100$	200	400	800	1,600
General matrix	396	796	1,596	3,196	6,396
Case A (constant)	97.7	98.6	98.1	96.8	95.7
Case B (periodic)	95.9	97.4	96.7	95.4	93.9
Case C (random 1)	97.8	99.7	98.8	97.5	96.0
Case D (random 2)	95.5	97.5	96.6	95.4	94.1
Case E (cracks)	95.7	98.1	97.7	96.8	95.5

Table 3 HSS ranks of the Schur complements for a matrix of size $6,396 \times 6,396$. The table shows the HSS-ranks (as described in Remark 5) of blocks in the solution operator for the different models. The reported rank was the average numerical rank (at precision 10^{-10}) over all HSS blocks of size N_{block} that arise in the compressed representation

	$N_{block} = 50$	100	200	400	800	1,600
General matrix	50	100	200	400	800	1,600
Case A (constant)	19.3	22.7	26.0	31.0	39.0	53.0
Case B (periodic)	18.8	21.6	24.8	29.3	37.0	50.0
Case C (random 1)	19.3	22.8	26.8	31.6	39.8	54.0
Case D (random 2)	18.7	21.9	25.5	30.8	38.8	52.5
Case E (cracks)	19.2	22.7	25.9	30.9	38.8	52.5

Once the compressed solution operator has been computed, it can be applied to a vector more or less instantaneously. For our simple implementation, we found the following for the time t_{solve} (in seconds) required for a single solve:

N_{side}	200	400	800	1,600	3,200
t_{solve} (sec)	4.4e-3	8.7e-3	1.8e-2	3.4e-2	7.1e-2

These numbers refer to a reduced model that is precise to within ten digits, and we would like to emphasize that the largest example reported, which requires 0.07 s for one solve, involves a problem whose micro-structure was originally resolved using $3\,200 \times 3\,200 \approx 10^7$ nodes.

The results reported in Tables 1–3 indicate that reduced models that are precise to within ten digits of accuracy in principle exist, even in the presence of the following complications:

- Solutions that are oscillatory on the boundary, even when the period of the oscillation is not very much smaller than the size of the domain (as in Case B).
- Solutions that are highly irregular on the boundary (as in Cases C, D, and E).
- Boundary loads that exhibit no smoothness. (We observe that the solution operator is constructed under no assumption on smoothness of the boundary data.)
- Solutions that involve significant discontinuities (as shown in Fig. 4b).

In Sects. 3.4, 3.5, and 3.6, we will describe practical techniques for inexpensively computing such reduced models.

3.4 Techniques for Computing the Solution Operator That Fully Resolve the Micro-Structure

Given a realization of a lattice model, the operator T defined by (12) can of course be computed with brute force. While Gaussian elimination has an $O(N_{\text{side}}^6)$ asymptotic cost that quickly becomes prohibitive, substantially more efficient techniques exist. Appendix describes a variation of the classical *nested dissection* method which in the present environment requires $O(N_{\text{side}}^3)$ floating point operations (flops) and $O(N_{\text{side}}^2)$ memory. This technique is exact up to rounding errors, and is very easy to implement. It was used to calculate the numbers reported in Sect. 3.3 and is sufficiently fast that the solution operator associated with an 800×800 lattice can be determined in 40 s via a Matlab implementation running on a standard desktop PC.

More recently, techniques have been developed that compute an operator such as T in $O(N_{\text{side}}^2)$ time (or possibly $O(N_{\text{side}}^2(\log N_{\text{side}})^\kappa)$ for a small integer κ), which is optimal since there are $O(N_{\text{side}}^2)$ links in the lattice [12,23,29]. These techniques are highly efficient, and enable the brute force calculation of a reduced model in many important environments in both two and three dimensions. For a brief introduction, see Section "Accelerations" in Appendix.

3.5 Techniques Accelerated by Collecting Statistics from a Representative Volume Element

In situations where there is a good separation of length-scales, variations of classical homogenization techniques can be used to dramatically accelerate the computation of a compressed boundary operator. To illustrate, let us investigate Case C in Sect. 3.2 (the case of random conductivities, drawn uniformly from the interval $[1, 2]$). The most basic "homogenized equation" is in this case a lattice in which all links have the same conductivity. Through experiments on an RVE, we determined that this conductivity should be

$$c_3 = 1.4718\cdots$$

We let T_{hom} denote the solution operator (*i.e.* the lattice Neumann-to-Dirichlet operator) for the homogenized lattice. We measured the discrepancy between the homogenized operator T_{hom}, and the operator associated with the original lattice T, using the measures:

$$E_{\text{N2D}} = \frac{||\mathsf{T}_{\text{hom}} - \mathsf{T}||}{||\mathsf{T}||}, \qquad \text{and} \qquad E_{\text{D2N}} = \frac{||\mathsf{T}_{\text{hom}}^{-1} - \mathsf{T}^{-1}||}{||\mathsf{T}^{-1}||}. \tag{16}$$

The first column of Table 4 gives the results for a particular realization of a 50×50 lattice. In addition to the discrepancies measured in operator norm, the table also provides the errors

$$E_{\text{smooth}} = \frac{||\left(\mathsf{T}_{\text{hom}} - \mathsf{T}\right) f_{\text{smooth}}||}{||\mathsf{T}\, f_{\text{smooth}}||}, \qquad \text{and} \qquad E_{\text{rough}} = \frac{||\left(\mathsf{T}_{\text{hom}} - \mathsf{T}\right) f_{\text{rough}}||}{||\mathsf{T}\, f_{\text{rough}}||}, \tag{17}$$

associated with two particular Neumann data vectors f_{smooth} and f_{rough}. The solutions associated with these data vectors are shown in Fig. 5. These examples show that as one would expect, the homogenized equation provides quite high accuracy for a smooth solution, and very poor accuracy for a rough one. (Table 4 also reports errors associated with improved "buffered" homogenization schemes, which will be introduced in Sect. 3.6.)

We next repeated all experiments for Case D (as defined in Sect. 3.2). In this case, numerical experiments indicated that the homogenized conductivity is

$$c_4 = 1 - \frac{1}{2}p + O(p^2).$$

The first column of Table 5 shows the errors associated with a realization of "Case D" on a 50×50 grid, with $p = 0.04$, and $c_4 = 0.98$.

Remark 6 (Computational cost). The solution operator T_{hom} associated with a constant coefficient lattice can be computed in time proportional to $O(N_{\text{side}})$ (in other words, in time proportional to the number of nodes on the boundary). This means that very large lattices can be handled rapidly. It was demonstrated in [22]

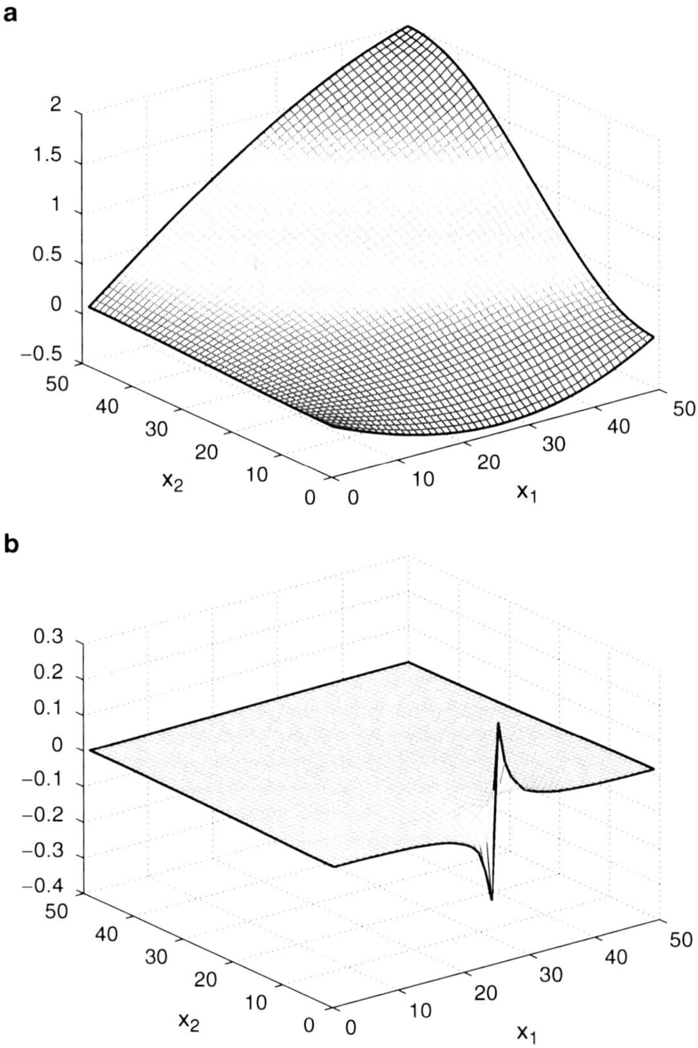

Fig. 5 Solutions for non-homogenized equation. (**a**) Solution resulting from the smooth boundary data f_{smooth}. (**b**) Solution resulting from the rough boundary data f_{rough}

Table 4 Errors in homogenized operator for "Case C". Discrepancy between the solution operator of an given lattice, and the homogenized solution operator. These numbers refer to the model described as "Case C" in Sect. 3.2 (random conductivities). The errors $E_{\rm D2N}$, $E_{\rm N2D}$, $E_{\rm smooth}$, and $E_{\rm rough}$ are defined in (16) and (17)

	No buffer	Homogenization with buffer of width b					
		$b=1$	$b=2$	$b=3$	$b=4$	$b=5$	$b=10$
$E_{\rm D2N}$	1.9e-01	5.4e-03	1.2e-03	3.9e-04	3.3e-04	1.3e-04	6.6e-05
$E_{\rm N2D}$	1.1e-02	7.5e-03	5.6e-03	5.7e-03	4.3e-03	4.9e-03	2.4e-03
$E_{\rm smooth}$	7.3e-03	4.1e-03	4.1e-03	4.1e-03	2.8e-03	2.6e-03	1.4e-03
$E_{\rm rough}$	1.5e-01	2.1e-02	1.1e-02	2.2e-03	8.8e-04	3.5e-03	9.2e-04

Table 5 Errors in homogenized operator for "Case D". Discrepancy between the solution operator of an given lattice, and the homogenized solution operator. These numbers refer to the model described as "Case D" in Sect. 3.2 (randomly cut bars). The errors $E_{\rm D2N}$, $E_{\rm N2D}$, $E_{\rm smooth}$, and $E_{\rm rough}$ are defined in (16) and (17)

	No buffer	Homogenization with buffer of width b					
		$b=1$	$b=2$	$b=3$	$b=4$	$b=5$	$b=10$
$E_{\rm D2N}$	4.4e-01	1.5e-02	4.5e-03	1.7e-03	1.2e-03	7.6e-04	3.3e-04
$E_{\rm N2D}$	8.7e-02	6.1e-02	5.6e-02	5.2e-02	4.5e-02	4.4e-02	2.8e-02
$E_{\rm smooth}$	7.4e-02	5.9e-02	5.4e-02	4.8e-02	4.2e-02	4.1e-02	2.7e-02
$E_{\rm rough}$	1.0e-01	7.0e-02	6.8e-02	6.2e-02	5.1e-02	5.0e-02	3.4e-02

that the solution operator associated with a lattice with 10^{10} nodes can be computed in less than two minutes on a standard desktop PC. (Observe that only the $4 \cdot 10^5$ nodes on the boundary actually need to enter the calculation.)

3.6 Fusing a Homogenized Model to a Locally Fully Resolved Region

In the environments under consideration here, domains are loaded only on the border. This of course raises the possibility of improving the accuracy in the homogenized model by preserving the actual micro-structure in a thin strip along the boundary, and use the homogenized equations only in the interior. In the framework proposed here, where the simplified model consists of a solution operator rather than a differential operator (or in the present case, difference operator), it is extra ordinarily simple to do so.

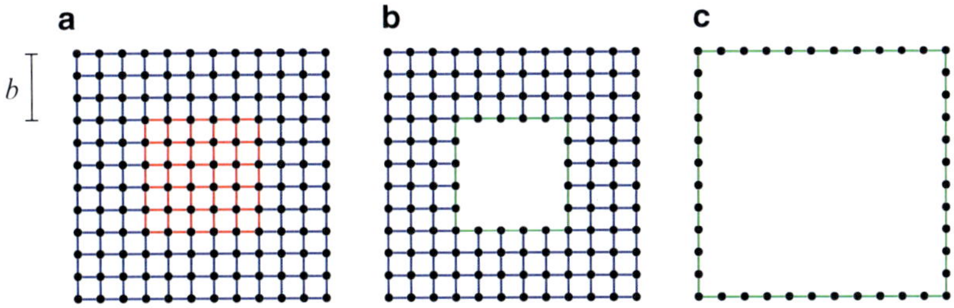

Fig. 6 Construction of a highly accurate reduced model by fusing a homogenized region with a region in which the micro-structure is fully resolved. (**a**) The blue links are within distance b of the boundary, and maintain their original conductivity. The red links are all assigned the "homogenized" conductivity. (**b**) All red links are eliminated from the model. This requires the construction of the solution operator for a constant coefficient lattice at cost $O(N_{\text{side}})$ (see Remark 6). (**c**) The few remaining links are eliminated to construct a highly approximate approximation to the solution operator

To illustrate, suppose that we are given a realization of an $N_{\text{side}} \times N_{\text{side}}$ lattice with heterogeneous conductivities. We fix a parameter b that indicates how broad of a band of cells we preserve, and then replace all bars that are more than b cells away from the boundary by bars with the homogenized conductivity, as illustrated in Fig. 6a. Then use the techniques of Sect. 3.5 to compute the Neumann-to-Dirichlet operator for the constant coefficient lattice of size $(N_{\text{side}}-2b) \times (N_{\text{side}}-2b)$ in the center. As observed in Remark 6, the cost is only $O(N_{\text{side}})$, and the new reduced model involves only $O(N_{\text{side}})$ degrees of freedom. As Tables 4 and 5 demonstrate, for our model problems ("Case C" and "Case D") keeping only five layers of the original lattice leads to a reduced model that is accurate to three or four digits.

Remark 7 (Accuracy of Neumann vs. Dirichlet problems). Tables 4 and 5 show that when "unbuffered" homogenization is used, the resulting error E_{D2N} associated with Dirichlet problems is significantly larger than the error E_{N2D} associated with Neumann problems. The tables also show that the accuracy of Dirichlet problems improve dramatically upon the introduction of even a very thin boundary layer. This is as one would expect since the Dirichlet-to-Neumann operator is dominated by short range interactions.

4 Case Study: Two-Phase Media

In this section, we briefly investigate the compressibility of the Neumann-to-Dirichlet operator for a two-phase material modeled by (2). The two geometries we consider are shown in Fig. 7, with the conductivity of the inclusions set to zero. In this case, the operator under consideration is a boundary integral operator T supported on the square outer boundary. Using techniques described in Remark 8, we constructed an $1,144 \times 1,144$ matrix T that approximated T. With this number of nodes, any Neumann data generated by point sources up to a distance of 0.5% of the side length of the square can be resolved to eight digits of accuracy. We compressed the matrix T into the HSS format described in Sect. 2 to a relative precision of 10^{-10}. The resulting data required 1.19KB of memory to store for the geometry shown in Fig. 7a, and 1.22KB of memory for the geometry shown in Fig. 7b. This corresponds to about 135 words of storage per row in the matrix. The HSS-ranks (as defined in Remark 5) are reported in Table 6. We make three observations:

- A compressed version of the boundary operator can in this case be stored using about the same amount of memory (100 words per degree of freedom) as the operators associated with the discrete problems described in Sect. 3.
- The two geometries shown in Fig. 7 require about the same amount of memory. This is note-worthy since the one labeled (b) corresponds to an almost singular geometry in which the domain is very close to being split in two halves. The effect is illustrated the solution shown in Fig. 8b where steep gradients are seen in middle of the piece. Standard assumptions used when homogenizing an elliptic differential operator are violated in this case.
- In Table 6, the ranks of HSS-blocks of size 143 are *larger* than those of HSS-blocks of size 286. We speculate that this unusual situation can be traced to the fact that the larger blocks are larger than the inclusions, and largely do not "see" the heterogeneities.

Table 6 Average ranks of HSS blocks for composite material example in Sect. 4. The average HSS-ranks (as defined in Remark 5) for the blocks in a data-sparse representation of the Neumann-to-Dirichlet operator for the geometries shown in Fig. 7

	$N_{\text{block}} = 36$	71	143	286
Geometry shown in Fig. 7a	18.2	27.0	39.5	25.8
Geometry shown in Fig. 7b	18.3	27.3	41.1	28.0

Remark 8 (Details of computation). To derive our approximation to the Neumann-to-Dirichlet operator, we recast the Neumann Laplace equation (2) as a BIE defined on the joint boundary $\Gamma \cup \Gamma_{\text{int}}$. In the present case with non-conducting inclusions, the boundary condition on all interior boundaries simplifies to a homogeneous Neumann condition. We represented the solution as a single layer representation supported on both the outer boundary Γ and the interior boundary Γ_{int}. In other words, we sought a solution of the form

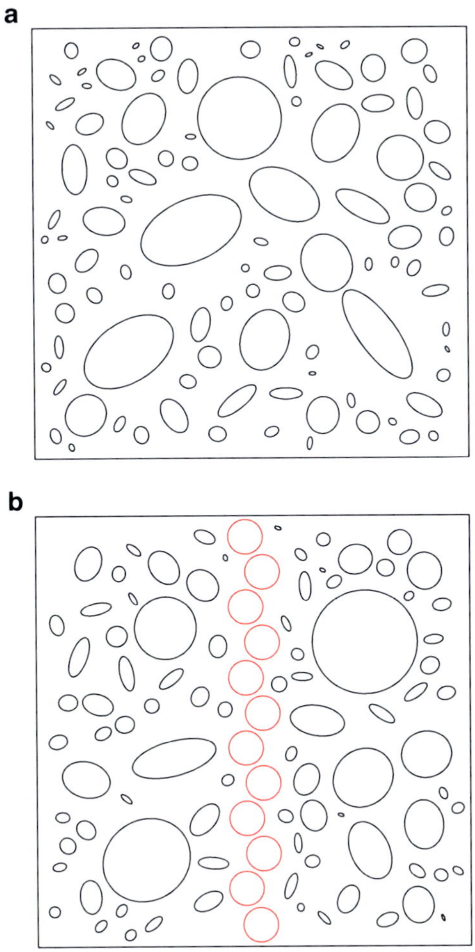

Fig. 7 Geometry for computations in Sect. 4. (**a**) A perforated material. (**b**) A perforated material with a chain of holes that almost line up

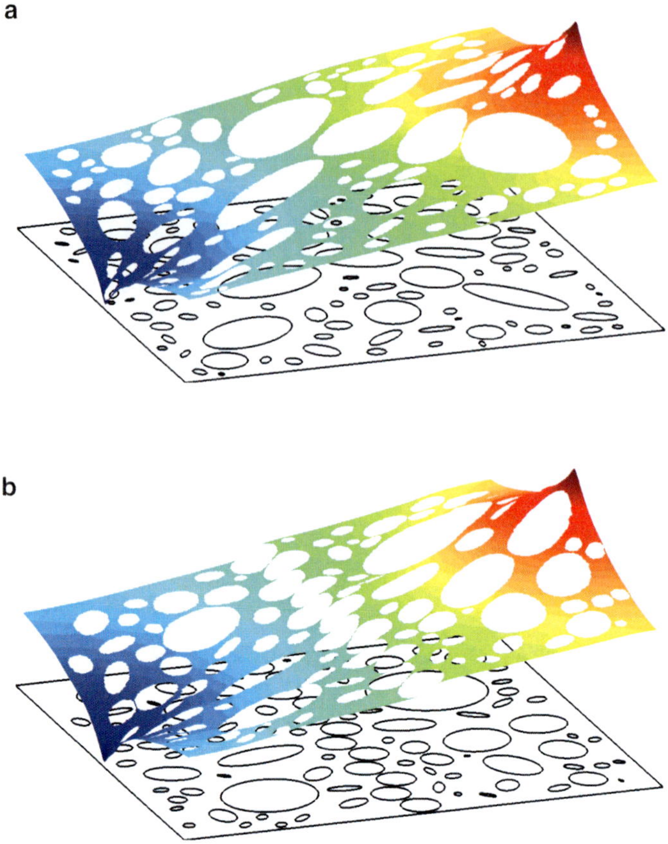

Fig. 8 Solutions to the Laplace's equation with Neumann boundary conditions on the geometries (**a**) and (**b**) shown in Fig. 7. The boundary flux is set to be identically zero, except for two point sources of strengths ± 1

$$u(x) = \int_{\Gamma} \log|x-y|\,\sigma(y)\,ds(y) + \int_{\Gamma_{\text{int}}} \log|x-y|\,\tau(y)\,ds(y). \tag{18}$$

The resulting BIE was discretized using a Nyström method combined with trapezoidal quadrature on the interior holes, and a Gaussian quadrature on the exterior boundary supported on 44 panels with 26 nodes each. The quadrature rule was locally modified as described in [8] to maintain eight digit accuracy in the presence of corners. This resulted in a large linear system from which all degrees of freedom associated with internal nodes (those associated with the density τ in (18)) were eliminated. The resulting Schur complement was multiplied by a matrix representing evaluation of a single layer potential on the boundary to produce the final discrete approximation T to the "true" analytic Neumann-to-Dirichlet operator T.

5 Generalizations

This report focused on problems modeled by simple Laplace-type problems in two dimensions involving no body loads. However, the techniques can be extended to much more general environments:

Other boundary conditions: While we focused on problems with Neumann boundary conditions, the extension to Dirichlet or mixed boundary conditions is trivial.

Other elliptic equations: The methods described extend readily to other elliptic equations whose kernels are non-oscillatory such as Stokes, elasticity, Yukawa, *etc*. The extension to wave problems modeled by Helmholtz equation, or the time-harmonic version of Maxwell, is more complicated for two reasons: (1) The presence of resonances (both true ones corresponding to the actual physics, and artificial ones present in the mathematical model only) must be dealt with. This can be done, but requires careful attention. (2) As the wave-number increases, the compressibility of the solution operator deteriorates, and eventually renders the proposed approach wholly unaffordable.

Body loads: The extension to problems involving body loads is in principle straightforward (see Remark 1). However, the compressed solution operator becomes more expensive to store.

Problems in three dimensions: In principle, the methodology proposed extends straight-forwardly to problems in three dimensions. However, the construction of the solution operator does become more expensive, and the method might be best suited for environments where a pre-computation is possible, or where the construction of the solution operator can be accelerated via the use of homogenized models in parts of the domain (as illustrated in Sect. 3.6). Moreover, for problems in three dimensions involving body loads, memory requirements may become prohibitive.

6 Conclusions

The purpose of this report is to attempt to draw attention to recent developments in numerical analysis that could be very useful in modeling heterogeneous media. Specifically, it has become possible to inexpensively compute an approximation to the solution operator associated with many elliptic PDEs, and to perform various operations involving such solution operators: addition, multiplication, inversion, merging operators for different sub-domains, *etc.* We argue that such solution operators form excellent "reduced models" for many problems that have proven difficult to handle using traditional homogenization techniques.

Constructing reduced models by approximating the solution operator is particularly advantageous in the following environments:

Domains that are loaded on the boundary only: For problems that involve no body load, the solution operator is defined on the boundary only. This reduction in dimensionality means that once it is computed, it can be stored very efficiently, and applied to vectors sufficiently fast that real time simulations become possible. For some problems in this category, the actual construction of the solution operator requires a large-scale (but very efficient) computation involving the entire micro-structure, but as shown in Sect. 3.6, the solution operator can sometime be dramatically accelerated by using a homogenized model in the interior of the domain.

Situations where a pre-computation is possible: When the entire micro-structure needs to be resolved (as happens when the problem involves a body load, or a micro-structure not suitable for homogenization methods), the initial construction of the solution operator can become somewhat expensive, in particular for problems in three dimensions. However, once it has been constructed, it can usually be applied to a vector very rapidly. This raises the possibility of pre-computing a library of compressed models which can then be used as building blocks in computational simulations.

Problems in two dimensions (whether involving volume loads or not): Given current trends in algorithmic and hardware development, we predict that for a great many problems in two dimensions, it will soon become entirely affordable to resolve the entire micro-structure, and computationally derive a reduced model of the solution operator. The automatic nature of such a procedure would save much human effort, and would be very robust in the sense that the computed model would be guaranteed to be accurate to whichever tolerance was requested.

Appendix: Efficient computation of the Neumann-to-Dirichlet operator

In this appendix, we describe an efficient technique for computing the Neumann-to-Dirichlet operator T defined by (12). It is a variation of the classical *nested dissection* techniques [21]. Throughout the appendix, Ω is a rectangular lattice, as defined by (8), and A is an associated discrete Laplace operator, as defined by (9).

To be precise, the technique we will describe does not compute the Neumann-to-Dirichlet operator T, but rather the *Schur complement* S, defined via

$$\mathsf{S} = \mathsf{A}_{\mathrm{b,b}} - \mathsf{A}_{\mathrm{b,i}}\,\mathsf{A}_{\mathrm{i,i}}^{-1}\,\mathsf{A}_{\mathrm{i,b}}. \tag{19}$$

Comparing (12) and (19), we see that $\mathsf{T} = \mathsf{S}^{-1}$.

Outline

The technique is a divide-and-conquer scheme in which the computational domain Ω is first split into $2^L \times 2^L$ roughly equisized small boxes. The parameter L is chosen so that each of the small boxes is sufficiently small that its Schur complement can be computed by evaluating (19) via brute force. (In practice, we found that letting the smallest boxes be of size roughly 50×50, or $L \approx \log_2(N_{\mathrm{side}}/50)$, works well.) Then it turns out to be possible to merge the Schur complements of two small adjacent boxes to form the Schur complement of the larger box; the process is described in Section "Merging of Two Schur Complements" in Appendix. The scheme proceeds by continuing the merge process to form the Schur complements of larger and larger boxes until eventually the entire box Ω has been processed. To illustrate, we describe the process graphically for a 24×24 domain that is originally split into 4×4 boxes, each containing 6×6 nodes.

Step 1: Partition the box Ω into 16 small boxes. For each box, identify the internal nodes (marked in blue) and eliminate them using formula (19).

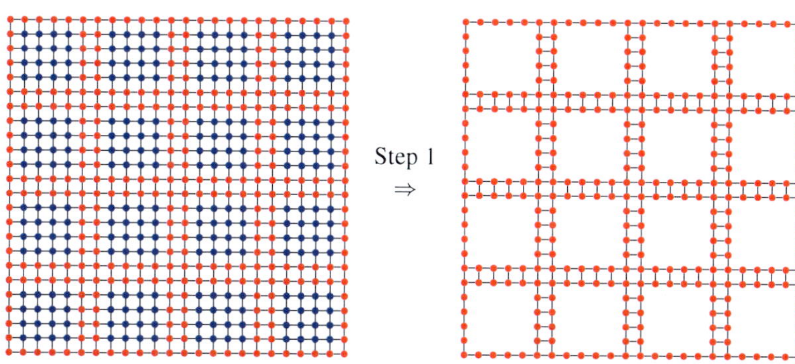

Step 2: Join the small boxes by pairs to form the Schur complements of boxes holding twice the number of nodes via the process to be described in Section "Merging of Two Schur Complements" in Appendix. The effect is to eliminate the interior nodes (marked in blue) of the newly formed larger boxes.

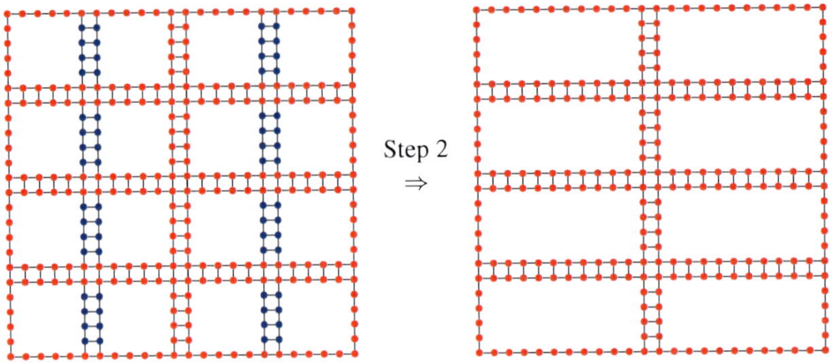

Step 3: Merge the boxes created in Step 2 in pairs, again via the process described in Section "Merging of Two Schur Complements" in Appendix.

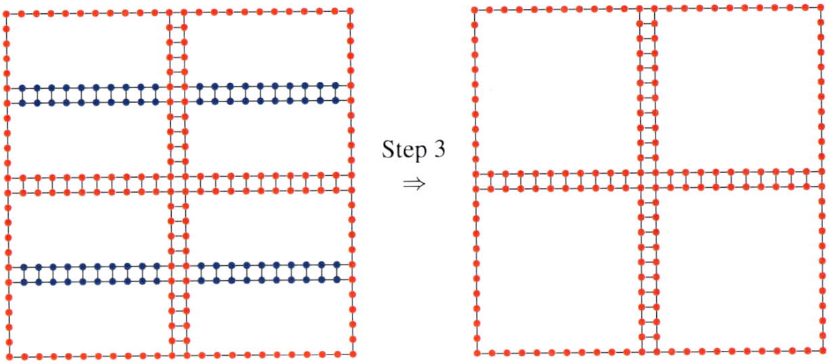

Step 4: Repeat the merge process once more.

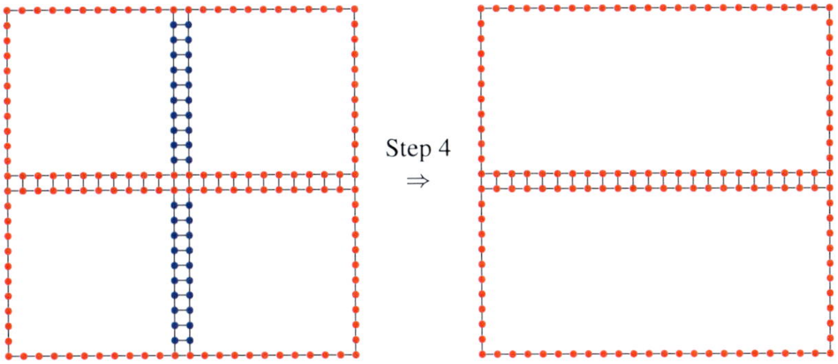

Step 5: Repeat the merge process one final time to obtain the Schur complement associated with the top level box Ω.

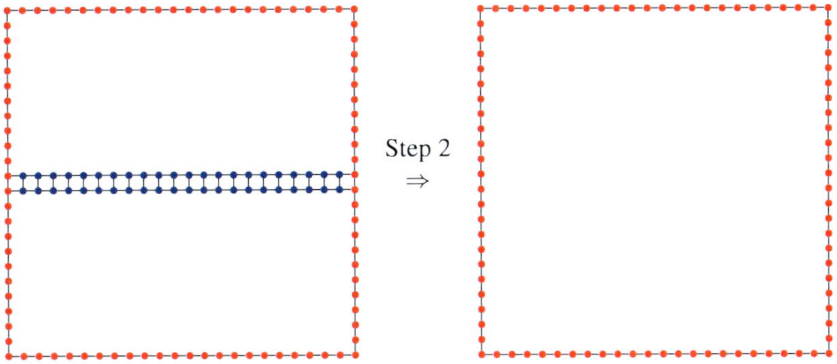

Merging of Two Schur Complements

Suppose that Ω is a box consisting of the two smaller boxes Ω_w and Ω_e (as in west and east):

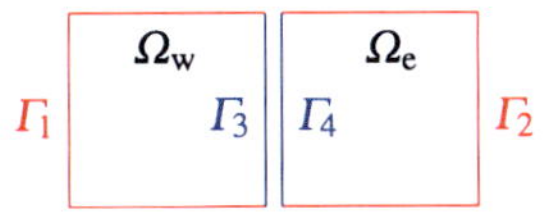

Suppose further that we know the corresponding Schur complements S_w and S_e and seek the Schur complement S of Ω. In effect, we need to remove the "interior" points along the middle lines (marked in blue in the figure).

First partition the nodes in Γ_w into the subsets Γ_1 and Γ_3, and partition Γ_e into Γ_2 and Γ_4 as shown in the figure. The Schur complements S_w and S_e are partitioned accordingly,

$$\mathsf{S}_w = \begin{bmatrix} \mathsf{S}_{11} & \mathsf{S}_{13} \\ \mathsf{S}_{31} & \mathsf{S}_{33} \end{bmatrix}, \qquad \text{and} \qquad \mathsf{S}_e = \begin{bmatrix} \mathsf{S}_{22} & \mathsf{S}_{24} \\ \mathsf{S}_{42} & \mathsf{S}_{44} \end{bmatrix}.$$

Since the interior edges are unloaded, the joint equilibrium equation for the two boxes now reads

$$\left[\begin{array}{cc|cc} \mathsf{S}_{11} & \mathsf{A}_{12} & \mathsf{S}_{13} & 0 \\ \mathsf{A}_{21} & \mathsf{S}_{22} & 0 & \mathsf{S}_{24} \\ \hline \mathsf{S}_{31} & 0 & \mathsf{S}_{33} & \mathsf{A}_{34} \\ 0 & \mathsf{S}_{24} & \mathsf{A}_{43} & \mathsf{S}_{44} \end{array}\right] \begin{bmatrix} u_1 \\ u_2 \\ u_3 \\ u_4 \end{bmatrix} = \begin{bmatrix} f_1 \\ f_2 \\ 0 \\ 0 \end{bmatrix}, \tag{20}$$

where A_{ij} are the relevant submatrices of the original discrete Laplacian A. To be precise, with A denoting the global discrete Laplace operator, and with J_i denoting an index vector marking the nodes in Γ_i, we have $\mathsf{A}_{ij} = \mathsf{A}(J_i, J_j)$. We observe that all matrices A_{ij} are very sparse (indeed, A_{12} and A_{21} have only two non-zero elements each). From (20), it is clear that the Schur complement of the large box is

$$\mathsf{S} = \begin{bmatrix} \mathsf{S}_{11} & \mathsf{A}_{12} \\ \mathsf{A}_{21} & \mathsf{S}_{22} \end{bmatrix} - \begin{bmatrix} \mathsf{S}_{13} & 0 \\ 0 & \mathsf{S}_{24} \end{bmatrix} \begin{bmatrix} \mathsf{S}_{33} & \mathsf{A}_{34} \\ \mathsf{A}_{43} & \mathsf{S}_{44} \end{bmatrix}^{-1} \begin{bmatrix} \mathsf{S}_{31} & 0 \\ 0 & \mathsf{S}_{42} \end{bmatrix}. \tag{21}$$

Accelerations

The scheme described in Sections "Outline" and "Merging of Two Schur Complements" in Appendix requires $O(N_{\text{side}}^3)$ floating point operations, and $O(N_{\text{side}}^2)$ storage, just like the original nested dissection scheme. This cost is incurred by the repeated evaluation of the formula (21) which involve matrices S_{ij} that are dense. However, as discussed at length in Sect. 3.3, these matrices have internal structure that allows operations such as matrix inversion, and matrix-matrix multiplication, to be evaluated in linear time. Incorporating such accelerated procedures reduces the overall cost (both floating point operations and memory) of the scheme to $O(N_{\text{side}}(\log N_{\text{side}})^{\kappa})$. For recent work in this direction, see, *e.g.* [12, 23, 29].

Remark 9. The process described in Section "Outline" in Appendix requires all Schur complements associated with one level to be kept in memory at one time. It is straight-forward to change the order in which the boxes are processed so that at most four Schur complements on each level must be kept in memory. When dense linear algebra is used, either approach requires $O(N_{\text{side}}^2)$ memory, but when data-sparse matrix formats are used, such an ordering reduces the memory requirement from $O(N_{\text{side}}^2)$ to $O(N_{\text{side}}(\log N_{\text{side}})^{\kappa})$.

Remark 10. Even without accelerations, the scheme described in Section "Outline" in Appendix can handle moderate size problems quite efficiently. For a rudimentary implementation in Matlab executed on a standard desktop (with an Intel i7 CPU running at 2.67 GHz), the time t required to compute T was:

N_{side}	100	200	400	800	1,600	3,200
t (sec)	2.6e-1	1.2e0	6.4e0	4.5e1	5.0e2	6.7e3

Note that less than a minute is required to process a lattice involving $800^2 = 640\,000$ nodes.

References

1. Ulf Andersson, Björn Engquist, Gunnar Ledfelt, and Olof Runborg, *A contribution to wavelet-based subgrid modeling*, Appl. Comput. Harmon. Anal. **7** (1999), no. 2, 151–164. MR MR1711012 (2000f:65130)
2. N. Bakhvalov and G. Panasenko, *Homogenisation: averaging processes in periodic media*, Kluwer Academic Publishers Group, Dordrecht, 1989, Mathematical problems in the mechanics of composite materials, Translated from the Russian by D. Leĭtes. MR 92d:73002
3. J. Barnes and P. Hut, *A hierarchical $O(n \log n)$ force-calculation algorithm*, Nature **324** (1986), no. 4.
4. G. Beylkin, R. Coifman, and V. Rokhlin, *Wavelets in numerical analysis*, Wavelets and their applications, Jones and Bartlett, Boston, MA, 1992, pp. 181–210. MR 93j:65215
5. S. Börm, *$\mathcal{H}^2$-matrix arithmetics in linear complexity*, Computing **77** (2006), no. 1, 1–28. MR MR2207953 (2006k:65111)
6. ———, *Approximation of solution operators of elliptic partial differential equations by $\mathcal{H}$- and $\mathcal{H}^2$–matrices*, Tech. Report 85/2007, Max Planck Institute, 2007.
7. ———, *Construction of data-sparse $\mathcal{H}^2$-matrices by hierarchical compression*, Tech. Report 92/2007, Max Planck Institute, 2007.
8. J. Bremer and V. Rokhlin, *Efficient discretization of Laplace boundary integral equations on polygonal domains*, J. Comput. Phys. **229** (2010), no. 7, 2507–2525. MR MR2586199
9. M. E. Brewster and G. Beylkin, *A multiresolution strategy for numerical homogenization*, Appl. Comput. Harmon. Anal. **2** (1995), no. 4, 327–349. MR 96h:65156
10. S. Chandrasekaran and M. Gu, *Fast and stable algorithms for banded plus semiseparable systems of linear equations*, SIAM J. Matrix Anal. Appl. **25** (2003), no. 2, 373–384 (electronic). MR MR2047424 (2005f:65039)
11. S. Chandrasekaran, M. Gu, X.S. Li, and J. Xia, *Fast algorithms for hierarchically semiseparable matrices*, Numer. Linear Algebra Appl., **17** (2010), 953–976.
12. ———, *Superfast multifrontal method for structured linear systems of equations*, SIAM J. Matrix Anal. Appl. **31** (2009), 1382–1411.
13. S. Chandrasekaran, M. Gu, and W. Lyons, *A fast adaptive solver for hierarchically semiseparable representations*, Calcolo **42** (2005), no. 3-4, 171–185. MR MR2191196 (2006i:65038)
14. Doina Cioranescu and Jeannine Saint Jean Paulin, *Homogenization of reticulated structures*, Applied Mathematical Sciences, vol. 136, Springer-Verlag, New York, 1999. MR MR1676922 (2000d:74064)
15. Patrick Dewilde and Shivkumar Chandrasekaran, *A hierarchical semi-separable Moore-Penrose equation solver*, Wavelets, multiscale systems and hypercomplex analysis, Oper. Theory Adv. Appl., vol. 167, Birkhäuser, Basel, 2006, pp. 69–85. MR MR2240291 (2007c:65039)

16. Mihai Dorobantu and Björn Engquist, *Wavelet-based numerical homogenization*, SIAM J. Numer. Anal. **35** (1998), no. 2, 540–559 (electronic). MR 99a:65183
17. Yalchin Efendiev and Thomas Y. Hou, *Multiscale finite element methods*, Surveys and Tutorials in the Applied Mathematical Sciences, vol. 4, Springer, New York, 2009, Theory and applications. MR MR2477579
18. Björn Engquist and Olof Runborg, *Wavelet-based numerical homogenization with applications*, Multiscale and multiresolution methods, Lect. Notes Comput. Sci. Eng., vol. 20, Springer, Berlin, 2002, pp. 97–148. MR MR1928565 (2003k:65180)
19. ———, *Wavelet-based numerical homogenization*, Highly oscillatory problems, London Math. Soc. Lecture Note Ser., vol. 366, Cambridge Univ. Press, Cambridge, 2009, pp. 98–126. MR MR2562507
20. Jacob Fish and Amir Wagiman, *Multiscale finite element method for a locally nonperiodic heterogeneous medium*, Computational Mechanics **12** (1993), 164–180, 10.1007/BF00371991.
21. A. George, *Nested dissection of a regular finite element mesh*, SIAM J. on Numerical Analysis **10** (1973), 345–363.
22. A. Gillman and P.G. Martinsson, *Fast and accurate numerical methods for solving elliptic difference equations defined on lattices*, J. Comput. Phys. **220**(24) (2010), 9026–9041.
23. Lars Grasedyck, Ronald Kriemann, and Sabine Le Borne, *Domain decomposition based $\mathscr{H}$-LU preconditioning*, Numer. Math. **112** (2009), no. 4, 565–600. MR MR2507619 (2010e:65200)
24. L. Greengard and V. Rokhlin, *A fast algorithm for particle simulations*, J. Comput. Phys. **73** (1987), no. 2, 325–348.
25. Leslie Greengard and Vladimir Rokhlin, *A new version of the fast multipole method for the Laplace equation in three dimensions*, Acta numerica, 1997, Acta Numer., vol. 6, Cambridge Univ. Press, Cambridge, 1997, pp. 229–269.
26. W. Hackbusch, *The panel clustering technique for the boundary element method (invited contribution)*, Boundary elements IX, Vol. 1 (Stuttgart, 1987), Comput. Mech., Southampton, 1987, pp. 463–474. MR MR965331 (89i:76011)
27. W. Hackbusch, B. Khoromskij, and S. Sauter, *On $\mathscr{H}^2$-matrices*, Lectures on Applied Mathematics, Springer Berlin, 2002, pp. 9–29.
28. Wolfgang Hackbusch, *A sparse matrix arithmetic based on H-matrices; Part I: Introduction to H-matrices*, Computing **62** (1999), 89–108.
29. P.G. Martinsson, *A fast direct solver for a class of elliptic partial differential equations*, J. Sci. Comput. **38** (2009), no. 3, 316–330. MR MR2475654 (2010c:65041)
30. P.G. Martinsson and V. Rokhlin, *A fast direct solver for boundary integral equations in two dimensions*, J. Comput. Phys. **205** (2005), no. 1, 1–23.
31. ———, *An accelerated kernel independent fast multipole method in one dimension*, SIAM Journal of Scientific Computing **29** (2007), no. 3, 1160–11178.
32. E. Michielssen, A. Boag, and W. C. Chew, *Scattering from elongated objects: direct solution in $O(N \log^2 N)$ operations*, IEE Proc. Microw. Antennas Propag. **143** (1996), no. 4, 277–283.
33. Zhifeng Sheng, Patrick Dewilde, and Shivkumar Chandrasekaran, *Algorithms to solve hierarchically semi-separable systems*, System theory, the Schur algorithm and multidimensional analysis, Oper. Theory Adv. Appl., vol. 176, Birkhäuser, Basel, 2007, pp. 255–294. MR MR2342902
34. Page Starr and Vladimir Rokhlin, *On the numerical solution of two-point boundary value problems. II*, Comm. Pure Appl. Math. **47** (1994), no. 8, 1117–1159. MR MR1288634 (95j:65090)
35. Jinchao Xu and Ludmil Zikatanov, *On an energy minimizing basis for algebraic multigrid methods*, Computing and Visualization in Science **7** (2004), 121–127, 10.1007/s00791-004-0147-y.

Adaptive Multilevel Monte Carlo Simulation

Håkon Hoel, Erik von Schwerin, Anders Szepessy, and Raúl Tempone

Abstract This work generalizes a multilevel forward Euler Monte Carlo method introduced in Michael B. Giles. (Michael Giles. Oper. Res. 56(3):607–617, 2008.) for the approximation of expected values depending on the solution to an Itô stochastic differential equation. The work (Michael Giles. Oper. Res. 56(3):607–617, 2008.) proposed and analyzed a forward Euler multilevel Monte Carlo method based on a hierarchy of uniform time discretizations and control variates to reduce the computational effort required by a standard, single level, Forward Euler Monte Carlo method. This work introduces an adaptive hierarchy of non uniform time discretizations, generated by an adaptive algorithm introduced in (Anna Dzougoutov et al. Raúl Tempone. Adaptive Monte Carlo algorithms for stopped diffusion. In *Multiscale methods in science and engineering*, volume 44 of *Lect. Notes Comput. Sci. Eng.*, pages 59–88. Springer, Berlin, 2005; Kyoung-Sook Moon et al. Stoch. Anal. Appl. 23(3):511–558, 2005; Kyoung-Sook Moon et al. An adaptive algorithm for ordinary, stochastic and partial differential equations. In *Recent advances in adaptive computation*, volume 383 of *Contemp. Math.*, pages 325–343. Amer. Math. Soc., Providence, RI, 2005.). This form of the adaptive algorithm generates stochastic, path dependent, time steps and is based on a posteriori error expansions first developed in (Anders Szepessy et al. Comm. Pure Appl. Math. 54(10):1169–1214, 2001). Our numerical results for a stopped diffusion problem, exhibit savings

E. von Schwerin (✉) · R. Tempone
Applied Mathematics and Computational Sciences, KAUST, Thuwal, Saudi Arabia
e-mail: erik.vonschwerin@kaust.edu.sa; raul.tempone@kaust.edu.sa

H. Hoel
Department of Numerical Analysis, CSC, Royal Institute of Technology (KTH), Stockholm, Sweden
e-mail: hhoel@csc.kth.se

A. Szepessy
Department of Mathematics, Royal Institute of Technology (KTH), Stockholm, Sweden
e-mail: szepessy@kth.se

B. Engquist et al. (eds.), *Numerical Analysis of Multiscale Computations*, Lecture Notes in Computational Science and Engineering 82, DOI 10.1007/978-3-642-21943-6_10,

in the computational cost to achieve an accuracy of $\mathcal{O}(\mathrm{TOL})$, from $\mathcal{O}\left(\mathrm{TOL}^{-3}\right)$ using a single level version of the adaptive algorithm to $\mathcal{O}\left(\left(\mathrm{TOL}^{-1}\log(\mathrm{TOL})\right)^2\right)$.

1 Introduction

This work develops a multilevel version of an adaptive algorithm for weak approximation of Itô stochastic differential equations (SDEs)

$$dX(t) = a(t, X(t))dt + b(t, X(t))dW(t), \qquad 0 < t < T, \tag{1}$$

where $X(t;\omega)$ is a stochastic process in $\mathbb{R}^d$, with randomness generated by a k-dimensional Wiener process with independent components, $W(t;\omega)$, on the probability space $(\Omega, \mathcal{F}, P)$; see [4, 7]. The functions $a(t,x) \in \mathbb{R}^d$ and $b(t,x) \in \mathbb{R}^{d\times k}$ are given drift and diffusion fluxes.

Our goal is to, for any given sufficiently well behaved function $g : \mathbb{R}^d \to \mathbb{R}$, approximate the expected value $E[g(X(T))]$ by adaptive multilevel Monte Carlo methods. A typical example of such an expected value is to compute option prices in mathematical finance; see [5] and [12]. Other models based on stochastic dynamics are used for example in molecular dynamics simulations at constant temperature and for stochastic climate prediction; cf. [6] and [2].

The multilevel Monte Carlo method based on uniform time stepping was introduced by Giles in [10]. He developed a clever control variate type variance reduction technique for a numerical method, denoted here by $\overline{X}$, that approximates the solution of the SDE (1). The key to the variance reduction in [10] is to compute approximate solutions, $\overline{X}_\ell$, on hierarchies of uniform time meshes with size

$$\Delta t_\ell = C^{-\ell}\Delta t_0, \qquad C \in \{2,3,\ldots\} \text{ and } \ell \in \{0,1,\ldots,L\}, \tag{2}$$

thereby generating sets of realizations on different mesh levels. After computing numerical approximations on a mesh hierarchy, the expected value $E[g(X(T))]$ is approximated by the multilevel Monte Carlo estimator

$$\mathcal{E}_{\{\mathcal{S}_\ell\}_{\ell=0}^L}\left(g(\overline{X}_L(T))\right) = \sum_{i=1}^{M_0} \frac{g(\overline{X}_0(T;\omega_{i,0}))}{M_0} + \sum_{\ell=1}^{L}\sum_{i=1}^{M_\ell} \frac{g(\overline{X}_\ell(T;\omega_{i,\ell})) - g(\overline{X}_{\ell-1}(T;\omega_{i,\ell}))}{M_\ell}. \tag{3}$$

Here $\{\mathcal{S}_\ell\}_{\ell=0}^L$ denotes mutually independent sample sets on the respective meshes, each with M_ℓ independent samples. To reduce the variance in the above estimator, the realization pairs $\overline{X}_\ell(T;\omega_{i,\ell})$ and $\overline{X}_{\ell-1}(T;\omega_{i,\ell})$ of the summands $g(\overline{X}_\ell(T;\omega_{i,\ell})) - g(\overline{X}_{\ell-1}(T;\omega_{i,\ell}))$ for each level $\ell > 0$ are generated by the same

Brownian path, $W_t(\omega_i)$, but they are realized on different temporal grids with uniform time steps, Δt_ℓ and $\Delta t_{\ell-1}$, respectively. The efficiency of this computation relies on an a priori known order of strong convergence for the numerical method employed on each level of the hierarchy.

Let TOL > 0 be a desired accuracy in the approximation of $E[g(X(T))]$. The main result of Giles' work [10] is that the computational cost needed to achieve the Mean Square Error (MSE)

$$E\left[\left(\mathscr{E}_{\{\mathscr{S}_\ell\}_{\ell=0}^L}(g(\overline{X}_L(T))) - E[g(X(T))]\right)^2\right] = \mathscr{O}\left(\mathrm{TOL}^2\right), \tag{4}$$

when using the Forward Euler method to create the approximate realizations $\overline{X}_\ell(T;\omega)$, can be reduced to

$$\mathscr{O}\left((\mathrm{TOL}^{-1}\log(\mathrm{TOL}^{-1}))^2\right),$$

with Giles' multilevel Monte Carlo method; the corresponding complexity using the standard Monte Carlo method is $\mathscr{O}\left(\mathrm{TOL}^{-3}\right)$ since the Forward Euler method has weak order of convergence 1 and the Monte Carlo sampling order $1/2$ by the Central Limit Theorem. Furthermore, whenever the function g is Lipschitz and for scalar Itô stochastic differential equations, the computational cost can be further reduced to $\mathscr{O}\left(\mathrm{TOL}^{-2}\right)$ using the first order strong convergence Milstein method. In addition, the work [11] shows how to apply the Milstein method for several scalar SDE cases where the Lipschitz condition is not fulfilled and still obtain the cost $\mathscr{O}\left(\mathrm{TOL}^{-2}\right)$.

In this work we use the Forward Euler method with non uniform time steps. Let $0 = t_0 < t_1 < \cdots < t_N = T$ denote a given time discretization, without reference to its place in the hierarchies, and $\{0 = W(t_0;\omega), W(t_1;\omega), \ldots, W(t_N;\omega)\}$ denote a generated sample of the Wiener process on that discretization. Then the Forward Euler method computes an approximate solution of (1) by the scheme

$$\begin{aligned} \overline{X}(t_0;\omega) &= X(0), \\ \overline{X}(t_{n+1};\omega) &= a(\overline{X}(t_n;\omega), t_n)\Delta t_n + b(\overline{X}(t_n;\omega), t_n)\Delta W(t_n;\omega), \qquad n \geq 0, \end{aligned} \tag{5}$$

where $\Delta t_n = t_{n+1} - t_n$ and $\Delta W(t_n;\omega) = W(t_{n+1};\omega) - W(t_n;\omega)$ are the time steps and Wiener increments, respectively.

The contribution of the present paper to the multilevel Monte Carlo method is the development of a novel algorithm with adaptive, non uniform time steps. The algorithm uses adaptive mesh refinements to stochastically create a path dependent mesh for each realization. The construction and analysis of the adaptive algorithm is inspired by the work on single level adaptive algorithms for weak approximation of ordinary stochastic differential equations [8], and uses the adjoint weighted global error estimates first derived in [1]. The goal of the adaptive algorithm is to choose the time steps and the number of realizations such that the event

$$\left|\mathscr{E}_{\{\mathscr{S}_\ell\}_{\ell=0}^L}\left(g(\overline{X}_L(T))\right) - E[g(X(T))]\right| \leq \mathrm{TOL}, \tag{6}$$

holds with probability close to one.

It should be noted that in the setting of adaptive mesh refinement there is no given notion of mesh size, so a hierarchy of meshes can no longer be described as in the constant time step case (2). Instead, we generate a hierarchy of meshes by successively increasing the accuracy in our computations: setting the tolerance levels

$$\mathrm{TOL}_\ell = \frac{\mathrm{TOL}_0}{2^\ell}, \quad \text{for} \quad \ell \in \{0, 1, \ldots, L\}, \tag{7}$$

and (by adaptive refinements based on error indicators) finding corresponding meshes so that for each level $\ell \in \{0, 1, \ldots, L\}$,

$$\left| E[g(X(T))] - E[g(\overline{X}_\ell(T))] \right| \lesssim \frac{\mathrm{TOL}_\ell}{2}.$$

The efficiency and accuracy of the multilevel adaptive Monte Carlo algorithm is illustrated by a numerical example, in the case of the stopped diffusion problems used to test the single level version of the algorithm in [3]. For this example multilevel Monte Carlo based on adaptive time steps requires a computational work $\mathcal{O}\left(\mathrm{TOL}^{-2}\log(\mathrm{TOL}^{-1})^2\right)$ while a direct application of the multilevel Monte Carlo method based on uniform time steps would be less efficient since the underlying Euler–Maruyama method has reduced orders of weak and strong convergence for the barrier problem.

The rest of this paper is organized as follows: Subsection 1.1 introduces the notion of error density and error indicators, and recalls useful results for single level adaptive forward Euler Monte Carlo methods. Section 2 describes the new adaptive multilevel Monte Carlo algorithm. Section 3 presents results from the numerical experiment.

1.1 A Single Level Posteriori Error Expansion

Here we recall previous single level results that are used for constructing the multilevel algorithm in Sect. 2. In particular, we recall adjoint based error expansions with computable leading order term. Assume that the process X satisfies (1) and its approximation, $\overline{X}$, is given by (5); then the error expansions in theorems 1.2 and 2.2 of [1] have the form

$$E[g(X(T)) - g(\overline{X}(T))] = E\left[\sum_{n=1}^{N} \rho_n \Delta t_n^2\right] + \text{higher order terms}, \tag{8}$$

where $\rho_n \Delta t_n^2$ are computable error indicators, that is they provide information for further improvement of the time mesh and ρ_n measures the density of the global error in (8). A typical adaptive algorithm does two things iteratively:

1. If the error indicators satisfy an accuracy condition then it stops; otherwise
2. The algorithm chooses where to refine the mesh based on the error indicators and then makes an iterative step to 1

In addition to estimating the global error $E[g(X(T)) - g(\overline{X}(T))]$ in the sense of equation (8), the error indicators $\rho_n \Delta t_n^2$ also give simple information on where to refine to reach an optimal mesh, based on the almost sure convergence of the density ρ_n as we refine the discretization, see Sect. 4 in [9].

In the remaining part of this section we state in Theorem 1 a single level error expansion from [1].

Given an initial time discretization $\Delta t[0](t)$ and, for the stochastic time steps algorithm, refining until[1]

$$|\rho(t,\omega)|\big(\Delta t(t)\big)^2 < \text{constant}, \tag{9}$$

we construct a partition $\Delta t(t)$ by repeated halving of intervals so that it satisfies

$$\Delta t(t) = \Delta t[0](t)/2^n \quad \text{for some natural number } n = n(t,\omega).$$

The criterion (9) uses an approximate error density function ρ, satisfying for $t \in [0,T]$ and all outcomes ω the uniform upper and lower bounds

$$\rho_{low}(\text{TOL}) \le |\rho(t,\omega)| \le \rho_{up}(\text{TOL}). \tag{10}$$

The positive functions ρ_{low} and ρ_{up} are chosen so that $\rho_{up}(\text{TOL}) \to +\infty$ as $\text{TOL} \to 0$ while $\rho_{low}(\text{TOL}) \to 0$ such that $\text{TOL}/\rho_{low}(\text{TOL}) \to 0$. We further make the assumption that for all $s,t \in [0,T]$ the sensitivity of the error density to values of the Wiener process can be bounded,

$$|\partial_{W(t)}\rho(s,\omega)| \le D\rho_{up}(\text{TOL}), \tag{11}$$

for some positive function $D\rho_{up}$ such that $D\rho_{up}(\text{TOL}) \to +\infty$ as $\text{TOL} \to 0$. For each realization successive subdivisions of the steps yield the largest time steps satisfying (9). The corresponding stochastic increments ΔW will have the correct distribution, with the necessary independence, if the increments ΔW related to the new steps are generated by Brownian bridges [7], that is the time steps are generated by conditional expected values of the Wiener process.

We begin now by stating in the next lemma the regularity conditions to be used in the analysis of the adaptive multilevel algorithms.

Lemma 1 (Regularity). (a) *Assume that the following regularity conditions hold:*

(1) *The functions $a(t,x)$ and $b(t,x)$ are continuous in (t,x) and are twice continuously differentiable with respect to x.*
(2) *The partial derivatives of first and second order with respect to x of the functions a and b are uniformly bounded.*

[1] The precise expression is given in (34) below.

(3) *The function g is twice continuously differentiable, and together with its partial derivatives of first and second order it is uniformly bounded.*

Then the cost to go *function, defined by*

$$u(t,x) = E\big[g(X(T)) \mid X(t) = x\big], \tag{12}$$

satisfies the Kolmogorov equation

$$\partial_t u(t,x) + a_k \partial_k u(t,x) + d_{kn} \partial_{kn} u(t,x) = 0, \qquad u(T,\cdot) = g, \tag{13}$$

where we have used Einstein summation convention[2], *and where* $d_{kn} = \frac{1}{2} b_k^l b_n^l$.

(b) *Furthermore, if the following regularity conditions are satisfied:*

(1) *The functions* $\partial_\beta a(t,\cdot)$ *and* $\partial_\beta b(t,\cdot)$ *are bounded uniformly in t for multi-indices* β *with* $1 \le |\beta| \le 8$*;*
(2) *The functions* $a(\cdot,x)$, $b(\cdot,x)$ *have continuous and uniformly bounded first order time derivatives;*
(3) *The function g has spatial derivatives* $\partial_\beta g$*, with polynomial growth for* $|\beta| \le 8$*;*

then the function u has continuous partial derivatives with respect to x up to the order 8*, satisfying the following polynomial growth condition: for all* $i \in \{0,1,2\}$ *and* $\alpha \in \mathbb{N}^d$ *with* $i + |\alpha| \le 8$ *there exists* $p_{\alpha,i} \in \mathbb{N}$ *and* $C_{\alpha,i} > 0$ *such that*

$$\max_{0 \le t \le T} \left| \partial_t^i \partial_\alpha u(t,x) \right| \le C_{\alpha,i} \left(1 + |x|^{p_{\alpha,i}}\right) \quad \forall x \in \mathbb{R}^d.$$

In what follows, Lemma 2 and Theorem 1 show that although the steps adaptively generated to satisfy (9)–(11) are not adapted to the natural Wiener filtration, the method indeed converges to the correct limit, which is the same as the limit of the forward Euler method with adapted time steps.

Lemma 2 (Strong Convergence). *For X the solution of* (1) *suppose that a, b, and g satisfy the assumptions in Lemma 1, that* $\overline{X}$ *is constructed by the forward Euler method, based on the stochastic time stepping algorithm defined in Sect. 2, with step sizes* Δt_n *satisfying* (9)–(11)*, and that their corresponding* ΔW_n *are generated by Brownian bridges. Then*

$$\sup_{0 \le t \le T} E[|X(t) - \overline{X}(t)|^2] = \mathcal{O}\left(\Delta t_{\text{sup}}\right) = \mathcal{O}\left(\frac{\text{TOL}}{\rho_{low}(\text{TOL})}\right) \longrightarrow 0 \tag{14}$$

as TOL $\to 0$*, where* $\Delta t_{\text{sup}} \equiv \sup_{n,\omega} \Delta t_n(\omega)$.

In Theorem 1 and the rest of this work, we will use Einstein summation convention with respect to functional and spatial indices, but not with respect to the temporal one (usually denoted t_n).

[2] When an index variable appears twice in a single term this means that a summation over all possible values of the index takes place; for example $a_k \partial_k u(t,x) = \sum_{k=1}^d a_k \partial_k u(t,x)$, where d is the space dimension of the SDE ($a, x \in \mathbb{R}^d$).

Theorem 1 (Single level stochastic time steps error expansion). *Given the assumptions in Lemma 2 and a deterministic initial value $X(0)$, the time discretization error in* (8) *has the following expansion, based on both the drift and diffusion fluxes and the discrete dual functions φ, φ', and φ'' given in* (17)–(22), *with computable leading order terms:*

$$
\begin{aligned}
E[g(X(T))]-E[g(\overline{X}(T))] &= E\left[\sum_{n=0}^{N-1}\tilde{\rho}(t_n,\omega)(\Delta t_n)^2\right] \\
&\quad + \mathcal{O}\left(\left(\frac{\mathrm{TOL}}{\rho_{low}(\mathrm{TOL})}\right)^{1/2}\left(\frac{\rho_{up}(\mathrm{TOL})}{\rho_{low}(\mathrm{TOL})}\right)^{\epsilon}\right)E\left[\sum_{n=0}^{N-1}(\Delta t_n)^2\right],
\end{aligned}
\tag{15}
$$

for any $\epsilon > 0$ and where

$$
\begin{aligned}
\tilde{\rho}(t_n,\omega) \equiv \frac{1}{2}\Big(&\big(\partial_t a_k + \partial_j a_k a_j + \partial_{ij} a_k d_{ij}\big)\varphi_k(t_{n+1}) \\
&+ \big(\partial_t d_{km} + \partial_j d_{km} a_j + \partial_{ij} d_{km} d_{ij} + 2\partial_j a_k d_{jm}\big)\varphi'_{km}(t_{n+1}) \\
&+ \big(2\partial_j d_{km} d_{jr}\big)\varphi''_{kmr}(t_{n+1})\Big)
\end{aligned}
\tag{16}
$$

and the terms in the sum of (16) *are evaluated at the a posteriori known points $(t_n,\overline{X}(t_n))$, i.e.,*

$$
\partial_\alpha a \equiv \partial_\alpha a(t_n,\overline{X}(t_n)),\quad \partial_\alpha b \equiv \partial_\alpha b(t_n,\overline{X}(t_n)),\quad \partial_\alpha d \equiv \partial_\alpha d(t_n,\overline{X}(t_n)).
$$

Here $\varphi \in \mathbb{R}^d$ is the solution of the discrete dual backward problem

$$
\begin{aligned}
\varphi_i(t_n) &= \partial_i c_j(t_n,\overline{X}(t_n))\varphi_j(t_{n+1}),\quad t_n < T, \\
\varphi_i(T) &= \partial_i g(\overline{X}(T)),
\end{aligned}
\tag{17}
$$

with

$$
c_i(t_n,x) \equiv x_i + \Delta t_n a_i(t_n,x) + \Delta W_n^{\ell} b_i^{\ell}(t_n,x)
\tag{18}
$$

and its first and second variation

$$
\varphi'_{ij} \equiv \partial_{x_j(t_n)}\varphi_i(t_n) \equiv \frac{\partial\varphi_i(t_n;\overline{X}(t_n)=x)}{\partial x_j},
\tag{19}
$$

$$
\varphi''_{ikm}(t_n) \equiv \partial_{x_m(t_n)}\varphi'_{ik}(t_n) \equiv \frac{\partial\varphi'_{ik}(t_n;\overline{X}(t_n)=x)}{\partial x_m},
\tag{20}
$$

which satisfy

$$
\begin{aligned}
\varphi'_{ik}(t_n) &= \partial_i c_j(t_n,\overline{X}(t_n))\partial_k c_p(t_n,\overline{X}(t_n))\varphi'_{jp}(t_{n+1}) \\
&\qquad + \partial_{ik} c_j(t_n,\overline{X}(t_n))\varphi_j(t_{n+1}),\quad t_n < T, \\
\varphi'_{ik}(T) &= \partial_{ik} g(\overline{X}(T)),
\end{aligned}
\tag{21}
$$

and

$$
\begin{aligned}
\varphi''_{ikm}(t_n) &= \partial_i c_j(t_n,\overline{X}(t_n))\partial_k c_p(t_n,\overline{X}(t_n))\partial_m c_r(t_n,\overline{X}(t_n))\varphi''_{jpr}(t_{n+1}) \\
&\quad + \partial_{im} c_j(t_n,\overline{X}(t_n))\partial_k c_p(t_n,\overline{X}(t_n))\varphi'_{jp}(t_{n+1}) \\
&\quad + \partial_i c_j(t_n,\overline{X}(t_n))\partial_{km} c_p(t_n,\overline{X}(t_n))\varphi'_{jp}(t_{n+1}) \\
&\quad + \partial_{ik} c_j(t_n,\overline{X}(t_n))\partial_m c_p(t_n,\overline{X}(t_n))\varphi'_{jp}(t_{n+1}) \\
&\quad + \partial_{ikm} c_j(t_n,\overline{X}(t_n))\varphi_j(t_{n+1}), \quad t_n < T, \\
\varphi''_{ikm}(T) &= \partial_{ikm} g(\overline{X}(T)),
\end{aligned} \tag{22}
$$

respectively.

Observe that the constant in $\mathcal{O}$ that appears in (15) may not be uniform with respect to the value ϵ. Thus, in practice one chooses $\epsilon = \epsilon(\text{TOL})$ to minimize the contribution of the remainder term to the error expansion (15).

Let us now discuss how to modify the error density $\tilde{\rho}(t_n,\omega)$ in (16) to satisfy the bounds (10) and at the same time guarantee that $\Delta t_{\sup} \to 0$ as $\text{TOL} \to 0$, see Lemma 2.

Consider, for $t \in [t_n, t_{n+1})$ and $n = 1,\ldots,N$, the piecewise constant function

$$\rho(t) \equiv \text{sign}(\tilde{\rho}(t_n)) \min\left(\max(|\tilde{\rho}(t_n)|, \rho_{low}(\text{TOL})), \rho_{max}(\text{TOL})\right), \tag{23}$$

where

$$
\begin{aligned}
\rho_{low}(\text{TOL}) &= \text{TOL}^{\bar{\gamma}}, \quad 0 < \bar{\gamma} < \tfrac{\alpha}{\alpha+2}, \quad 0 < \alpha < \tfrac{1}{2}, \\
\rho_{max}(\text{TOL}) &= \text{TOL}^{-r}, \; r > 0,
\end{aligned} \tag{24}
$$

and with the standard notation for the function sign, that is $\text{sign}(x) = 1$ for $x \geq 0$ and -1 for $x < 0$. The function ρ defined by (23) measures the density of the time discretization error; it is used in (33) and (34) to guide the mesh refinements. From now on, with a slight abuse of notation, $\rho(t_n) = \rho_n$ denotes the modified density (23).

Following the error expansion in Theorem 1, the time discretization error is approximated by

$$|\mathcal{E}_T| = |E[g(X(T)) - g(\overline{X}(T))]| \lesssim E\left[\sum_{n=1}^{N} r(n)\right], \tag{25}$$

using the error indicator, $r(n)$, defined by

$$r(n) \equiv |\rho(t_n)|\Delta t_n^2, \tag{26}$$

with the modified error density defined by (23). According to Corollary 4.3 and Theorem 4.5 in [9], we have the almost sure convergence of the error density to a limit density denoted by $\hat{\rho}$, $\rho \to \hat{\rho}$ as $\text{TOL} \to 0$.

2 Adaptive Algorithms and Multilevel Variance Reduction

In this section we describe the multilevel Monte Carlo algorithm with adaptive stochastic time steps for approximating $E[g(X(T))]$.

Given a tolerance $\mathrm{TOL} > 0$ for which we want the estimate (6) to be fulfilled, we split the tolerance into a time discretization tolerance and a statistical error tolerance,

$$\mathrm{TOL} = \mathrm{TOL_T} + \mathrm{TOL_S}.$$

The optimal way of choosing $\mathrm{TOL_T}$ and $\mathrm{TOL_S}$ in terms of minimizing the computational work can be approximated by Lagrangian optimization. The basis of the error control is to choose the number of samples large enough to make the estimated statistical error smaller than $\mathrm{TOL_S}$ and adaptively refining the time steps, for each realization, until the estimated time discretization error is smaller than $\mathrm{TOL_T}$.

The stochastic time stepping algorithm uses criteria related to (9) with an outer and an inner loop, described below. Given the value of M_0, and mutually independent sample sets $\{\mathscr{S}_\ell\}_{\ell=0}^{L}$ where each $\mathscr{S}_\ell$ consists of

$$M_\ell = \left\lceil M_0 \frac{\rho_{low}(\mathrm{TOL}_0)\mathrm{TOL}_\ell}{\rho_{low}(\mathrm{TOL}_\ell)\mathrm{TOL}_0} \right\rceil \tag{27}$$

independent realisations of the underlying Wiener process, the *outer loop* uses a multilevel Monte Carlo technique to estimate $E[g(X(T))]$ and, if necessary, update the value M_0. Recall that the lower bound for the error density, ρ_{low}, was introduced in (24). We use the enforced deterministic lower bound

$$M_0 \geq M_{-1} = \mathrm{const} \cdot \mathrm{TOL}^{-1}. \tag{28}$$

The sample set independence makes it possible to estimate $E[g(X(T))]$ by the sum of sample averages

$$\mathscr{E}_{\{\mathscr{S}_\ell\}_{\ell=0}^{L}}\left(g(\overline{X}_L(T))\right) = \mathscr{A}_{\mathscr{S}_0}\left[g(\overline{X}_0(T))\right] + \sum_{\ell=1}^{L} \mathscr{A}_{\mathscr{S}_\ell}\left[g(\overline{X}_\ell(T)) - g(\overline{X}_{\ell-1}(T))\right],$$

$$\mathscr{A}_{\mathscr{S}_\ell}[f] := M_\ell^{-1} \sum_{\omega \in \mathscr{S}_\ell}^{M_\ell} f(\omega),$$

where the algorithm for constructing $g(\overline{X}_{\ell-1}(T))$ must be identical on levels ℓ and $\ell - 1$ for the telescoping sum to work perfectly; this is described in detail later in this section and explicitly in Algorithm 2.1. Approximate the variance of $\mathscr{E}_{\{\mathscr{S}_\ell\}_{\ell=0}^{L}}\left(g(\overline{X}_L(T))\right)$ by the sum of sample variances

$$\sigma^2 = \frac{\mathscr{V}_{\mathscr{S}_0}\left[g(\overline{X}_0(T))\right]}{M_0} + \sum_{\ell=1}^{L} \frac{\mathscr{V}_{\mathscr{S}_\ell}\left[g(\overline{X}_\ell(T)) - g(\overline{X}_{\ell-1}(T))\right]}{M_\ell} \tag{29}$$

and aim to control this variance by choosing M_0 sufficiently large so that

$$\sigma < \frac{TOL_S}{C_C}. \tag{30}$$

If $\sigma > \frac{TOL_S}{C_C}$, the number of samples M_0 is increased in the next batch; in the numerical examples of Sect. 3 the size of the new sample set was set to

$$\left\lceil M_{0,old} \max\left\{2, \min\left\{\sigma^2 \left(C_C/\mathrm{TOL_S}\right)^2, \mathrm{MCH}\right\}\right\}\right\rceil, \tag{31}$$

with $\mathrm{MCH} = 10$, but we may use the rule $M_{0,new} = 2M_{0,old}$ as well. The parameter MCH should not be taken too close to one in order to avoid a large number of iterations with similar M_0 before convergence, yielding a total computational work much larger than the computational work corresponding to the accepted M_0. On the other hand, MCH should not be too large in order to avoid using an excessively large M_0.

The *inner loop*, with iteration index ℓ representing a level in the adaptive mesh hierarchy, generates M_ℓ realization pairs[3], $(\overline{X}_{\ell-1}(T), \overline{X}_\ell(T))$, of (5) approximating (1) to the accuracy tolerances $\mathrm{TOL}_{\ell-1}$ and TOL_ℓ. These pairs are constructed by successive subdivision of an initial grid Δt_{-1}. First, the algorithm determines the grid $\Delta t_{\ell-1}$ from the initial grid Δt_{-1} by starting out with the tolerance $\mathrm{TOL}_0 = 2^L \mathrm{TOL_T}$ for the time discretization error and successively halving that tolerance until it becomes $\mathrm{TOL}_{\ell-1} = 2^{(L-\ell+1)}\mathrm{TOL_T}$ while for each new tolerance constructing the new grid by repeated adaptive subdivision of the previously constructed mesh. This iterative procedure in Algorithm 2.1, with index $\tilde{\ell} = 0, \ldots, \ell-1$, ensures that a grid $\Delta t_{\ell-1}$ on level ℓ is generated in the same way as a grid $\Delta t_{\ell-1}$ on level $\ell-1$ and consequently that $E[\overline{X}_\ell(T)]$ when computed as the coarser approximation in a pair $(\overline{X}_\ell(T), \overline{X}_{\ell+1}(T))$ is the same as when computed as the finer approximation in a pair $(\overline{X}_{\ell-1}(T), \overline{X}_\ell(T))$. The above mentioned property is necessary for the telescopic expansion of the time discretization error introduced by Giles in [10]. Second, the algorithm determines the grid Δt_ℓ by successively subdividing the recently determined $\Delta t_{\ell-1}$ according to the refinement criterion (34) until the stopping criterion (33) is satisfied.

Due to the stochastic nature of SDEs, each realization pair of $(\overline{X}_{\ell-1}(T), \overline{X}_\ell(T))$ may refine the initial grid Δt_{-1} differently. In particular, grids corresponding to different realizations on the same level ℓ may be different. To take this feature into account in the grid refinement, we introduce some notation. Let N_ℓ and $\overline{\mathcal{N}}_\ell$ denote the number of time steps and the approximate average number of time steps for realizations at level ℓ, respectively; see Algorithm 2.2 for details on the approximation technique and its update through the iteration. Further, denote the grid corresponding to one realization at level ℓ by

$$\Delta t_\ell = [\Delta t_\ell(0), \ldots, \Delta t_\ell(N_\ell - 1)], \tag{32}$$

[3] Observe that for the level $\ell = 0$ only the realisation of $\overline{X}_0(T)$ is generated.

and its corresponding Wiener increments by

$$\Delta W_\ell = [\Delta W_\ell(0), \ldots, \Delta W_\ell(N_\ell - 1)].$$

The refinement condition is based on the error indicator $r_{[\ell]}$, defined in (26), and uses similar refinements to those defined for the single level method. The stopping condition for refinement of Δt_ℓ is

$$\max_{1 \le n \le N_\ell} r_{[\ell]}(n) < C_S \frac{\mathrm{TOL}_\ell}{\overline{\mathcal{N}_\ell}}. \tag{33}$$

When inequality (33) is violated, the nth time step of Δt_ℓ is refined if

$$r_{[\ell]}(n) \ge C_R \frac{\mathrm{TOL}_\ell}{\overline{\mathcal{N}_\ell}}. \tag{34}$$

Normally, the value for C_R would be around 2, and $C_S > C_R$ following the theory developed in [8, 9].

The inputs in Algorithm 2.1 are: TOL_S, TOL_T, initial number of sample realisations M_0, L, Δt_{-1}, initial guesses for the mean number of time steps $(\overline{\mathcal{N}_\ell})_{\ell=0}^{L}$ needed for fulfillment of (33), and the three parameters C_R, C_C, and C_S used in the refinement condition (34) and stopping conditions (30) and (33), respectively. In this algorithm the mean number of initial time steps are chosen as $\overline{\mathcal{N}_\ell} = c\mathrm{TOL}_\ell^{-1}$, for $\ell = 0, \ldots, L$ and a small constant c.

Algorithm 2.1: Multilevel Monte Carlo with stochastic time stepping

Input : $\mathrm{TOL_S}$, $\mathrm{TOL_T}$, M_0, Δt_{-1}, $\{\overline{\mathcal{N}}_\ell\}_{\ell=0}^{L}$, C_R, C_S, C_C
Output: $\mu \approx E\left[g(X(T))\right]$

Set $k = 0$.
while $k < 1$ **or** (30) is violated **do**
  Compute M_0 new realizations of $g\left(\overline{X}_0(T)\right)$
  and their corresponding number of time steps, $\{N_0\}_1^{M_0}$,
  by generating Wiener increments ΔW_{-1} on the mesh Δt_{-1} (independently for each realization) and calling Algorithm 2.3:
  ATSSE($\Delta t_{-1}, \Delta W_{-1}, \mathrm{TOL_T} 2^L, \overline{\mathcal{N}}_0$).
  Set $\mu = \mathcal{A}_{\mathcal{S}_0}\left[g\left(\overline{X}_0(T)\right)\right]$ and $\sigma^2 = \frac{\mathcal{V}_{\mathcal{S}_0}[g(\overline{X}_0(T))]}{M_0}$.
  Compute the average number of time steps $\mathcal{A}_{\mathcal{S}_0}[N_0]$.
  for $\ell = 1, \ldots, L$ **do**
    Set M_ℓ as in (27)

    Compute M_ℓ new realizations of $g\left(\overline{X}_{\ell-1}(T)\right)$,
    their corresponding number of time steps, $\{N_{\ell-1}\}_1^{M_\ell}$, and Wiener increments, $\Delta W_{\ell-1}$, by generating Wiener steps ΔW_{-1} on the mesh Δt_{-1} (independently for each realization) and using the loop
    for $\hat{\ell} = 0, \ldots, \ell - 1$ **do**
      compute $\Delta t_{\hat{\ell}}$ and $\Delta W_{\hat{\ell}}$ by calling Algorithm 2.3:
      ATSSE($\Delta t_{\hat{\ell}-1}, \Delta W_{\hat{\ell}-1}, \mathrm{TOL}_T 2^{L-\hat{\ell}}, \overline{\mathcal{N}}_{\hat{\ell}}$).
    end

    Compute the corresponding M_ℓ realizations of $g\left(\overline{X}_\ell(T)\right)$
    and their number of time steps, N_ℓ, by calling Algorithm 2.3:
    ATSSE($\Delta t_{\ell-1}, \Delta W_{\ell-1}, \mathrm{TOL}_T 2^{L-\ell}, \overline{\mathcal{N}}_\ell$).
    Set $\mu = \mu + \mathcal{A}_{\mathcal{S}_\ell}\left[g\left(\overline{X}_\ell(T)\right) - g\left(\overline{X}_{\ell-1}(T)\right)\right]$ and
    $\sigma^2 = \sigma^2 + \frac{\mathcal{V}_{\mathcal{S}_\ell}[g(\overline{X}_\ell(T)) - g(\overline{X}_{\ell-1}(T))]}{M_\ell}$.
    Compute the average number of time steps $\mathcal{A}_{\mathcal{S}_\ell}[N_{\ell-1}]$ and $\mathcal{A}_{\mathcal{S}_\ell}[N_\ell]$.
  end

  if σ violates (30) **then**
    Update the number of samples by
    $\left\lceil M_0 \max\left\{2, \min\left\{\sigma^2 \left(C_\mathrm{C}/\mathrm{TOL_S}\right)^2, \mathrm{MCH}\right\}\right\}\right\rceil$.
    Update the values of $\{\overline{\mathcal{N}}_\ell\}_{\ell=0}^{L}$ by calling Algorithm 2.2:
    UMNT ($\{M_\ell\}_{\ell=0}^{L}$, $\{\mathcal{A}_{\mathcal{S}_\ell}[N_\ell]\}_{\ell=0}^{L}$, $\{\mathcal{A}_{\mathcal{S}_\ell}[N_{\ell-1}]\}_{\ell=1}^{L}$).
  end
  Increase k by 1.
end

Algorithm 2.2: Update for the mean number of time steps, **(UMNT)**

Input : $\{M_\ell\}_{\ell=0}^L$, $\{\mathscr{A}_{\mathscr{S}_\ell}[N_\ell]\}_{\ell=0}^L$, $\{\mathscr{A}_{\mathscr{S}_\ell}[N_{\ell-1}]\}_{\ell=1}^L$
Output: $\{\overline{\mathscr{N}_\ell}\}_{\ell=0}^L$

for $\ell = 0, 1, \ldots, L$ **do**
  if $\ell < L$ **then**
    Set $\overline{\mathscr{N}}_\ell = \frac{M_\ell \mathscr{A}_{\mathscr{S}_\ell}[N_\ell] + M_{\ell+1}\mathscr{A}_{\mathscr{S}_{\ell+1}}[N_\ell]}{M_\ell + M_{\ell+1}}$.
  else
    Set $\overline{\mathscr{N}}_L = \mathscr{A}_{\mathscr{S}_L}[N_L]$.
  end
end

Algorithm 2.3: Adaptive Time Step Stochastic Euler **(ATSSE)**

Input : $\Delta t_{in}, \Delta W_{in}$, TOL, $\overline{\mathscr{N}}_{in}$
Output: $\Delta t_{out}, \Delta W_{out}, N_{out}, g_{out}$
Set $m = 0$, $\Delta t_{[0]} = \Delta t_{in}$, $\Delta W_{[0]} = \Delta W_{in}$, $N_{[0]}$ = number of steps in Δt_{in}
while $m < 1$ **or** $(r_{[m-1]};$ TOL, $\overline{\mathscr{N}}_{in})$ violates (33) **do**
  Compute the Euler approximation $\overline{X}_{[m]}$ and the error indicators $r_{[m]}$ on $\Delta t_{[m]}$ with the known Wiener increments $\Delta W_{[m]}$.
  if $(r_{[m]};$ TOL, $\overline{\mathscr{N}}_{in})$ violates (33) **then**
    Refine the grid $\Delta t_{[m]}$ by
    forall intervals $n = 1, 2, \ldots, N_{[m]}$ **do**
      if $r_{[m]}(n)$ satisfies (34) **then**
        divide the interval n into two equal parts
      end
    end
    and store the refined grid in $\Delta t_{[m+1]}$.
    Compute $\Delta W_{[m+1]}$ from $\Delta W_{[m]}$ using Brownian bridges on $\Delta t_{[m+1]}$.
    Set $N_{[m+1]}$ = number of steps in $\Delta t_{[m+1]}$.
  end
  Increase m by 1.
end
Set $\Delta t_{out} = \Delta t_{[m-1]}$, $\Delta W_{out} = \Delta W_{[m-1]}$, $N_{out} = N_{[m-1]}$, $g_{out} = g(\overline{X}_{[m-1]})$.

3 A Stopped Diffusion Example

This section presents numerical results from an implementation of the algorithm of Sect. 2. We apply the algorithm to a challenging problem where the computational work of multilevel Monte Carlo based on uniform meshes is larger than the optimal $\mathscr{O}\left((\mathrm{TOL}^{-1}\log(\mathrm{TOL}))^2\right)$, which is still attained by the adaptive multilevel Monte Carlo algorithm. This motivates the use of stochastic time steps that are adaptively refined for each sample path.

The additional difficulty of the problem is that we now wish to compute approximations of an expected value

$$E[g(X(\tau),\tau)], \tag{35}$$

where $X(t)$ solves the SDE (1), but where the function $g : D \times [0,T] \to \mathbb{R}$ is evaluated at the first exit time

$$\tau := \inf\{t > 0 : (X(t),t) \notin D \times (0,T)\}$$

from a given open domain $D \times (0,T) \subset \mathbb{R}^d \times (0,T)$. This kind of stopped (or killed) diffusion problems arise for example in mathematical finance when pricing barrier options and for boundary value problems in physics.

The main difficulty in the approximation of the stopped diffusion on the boundary ∂D is that a continuous sample path may exit the given domain D even though a discrete approximate solution does not cross the boundary of D. Due to this hitting of the boundary the order of weak convergence of the Euler–Maruyama method is reduced from 1 to $1/2$, in terms of the step size of uniform meshes, and the order of strong convergence is less than $1/2$ so that the complexity estimate in Theorem 1 of [10] for uniform multilevel simulations can not be applied.

We combine the adaptive multilevel algorithm of Sect. 2 with an error estimate derived in [3] that takes into account also the hitting error. The hitting error is accounted for by an extra contribution to the error density in (23); this contribution can be expressed in terms of exit probabilities for individual time steps, conditioned on the computed path at the beginning and the end of the time steps, and of the change in the goal function, g, when evaluated at a possible exit point within the time step instead of the actually computed exit $(\overline{X}(\bar{\tau}),\bar{\tau})$. The full expression of the resulting error indicators is given in equation (50) of [3]. Since the differential $\partial_i g(\overline{X}(T),T)$ in the discrete dual backward problem (17) does not exist if T is replaced by $\bar{\tau} < T$ this initial value must be alternatively defined; this can be done using difference quotients with restarted computed trajectories as described, both for the discrete dual and for its first and second variations, in (20)–(25) of [3]. Note that for this modified error density the proof in [9] of almost sure convergence to a limit density does not apply.

The results in this section are on the accuracy and cost of the adaptive multilevel algorithm of Sect. 2, applied to (35)–(36), with the error estimate modified for the barrier problem.

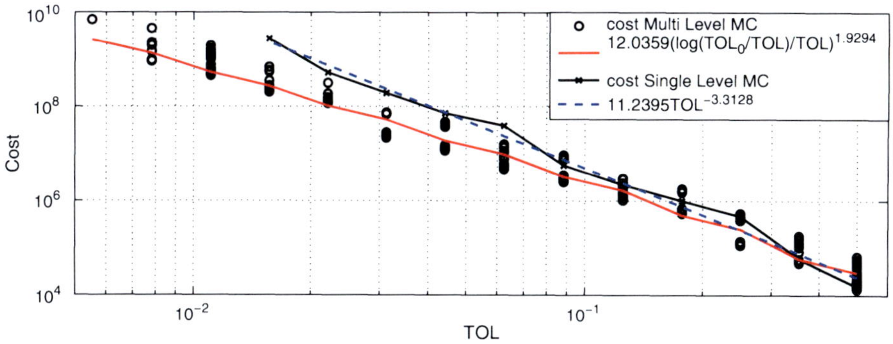

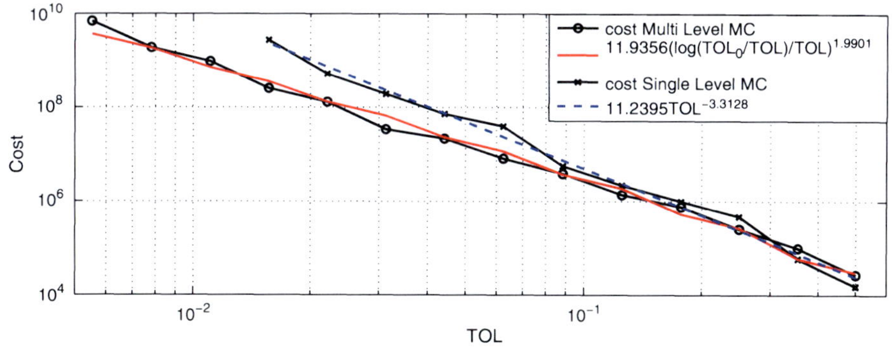

Fig. 1 Experimental complexity for the barrier example. On top, the computational cost of the multilevel adaptive algorithm is shown for varying tolerances using different initial states in the pseudo random number algorithm. A least squares fit, in $\log_2-\log_2$-scale, of the model $cost = c_1\left(\log(\mathrm{TOL}_0/\mathrm{TOL})/\mathrm{TOL}\right)^{c_2}$ with equal weight on all observations results in $c_1 = 12$ and $c_2 = 1.9$. One realisation of the corresponding cost using a single level implementation of the same adaptive Monte Carlo method is included for reference. At the bottom is shown the mean computational cost over all observations where the values for large tolerances are based on more observations than those for small tolerances. When the least square fit is performed on the average values the resulting coefficients are $c_1 = 12$ and $c_2 = 2.0$

For the numerical example we consider the stopped diffusion problem

$$\begin{aligned} dX(t) &= \frac{11}{36}X(t)\,dt + \frac{1}{6}X(t)\,dW(t), \text{for } t \in [0,2] \text{ and } X(t) \in (-\infty, 2), \\ X(0) &= 1.6. \end{aligned} \tag{36}$$

For $g(x,t) = x^3 e^{-t}$ with $x \in \mathbb{R}$, this problem has the exact solution $E[g(X_\tau, \tau)] = u(X(0), 0) = X(0)^3$, where the solution, u, of the Kolmogorov backward equation is $u(x,t) = x^3 e^{-t}$. We chose an example in one space dimension for simplicity, although it is only in high dimension that Monte Carlo methods are more efficient than deterministic finite difference or finite element methods to solve stopped diffusion problems. The comparison here between the standard Monte Carlo and the Multilevel Monte Carlo methods in the simple one dimensional example indicates

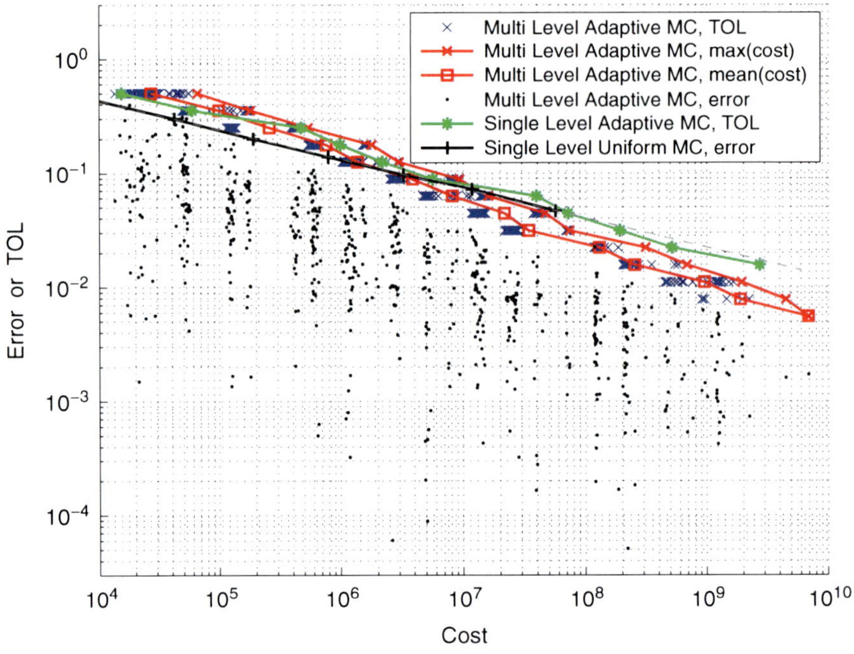

Fig. 2 The multilevel adaptive Monte Carlo algorithm with stochastic time steps has been tested on the barrier problem using a sequence of tolerances and different initial states in the pseudo random number generator. For each tolerance and each sample the computational cost is marked by an ×; the maximal cost and the average cost for a given tolerance have been chosen as representative measures. One realisation of the computational cost using a single level implementation of the adaptive algorithm for a sequence of tolerances is included as a reference, showing that the multilevel version is more efficient for small tolerances. A further comparison can be made with a basic single level, constant time step, Monte Carlo algorithm. This algorithm lacks error control; instead the statistical error was balanced against the time discretisation error in two steps: first the statistical error was over killed to reveal the time discretisation error for each time step size, and then the number of samples needed to make the statistical error the same size as the time discretisation error was estimated using variance estimates from the computation. This represents an ideal situation and it explains the very regular decay of the error with increasing cost seen in the graph; the least square fit, shown as a dashed line, has the slope –0.26, consistent with the $\mathcal{O}(\sqrt{\Delta t})$ time discretisation error for the barrier problem, and a $\mathcal{O}(1/\sqrt{N})$ statistical error

that the Multilevel Monte Carlo method will also be more efficient in high dimensional stopped diffusion problems, where a Monte Carlo method is a good choice.

In the simulations the tolerance levels were chosen as $\mathrm{TOL_S} = \mathrm{TOL}/2$ and $\mathrm{TOL_T} = \mathrm{TOL}/4$. We used for the stopping and refinement constants the values $C_S = 5$ and $C_R = 2$. The computations were performed in `Matlab 7` using the built in pseudo random number generator `randn` for simulating sampling from the normal distribution.

In the numerical complexity results the cost is measured by counting the total number of time steps in all batches and on all levels. The complexity study in Fig. 1

is based on multiple simulations for each tolerance using different initial states in the pseudo random number generator, with more data on the large tolerances than on the smallest ones. A least squares fit of the model[4]

$$cost = c_1 \left(\log\left(\frac{\mathrm{TOL}_0}{\mathrm{TOL}}\right) \frac{1}{\mathrm{TOL}} \right)^{c_2} \tag{37}$$

in the $\log_2$-$\log_2$-scale of the graph using equal weights on all data points gives $c_2 = 1.9$ where the value 2 is predicted by theory. When the least squares fit is made on the mean cost for each tolerance the parameter in the cost model is $c_2 = 2.0$. The corresponding cost using the single level adaptive algorithm with just one data point per tolerance used grows faster than TOL^3 in this example.

In Fig. 2 the data on cost versus tolerance of Fig. 1 is shown together with the corresponding errors. The observed errors are scattered below the corresponding tolerances showing that the algorithm achieves the prescribed accuracy. It was already observed above that the multilevel version of the adaptive algorithm improves on the convergence of the single level version; this figure also shows that the error using a basic single level Monte Carlo method with uniform time steps for the stopped diffusion problem decreases only like $cost^{-0.26}$, which is worse than the convergence rate of the single level version of the adaptive algorithm.

We remark that we present the error versus cost results for the basic Monte Carlo algorithm in a way that slightly favours it over the adaptive methods. To explain this we note that the adaptive algorithms aim to balance the contributions to the total error made by the statistical and by time discretization errors; since the constant time step algorithm was implemented without time discretization error estimates this balancing could not be made in the computations. Instead, for each step size, the cost and error pair displayed in the graph was obtained indirectly by first overkilling the statistical error using a large number of samples and then by, knowing that the resulting error was dominated by the time discretization error, using the computed sample variance to get an estimate of the number of samples that would have been sufficient for obtaining a statistical error of the same size as the time discretization error. This procedure favours the constant time step method over the adaptive methods in that it gives an ideal constant factor in the cost, but the order of convergence is not affected. On the other hand the computational overhead in the implementation of the adaptive time stepping algorithm is significantly greater than in the naive Monte Carlo algorithm; again the order of convergence is not changed.

In conclusion the observed convergence of the adaptive multilevel Monte Carlo method applied to the barrier problem (36) is close to the predicted

$$cost = c_1 \left(\log\left(\frac{\mathrm{TOL}_0}{\mathrm{TOL}}\right) \frac{1}{\mathrm{TOL}} \right)^{2}.$$

[4] The number of levels is $1 + L = 1 + \log_2\left(\frac{\mathrm{TOL}_0}{\mathrm{TOL}_\mathrm{T}}\right) = \log_2\left(\frac{\mathrm{TOL}_0}{\mathrm{TOL}}\right)$.

This shows an improved convergence compared to the single level version of the adaptive Monte Carlo algorithm where the cost grows approximately like TOL^{-3}, which in itself is a better order of weak convergence than the one obtained using a single level Monte Carlo method with constant time steps where the cost grows like $error^{-4}$.

References

1. Anders Szepessy, Raúl Tempone, and Georgios E. Zouraris. Adaptive weak approximation of stochastic differential equations. *Comm. Pure Appl. Math.*, 54(10):1169–1214, 2001.
2. Andrew J. Majda, Ilya Timofeyev, and Eric Vanden Eijnden. A mathematical framework for stochastic climate models. *Comm. Pure Appl. Math.*, 54(8):891–974, 2001.
3. Anna Dzougoutov, Kyoung-Sook Moon, Erik von Schwerin, Anders Szepessy, and Raúl Tempone. Adaptive Monte Carlo algorithms for stopped diffusion. In *Multiscale methods in science and engineering*, volume 44 of *Lect. Notes Comput. Sci. Eng.*, pages 59–88. Springer, Berlin, 2005.
4. Bernt Øksendal. *Stochastic differential equations*. Universitext. Springer-Verlag, Berlin, fifth edition, 1998. An introduction with applications.
5. Elyes Jouini, Jaksa Cvitanić, and Marek Musiela, editors. *Option pricing, interest rates and risk management*. Handbooks in Mathematical Finance. Cambridge University Press, Cambridge, 2001.
6. Eric Cancès, Frédéric Legoll, and Gabriel Stoltz. Theoretical and numerical comparison of some sampling methods for molecular dynamics. *M2AN Math. Model. Numer. Anal.*, 41(2):351–389, 2007.
7. Ioannis Karatzas and Steven E. Shreve. *Brownian motion and stochastic calculus*, volume 113 of *Graduate Texts in Mathematics*. Springer-Verlag, New York, second edition, 1991.
8. Kyoung-Sook Moon, Anders Szepessy, Raúl Tempone, and Georgios E. Zouraris. Convergence rates for adaptive weak approximation of stochastic differential equations. *Stoch. Anal. Appl.*, 23(3):511–558, 2005.
9. Kyoung-Sook Moon, Erik von Schwerin, Anders Szepessy, and Raúl Tempone. An adaptive algorithm for ordinary, stochastic and partial differential equations. In *Recent advances in adaptive computation*, volume 383 of *Contemp. Math.*, pages 325–343. Amer. Math. Soc., Providence, RI, 2005.
10. Michael B. Giles. Multilevel Monte Carlo path simulation. *Oper. Res.*, 56(3):607–617, 2008.
11. Mike Giles. Improved multilevel Monte Carlo convergence using the Milstein scheme. In *Monte Carlo and quasi-Monte Carlo methods 2006*, pages 343–358. Springer, Berlin, 2008.
12. Paul Glasserman. *Monte Carlo methods in financial engineering*, volume 53 of *Applications of Mathematics (New York)*. Springer-Verlag, New York, 2004. Stochastic Modelling and Applied Probability.

Coupled Coarse Graining and Markov Chain Monte Carlo for Lattice Systems

Evangelia Kalligiannaki*, Markos A. Katsoulakis†, and Petr Plecháč‡

Abstract We propose an efficient Markov Chain Monte Carlo method for sampling equilibrium distributions for stochastic lattice models. The method is capable of handling correctly and efficiently long and short-range particle interactions. The proposed method is a Metropolis-type algorithm with the proposal probability transition matrix based on the coarse-grained approximating measures introduced in (Katsoulakis et al. Proc. Natl. Acad. Sci. 100(3):782–787, 2003; Katsoulakis et al. ESAIM-Math. Model. Numer. Anal. 41(3):627–660, 2007). The proposed algorithm reduces the computational cost due to energy differences and has comparable mixing properties with the classical microscopic Metropolis algorithm, controlled

* The research of E.K. was supported by the National Science Foundation under the grant NSF-CMMI-0835582.
† The research of M.A. K. was supported by the National Science Foundation through the grants NSF-DMS-0715125 and the CDI – Type II award NSF-CMMI-0835673.
‡ The research of P.P. was partially supported by the National Science Foundation under the grant NSF-DMS-0813893 and by the Office of Advanced Scientific Computing Research, U.S. Department of Energy under DE-SC0001340; the work was partly done at the Oak Ridge National Laboratory, which is managed by UT-Battelle, LLC under Contract No. DE-AC05-00OR22725.

P. Plecháč (✉)
Department of Mathematics, University of Tennessee, 1900 Laurel Ave, Knoxville, TN 37916, USA
e-mail: plechac@math.utk.edu

E. Kalligiannaki
Joint Institute for Computational Sciences, University of Tennessee and Oak Ridge National Laboratory, P.O. Box 2008, Oak Ridge, TN 37831, USA
e-mail: evy@ornl.gov

M.A. Katsoulakis
Department of Applied Mathematics, University of Massachusetts, University of Crete and Foundation of Research and Technology-Hellas, Nikolaou Plastira 100, Vassilika Vouton, GR – 700 13 Heraklion, Crete, Greece
e-mail: markos@math.umass.edu

B. Engquist et al. (eds.), *Numerical Analysis of Multiscale Computations*, Lecture Notes in Computational Science and Engineering 82, DOI 10.1007/978-3-642-21943-6_11,

by the level of coarsening and reconstruction procedure. The properties and effectiveness of the algorithm are demonstrated with an exactly solvable example of a one dimensional Ising-type model, comparing efficiency of the single spin-flip Metropolis dynamics and the proposed coupled Metropolis algorithm.

1 Introduction

Microscopic, *extended* (many-particle) systems with *complex interactions* are ubiquitous in science and engineering applications in a variety of physical and chemical systems, exhibiting rich mesoscopic morphologies. For example, nano-pattern formation via self-assembly, arises in surface processes e.g., in heteroepitaxy, induced by competing short and long-range interactions [6]. Other examples include macromolecular systems such as polymers, proteins and other soft matter systems, quantum dots and micromagnetic materials. Scientific computing for this class of systems can rely on molecular simulation methods such as Kinetic Monte Carlo (KMC) or Molecular Dynamics (MD), however their extensivity, their inherently complex interactions and stochastic nature, severely limit the spatio-temporal scales that can be addressed by these direct numerical simulation methods.

One of our primary goals is to develop systematic mathematical and computational strategies for the speed-up of microscopic simulation methods by developing *coarse-grained* (CG) approximations, thus reducing the extended system's degrees of freedom. To date coarse-graining methods have been a subject of intense focus, mainly outside mathematics and primarily in the physics, applied sciences and engineering literatures [10, 18, 26, 29, 31]. The existing approaches can give unprecedented speed-up to molecular simulations and can work well in certain parameter regimes, for instance, at high temperatures or low density. On the other hand, in many parameter regimes, important macroscopic properties may not be captured properly, e.g. [1, 31, 32]. Here we propose to, develop *reliable* CG algorithms for stochastic lattice systems with complex, and often competing particle interactions in equilibrium. Our proposed methodologies stem from the synergy of stochastic processes, statistical mechanics and statistics sampling methods.

Monte Carlo algorithms provide a computational tool capable of estimating observables defined on high-dimensional configuration spaces that are typical for modeling of complex interacting particle systems at or out of equilibrium. Markov Chain Monte Carlo (MCMC) simulation methods such as the Metropolis algorithm, were first proposed in 1953 by Metropolis and his coauthors [30] for the numerical calculation of the equation of state for a system of rigid spheres. It was generalized in 1970 by Hastings [14] and it is commonly referred to as the Metropolis-Hastings (MH) Monte Carlo method. This method belongs to the family of MCMC methods which generate ergodic Markovian chains with the stationary distribution being the desired sampled probability measure. Metropolis algorithm consists of two main ingredients: (a) the probability transition kernel q; the *proposal*, that generates trial states and (b) the acceptance probability α according to which the proposed trial is accepted or rejected. There are though some drawbacks of this method when applied to large systems, such as a small acceptance probability α, that leads to costly calculations of a large number of samples that are discarded. A way to reduce these costs is to *predict* efficient proposal measures such that the computational cost of calculating a sample is lower and, if possible, increase the acceptance probability. Convergence and ergodicity properties of Metropolis type algorithms are studied extensively in a series of works [7, 8, 33]. The rate of convergence to stationarity is strongly dependent on the proposal distribution and its relation to the stationary measure [33, Chap. 7]. A quantity that measures the speed of convergence in distribution to stationarity is the spectral gap. In order to improve an MCMC method one has to increase its spectral gap by smartly constructing a good proposal.

In this work we propose the Coupled Coarse Graining Monte Carlo (Coupled CGMC) method, a new method of constructing efficient *proposal measures* based on coarse-graining properties of the sampling models. We prove that such approach is suitable for models that include both short and long-range interactions between particles. Long-range interactions are well-approximated by coarse graining techniques [18, 19, 21], and Coarse Graining Monte Carlo (CGMC) are adequate simulation methods [20, 22]. Furthermore, models where only short-range interactions appear are inexpensive to simulate, for example with a single spin-flip Metropolis method. However, when both short and long-range interactions are present the classical MH algorithm becomes prohibitively expensive due to the high cost of calculating energy differences arising from the long-range interaction potential. In [16] we extend our framework for coupled CGMC to the dynamics case, developing kinetic Monte Carlo algorithms based on coarse-level rates.

Section 2 describes the classical Metropolis-Hastings algorithm and some known mathematical theory for the convergence and the rate of convergence for MCMC methods. In Sect. 3 we present the proposed Coupled CGMC method in a general framework describing its mathematical properties. We state the main theorem that compares the rate of convergence to equilibrium with the rate of the classical MH method. In Sect. 4 we describe stochastic lattice systems and the coarse-graining procedure in order to prepare for the application of the proposed method in Sect. 5 to a generic model of lattice systems in which both short and long-range interactions are present.

2 MCMC Methods

Before describing the Metropolis-Hastings method we need to introduce some necessary definitions and theoretical facts.

Let $\{X_n\}$ be a Markov chain on space Σ with transition kernel $\mathscr{K}$.

Definition 1. A transition kernel $\mathscr{K}$ has the *stationary measure* μ if

$$\mathscr{K}\mu = \mu .$$

Definition 2. $\mathscr{K}$ is called reversible with respect to μ if

$$(g, \mathscr{K} h)_\mu = (\mathscr{K} g, h)_\mu, \quad \text{for all } g, h \in L^2(\mu).$$

where $(g,h)_\mu = \int_\Sigma \overline{g(\sigma)} h(\sigma)\mu(d\sigma)$ and $\mathscr{K} g(\sigma) = \int_\Sigma \mathscr{K}(\sigma, d\sigma')g(\sigma'), \forall \sigma \in \Sigma$.

A sufficient condition for μ being a stationary measure of $\mathscr{K}$ is the detailed balance (DB) condition, that is often easy to check.

Definition 3. A Markov chain with transition kernel $\mathscr{K}$ satisfies the detailed balance condition if there exists a function f satisfying

$$\mathscr{K}(\sigma, \sigma') f(\sigma) = \mathscr{K}(\sigma', \sigma) f(\sigma'). \tag{1}$$

Here we focus on the Metropolis-Hastings algorithm [33]. The algorithm generates an ergodic Markov chain $\{X_n\}$ in the state space Σ, with stationary measure $\mu(d\sigma)$. Let $f(\sigma)$ be the probability density corresponding to the measure μ and $X_0 = \sigma_0$ be arbitrary. The n-th iteration of the algorithm consists of the following steps

Algorithm 1 (Metropolis-Hastings algorithm)

Given $X_n = \sigma$

Step 1 *Generate* $Y_n = \sigma' \sim q(\sigma', \sigma)$
Step 2 *Accept-Reject*

$$X_{n+1} = \begin{cases} Y_n = \sigma' \text{ with probability } \alpha(\sigma, \sigma'), \\ X_n = \sigma_n \text{ with probability } 1 - \alpha(\sigma, \sigma'), \end{cases}$$

where

$$\alpha(\sigma, \sigma') = \min\left\{1, \frac{f(\sigma') q(\sigma', \sigma)}{f(\sigma) q(\sigma, \sigma')}\right\}.$$

We denote $q(\sigma', \sigma)$ *the proposal probability transition kernel, and* $\alpha(\sigma, \sigma')$ *the probability of accepting the proposed state* σ'. *The transition kernel associated to MH algorithm is*

$$\mathscr{K}_c(\sigma, \sigma') = \alpha(\sigma, \sigma') q(\sigma, \sigma') + \left[1 - \int \alpha(\sigma, \sigma') q(\sigma, \sigma') d\sigma'\right] \delta(\sigma' - \sigma). \tag{2}$$

where δ denotes the Dirac function.

Convergence and ergodicity properties of the chain $\{X_n\}$ depend on the proposal kernel $q(\sigma, \sigma')$, and they are studied extensively in [33]. $\mathscr{K}_c$ satisfies the DB condition with f ensuring that it has stationary measure μ. $\mathscr{K}_c$ is irreducible and aperiodic, nonnegative definite, and reversible, thus the Markov chain with transition kernel $\mathscr{K}_c$ converges in distribution to μ.

2.1 Mixing Times and Speed of Convergence

It is known [7] that for a discrete-time Markov chain $\{X_n\}$ with the transition kernel $\mathscr{K}$ and stationary distribution f, the rate of convergence to its stationarity can be measured in terms of the kernel's second largest eigenvalue, according to inequality

$$2||\mathscr{K}^n(\sigma,\cdot) - f||_{TV} \le \frac{1}{f(\sigma)^{1/2}}\beta^n,$$

where $||\cdot||_{TV}$ denotes the total variance norm and $\beta = max\{|\beta_{min}|, \beta_1\}$ with $-1 \le \beta_{min} \le \cdots \le \beta_1 \le \beta_0 = 1$ are the real eigenvalues of $\mathscr{K}$. The spectral gap of kernel $\mathscr{K}$ is defined by

$$\lambda(\mathscr{K}) = \min\left\{\frac{\mathscr{E}(h,h)}{\mathrm{Var}(h)}; \mathrm{Var}(h) \ne 0\right\},$$

which for a self-adjoint, because of reversibility, kernel $\mathscr{K}$ is $\lambda(\mathscr{K}) = 1 - \beta_1$. With the Dirichlet form $\mathscr{E}$ and the variance defined by

$$\mathscr{E}(h,h) = \frac{1}{2}\sum_{\sigma,\sigma'}|h(\sigma) - h(\sigma')|^2 \mathscr{K}(\sigma,\sigma')f(\sigma),$$

$$\mathrm{Var}(h) = \frac{1}{2}\sum_{\sigma,\sigma'}|h(\sigma) - h(\sigma')|^2 f(\sigma')f(\sigma).$$

Between two algorithms producing Markov chains with identical equilibrium distributions *better* in terms of the speed of convergence is the one with the *smaller second eigenvalue in absolute value* or equivalently with the *larger spectral gap.*

3 The Coupled CGMC Method

The proposed algorithm is designed to generate samples from the microscopic probability measure μ with density f on a space Σ, coupling properly states of the microscopic space Σ with states on a *coarse* space $\bar{\Sigma}$ having less degrees of freedom. A properly constructed coarse measure on $\bar{\Sigma}$ will be the basis for constructing efficient proposal kernels for MH algorithms sampling large systems.

The *coarsening* procedure is based on the expansion of the target measure μ to a coarse and a finer part. Abstractly we write $f(\sigma) = f(\eta,\xi)$ and $\Sigma = \bar{\Sigma} \times \bar{\Sigma}'$, where $\eta \in \bar{\Sigma}$ represents the coarse variables.

We denote the projection operator on the coarse variables

$$T : \Sigma \to \bar{\Sigma}, \quad T\sigma = \eta.$$

The exact coarse marginal is

$$\bar{f}(\eta) = \int_{\bar{\Sigma}'} f(\eta,\xi)d\xi\,.$$

To obtain an explicit formula of the coarse marginal is as difficult as sampling the original target distribution since space $\bar{\Sigma}'$ remains high dimensional. Therefore use of distributions approximating $\bar{f}$ becomes necessary. Such approximations have been proposed in [18, 22] for stochastic lattice systems and are abstractly described in Sect. 4 and for complex macromolecular systems see [4, 11, 13, 36].

Denote $\bar{f}_0$ an approximation of $\bar{f}$ on $\bar{\Sigma}$. This distribution, combined with a reconstruction distribution $f_r(\xi|\eta)$ corresponding to the finer variables ξ, will construct a candidate for proposal distribution in MH algorithms performed in order to sample from f at the original space Σ. An example of a "good" proposal distribution is $f_0(\sigma) := \bar{f}_0(\eta) f_r(\xi|\eta)$. For notational simplicity we write $f_r(\sigma|\eta)$ instead of $f_r(\xi|\eta)$. In terms of the Metropolis-Hastings algorithm this means that $q(\sigma,\sigma') = f_0(\sigma')$, or that f_0 is the stationary measure of the proposal kernel $q(\sigma,\sigma')$.

The coupled CGMC algorithm is composed of two coupled Metropolis iterations, the first generating samples from the proposal distribution and the second samples from the target measure. The first Metropolis step samples the coarse approximating marginal $\bar{f}_0(\eta)$, using an arbitrary proposal transition kernel $\bar{q}_0(\eta,\eta')$ to produce trial samples η'. The second step is performed if the coarse trial sample is accepted, and consists of the reconstruction from the coarse trial state and a Metropolis acceptance criterion designed to ensure sampling from the correct microscopic density f. If a trial coarse sample is rejected, then we go back to the first step to rebuild a new coarse trial, so that the fine Metropolis step is not performed and no computational time is wasted on checking fine trial samples that are most likely to be rejected.

In [9] Efendiev et al. propose the Preconditioning MCMC, a two stage (coarse and fine) Metropolis MCMC method, applied to inverse problems of subsurface characterization. The coarse and fine models are finite volume schemes of different resolutions for a PDE two-phase flow model. Our algorithm shares the same idea and structure with the Preconditioning MCMC of constructing a proposal density based on meso/macro-scopic properties of the model studied and taking advantage of the first stage rejections. In terms of the MC method "coarsening" corresponds to enriching the range of the sampling measure based on coarse-scale models proposed by multiscale finite volume methods. The major difference of the Preconditioning MCMC and the proposed algorithm is that the latter alternates between different state spaces during each MC iteration, the coarse and the finer, whether in the former the state space remains the same since coarse and fine problems are solved independently. Thus, at the end of a simulation we will have both fine-scale and "compressed", coarse-grained, data. The performance of the coarse proposals in our case can be further estimated based on a systematic error analysis such as in (14).

The proposed procedure has also some common features with the modified Configurational bias Monte Carlo (CBMS) where the trial density is built up sequentially with stage-wise rejection decision, described in [28], applied effectively in quantum mechanical systems [5]. There are also some similarities with simulated sintering and transdimensional MCMC, see [28] and references therein. However, in our method, the construction of the variable dimensionality (and level of coarse-graining) state spaces and the corresponding Gibbs measures relies on statistical mechanics tools that allow a systematic control of the error from one level of coarse-graining to the next, e.g. (14).

3.1 The Algorithm

We describe in detail the coupled CGMC Metropolis algorithm outlined above.

Algorithm 2 (Coupled CGMC algorithm)
Let $X_0 = \sigma_0$ *be arbitrary. For* $n = 0, 1, 2, \ldots$
given $X_n = \sigma$

Step 1 *Compute the coarse variable* $\eta = T\sigma$.
Step 2 *Generate a coarse sample* $\eta' \sim \bar{q}_0(\eta, \eta')$.
Step 3 *Coarse Level Accept-Reject.*
Accept η' *with probability:*

$$\alpha_{CG}(\eta, \eta') = \min\left\{1, \frac{\bar{f}_0(\eta')\bar{q}_0(\eta', \eta)}{\bar{f}_0(\eta)\bar{q}_0(\eta, \eta')}\right\}.$$

If η' **is accepted then proceed to Step 4**
else generate a new coarse sample Step 2

Step 4 *Reconstruct* σ' *given the coarse trial* η',

$$\sigma' \sim f_r(\cdot|\eta').$$

Step 5 *Fine Level Accept-Reject.*
Accept σ' *with probability*

$$\alpha_f(\sigma, \sigma') = \min\left\{1, \frac{f(\sigma')\,\bar{f}_0(\eta)\,f_r(\sigma|\eta)}{f(\sigma)\,\bar{f}_0(\eta')\,f_r(\sigma'|\eta')}\right\}.$$

Steps 2 and 3 generate a Markov chain $\{Z_n\}$ in the coarse space $\bar{\Sigma}$ with the transition kernel

$$\mathcal{Q}(\eta, \eta') = \alpha_{CG}(\eta, \eta')\bar{q}_0(\eta, \eta') + \left[1 - \int \alpha_{CG}(\eta, z)\bar{q}_0(\eta, z)\right]\delta(\eta' - \eta).$$

The stationary measure of kernel $\mathcal{Q}$ is $\bar{f}_0(\eta)$. Combination of this kernel and Steps 1 and 4 constructs the desired proposal transition kernel $q_0(\sigma, \sigma')$ on the fine level space Σ,

$$q_0(\sigma, \sigma') = \mathcal{Q}(\eta, \eta')\,f_r(\sigma'|\eta').$$

According to the MH algorithm in order to sample from f, the fine level acceptance probability should be $\alpha_f(\sigma,\sigma') = \min\left\{1, \frac{f(\sigma')q_0(\sigma',\sigma)}{f(\sigma)q_0(\sigma,\sigma')}\right\}$, but since $\mathscr{Q}$ satisfies the detailed balance condition $\mathscr{Q}(\eta,\eta')\bar{f}_0(\eta) = \mathscr{Q}(\eta',\eta)\bar{f}_0(\eta')$, α_f is equal to

$$\alpha_f(\sigma,\sigma') = \min\left\{1, \frac{f(\sigma')\mathscr{Q}(\eta',\eta)f_r(\sigma|\eta)}{f(\sigma)\mathscr{Q}(\eta,\eta')f_r(\sigma'|\eta')}\right\}$$
$$= \min\left\{1, \frac{f(\sigma')\bar{f}_0(\eta)f_r(\sigma|\eta)}{f(\sigma)\bar{f}_0(\eta')f_r(\sigma'|\eta')}\right\}.$$

Chain $\{X_n\}$ generated by the Coupled CGMC algorithm is a Markov chain on the fine space Σ, with the transition kernel

$$\mathscr{K}_{CG}(\sigma,\sigma') = \alpha_f(\sigma,\sigma')q_0(\sigma,\sigma') + \left[1 - \int \alpha_f(\sigma,\sigma')q_0(\sigma,\sigma')d\sigma'\right]\delta(\sigma'-\sigma). \quad (3)$$

The Markov chain $\{X_n\}$ converges to the correct stationary distribution f and is ergodic, which ensures that $\frac{1}{n}\sum_{j=1}^{n} h(X_j)$ is a convergent approximation of the averages $\int h(\sigma)f(\sigma)d\sigma$ for any $h \in L^1(f)$. Ergodicity and reversibility properties are satisfied ensuring that the algorithm generates samples from the correct measure.We state this fact as a separate theorem proof of which is given in detail in [15].

We denote $E = \{\sigma \in \Sigma; f(\sigma) > 0\}$, $\bar{E} = \{\eta \in \bar{\Sigma}; \bar{f}_0(\eta) > 0\}$.

Theorem 1. *For every conditional distribution $\bar{q}_0$, and f_r such that the support of $q_0 f_r$ includes E,*

(i) The transition kernel satisfies the detailed balance (DB) condition with f.

$$\mathscr{K}_{CG}(\sigma,\sigma')f(\sigma) = \mathscr{K}_{CG}(\sigma',\sigma)f(\sigma').$$

(ii) f is a stationary distribution of the chain.
(iii) If $q_0(\sigma,\sigma') > 0$, $\forall \sigma,\sigma' \in E$ and $E \subseteq supp(f_0)$ then $\{X_n\}$ is f-irreducible.
(iv) $\{X_n\}$ is aperiodic.

3.2 The Rate of Convergence

The calculation of the rate of convergence to stationarity is a hard problem since it is model dependent. Here we can prove for the proposed method that its rate of convergence is comparable to the classical Metropolis-Hastings algorithm described in Algorithm 1. This fact is stated rigorously in the following theorem which we prove in [15].

Let $\lambda(\mathscr{K}_{CG}), \lambda(\mathscr{K}_c)$ be the spectral gap corresponding to the coupled CGMC $\mathscr{K}_{CG}$, (3), and the classical MH $\mathscr{K}_c$, (2), transition kernels respectively.

Theorem 2. *Let $q(\sigma,\sigma')$ be a symmetric proposal transition probability for the classical MH algorithm and $\bar{q}_0(\eta,\eta')$ a symmetric proposal transition probability on* *the coarse space $\bar{\Sigma}$ for the coupled CGMC algorithm, then for any reconstruction conditional probability $f_r(\sigma|\eta)$*

(i)

$$\mathscr{K}_{CG}(\sigma,\sigma') = \mathscr{A}(\sigma,\sigma')\mathscr{B}(\sigma,\sigma')\mathscr{K}_c(\sigma,\sigma'), \qquad \sigma \neq \sigma', \tag{4}$$

$$\mathscr{B}(\sigma,\sigma') = \begin{cases} \frac{\bar{q}_0(\eta,\eta') f_r(\sigma'|\eta')}{q(\sigma,\sigma')}, & \text{if } \alpha_f = 1, \\ \frac{\bar{q}_0(\eta',\eta) f_r(\sigma|\eta)}{q(\sigma',\sigma)}, & \text{if } \alpha_f < 1. \end{cases}$$

Furthermore we define the subsets

$$C_1 = \{(\sigma,\sigma') \in \Sigma \times \Sigma : \{\alpha < 1, \alpha_{CG} < 1, \alpha_f < 1\} \text{ or } \{\alpha = 1, \alpha_{CG} = 1, \alpha_f = 1\}\},$$
$$C_2 = \{(\sigma,\sigma') \in \Sigma \times \Sigma : \{\alpha = 1, \alpha_{CG} < 1, \alpha_f = 1\} \text{ or } \{\alpha < 1, \alpha_{CG} = 1, \alpha_f < 1\}\},$$
$$C_3 = \{(\sigma,\sigma') \in \Sigma \times \Sigma : \{\alpha = 1, \alpha_{CG} = 1, \alpha_f < 1\} \text{ or } \{\alpha < 1, \alpha_{CG} < 1, \alpha_f = 1\}\},$$
$$C_4 = \{(\sigma,\sigma') \in \Sigma \times \Sigma : \{\alpha < 1, \alpha_{CG} = 1, \alpha_f = 1\} \text{ or } \{\alpha = 1, \alpha_{CG} < 1, \alpha_f < 1\}\},$$

$$\mathscr{A}(\sigma,\sigma') = \begin{cases} 1, & \text{if } (\sigma,\sigma') \in C_1, \\ \min\{\frac{\bar{f}_0(\eta')}{\bar{f}_0(\eta)}, \frac{\bar{f}_0(\eta)}{\bar{f}_0(\eta')}\}, & \text{if } (\sigma,\sigma') \in C_2, \\ \min\{\frac{f(\sigma')\bar{f}_0(\eta)}{f(\sigma)\bar{f}_0(\eta')}, \frac{f(\sigma)\bar{f}_0(\eta')}{f(\sigma')\bar{f}_0(\eta)}\}, & \text{if } (\sigma,\sigma') \in C_3, \\ \min\{\frac{f(\sigma')}{f(\sigma)}, \frac{f(\sigma)}{f(\sigma')}\}, & \text{if } (\sigma,\sigma') \in C_4. \end{cases}$$

(ii)

$$\mathscr{A}\underline{\gamma}\lambda(\mathscr{K}_c) \leq \lambda(\mathscr{K}_{CG}) \leq \bar{\gamma}\lambda(\mathscr{K}_c) \tag{5}$$

where $\mathscr{A} = \inf_{\sigma,\sigma'} \mathscr{A}(\sigma,\sigma')$ *and* $\underline{\gamma} > 0, \bar{\gamma} > 0$ *such that* $\underline{\gamma} \leq \mathscr{B}(\sigma,\sigma') \leq \bar{\gamma}$.

4 Extended Lattice Systems

This class of stochastic processes is employed in the modeling of adsorption, desorption, reaction and diffusion of chemical species in numerous applied science areas such as catalysis, microporous materials, biological systems, etc. [3, 27]. To demonstrate the basic ideas, we consider an Ising-type system on a periodic d-dimensional lattice Λ_N with $N = n^d$ lattice points. At each $x \in \Lambda_N$ we can define an order parameter $\sigma(x)$; for instance, when taking values 0 and 1, it can describe vacant and occupied sites. The energy H_N of the system, at the configuration $\sigma = \{\sigma(x) : x \in \Lambda_N\}$ is given by the Hamiltonian,

$$H_N(\sigma) = -\frac{1}{2} \sum_{x \in \Lambda_N} \sum_{y \neq x} [K(x-y) + J(x-y)]\sigma(x)\sigma(y) + \sum h\sigma(x), \tag{6}$$

where h, is the external field and K, J are the inter-particle potentials. Equilibrium states at the temperature $\sim \beta^{-1}$ are described by the (canonical) Gibbs probability measure,

$$\mu_{N,\beta}(d\sigma) = Z_N^{-1} \exp\big(-\beta H_N(\sigma)\big) P_N(d\sigma), \tag{7}$$

and Z_N is the normalizing constant (partition function). Furthermore, the product Bernoulli distribution $P_N(\sigma)$ is the *prior distribution* on Λ_N.

The inter-particle potentials K, J account for interactions between occupied sites. We consider K corresponding to the short and J to the long-range interactions discussed in detail in Sect. 4.2. General potentials with combined short and long-range interactions are discussed here, while we can also address potentials with suitable decay/growth conditions [2].

The prior $P_N(d\sigma)$ is typically a product measure, describing the system at $\beta = 0$, when interactions in H_N are unimportant and thermal fluctuations-disorder-associated with the product structure of $P_N(d\sigma)$ dominates. By contrast at zero temperature, $\beta = \infty$, interactions and hence order, prevail. Finite temperatures, $0 < \beta < \infty$, describe intermediate states, including possible phase transitions between ordered and disordered states. For both on-lattice or off-lattice particle systems, the finite-volume equilibrium states of the system have the structure (7).

4.1 Coarse-Graining of Microscopic Systems

Coarse-graining (CG) of microscopic systems is essentially an approximation theory and a numerical analysis question. However, the presence of *stochastic fluctuations* on one hand, and the *extensivity* of the models (the system size scales with the number of particles) on the other, create a new set of challenges. We discuss all these issues next, in a general setting that applies to both on-lattice and off-lattice systems.

First, we write the microscopic configuration σ in terms of coarse variables η and corresponding fine ones ξ so that $\sigma = (\eta, \xi)$. We denote by T the coarse-graining map $T\sigma = \eta$.

The CG system size is denoted by M, while the microscopic system size is $N = Mq$, where we refer to q as the level of coarse graining, and $q = 1$ corresponds to no coarse graining. The exact CG Gibbs measure is given (with a slight abuse of notation) by $\bar{\mu}_{M,\beta} = \mu_{N,\beta} \circ T^{-1}$. In order to write $\bar{\mu}_{M,\beta}$ in a more convenient form we first define the CG prior $\bar{P}_M(d\eta) = P_N \circ T^{-1}$. The conditional prior $P_N(d\sigma|\eta)$ is the probability of having a microscopic configuration σ, given a coarse configuration η. We now rewrite $\bar{\mu}_{M,\beta}$ using the *exact coarse-grained Hamiltonian*:

$$e^{-\beta \bar{H}_M(\eta)} = \mathbb{E}[e^{-\beta H_N}|\eta] = \int e^{-\beta H_N(\sigma)} P_N(d\sigma|\eta), \tag{8}$$

a procedure known as the *renormalization group map*, [12]; $\bar{\mu}_{M,\beta}(d\eta)$ is now re-written using (8) as

$$\bar{\mu}_{M,\beta}(d\eta) = \frac{1}{\bar{Z}_M} e^{-\beta \bar{H}_M(\eta)} \bar{P}_M(d\eta). \tag{9}$$

Although typically $\bar{P}_M(d\eta)$ is easy to calculate, even for moderately small values of N, the exact computation of the coarse-grained Hamiltonian $\bar{H}_M(\eta)$ given by (8) is, in general, impossible.

We have shown in [22] that there is an expansion of $\bar{H}_M(\eta)$ into a convergent series

$$\bar{H}_M(\eta) = \bar{H}_M^{(0)}(\eta) + \bar{H}_M^{(1)}(\eta) + \bar{H}_M^{(2)}(\eta) + \cdots + \bar{H}_M^{(p)}(\eta) + N \times \mathcal{O}(\epsilon^p), \tag{10}$$

by constructing a suitable first approximation $\bar{H}_M^{(0)}(\eta)$ and identifying a suitable small parameter ϵ to control the higher order terms in the expansions. Truncations including the first terms in (10) correspond to coarse-graining schemes of increasing accuracy. In order to obtain this expansion we rewrite (8) as

$$\bar{H}_M(\eta) = \bar{H}_M^{(0)}(\eta) - \frac{1}{\beta} \log \mathbb{E}[e^{-\beta(H_N - \bar{H}_M^{(0)}(\eta))} | \eta]. \tag{11}$$

We need to show that the logarithm can be expanded into a convergent series, yielding eventually (10), however, two interrelated difficulties emerge immediately: first, the stochasticity of the system in the finite temperature case, yields the nonlinear log expression which in turn will need to be expanded into a series. Second, the extensivity of the microscopic system, i.e., typically the Hamiltonian scales as $H_N = O(N)$, does not allow the expansion of the logarithm and exponential functions into a Taylor series. For these two reasons, one of the mathematical tools we employed is the *cluster expansion method*, see [34] for an overview. Cluster expansions allow us to identify uncorrelated components in the expected value $\mathbb{E}[e^{-\beta(H_N - \bar{H}_M^{(0)}(\eta))} | \eta]$, which in turn will permit us to factorize it, and subsequently expand the logarithm.

The coarse-graining of systems with purely long- or intermediate-range interactions of the form

$$J(x-y) = L^{-1} V\Big((x-y)/L\Big), \quad x, y \in \Lambda_N, \tag{12}$$

where $V(r) = V(-r)$, $V(r) = 0, |r| > 1$, was studied using cluster expansions in [2,21,22]. The corresponding CG Hamiltonian is

$$\bar{H}^0(\eta) = -\frac{1}{2} \sum_{l \in \bar{\Lambda}_M} \sum_{\substack{k \in \bar{\Lambda}_M \\ k \neq l}} \bar{J}(k,l) \eta(k) \eta(l) - \frac{\bar{J}(0,0)}{2} \sum_{l \in \bar{\Lambda}_M} \eta(l)\big(\eta(l)-1\big) + \sum_{k \in \bar{\Lambda}_M} \bar{h} \eta(k), \tag{13}$$

$$\bar{J}(k,l) = \frac{1}{q^2} \sum_{x \in C_k} \sum_{y \in C_l} J(x-y), \quad \bar{J}(k,k) = \frac{1}{q(q-1)} \sum_{x \in C_k} \sum_{y \in C_k, y \neq x} J(x-y).$$

One of the results therein is on deriving error estimates in terms of the specific relative entropy $\mathscr{R}(\mu|\nu) := N^{-1}\sum_{\sigma}\log\{\mu(\sigma)/\nu(\sigma)\}\mu(\sigma)$ between the corresponding equilibrium Gibbs measures. Note that the scaling factor N^{-1} is related to the extensivity of the system, hence the proper error quantity that needs to be tracked is the loss of information *per particle*. Using this idea we can assess the *information compression* for the same level of coarse graining in schemes differentiated by the truncation level p in (10)

$$\mathscr{R}\left(\bar{\mu}^{(p)}_{M,\beta}|\mu_{N,\beta}\circ T^{-1}\right) = \mathscr{O}\left(\epsilon^{p+1}\right), \qquad \epsilon \equiv \beta\|\nabla V\|_1\left(\frac{q}{L}\right), \tag{14}$$

where $\bar{H}^{(0)}_M(\eta)$ in (10) is given by (13). The role of such higher order schemes was demonstrated in nucleation, metastability and the resulting switching times between phases, [2].

Although CGMC and other CG methods can provide a powerful computational tool in molecular simulations, it has been observed that in some regimes, important macroscopic properties may not be captured properly. For instance, (over-)coarse graining in polymer systems may yield wrong predictions in the melt structure [1]; similarly wrong predictions on crystallization were also observed in the CG of complex fluids, [32]. In CGMC for lattice systems, hysteresis and critical behavior may also not be captured properly for short and intermediate range potentials, [20, 22]. Motivated by such observations, in our recent work we studied when CG methods perform satisfactorily, and how to quantify the CG approximations from a *numerical analysis* perspective, where error is assessed in view of a specified tolerance. Next, we discuss systems with *long range* interactions, i.e., $L \gg 1$ in (12). These systems can exhibit complex behavior such as phase transitions, nucleation, etc., however, they are more tractable analytically. At the same time they pose a serious challenge to conventional MC methods due to the large number of neighbors involved in each MC step.

Here we adopt this general approach, however, the challenges when both short and long-range interactions are present, require a new methodology. Short-range interactions induce strong "sub-coarse grid" fine-scale correlations between coarse cells, and need to be explicitly included in the initial approximation $\bar{H}^{(0)}_M(\eta)$. For this reason we introduced in [25] a *multi-scale decomposition* of the Gibbs state (7), into fine and coarse variables, which in turn allows us to describe in an explicit manner the communication across scales, for both short and long-range interactions.

4.2 Multiscale Decomposition and Splitting Methods for MCMC

We first focus on general lattice systems, and subsequently discuss related applications in later sections. We consider (6) where in addition to the long-range potential (12), we add the short-range $K(x-y) = S^{-1}U\left(N|x-y|/S\right)$, where $S \ll L$ and

U has similar properties as V in (12); for $S = 1$ we have the usual nearest neighbor interaction. The new Hamiltonian includes both long and short-range interactions: $H_N = H_N^l + H_N^s$.

The common theme is the observation that long-range interactions $L \gg 1$ can be handled very efficiently by CGMC, (14). On the other hand short-range interactions are relatively inexpensive and one could simulate them with Direct Numerical Simulation (DNS) provided there is a suitable *splitting* of the algorithm in short and long-range parts, that can reproduce within a given tolerance equilibrium Gibbs states and dynamics. We return to the general discussion in (10) and outline the steps we need in order to construct the CG Hamiltonian for the combined short and long-range interactions.

Step 1: Semi-analytical splitting schemes. Here we take advantage of CG approximations developed in (14) in order to decompose our calculation into analytical and numerical components, the latter involving only short-range interactions:

$$\begin{aligned}\mu_{N,\beta}(d\sigma) &\sim e^{-\beta H_N(\sigma)} P_N(d\sigma) \\ &= e^{-\left(\beta H_N^l(\sigma) - \bar{H}_M^{l,0}(\eta)\right)} \left[e^{-\beta H_N^s(\sigma)} P_N(d\sigma|\eta) \right] e^{-\bar{H}_M^{l,0}(\eta)} \bar{P}_M(\eta),\end{aligned}$$

where $\bar{H}_M^{l,0}$ is the analytical CG formula (13) constructed for the computationally expensive, for conventional MC, long-range part; due to the estimates (14), the first term has controlled error. Furthermore, the dependence of ϵ on ∇V in these estimates suggests a *rearrangement* of the overall combined short- and long-range potential, into a new short-range interaction that includes possible singularities originally in the long-range component (12), e.g., the singular part in a Lennard-Jones potential, and a locally integrable (or smooth) long-range decaying component that can be analytically coarse-grained using (13), with a small error due to (14). This breakdown allows us to isolate the short-range interactions (after a possible re-arrangement!), and suggests the two alternative computational approaches: either seek an approximation $e^{-\beta \bar{H}_M^s(\eta)} = \int e^{-\beta H_N^s} P_N(d\sigma|\eta)$, or use sampling methods to account for the short-range "unresolved" terms.

4.3 Microscopic Reconstruction

The reverse procedure of coarse-graining, i.e., reproducing "atomistic" properties, directly from CG simulations is an issue that arises extensively in the polymer science literature, [31, 37]. The principal idea is that computationally inexpensive CG simulations will reproduce the large scale structure and subsequently microscopic information will be added through *microscopic reconstruction*, e.g., the calculation of diffusion of penetrants through polymer melts, reconstructed from CG simulation, [31]. In this direction, CGMC provides a simpler lattice framework to mathematically formulate microscopic reconstruction and study related numerical and computational issues. Interestingly this issue arised also in the mathematical error analysis in [19, 23].

The mathematical formulation for the reconstruction of the microscopic equilibrium follows trivially when we rewrite the Gibbs measure (7) in terms of the exact CG measure corresponding to (8), defined in (9), [21]:

$$\mu_N(d\sigma) \sim e^{-\beta(H(\sigma)-\bar{H}(\eta))} P_N(d\sigma|\eta)\bar{\mu}_M(d\eta) \equiv \mu_N(d\sigma|\eta)\bar{\mu}_M(d\eta)\,.$$

We can define the conditional probability $\mu_N(d\sigma|\eta)$ as the *exact reconstruction* of $\mu_N(d\sigma)$ from the exact CG measure $\bar{\mu}_M(d\eta)$. Although many fine-scale configurations σ correspond to a single CG configuration η, the "reconstructed" conditional probability $\mu_N(d\sigma|\eta)$ is *uniquely* defined, given the microscopic and the coarse-grained measures $\mu_N(d\sigma)$ and $\bar{\mu}_M(d\eta)$ respectively.

A coarse-graining scheme provides an approximation $\bar{\mu}_M^{\mathrm{app}}(d\eta)$ for $\bar{\mu}_M(d\eta)$, at the coarse level. The approximation $\bar{\mu}_M^{\mathrm{app}}(d\eta)$ could be, for instance, any of the schemes discussed in Sect. 4.2. To provide a reconstruction we need to lift the measure $\bar{\mu}_M^{\mathrm{app}}(d\eta)$ to a measure $\mu_N^{\mathrm{app}}(d\sigma)$ on the microscopic configurations. That is, we need to specify a conditional probability $\nu_N(d\sigma|\eta)$ and set $\mu_N^{\mathrm{app}}(d\sigma) := \nu_N(d\sigma|\eta)\bar{\mu}_M^{\mathrm{app}}(d\eta)$. In the spirit of our earlier discussion, it is natural to measure the efficiency of the reconstruction by the relative entropy,

$$\mathscr{R}\left(\mu_N^{\mathrm{app}}\,|\,\mu_N\right) = \mathscr{R}\left(\bar{\mu}_M^{\mathrm{app}}\,|\,\bar{\mu}_M\right) + \int \mathscr{R}\left(\nu_N(\cdot|\eta)\,|\,\mu_N(\cdot\,|\,\eta)\right)\bar{\mu}_M^{\mathrm{app}}(d\eta)\,, \qquad (15)$$

i.e., relative entropy splits the total error at the microscopic level into the sum of the error at the coarse level and the error made during reconstruction, [21, 35].

The first term in (15) can be controlled via CG estimates, e.g., (14). However, (15) suggests that in order to obtain a successful reconstruction we then need to construct $\nu_N(d\sigma\,|\,\eta)$ such that (a) $\mathscr{R}\left(\nu_N(d\sigma\,|\,\eta)\,|\,\mu_N(d\sigma\,|\,\eta)\right)$ should be of the same order as the first term in (15), and (b) it is easily computable and implementable.

The simplest example of reconstruction is obtained by considering a microscopic system with intermediate/long-range interactions (12)

$$\bar{\mu}_M^{\mathrm{app}}(d\eta) = \bar{\mu}_M^{(0)}(d\eta)\,, \quad \nu_N(d\sigma\,|\,\eta) = P_N(d\sigma\,|\,\eta)\,. \qquad (16)$$

Thus we first sample the CG variables η involved in $\bar{\mu}_M^{(0)}$, using a CGMC algorithm; then we reconstruct the microscopic configuration σ by distributing the particles uniformly on the coarse cell, conditioned on the value of η. Since $P_N(d\sigma|\eta)$ is a product measure this can be done numerically in a very easy way, without communication between coarse cells and only at the coarse cells where an update has occurred in the CGMC algorithm. In this case the analysis in [24] yields the estimates

$$\mathscr{R}\left(\bar{\mu}_M^{(0)}\,|\,\bar{\mu}_M\right) = O(\epsilon^2)\,,\, \mathscr{R}\left(\mu_N(\cdot\,|\,\eta)|\, P_N(\cdot\,|\,\eta)\right) = \frac{\beta}{N}\left(\bar{H}^{(0)}(\eta) - \bar{H}(\eta)\right) = O(\epsilon^2)\,.$$

Hence the reconstruction is second order accurate and of the same order as the coarse-graining given by (13).

5 Example: Short and Long-Range Interactions

Short and long-range interactions pose a formidable computational challenge. We consider an example that has been explicitly solved by Kardar in [17]. The model considered has state space $\Sigma_N = \{0,1\}^{\Lambda_N}$, where Λ_N is a 1-dimensional lattice with N sites. The energy of the system at configuration $\sigma = \{\sigma(x), x \in \Lambda_N\}$ is

$$\begin{aligned}\beta H_N(\sigma) &= -\frac{K}{2}\sum_x \sum_{|x-y|=1} \sigma(x)\sigma(y) - \frac{J}{2N}\sum_x \sum_{y\neq x} \sigma(x)\sigma(y) - h\sum \sigma(x)\\ &\equiv H_N^s(\sigma) + H_N^l(\sigma) + E(\sigma)\,.\end{aligned}$$

Hamiltonian $H_N(\sigma)$ consists of the short-range term H_N^s, the long-range term H_N^l and an external field E. The interactions involved in H_N^s are of the nearest-neighbor type with strength K, while H_N^l represents a mean-field approximation or the Curie-Weiss model defined by the potential J averaged over all lattice sites. For this generic model Kardar gave in [17] a closed form solution for magnetization $M_\beta(K,J,h)$, for the state space $\{-1,1\}$

$$\begin{aligned}M_\beta(K,J,h) = \arg\min_m \Big(\frac{J}{2}m^2 - \log\Big[e^K\cosh(h+Jm)\\ + \sqrt{e^{2K}\sin^2(h+Jm)+e^{-2K}}\Big]\Big),\end{aligned}$$

a simple rescaling of which gives the exact average coverage $m_\beta(K,J,h)$ for the lattice-gas model considered here,

$$m_\beta(K,J,h) = \frac{1}{2}\left(M_\beta\left(\frac{1}{4}K, \frac{1}{4}J, \frac{1}{2}h - \frac{1}{4}J - \frac{1}{4}K\right) + 1\right). \tag{17}$$

We have constructed the classical single spin-flip MH algorithm and the coupled MH CGMC for the single spin-flip algorithm, both generating samples from the Gibbs measure

$$\mu_{N,\beta} = \frac{1}{Z_N} e^{-\beta H_N(\sigma)} P_N(d\sigma)\,.$$

We denote σ^x the state that differs from σ only at the site x, $\sigma^x(y) = \sigma(y)$, $y \neq x$, $\sigma^x(x) = 1 - \sigma(x)$. The proposal transition kernel is $q(\sigma'|\sigma) = \frac{1}{N}\sum_x \delta(\sigma' - \sigma^x)$, proposing a spin-flip at the site x with the probability $\frac{1}{N}$.

We apply the coupled CGMC method with coarse updating variable

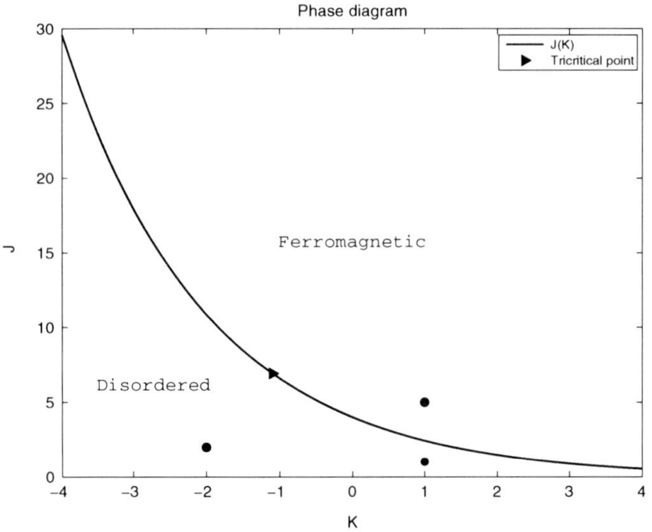

Fig. 1 Phase Diagram [17]. The points marked by (•) depict the choice of the parameters K, J in the presented simulations. The curve $J(K)$ indicates the phase transition

Table 1 Operations count for evaluating energy differences for n iterations

Cost	Metropolis hastings	Coupled CGMC $q < N$	Coupled CGMC $q = N$
Coarse A-R	–	$n \times O(M)$	$n \times O(1)$
Fine A-R	$n \times O(N)$	$n_1 \times O(1)$	$n_1 \times O(1)$

$$\eta := T\sigma = \{\eta(k), k = 1, \dots, M\},$$

$\eta(k) := \sum_{x \in C_k} \sigma(x)$, $qM = N$ with a coarsening level $q < N$. For the maximum coarsening $q = N$ the coarse variable is total magnetization $\eta = \sum_{x \in \Lambda_N} \sigma(x)$, this can be thought as a coarsening procedure constructing a system consisting of one big coarse cell $M = 1$ with $q = N$ sites. Since we want to consider only single spin-flip updates, for the sake of comparison to the classical Metropolis method, the cell updating can take only the values ± 1. The reconstruction is chosen uniform in each cell, in the sense described in example at Sect. 4.3, though for implementation ease and to demonstrate the importance of the reconstruction procedure, a simplified reconstruction is used in Figs. 1–3 and Tables 2, 3 while the exact reconstruction is used in Fig. 4 and Table 4.

The simplified reconstruction is a linear search over the cell sites, picking the first site that is appropriate for the adsorption/desorption avoiding the use of a random number. This simplification introduces error dependent on the cell size q as is evident in the numerical results.

Table 2 $N = 4096$

	CG level q	Error_c	CPU(min)
$K = -2, J = 2$	4	0.089	93.5
	8	0.302	45.8
$K = 1, J = 5$	4	0.003	93.6
	8	0.003	45.9
$K = 1, J = 1$	4	0.027	91.6
	8	0.100	45.5

Table 1 gives a comparison of the classical single-site updating Metropolis Hastings algorithm with the proposed coupled Metropolis CGMC algorithm, in terms of computational complexity per iteration. By computational complexity here we mean the cost of calculating energy differences involved at the acceptance probabilities. Consider the case that both the microscopic single-site updating Metropolis and the two-step CGMC are run n times. This is reasonable to consider since as stated at Theorem 2 the two methods have comparable mixing times, therefore the number of iterations needed to achieve stationarity are comparable. We denote $\mathrm{E}(\alpha_{CG}) := \int\int \alpha_{CG}(\eta, \eta')\bar{q}_0(\eta, \eta')\bar{f}_0(\eta)d\eta d\eta'$ the average acceptance rate of the coarse proposal. The average number of accepted coarse samples is $n_1 := [\mathrm{E}(\alpha_{CG})n]$, for which $n_1 < n$ since $\mathrm{E}(\alpha_{CG}) < 1$. This means that the reconstruction and fine step acceptance criterion are performed in average only for n_1 iterations.

Results of computational implementation are shown in Figs. 2–4 and Tables 2–4. Fig. 2a represents the average coverage versus the external field h for the exact solution m_{ex}, the classical MH result $< m_{cl} >$ and the coupled CGMC $< m >$, for a choice of interaction parameters $K = 1,\ J = 5$ in the ferromagnetic region as is stated at the phase diagram depicted in Fig. 1. The exact solution m_{ex} as is plotted in Fig. 2a corresponds to the part of the full solution (17) up to the point it jumps. Fig. 2b is a graph of the average acceptance rates for the classical MH algorithm and the coupled CGMC algorithm, that verifies the theoretical proof of the fact that the two algorithms have comparable mixing times since the acceptance rate is strongly related to mixing times. In the same figure we also give the average acceptance rates of the coarse and fine step of the coupled method, noting that the fine acceptance rate is large proving that a relative high number of the trial samples entering the fine step are accepted.

Table 2 reports the error between the exact solution and the average coverage obtained from the coupled CGMC algorithm with the simplified reconstruction for a lattice of size $N = 4096$. Error is measured in terms of the pointwise solutions as

$$\text{Error}_c = \left(\sum_i (m_{ex}(h_i) - < m > (h_i))^2 \right)^{1/2},$$

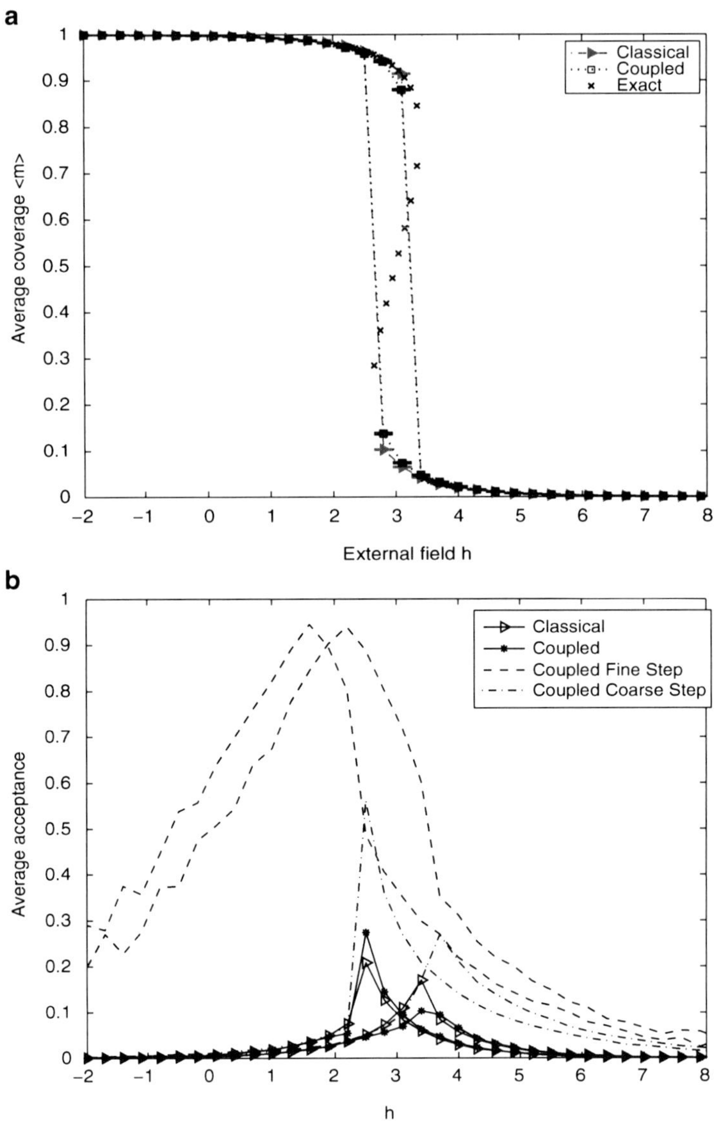

Fig. 2 $N = 1024$, $q = 8$, $K = 1$, $J = 5$: (**a**) Coverage ; (**b**) Average acceptance

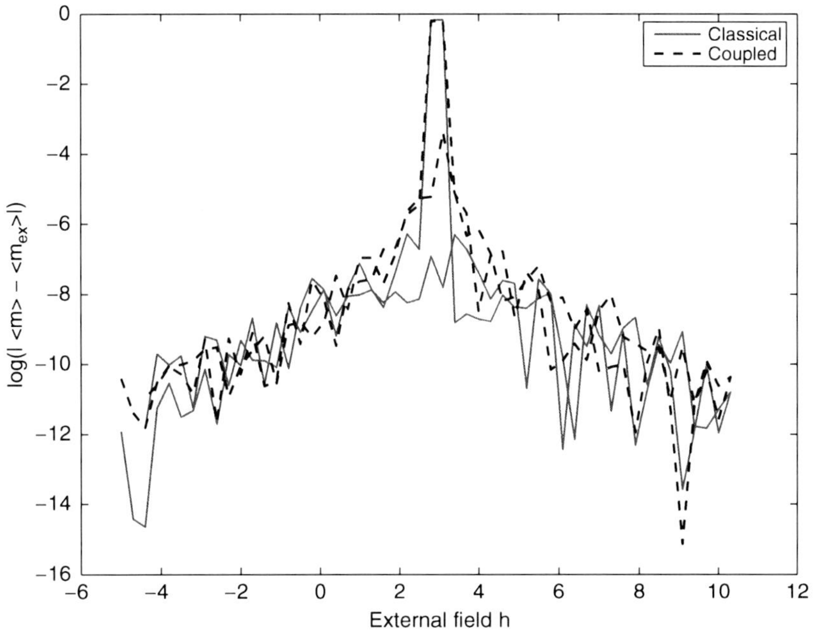

Fig. 3 $N = 1024, q = 8, K = 1, J = 5$: Local error $\log(| < m > - < m_{ex} > |)$

and

$$\text{Error}_{cl} = \left(\sum_i (m_{ex}(h_i) - < m_{cl} > (h_i))^2 \right)^{1/2},$$

for the coupled and the classical method respectively, where h_i are the different external field parameters for which the average coverages are computed. CPU times are compared for the coarse-graining levels $q = 4$ and $q = 8$. To demonstrate the robustness of the algorithm we present simulations at three different points of the phase diagram plane $K - J$: in the disordered $((K = -2, J = 2)$ and $(K = 1, J = 1))$ and ferromagnetic $(K = 1, J = 5)$ regions. In Table 3 we compare the error between the coupled CGMC average coverage with the exact solution and the corresponding CPU time for $q = 4$ and $q = 8$, in the ferromagnetic region $(K = 1, J = 5)$ and compare with the classical spin-flip MH results error and computational time for a smaller lattice of size $N = 1024$.

These results demonstrate the efficiency of the coupled CGMC methods in terms of computational time, the run time gain scales almost linearly with the coarsening level. We should also mention that a large number of samples (10^5) were considered ensuring the statistical error is small enough.

The example studied here demonstrates the effectiveness of the proposed algorithm due to the splitting of the long and short range interactions into the coarse and the fine space respectively. The error of approximating the long range interactions Hamiltonian is not apparent, since the coarse grained Hamiltonian is exact in this

Table 3 $N = 1024$, $K = 1$, $J = 5$, Error$_{cl} = 0.003$, Classical CPU = 94.5 min

CG level	Error$_c$	Coupled CPU(min)
$q = 4$	0.01	23.1
$q = 8$	0.04	12.1

case, that is $\bar{H}^l_M(\eta) = H^l_N(\sigma)$. When $\bar{H}^l_N(\eta)$ is not exact this approximating error is controlled by the estimates given section in 4.1 and extensively studied in [20, 22], if no correction terms are included in the reconstruction procedure as described in Sect. 4.3. On the other hand if the reconstruction procedure is perfect, being the exact marginal of the microscopic distribution, we expect that the method will be independent of the coarse graining parameter.

The coarse graining parameter q dependent error appearing in Tables 2 and 3 is due to the simplification of the reconstruction procedure. Figure 4 and Table 4 shows a comparison of the average coverage for the method implemented with the simplified and the exact reconstruction where we used a larger number of samples to 5×10^6 to reduce the statistical error. The computational time gain, with respect to the traditional spin-flip MH, corresponding to the simplified reconstruction is small compared to the overall gain, as the comparison in Table 4 shows, ensuring that for the perfect reconstruction we indeed have an overall computational time reduction of the order of the coarse graining level q.

The direct numerical simulation also yields an error comparable to the coupled method with the perfect reconstruction, both errors depending on the finite lattice size effect and statistical errors.

Table 4 L^1 error Coupled – Exact: $N = 1024$, $K = 1$, $J = 5$, L^1 error Classical – Exact = 4.0e-05

CG level q	Simple reconstruction CPU time = 21.5 min	Perfect reconstruction CPU time = 23.9 min
2	3.55e-04	4.0e-05
4	3.62e-04	5.1e-05
8	4.23e-04	1.05e-04
16	2.93e-04	3.0e-05
32	5.08e-04	3.26e-04
64	7.15e-04	3.2e-05
128	1.18e-03	1.01e-04
256	1.89e-03	3.7e-05
512	1.60e-03	2.7e-05

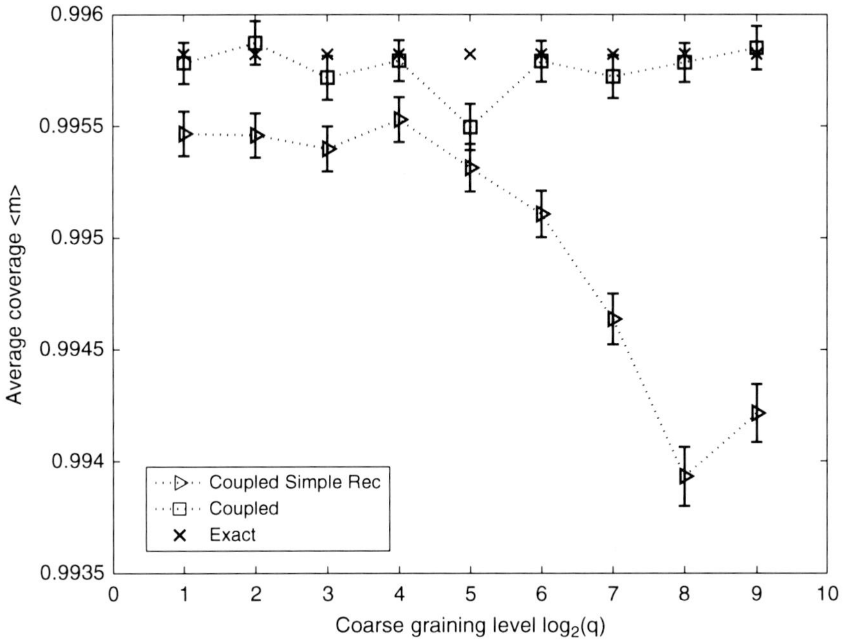

Fig. 4 $N = 1024$, $K = 1$, $J = 5$: Coverage

6 Conclusions

An advantage of the Coupled CGMC approach over the asymptotics methodology discussed in Sect. 4.2 is that the trial distribution may even be order one away from the target distribution, however, the method can still perform well. On the other hand, the methods can *complement* each other; for example, for equilibrium sampling considered in this work we use as a trial reconstructed distribution, the conditional measure $\nu(d\sigma|\eta)$ in the multiscale decomposition in [25], see also Sect. 4.3. Such proposals based on careful statistical mechanics-based approximations provide better trial choices for the MH methods and more efficient sampling, as is proved theoretically and numerically. The example illustrated makes clear that the coupled CGMC method implements a splitting of the short and long-range interaction terms, into the two Metropolis acceptance criteria involved. The long-range part which is responsible for the expensive calculations at a fully microscopic method, now enters only in the coarse approximation measure where its computational cost is much lower.

Coupling of a coarse and fine step is also effective in the study of dynamic processes of stochastic lattice systems with kinetic Monte Carlo methods, a topic studied in detail in [16].

References

1. Abrams, C.F., Kremer, K.: The effect of bond length on the structure of dense bead-spring polymer melts. J. Chem. Phys. **115**, 2776 (2001)
2. Are, S., Katsoulakis, M.A., Plecháč, P., Rey-Bellet, L.: Multibody interactions in coarse-graining schemes for extended systems. SIAM J. Sci. Comput. **31**(2), 987–1015 (2008)
3. Auerbach, S.M.: Theory and simulation of jump dynamics, diffusion and phase equilibrium in nanopores. Int. Rev. Phys. Chem. **19**(155) (2000)
4. Briels, W.J., Akkermans, R.L.C.: Coarse-grained interactions in polymer melts: a variational approach. J. Chem. Phys. **115**, 6210 (2001)
5. Ceperley, D.M.: Path integrals in the theory of condensed helium. Rev. Mod. Phys. **67**(2), 279–355 (1995)
6. Chatterjee, A., Vlachos, D.: Systems tasks in nanotechnology via hierarchical multiscale modeling: Nanopattern formation in heteroepitaxy. Chem. Eng. Sci. **62(18-20)**, 4852–4863 (2007)
7. Diaconis, P., Saloff-Coste, L.: Logarithmic Sobolev inequalities for finite Markov Chains. Ann. Appl. Prob. **6**(3), 695–750 (1996)
8. Diaconis, P., Saloff-Coste, L.: What Do We Know about the Metropolis Algorithm? Journal of Computer and System Sciences **57**, 20–36 (1998)
9. Efendiev, Y., Hou, T., Luo, W.: Preconditioning Markov chain Monte Carlo simulations using coarse-scale models. SIAM J. Sci. Comput. **28**(2), 776–803 (2006)
10. Espanol, P., Warren, P.: Statistics-mechanics of dissipative particle dynamics. Europhys. Lett. **30**(4), 191–196 (1995)
11. Fukunaga, H., J. Takimoto, J., Doi, M.: A coarse-grained procedure for flexible polymer chains with bonded and nonbonded interactions. J. Chem. Phys. **116**, 8183 (2002)
12. Goldenfeld, N.: Lectures on Phase Transitions and the Renormalization Group, vol. 85. Addison-Wesley, New York (1992)
13. Harmandaris, V.A., Adhikari, N.P., van der Vegt, N.F.A., Kremer, K.: Hierarchical modeling of polysterene: From atomistic to coarse-grained simulations. Macromolecules **39**, 6708 (2006)
14. Hastings, W.K.: Monte Carlo sampling methods using Markov chains and their applications. Biometrika **57**(1), 97–109 (1970)
15. Kalligiannaki, E., Katsoulakis, M.A., Plecháč, P., Vlachos D.G.: Multilevel coarse graining and nano–pattern discovery in many particle stochastic systems. Submitted to J. Comp. Physics, preprint: arXiv:1109.0459
16. Kalligiannaki, E., Katsoulakis, M.A., Plechac, P.: Multilevel kinetic Coarse Graining Monte Carlo methods for stochastic lattice dynamics. in preparation
17. Kardar, M.: Crossover to equivalent-neighbor multicritical behavior in arbitrary dimensions. Phys. Rev. B **28**(1), 244–246 (1983)
18. Katsoulakis, M., Majda, A., Vlachos, D.: Coarse-grained stochastic processes for microscopic lattice systems. Proc. Natl. Acad. Sci. **100**(3), 782–787 (2003)
19. Katsoulakis, M., Trashorras, J.: Information loss in coarse-graining of stochastic particle dynamics. J. Stat. Phys. **122**(1), 115–135 (2006)
20. Katsoulakis, M.A., Majda, A.J., Vlachos, D.G.: Coarse-grained stochastic processes and Monte Carlo simulations in lattice systems. J. Comp. Phys. **186**(1), 250–278 (2003)
21. Katsoulakis, M.A., Plechac, P., Rey-Bellet, L.: Numerical and statistical methods for the coarse-graining of many-particle stochastic systems. J. Sci. Comput. **37**(1), 43–71 (2008)
22. Katsoulakis, M.A., Plechac, P., Rey-Bellet, L., Tsagkarogiannis, D.K.: Coarse-graining schemes and a posteriori error estimates for stochastic lattice systems. ESAIM-Math. Model. Numer. Anal. **41**(3), 627–660 (2007)
23. Katsoulakis, M.A., Plecháč, P., Sopasakis, A.: Error analysis of coarse-graining for stochastic lattice dynamics. SIAM J. Numer. Anal. **44**(6), 2270–2296 (2006)
24. Katsoulakis, M.A., Rey-Bellet, L., Plecháč, P., K.Tsagkarogiannis, D.: Mathematical strategies in the coarse-graining of extensive systems: error quantification and adaptivity. J. Non Newt. Fluid Mech. (2008)

25. Katsoulakis, M.A., Rey-Bellet, L., Plecháč, P., Tsagkarogiannis, D.K.: Coarse-graining schemes for stochastic lattice systems with short and long range interactions. submitted to Math. Comp.
26. Kremer, K., Müller-Plathe, F.: Multiscale problems in polymer science: simulation approaches. MRS Bull. p. 205 (March 2001)
27. Landau, D., Binder, K.: A Guide to Monte Carlo Simulations in Statistical Physics. Cambridge University Press (2000)
28. Liu, J.S.: Monte Carlo Strategies in Scientific Computing. Springer, New York, Berlin (2001)
29. Lyubartsev, A.P., Karttunen, M., Vattulainen, P., Laaksonen, A.: On coarse-graining by the inverse Monte Carlo method: Dissipative particle dynamics simulations made to a precise tool in soft matter modeling. Soft Materials **1**(1), 121–137 (2003)
30. Metropolis, N., Rosenbluth, A.W., Rosenbluth, M.N., Teller, A.H., Teller, E.: Equation of state calculations by fast computing machines. J. Chem. Phys. **21**(6), 1087–1092 (1953)
31. Müller-Plathe, F.: Coarse-graining in polymer simulation: from the atomistic to the mesoscale and back. Chem. Phys. Chem. **3**, 754 (2002)
32. Pivkin, I., Karniadakis, G.: Coarse-graining limits in open and wall-bounded dissipative particle dynamics systems. J. Chem. Phys. **124**, 184,101 (2006)
33. Robert, C.P., Casella, G.: Monte Carlo Statistical Methods. Springer, New York (2004)
34. Simon, B.: The Statistical Mechanics of Lattice Gases, vol. I. Princeton series in Physics (1993)
35. Trashorras, J., Tsagkarogiannis, D.K.: Reconstruction schemes for coarse-grained stochastic lattice systems (2008). Submitted to SIAM J. Num. Anal.
36. Tschöp, W., Kremer, K., Hahn, O., Batoulis, J., Bürger, T.: Simulation of polymer melts. I. coarse-graining procedure for polycarbonates. Acta Polym. **49**, 61 (1998)
37. Tschöp, W., Kremer, K., Hahn, O., Batoulis, J., Bürger, T.: Simulation of polymer melts. II. from coarse-grained models back to atomistic description. Acta Polym. **49**, 75 (1998)

Calibration of a Jump-Diffusion Process Using Optimal Control

Jonas Kiessling

Abstract A method for calibrating a jump-diffusion model to observed option prices is presented. The calibration problem is formulated as an optimal control problem, with the model parameters as the control variable. It is well known that such problems are ill-posed and need to be regularized. A Hamiltonian system, with non-differentiable Hamiltonian, is obtained from the characteristics of the corresponding Hamilton-Jacobi-Bellman equation. An explicit regularization of the Hamiltonian is suggested, and the regularized Hamiltonian system is solved with a symplectic Euler method. The paper is concluded with some numerical experiments on real and artificial data.

1 Introduction

Jump-diffusion models are increasingly popular in financial mathematics. They present many new and challenging problems, for instance the design of efficient and stable calibration algorithms. One interesting aspect of such models are the different scales present: To price a contingent claim when the underlying is driven by a jump-diffusion process one needs to solve a partial-integral equation where the integral kernel typically has non-compact support.

Consider a stock $S = S_t$ priced in a market with risk-free interest rate r. Let $C = C(t, S; T, K)$ denote the price of an ordinary European call option on S with strike price K and maturity T. Under certain assumptions (see any book on financial mathematics, for instance [4, Chap. 9]) there is a probability measure $\mathcal{Q}$, on the set of all stock price trajectories, such that the price of the call option equals its discounted expected future payoff,

J. Kiessling (✉)
Department of Mathematics, Royal Institute of Technology (KTH), Valhallavägen 79, Kungl Tekniska Högskolan, SE-100 44, Stockholm, Sweden
e-mail: jonkie@kth.se

B. Engquist et al. (eds.), *Numerical Analysis of Multiscale Computations*, Lecture Notes in Computational Science and Engineering 82, DOI 10.1007/978-3-642-21943-6_12,

$$C(t, S_t; T, K) = e^{-r(T-t)} E^{\mathcal{Q}}[\max(S_T - K)|S_t]. \tag{1}$$

A priori not much is given about this *pricing measure* $\mathcal{Q}$, except that $e^{-rt} S_t$ is a martingale under $\mathcal{Q}$,

$$e^{-rt} S_t = e^{-rT} \mathrm{E}^{\mathcal{Q}}[S_T | S_t]. \tag{2}$$

Model calibration is the process of calibrating (that is, determine) $\mathcal{Q}$ from market data. The purpose of this work is to explain how the calibration problem can be solved using optimal control and the theory of Hamilton-Jacobi-Bellman equations. The idea is as follows: By (1) there is a price function of call options corresponding to each choice of measure $\mathcal{Q}$, $C = C(T, K; \mathcal{Q})$. Regarding $\mathcal{Q}$ as the control, we try to minimize

$$\int_0^{\hat{T}} \int_{\mathcal{R}_+} (C(T, K; \mathcal{Q}) - C_m(T, K))^2 \mathrm{d}T \mathrm{d}K, \tag{3}$$

where C_m denotes the market price of call options.

As stated, the problem of calibrating $\mathcal{Q}$ is too ill-posed. There are simply too many possible choices of pricing measures that would fit data accurately. The usual approach is to parametrize the dynamics of S_t under $\mathcal{Q}$, its *risk-neutral dynamics*. Concretely, one assumes that the price process S_t solves a stochastic differential equation, parametrized by some parameter. Calibration now amounts to choosing the parameter resulting in the best fit to market data. One could for instance assume that there is a number σ such that

$$\frac{\mathrm{d}S_t}{S_t} = r\mathrm{d}t + \sigma \mathrm{d}B_t. \tag{4}$$

Here and for the rest of the paper we let B_t denote Brownian motion. This was the approach taken by Black and Scholes in [3], and many others. The calibration problem is now reduced to determining one number σ. This simple model is probably still the most widely used model, especially in the day-to-day pricing and hedging of vanilla options. The problem with this approach is its poor ability to reproduce market prices. So poor, in fact, that different numbers σ are needed for options on the same underlying with different strikes, a clear violation of the original model.

There are many ways people have refined the model suggested by Black and Scholes. One popular approach is to assume *stochastic volatility*, i.e. σ is no longer a number, but a stochastic process, see for instance [10] or [2].

Following Dupire in [7], a second approach is to assume that σ is a deterministic function of time and price, $\sigma = \sigma(t, S_t)$, the so-called "local volatility" function. One nice feature of this model is that there is a closed formula for $\sigma(t, S)$ in terms of quoted option prices, see [7, p. 5].

Finally a popular approach is to introduce discontinuities (i.e. jumps) in the price process. This was initiated with the work of Merton in 1976 in [9]. A good reference for jump processes in finance is the book [4].

The model we choose to calibrate in this paper is a jump-diffusion model with state and time dependent volatility and time dependent jump intensity. It should be noted however that the techniques used in this paper are more widely applicable.

Even after making restrictions on the pricing measure, model calibration faces one major problem. It is typically *ill-posed* in the sense that the result is *highly* sensitive to changes in the data. One reason that the standard Black and Scholes pricing model is still so widely used is probably that a constant volatility is so easy to determine. One benefit of the optimal control approach described in this work, is that the ill-posedness is made explicit in terms of the non-smoothness of the Hamiltonian, see (36). Well-posedness is obtained after a regularization.

The focus of this work is to develop a method, first introduced in [12], for calibrating the pricing measure from quoted option prices. The method is summarized in Algorithm 4.1. As can be seen in the final sections of the paper, the method works in the sense that we can determine a measure $\mathcal{Q}$, such that pricing under $\mathcal{Q}$ results in prices in accordance with observed market prices. Of course, to apply the method in a real life situation would require more work. One challenge is to obtain a pricing measure that is exact enough to price exotic options. The work [13] suggests that the sensitivity of certain exotic contracts to even small changes in the pricing measure makes the procedure of calibrating on vanilla to price exotics rather dangerous. Another important challenge is to obtain a pricing measure that, not only gives reasonable prices, but also good values for *the greeks*. One can argue that it is more important to obtain good values for the greeks than for the price, as the greeks determine the hedging strategy whereas the market determines the price.

The outline for the rest of this paper is as follows: In the next section we introduce in more detail the SDE we wish to calibrate. Following Dupire, we also deduce a forward integro-partial differential equation satisfied by the call option in the strike and maturity variables T and K. This makes the numerical solution scheme much more efficient. In Sect. 3, we give a quick introduction to the theory of optimal control. In the following section, we develop a scheme for calibrating the local volatility and jump intensity. This is done by first formulating the problem as an optimal control problem, and then solving the corresponding regularized Hamiltonian system. We conclude the paper with some numerical experiments. We first try the method on artificial data obtained by solving the forward problem (1) with prescribed local volatility and jump intensity, thus obtaining a price function $C = C(T, K)$. The local volatility and jump intensity are then reconstructed from $C(T, K)$ using Algorithm 4.1. Finally we calibrate using data from the S&P 500-market. To start the procedure we need information on the jump distribution. In [1] this problem is solved by first calibrating a Lévy process to observed prices, then refining the calibration by allowing the volatility (and in our case, the jump intensity) to vary. We use their calibration of the jump distribution. The calibration scheme results in a volatility surface that is roughly constant in time, but varying in price with lower price implying higher volatility. The result is positive in the sense that we had no problem of convergence, once the jump distribution had been specified.

2 The Forward Equation

We begin this section by introducing a model for the risk-neutral stock price dynamics. Consider a stock S paying no dividend which is, under the pricing measure, affected by two sources of randomness: ordinary Brownian motion $B(t)$, and a compound Poisson process with deterministic time dependent jump intensity. These assumptions leads to the following evolution of S:

$$dS(t)/S(t-) = \Big(r - \mu(t)m(t)\Big)\mathrm{d}t + \sigma(t, S(t-))dB(t) + \Big(J(t) - 1\Big)d\pi(t), \quad (5)$$

where the relative jump-sizes $\{J(t)\}_{t>0}$ consists of a family of independent stochastic variables with at most time-dependent densities $\{\chi(t)\}_{t>0}$. Note that by definition $J(t) > 0$. The jump times are governed by a Poisson counting process $\pi(t)$ with time dependent intensity $\mu(t)$. As usual σ denotes the (state and time dependent) volatility function and r denotes the risk-free interest rate. The drift term is determined by the fact that $e^{-rt}S(t)$ is a martingale, forcing $m(t)$ to be $m(t) = \mathrm{E}[J(t) - 1]$. In (5), $t-$ is the usual notation for the limit of $t - |\epsilon|$ as $\epsilon \to 0$.

The price $C = C\big(t, S(t)\big)$ of any European style contingent claim, written on S and with payoff $g(S)$, equals its discounted future expected payoff,

$$C\big(t, S(t)\big) = e^{-r(t-T)} E^{\mathcal{Q}}[g\big(S(T)\big)|S(t)]. \quad (6)$$

Standard arguments (see for instance [4, Chap. 12]) show that C satisfies the backward integro partial-differential equation

$$\begin{aligned} rC &= C_t - \mu\Big(C + mSC_S + \mathrm{E}[C\big(t, J(t)S\big)]\Big) + \frac{1}{2}\sigma^2(t,S)S^2C_{SS} \\ &\quad + rSC_S, \\ C(T,S) &= g(S), \end{aligned} \quad (7)$$

where

$$\mathrm{E}[C(t, J(t)S)] = \int_{\mathcal{R}_+} C(t, Sx)\chi(x;t)\mathrm{d}x. \quad (8)$$

We use the notation C_t, C_S etc. to indicate the derivatives of $C = C(t,S)$ with respect to its first and second variable.

As it stands, in order to calculate call option prices for different strikes and maturities T and K, we need to solve the above equation once *for each different pair* (K,T). However, following Dupire ([7]) one can show that, due to the specific structure of the payoff function of a call option, C satisfies a similar forward equation *in the variables* T *and* K. A similar result is obtained in [1].

Proposition 1. *Assuming stock price dynamics given by (5), european call options* $C(T,K) = C(0,S;K,T)$, *at fixed* $t = 0$ *and* S, *satisfy the equation:*

$$\begin{aligned} C_T &= \mu(T)\Big(mKC_K - (m+1)C + E\big[J(T)C\big(T, J(T)^{-1}K\big)\big]\Big) \\ &\quad + \tfrac{1}{2}\sigma^2 K^2 C_{KK} - rKC_K, \end{aligned} \tag{9}$$

where

$$\begin{aligned} E\big[J(T)C\big(T, J(T)^{-1}K\big)\big] &= \textstyle\int_{\mathscr{R}_+} zC(T, K/z)\chi(z;T)dz \\ C(0,K) &= max(S-K,0). \end{aligned} \tag{10}$$

Proof. Let $f = f(t,x)$ denote some function defined on $\mathscr{R}_+ \times \mathscr{R}_+$. We begin by introducing the adjoint operators L^* and L:

$$\begin{aligned} L^* f(t,x) &= f_t(t,x) - \mu\Big(f(t,x) + mxf_x(t,x) + \textstyle\int_{\mathscr{R}_+} f(t,zx)\chi(z)\mathrm{d}z\Big) \\ &\quad + \tfrac{1}{2}\sigma^2 x^2 f_{xx}(t,x) + r\Big(xf_x(t,x) - f(t,x)\Big), \\ Lf(t,x) &= -f_t(t,x) + \mu\Big(f(t,x) - m\partial_x(xf) + \textstyle\int_{\mathscr{R}_+} z^{-1} f(t,z^{-1}x)\chi(z)\mathrm{d}z\Big) \\ &\quad + \tfrac{1}{2}\sigma^2 x^2 f_{xx}(t,x) - r\Big(\partial_x\big(xf(t,x)\big) + f\Big). \end{aligned} \tag{11}$$

From (7) we see that in its first two variables $C(t,S) = C(t,S;T,K)$ satisfies

$$L^* C(t,x) = 0. \tag{12}$$

We let $P = P(t,x;s,y)$ denote the solution to

$$\begin{aligned} L_{(t,x)} P(t,x;s,y) &= 0, \quad t > s, \\ P(s,x;s,y) &= \delta(x-y). \end{aligned} \tag{13}$$

Where the subscript in $L_{(t,x)}$ indicates that the operator is acting in the variables t and x.

Integration by parts yields:

$$\begin{aligned} 0 &= \textstyle\int_s^T \int_{\mathscr{R}} \Big(L^* C(t,x)\Big) P(t,x;s,y)\mathrm{d}x\mathrm{d}t \\ &= \Big[\textstyle\int_{\mathscr{R}} C(t,x)P(t,x;s,y)dx\Big]_{t=s}^{t=T} + \int_s^T \int_{\mathscr{R}} C(t,x)\big(L_{(t,x)} P(t,x;s,y)\big)\mathrm{d}x\mathrm{d}t \\ &= \textstyle\int_{\mathscr{R}} C(T,x)P(T,x;s,y)\mathrm{d}x - \int_{\mathscr{R}} C(s,x)P(s,x;s,y)\mathrm{d}x \\ &= \textstyle\int_{\mathscr{R}} C(T,x)P(T,x;s,y)\mathrm{d}x - C(s,y). \end{aligned} \tag{14}$$

This gives us the equality:

$$C(s,y) = \int_{\mathscr{R}} C(T,x)P(T,x;s,y)dx. \tag{15}$$

The payoff of a call option is $C(T,x) = \max(x-K,0)$ so:

$$C(s,y) = \int_{x=K}^{x=\infty} (x-K)P(T,x;s,y)dx. \tag{16}$$

Fixing s and y and differentiating twice with respect to K yields:

$$C_{KK}(s,y;T,K) = P(T,K;s,y), \tag{17}$$

and consequently, using the fact that P satisfies (13):

$$L_{(T,K)}C_{KK}(t,S;T,K) = 0. \tag{18}$$

We observe that $KC_{KK}(T,K) = \partial_K\big(KC_K(T,K) - C\big)$ and

$$C_{KK}(T,z^{-1}K) = z^2\partial_{KK}C(T,z^{-1}K).$$

Recall our choice of notation: $C_{KK}(T,z^{-1}K)$ denotes the derivative of $C(T,K)$ with respect to its second variable evaluated at $(T,z^{-1}K)$.

These observations and the above equation yield:

$$\partial_{KK}\Big(-C_T + \mu(T)\Big(m(T)KC_K - (m(T)+1)C + E\big[J(T)C(T,J(T)^{-1}K)\big]\Big) \\ +\tfrac{1}{2}\sigma^2(T,K)K^2C_{KK} - rKC_K\Big) = 0.$$

We integrate twice and observe that the left hand-side and its derivate with respect to K goes to zero as K tends to infinity. This forces the integrating constants to be zero and finishes the proof. □

For ease of notation we assume from now on that, unless otherwise is explicitly stated, the risk-free interest rate is zero, $r = 0$. Moreover we assume that the density of the jump-size $\chi(t)$ is constant over time. For this reason we use only symbols χ and J to denote χ_t and $J(t)$ respectively.

We conclude this section with introducing the two operators ψ_1, ψ_2 and their adjoints:

$$\begin{aligned} \psi_1(C) &= (m+1)C - mKC_K + E[JC(T,J^{-1}K)], \\ \psi_2(C) &= \tfrac{1}{2}K^2C_{KK}, \\ \psi_1^*(C) &= (m+1)C + m\partial_K(KC) + E[J^2C(T,JK)], \\ \psi_2^*(C) &= \tfrac{1}{2}\partial_{KK}(K^2C). \end{aligned} \tag{19}$$

The forward equation satisfied by the call options can now be written as

$$\begin{aligned} C_T &= \psi_1(C) + \psi_2(C), \\ C(0,K) &= \max(S-K,0). \end{aligned} \tag{20}$$

3 The Optimal Control Problem

Consider an open set $\Omega \subset \mathscr{R}^n$ and let V be some Hilbert space of functions on Ω, considered as a subspace of $L^2(\Omega)$ with its usual inner product. For a given *cost functional* $h: V \times V \to \mathscr{R}$, the *optimal control problem* consists of finding

$$\inf_{\sigma:\Omega\times[0,\hat{\mathrm{T}}]\to\mathscr{R}} \int_0^{\hat{\mathrm{T}}} h(\varphi,\sigma)\mathrm{dt}, \tag{21}$$

where $\varphi : \Omega \times [0,\hat{\mathrm{T}}] \to \mathscr{R}$ is the solution a differential equation

$$\varphi_t = f(\varphi;\sigma), \tag{22}$$

with a given initial function $\varphi(\cdot,0) = \varphi^0$. We call f the *flux*. For each choice of σ it is a function $f : V \to \mathscr{R}$. Recall that φ_t denotes the partial derivative with respect to t.

We refer to σ as the *control*, and the minimizer of (21), if it exists, is called the *optimal control*. We assume that σ takes values in some compact set $B \subset \mathscr{R}$.

There are different methods for solving optimal control type problems. In this work we study the characteristics associated to the non-linear Hamilton-Jacobi-Bellman equation. The first step is to introduce the value function U:

$$\begin{aligned} U(\phi,\tau) = \inf_{\sigma:\Omega\times[\tau,T]\to B}\Big\{ \textstyle\int_\tau^T h(\varphi^\phi,\sigma)dt \; : \; \varphi_t = f(\varphi;\sigma) \text{ for } \tau < t < T,\\ \varphi(\cdot,\tau) = \phi \in V\Big\}. \end{aligned} \tag{23}$$

The associated non-linear Hamilton-Jacobi-Bellman equation becomes:

$$\begin{cases} U_t + H(U_\phi,\phi) = 0, \\ \qquad U(\phi,T) = 0, \end{cases} \tag{24}$$

where $H : V \times V \to \mathscr{R}$ is the *Hamiltonian* associated to the above optimal control problem

$$H(\lambda,\varphi) = \inf_{a:\Omega\to B}\big\{\langle\lambda, f(\varphi,a)\rangle + h(\varphi,a)\big\}. \tag{25}$$

Here $\langle\cdot,\cdot\rangle$ is the inner-product in the Hilbert space V.

Crandall's, Evans and Lions proved that Hamilton-Jacobi-Bellman type equations often have well-posed viscosity solutions, see [5]. Constructing a viscosity solution to (24) directly is however computationally very costly. We shall instead construct a regularization of the characteristics of (24) and solve the corresponding coupled system of differential equations.

The well known method of characteristics associated to (24) yields the Hamiltonian system:

$$\begin{aligned} \varphi_t &= H_\lambda(\varphi,\lambda), \\ \lambda_t &= -H_\varphi(\varphi,\lambda), \\ \varphi(\cdot,0) &= \varphi^0, \\ \lambda(\cdot,T) &= 0, \end{aligned} \tag{26}$$

where H_λ and H_φ denote the Gâteaux derivatives of H w.r.t. λ and φ respectively.

Recall that, by definition of the Gâteaux derivative, $H_\varphi(\varphi,\lambda)$ satisfies

$$\left.\frac{\mathrm{d}}{\mathrm{d}t}\right|_{t=0} H(\varphi + tg,\lambda) = \int_\Omega g H_\varphi(\varphi,\lambda)dx, \tag{27}$$

for all $g \in V$, and similarly for H_λ.

In the applications in this work, the Hamiltonian H is not differentiable. In order for (26) to make sense, we first need to regularize H.

4 Reconstructing the Volatility Surface and Jump Intensity

4.1 The Hamiltonian System

Recall our model of the stock price and the corresponding integro-partial differential equation for call options, (20). For now we assume that the density of the jump-size is *known*, i.e. $\chi = \chi(x)$ is some given function. In Sect. 5 we indicate how χ can be determined in concrete examples. As will be clear later, we are forced to treat the jump size density separately from the local volatility and jump intensity. We use the explicit expression of the Hamiltonian, obtained in (36), to determine σ^2 and μ. There is no corresponding simple expression for the Hamiltonian if the jump density is unknown.

The remaining unknown quantities in (20) are: the local volatility function $\sigma = \sigma(t, S)$, and the jump intensity $\mu = \mu(t)$. The problem of calibrating these from option prices can be formulated as an optimal control problem. Recall the operators ψ_1, ψ_2 and their adjoints introduced in (19).

Suppose that $C_m = C_m(T, K)$ are call options priced in the market, for different strikes $K \geq 0$ and maturities $0 \leq T \leq \hat{\mathrm{T}}$. We wish to the determine the control (σ^2, μ) minimizing

$$\int_0^{\hat{\mathrm{T}}} \int_{\mathscr{R}_+} (C - C_m)^2 \mathrm{d}T \mathrm{d}K, \tag{28}$$

given that $C = C(T, K)$ satisfies

$$C_T = \mu \psi_1(C) + \sigma^2 \psi_2(C), \tag{29}$$

with boundary conditions

$$\begin{aligned} C(K,0) &= \max(S - K,0), \\ C(0,T) &= S. \end{aligned} \tag{30}$$

We further assume that for all T and K, $\sigma^2 \in [\sigma_-^2, \sigma_+^2]$ and $\mu \in [\mu_-, \mu_+]$, for constants σ_-, σ_+, μ_- and μ_+.

The problem as stated here is typically ill-posed as the solution often is very sensitive small changes in C_m.

A common way to impose well-posedness is to add a Tikhonov regularization term to (28), e.g. for some $\delta > 0$ one determines

$$\arg\min_{(\sigma,\mu)} \int_0^T \int_{\mathscr{R}_+} (C - C_m)^2 \mathrm{d}T \mathrm{d}K + \delta(\|\sigma^2\|^2 + \|\mu\|^2), \tag{31}$$

with C subject to (29). Minimizing (31) under the constraint (29) leads to a C^2-Hamiltonian

$$H^{\delta}(C,\lambda,T) = \min_{(\sigma^2,\mu)}\Big\{\mu\left(\delta\mu + \int_{\mathscr{R}_+}\lambda\psi_1(C)\mathrm{d}K\right) + \int_{\mathscr{R}_+}\sigma^2\left(\delta\sigma^2 + \lambda\psi_2(C)\right)\mathrm{d}K + \int_{\mathscr{R}_+}(C - C_m(T,K))^2\mathrm{d}K\Big\}. \tag{32}$$

A rigorous study of Tikhonov regularization for calibration of the local volatility can be found in the work of Crépey, see [6].

We take a different approach. Using the material presented in the previous section, we construct an explicit regularization of the Hamiltonian associated with (28) and (29), thus imposing well-posedness on the value function. As can be seen in (32), Tikhonov regularization corresponds to a different choice of regularization of the Hamiltonian. In this work we choose a slightly different regularization of the Hamiltonian, see (38). The particular choice of regularization in (38) is out of convenience.

The Hamiltonian associated with the optimal control problem (28) and (29) becomes

$$H(C,\lambda,T) = \inf_{(\sigma,\mu)}\Big\{\mu\int_{\mathscr{R}_+}\lambda\psi_1(C)\mathrm{d}K + \int_{\mathscr{R}_+}\sigma^2\lambda\psi_2(C)\mathrm{d}K + \int_{\mathscr{R}_+}(C - C_m(T,K))^2\mathrm{d}K\Big\}. \tag{33}$$

Only the sign of the terms

$$\int_{\mathscr{R}_+}\lambda\psi_1(C)\mathrm{d}K \quad \text{and} \quad \lambda\psi_2(C) \tag{34}$$

are important in solving the above optimization problem. This leads us to define the function

$$s_{[a,b]}(x) = \begin{cases} ax \text{ if } x < 0, \\ bx \text{ if } x > 0. \end{cases} \tag{35}$$

We can express the Hamiltonian using the function s in the following way:

$$H(C,\lambda,T) = s_{[\mu_-,\mu_+]}\left(\int_{\mathscr{R}_+}\lambda\psi_1(C)\mathrm{d}K\right) + \int_{\mathscr{R}_+}s_{[\sigma_-^2,\sigma_+^2]}\left(\lambda\psi_2(C)\right)\mathrm{d}K + \int_{\mathscr{R}_+}(C - C_m(T,K))^2\mathrm{d}K.$$

Recall that we assumed $\mu = \mu(T)$ to be *independent* of K, whereas $\sigma = \sigma(K,T)$ is a function of *both* T and K. This explains the different positions of s and the integral in the expression for H above.

It is clear from (36) and (35) that the Hamiltonian is not differentiable. We proceed by constructing an explicit regularization of the Hamiltonian H. A straightforward regularization of the Hamiltonian is to approximate $s(x)$ by

$$s_{\delta,[a,b]}(x) = x\frac{b-a}{2} - \frac{b+a}{2}\int_0^x \tanh(y/\delta)\mathrm{d}y, \tag{36}$$

for some $\delta > 0$. The derivative of s_δ,

$$s'_{\delta,[a,b]}(x) = \frac{b-a}{2} - \frac{b+a}{2}\tanh(x/\delta), \tag{37}$$

approaches a step function as δ tends to zero, see Fig. 1.

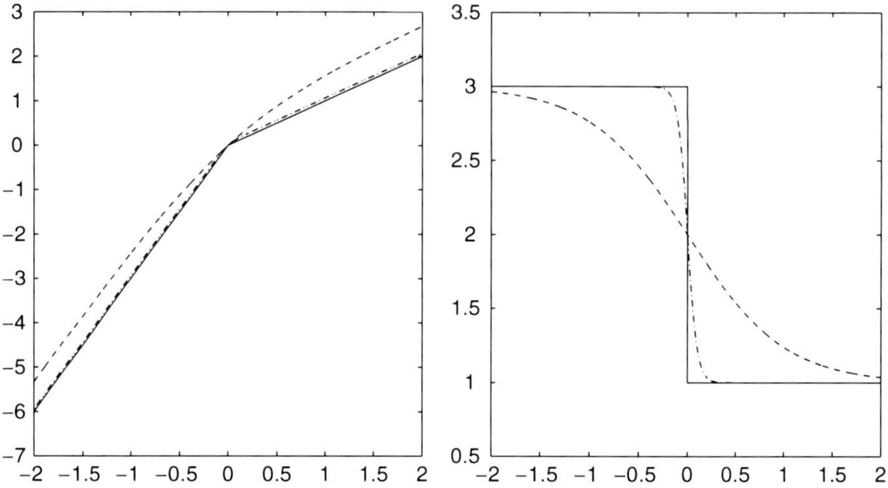

Fig. 1 Here $a = 1$ and $b = 3$. To the left: s_δ for $\delta = 1$(*dashed*), 0.1 (*dot-dashed*) and 0.0001 (*solid*) respectively. To the right: s'_δ for the same δ

We define the regularized Hamiltonian $H^\delta(C,\lambda,T)$ by

$$H^\delta(C,\lambda,T) = s_{\delta,[\mu_-,\mu_+]}\left(\int_{\mathscr{R}_+}\lambda\psi_1(C)\mathrm{d}K\right) + \int_{\mathscr{R}_+} s_{\delta,[\sigma_-^2,\sigma_+^2]}\left(\lambda\psi_2(C)\right)\mathrm{d}K + \int_{\mathscr{R}_+}(C - C_m(T,K))^2\mathrm{d}K. \tag{38}$$

The Hamiltonian system (26) associated to the regularized optimal control problem is

$$
\begin{aligned}
C_T &= s'_{\delta,[\mu_-,\mu_+]}\left(\int_{\mathscr{R}_+}\lambda\psi_1(C)\mathrm{d}K\right)\psi_1(C)+s'_{\delta,[\sigma_-^2,\sigma_+^2]}\left(\lambda\psi_2(C)\right)\psi_2(C),\\
-\lambda_T &= s'_{\delta,[\mu_-,\mu_+]}\left(\int_{\mathscr{R}_+}\lambda\psi_1(C)\mathrm{d}K\right)\psi_1^*(\lambda)\\
&\quad+\psi_2^*\left(\lambda s'_{\delta,[\sigma_-^2,\sigma_+^2]}\left(\psi_2(C)\lambda\right)\right)+2(C-C_m),\\
C(K,0) &= \max(S-K,0),\\
C(0,T) &= S,\\
\lambda(K,\hat{\mathrm{T}}) &= 0,\\
\lambda(0,T) &= 0,\\
\mu_\delta &= s'_{\delta,[\mu_-,\mu_+]}\left(\int_{\mathscr{R}_+}\lambda\psi_1(C)\mathrm{d}K\right),\\
\sigma_\delta &= s'_{\delta,[\sigma_-^2,\sigma_+^2]}\left(\lambda\psi_2(C)\right).
\end{aligned}
\tag{39}
$$

4.2 Discretization

We proceed by solving the Hamiltonian system (39). We suggest a discretization in the time dimension based on an implicit symplectic Pontryagin scheme introduced in [11]. The details are as follows:

We introduce a uniform partition of the time interval $[0,\hat{\mathrm{T}}]$ with $\Delta t = \hat{\mathrm{T}}/N$ for some integer N. We write $C^{(j)}(K) = C(K, j\Delta T)$ and $\lambda^{(j)}(K) = \lambda(K, j\Delta T)$ and demand that they satisfy a symplectic implicit Euler scheme:

$$
\begin{aligned}
C^{(j+1)} - C^{(j)} &= \Delta T H_\lambda^\delta(C,\lambda,T)^{(j)},\\
\lambda^{(j)} - \lambda^{(j+1)} &= \Delta T H_C^\delta(C,\lambda,T)^{(j)},
\end{aligned}
\tag{40}
$$

where $H^\delta(C,\lambda)^{(j)} = H^\delta(C^{(j)},\lambda^{(j+1)}, j\Delta t)$. Notice that we evaluate the Hamiltonian at different times for C and λ.

Remark 1. Symplecticity here means that the gradient of the discrete value function coincides with the discrete dual:

$$\overline{U}_C(C^{(i)},t) = \lambda^{(i)}. \tag{41}$$

Symplectic Euler is an example of a symplectic scheme. See chapter 6 in [8] for more examples and a more thorough discussion of symplectic methods and their use. An important property of the symplectic Euler method is that the numerical solution is an *exact* solution of a perturbed Hamiltonian system. See [11] for a detailed description of the perturbed Hamiltonian.

The main result of [12] (see Theorem 4.1) states that if the Hamiltonian is Lipschitz, and if $\lambda^{(j+1)}$ has uniformly bounded variation with respect to $C^{(j)}$ for all j and ΔT, the optimal solution to the Pontryagin problem (40), $(C^{(j)}, \lambda^{(j)})$, satisfies the error estimate (for $\delta \sim \Delta T$)

$$|\inf \int_0^{\hat{T}} \int_{\mathscr{R}+} (C - C_m)^2 \mathrm{d}K \mathrm{d}T - \Delta T \sum_j \int_{\mathscr{R}+} (C^{(j)} - C_m^{(j)})^2 \mathrm{d}K| = \mathscr{O}(\Delta T). \quad (42)$$

We now turn to the strike variable K. We truncate for large values of K: $C(T, K) = 0$ for $K > \hat{K}$, for some large $\hat{K}$, and introduce a uniform grid on $[0, \hat{K}]$, $\Delta K = \hat{K}/M$, for some integer M. We use the notation

$$\begin{aligned} C_i^{(j)} &= C(i\Delta K, j\Delta T), \\ \lambda_i^{(j)} &= \lambda(i\Delta K, j\Delta T). \end{aligned} \quad (43)$$

The next step is to discretize the operators (19). We use the standard central difference quotients to approximate the derivatives

$$D_i C = \frac{C_{i+1} - C_{i-1}}{2\Delta K}, \qquad D_i^2 C = \frac{C_{i+1} - 2C_i + C_{i-1}}{\Delta K^2}. \quad (44)$$

The integral $E[JC(T, J^{-1}K)] = \int_{\mathscr{R}_+} z\chi(z) C(T, z^{-1}K) \mathrm{d}x$ is calculated by first truncating for large values of z, say $z > \hat{z}$, then using the trapezoidal rule:

$$E\big[JC(T, J^{-1}K)\big] \approx E(C)_i := \Delta z$$

$$\sum_{k=0}^{P} \frac{f(z_k)C(T, i\Delta K/z_k) + f(z_{k+1})C(T, i\Delta K/z_{k+1})}{2},$$

where $\Delta z = \hat{z}/P$, $z_k = k\Delta z$ and $f(z) = z\chi(z)$. The value of $C(T, i\Delta K/z_{k+1})$ is approximated using linear interpolation. Define the integers $\gamma(i,k)$ by the rule $\gamma(i,k) \leq i/z_k < \gamma(i,k) + 1$. It is then possible to estimate

$$C(T, i\Delta K/z_k) \approx (C_{\gamma(i,k)+1} - C_{\gamma(i,k)})(i/z_k - \gamma(i,k)) + C_{\gamma(i,k)}. \quad (45)$$

We treat

$$E^*(C)_i \approx E[J^2 C(T, Ji\Delta K)] \quad (46)$$

in the same way.

This yields the discretization

$$\begin{aligned} \psi_1(C)_i &= m(i\Delta K) D_i C - (m+1)C_i + \mathrm{E}(C)_i, \\ \psi_2(C)_i &= \tfrac{1}{2}(i\Delta K)^2 D_i^2 C, \\ \psi^*(C)_i &= (m+1)C_i + m D_i(KC) + \mathrm{E}_i^*(C), \\ \psi^*(C)_i &= \tfrac{1}{2} D_i^2(K^2 C). \end{aligned} \quad (47)$$

We can now approximate H_λ^δ and H_C^δ by

$$
\begin{aligned}
H^{\delta}_{\lambda}(C,\lambda,T)^{(j)}_{i} &= s'_{\delta,[\mu_-,\mu_+]}\Big(\Delta K \textstyle\sum_k \lambda^{(j+1)}_k \psi_1(C^{(j)})_k\Big)\psi_1(C^{(j)})_i \\
&\quad + s'_{\delta,[\sigma^2_-,\sigma^2_+]}\Big(\lambda^{(j+1)}_i \psi_2(C^{(j)})_i\Big)\psi_2(C^{(j)})_i, \\
H^{\delta}_{C}(C,\lambda,T)^{(j)}_{i} &= s'_{\delta,[\mu_-,\mu_+]}\Big(\textstyle\int_{\mathscr{R}_+} \lambda\psi_1(C^{(j)})\mathrm{d}K\Big)\psi^*_1(\lambda)_i \\
&\quad + \psi^*_2\Big(\lambda s'_{\delta,[\sigma^2_-,\sigma^2_+]}\Big(\psi_2(C^{(j)})\lambda\Big)\Big)_i + 2(C^{(j)} - C^{(j)}_m)_i.
\end{aligned}
\tag{48}
$$

Finally we summarize the above and obtain the completely discretized Hamiltonian system

$$
\begin{aligned}
C^{(j+1)}_i - C^{(j)}_i &= \Delta T H^{\delta}_{\lambda}(C,\lambda)^{(j)}_i, \\
\lambda^{(j)}_i - \lambda^{(j+1)}_i &= \Delta T H^{\delta}_{C}(C,\lambda)^{(j)}_i, \\
C^{(0)}_i &= \max(S - i\Delta K, 0), \\
C^{(j)}_0 &= S, \\
C^{(j)}_M &= 0, \\
\lambda^{(N)}_i &= 0, \\
\lambda^{(j)}_0 &= 0, \\
\lambda^{(j)}_M &= 0, \\
\mu^{(j)} &= s'_{\delta,[\mu_-,\mu_+]}\Big(\Delta K \textstyle\sum_k \lambda^{(j+1)}_k \psi_1(C^{(j)})_k\Big), \\
\sigma^{(j)}_i &= s'_{\delta,[\sigma^2_-,\sigma^2_+]}\Big(\lambda^{(j+1)}_i \psi_2(C^{(j)})_i\Big).
\end{aligned}
\tag{49}
$$

Recall that μ and σ depend on the parameter δ.

4.3 The Newton Method

In order to solve the Hamiltonian system (49), one could use some fixed-point scheme that in each iteration removed the coupling by solving the equations for C and λ separately. This method has the advantage of being easy to implement but the major drawback of very slow (if any) convergence to the optimal solution.

We instead use information about the Hessian and solve (49) with the Newton method. The details are as follows:

We let the functions $F^{\delta}, G^{\delta} : \mathscr{R}^{MN} \to \mathscr{R}^{MN}$ be given by

$$
\begin{aligned}
F^{\delta}(C,\lambda)_{i+j*N} &= C^{(j+1)}_i - C^{(j)}_i - \Delta T H^{\delta}_{\lambda,ij}, \\
G^{\delta}(C,\lambda)_{i+j*N} &= C^{(j)}_i - C^{(j+1)}_i - \Delta T H^{\delta}_{C,ij}.
\end{aligned}
\tag{50}
$$

We seek (C,λ) such that $F^{\delta}(C,\lambda) = G^{\delta}(C,\lambda) = 0$.

Starting with some initial guess $(C[0],\lambda[0])$, the Newton method gives

$$\begin{pmatrix} C[k+1] \\ \lambda[k+1] \end{pmatrix} = \begin{pmatrix} C[k] \\ \lambda[k] \end{pmatrix} - \begin{pmatrix} X[k] \\ Y[k] \end{pmatrix}, \tag{51}$$

where $(X[k], Y[k])$ is the solution to the following system of linear equations

$$J_k \begin{pmatrix} X[k] \\ Y[k] \end{pmatrix} = \begin{pmatrix} F(C[k], \lambda[k]) \\ G(C[k], \lambda[k]) \end{pmatrix}. \tag{52}$$

We let J_k denote the Jacobian of $(F, G) : \mathscr{R}^{2MN} \to \mathscr{R}^{2MN}$ evaluated at $(C[k], \lambda[k])$.

As expected, the smaller the value of the regularizing parameter δ, the harder for (51) to converge. In particular, a small δ requires a good initial guess. Since ultimately we wish to solve the Hamiltonian system for very small δ, we are led to a iterative Newton scheme that brings δ down successively. The scheme is summarized in Algorithm 4.1.

Algorithm 4.1: Newton method

Input: Tolerance TOL, final regularization parameter δ_0, observed prices C_m.
Output: μ and σ.

Let δ be not too small (usually $\delta \approx 1$ will do).
Let β be some number $0 < \beta < 1$ (typically $\beta \approx 0.7$ will do).
Set $C_i^{(j)}[0] = \max(S - i\Delta K, 0)$ and $\lambda_i^{(j)}[0] = 0$.
while $\delta > \delta_0$ **do**
  Let $k = 0$.
  while $\|(F(C[k], \lambda[k]), G(C[k], \lambda[k]))\| > \text{TOL}$ **do**

$$\begin{pmatrix} C[k+1] \\ \lambda[k+1] \end{pmatrix} = \begin{pmatrix} C[k] \\ \lambda[k] \end{pmatrix} - \begin{pmatrix} X[k] \\ Y[k] \end{pmatrix} \tag{12.53}$$

    $k = k + 1$.
  end while
  Let $(C[0], \lambda[0]) = (C[k], \lambda[k])$.
  Put $\delta = \beta\delta$.
end while
Define σ and μ by:

$$\begin{aligned} \mu^{(j)} &= s'_{\delta,[\mu_-,\mu_+]}\left(\Delta K \sum_k \lambda_k^{(j+1)} \psi_1(C^{(j)})_k\right), \\ \sigma_i^{(j)} &= s'_{\delta,[\sigma_-^2,\sigma_+^2]}\left(\lambda_i^{(j+1)} \psi_2(C^{(j)})_i\right). \end{aligned} \tag{12.54}$$

5 Numerical Examples

5.1 Artificial Data No Jumps

As a first example we apply the method presented above, and summarized in Algorithm 4.1, to solve the calibration problem (28) without jumps. That is, from a set of solutions $\{C_m(K,T)\}$ we deduce $\sigma(K,T)$ by minimizing

$$\int_0^{\hat{T}} \int_{\mathscr{R}_+} (C - C_m)^2 \mathrm{d}T \mathrm{d}K, \tag{55}$$

where $C(T,K)$ solves

$$\begin{aligned} C_T &= \tfrac{1}{2}\sigma^2 C_{KK}, \\ C(K,0) &= \max(S-K,0). \end{aligned} \tag{56}$$

To test that the method does indeed converge to the correct solution, we assign a value to $\sigma(S,t)$, and by solving (56), we obtain a solution $C_m(K,T)$. Using this solution, we reconstruct $\sigma(K,T)$ with Algorithm 4.1. The result is presented in Fig. 2. As can be seen, away from the boundary, one can reconstruct $\sigma(T,K)$ to a very high degree of accuracy. It should also be noted that the regularizing parameter δ can virtually be eliminated, thus obtaining a nearly perfect fit of calibrated prices $C(T,K)$ to market prices $C_m(T,K)$.

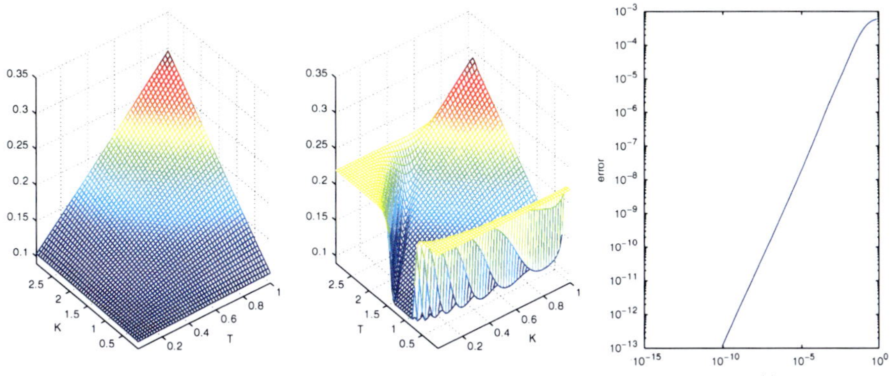

Fig. 2 Reconstruction of volatility $\sigma^2_{\text{true}}(K,T) = 0.1 + 0.2TK/3$ with $\delta = 10^{-9}$, $\sigma^2_- = 0.1$ and $\sigma^2_+ = 0.35$ with no jump present. In this experiment $S = 1$, $\hat{K} = 3$ and $\hat{T} = 1$. We use the grid size of $M = N = 50$. The three plots shows, from left to right: 1. The true volatility σ^2_{true} used to generate "quoted" option prices C_m, 2. The reconstructed volatility σ^2 for $\delta = 10^{-10}$ and, 3. The L^2-error in option prices as a function of the regularizing parameter δ: $\|C - C_m\|_{L^2}(\delta)$

5.2 Artificial Data with Jumps

A more interesting example is obtained by generating option prices using the full jump-diffusion model (5). We assume that the relative jump sizes J are log-normally distributed with mean 0 and variance 1, i.e.

$$\log J \sim N(0,1). \tag{57}$$

Option prices C_m are generated by solving (20) with prescribed functions σ and μ. We then reconstruct σ and μ using Algorithm 4.1. The result is presented in Fig. 3 below. Again the quality of the reconstructed data is very good. The calibrated volatility σ and jump intensity μ can be brought arbitrarily close to its "true" prescribed values.

5.3 Real Data

We conclude this section and the paper with an example from the S&P - 500 market.

In order to compare the described calibration method with existing methods we decided to re-calibrate the model calibrated in [1]. Andersen and Andreasen collected a set of bid and ask prices for call options on the S&P-500 index in April 1999. At page 11 in [1] a table of bid and ask volatilities is presented. We will focus only on data for options with maturities no more than 12 months.

The first step in the calibration is to determine the distribution of the jump sizes. In [1] the authors assumes that the jumps are log-normally distributed with unknown

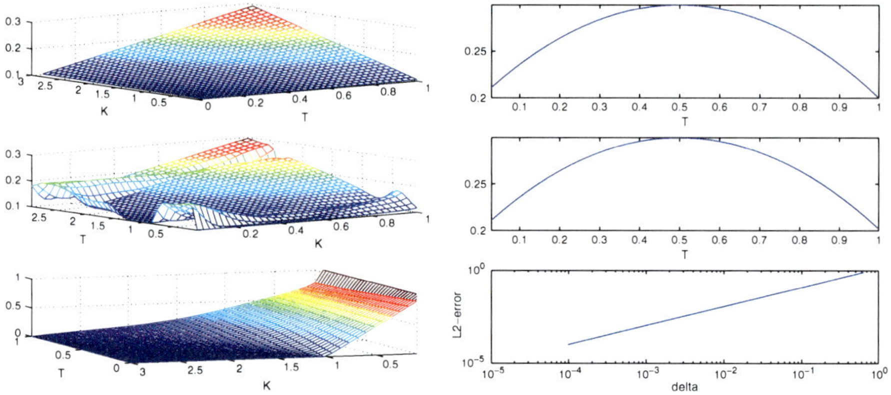

Fig. 3 Reconstruction of volatility and jump intensity when $\sigma^2_{\text{true}}(K,T) = 0.1 + 0.2TK/3$ and $\mu = 0.2 - 0.4((t-0.5)^2 - 0.25)$. Here we show results for $\delta = 10^{-4}$, $\sigma^2_- = 0.1$, $\sigma^2_+ = 0.35$, $\mu_- = 0.2$ and $\mu_+ = 0.3$. In this experiment $S = 1$, $\hat{K} = 3$ and $\hat{T} = 1$. We used a grid size of $M = 80$ and $N = 20$. The six plots from top left to bottom right represent respectively: 1. The true volatility σ^2_{true}. 2. True jump intensity. 3. Reconstructed volatility σ^2 for $\delta = 10^{-10}$. 4. Reconstructed jump intensity. 5. The price surface obtained using reconstructed prices. 6. The L^2-error in option prices as a function of the regularizing parameter δ: $\|C - C_m\|_{L^2}(\delta)$

mean α and variance β^2. They determine α and β by assuming that also σ and μ are constant. That is, they calibrate the Levy-process (q denotes the dividend yield)

$$dS_t/S_{t-} = (r - q - \mu m)dt + \sigma \mathrm{d}B(t) + (J-1)\mathrm{d}\pi_t. \tag{58}$$

A best fit, in the least-square sense, of the above parameters, to mid-implied volatilities, results in

$$\begin{aligned} \sigma &= 17.65\%, \\ \mu &= 8.90\%, \\ \alpha &= -88.98\%, \\ \beta &= 45.05\%. \end{aligned} \tag{59}$$

We assume that the above parameters determine the jump size distribution and proceed by calibrating the state and time dependent volatility and time dependent intensity using Algorithm 4.1. Note that the interest rate is non-zero and that there is a dividend yield. It is straightforward to obtain the forward equation with a yield term present, corresponding to (9) (see for instance Equation 4 in [1]).

As before, we let $C_m = C_m(T, K)$ denote the market price of options. The optimal control problem consists of minimizing

$$\int_0^{\hat{T}} \int_{\mathscr{R}_+} w(T, K)(C - C_m)^2 \mathrm{d}T \mathrm{d}K, \tag{60}$$

where we have introduced a weight function w to accommodate for the fact the C_m is not known everywhere. The specific weight function used in the calibration is

$$w(T, K) = \sum_{(T_i, K_i) \in I} \delta(T - T_i)\delta(K - K_i), \tag{61}$$

with the sum taken over all values (T_i, K_i) for which we have a market price.

We are now in a position to apply the technique explained in the previous section. The jump intensity was found to be roughly constant over time and equal to

$$\mu(T) = 16.5\%. \tag{62}$$

The resulting local volatility σ is plotted in Fig. 4. We used the constant values in (59) as starting values of σ and μ. The method worked well in the sense that we had no problems with convergence and the resulting volatility surface and intensity function were reasonable. Using the calibrated measure we could reproduce the option prices to within the bid-ask spread.

One drawback with the method presented in this work is that one needs an explicit Hamiltonian, and preferably an explicit expression of the Hessian. Otherwise, the method becomes more involved, and potentially more computationally costly. This is the reason why we determine the jump distribution as described in this section.

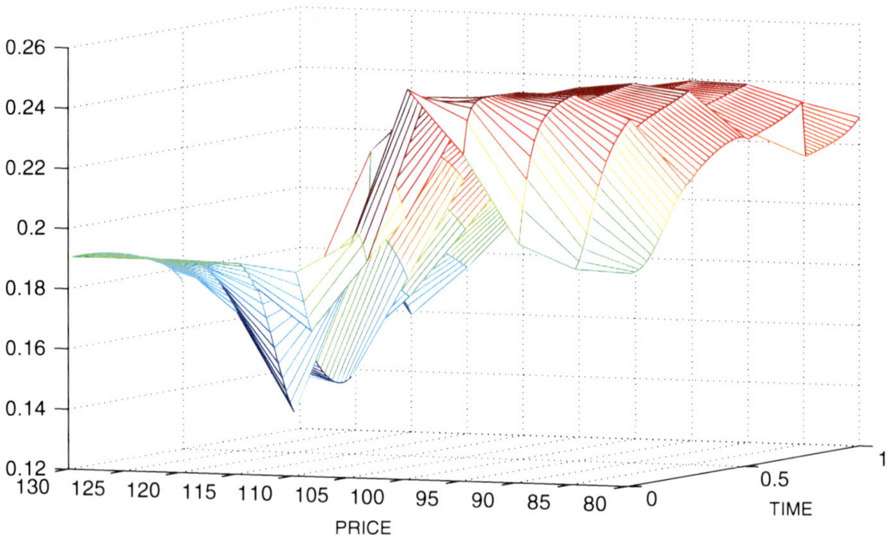

Fig. 4 Local diffusion volatilities for the S&P500 index, April 1999. Local volatilities for jump-diffusion model when fitted to S&P500 option prices. First axis is future spot relative current and second axis is time in years. Jump parameters are $\alpha = -88.89\%$ and $\beta = 45.05\%$. The jump intensity was calibrated to $\mu(t) = 16.5\%$

References

1. Leif Andersen and Jesper Andreasen. Jump-diffusion processes: Volatility smile fitting and numerical methods for option pricing. *Review of Derivatives Research*, 4(4):231–262, 2000.
2. Davis Bates. Jumps and stochastic volatility: Exchange rate processes implicit in deutsche mark options. *The Review of Financial Studies*, 9:69–107, 1996.
3. F. Black and M. Scholes. The pricing of options and corporate liabilities. *Journal of Political Economy*, 81:637–654, 1973.
4. Rama Cont and Peter Tankov. *Financial modelling with jump processes*. Chapman & Hall/CRC Financial Mathematics Series. Chapman & Hall/CRC, Boca Raton, FL, 2004.
5. M. G. Crandall, L. C. Evans, and P.-L. Lions. Some properties of viscosity solutions of Hamilton-Jacobi equations. *Trans. Amer. Math. Soc.*, 282(2):487–502, 1984.
6. S. Crépey. Calibration of the local volatility in a generalized Black-Scholes model using Tikhonov regularization. *SIAM J. Math. Anal.*, 34(5):1183–1206 (electronic), 2003.
7. Bruno Dupire. Pricing and hedging with smiles. In *Mathematics of derivative securities (Cambridge, 1995)*, volume 15 of *Publ. Newton Inst.*, pages 103–111. Cambridge Univ. Press, Cambridge, 1997.
8. Ernst Hairer, Christian Lubich, and Gerhard Wanner. *Geometric numerical integration*, volume 31 of *Springer Series in Computational Mathematics*. Springer-Verlag, Berlin, second edition, 2006. Structure-preserving algorithms for ordinary differential equations.
9. Robert Merton. Option pricing when underlying stock returns are discontinuous. *Journal of Financial Economics*, 3:125–144, 1976.
10. Steven L. Heston. A closed-form solution for options with stochastic volatility with applications to bond and currency options. *The Review of Financial Studies*, 6(2):327–342, 1993.

11. Mattias Sandberg. Extended applicability of the symplectic pontryagin method. arXiv:0901.4805v1.
12. Mattias Sandberg and Anders Szepessy. Convergence rates of symplectic Pontryagin approximations in optimal control theory. *M2AN Math. Model. Numer. Anal.*, 40(1):149–173, 2006.
13. W Schoutens, E Simons, and J Tistaert. A perfect calibration! Now what? *Wilmott*, 2004(2):66–78, 2004.

Some Remarks on Free Energy and Coarse-Graining

Frédéric Legoll and Tony Lelièvre

Abstract We present recent results on coarse-graining techniques for thermodynamic quantities (canonical averages) and dynamical quantities (averages of path functionals over solutions of overdamped Langevin equations). The question is how to obtain reduced models to compute such quantities, in the specific case when the functional to be averaged only depends on a few degrees of freedom. We mainly review, numerically illustrate and extend results from (Blanc et al. Journal of Nonlinear Science 20(2):241–275, 2010; Legoll and Lelièvre Nonlinearity 23(9):2131–2163, 2010.), concerning the computation of the stress-strain relation for one-dimensional chains of atoms, and the construction of an effective dynamics for a scalar coarse-grained variable when the complete system evolves according to the overdamped Langevin equation.

1 Motivation

In molecular simulation, two types of quantities are typically of interest: averages with respect to the canonical ensemble (*thermodynamic quantities*, such as stress, root-mean-square distance, ...), and averages of functionals over paths (*dynamic quantities*, like viscosity, diffusion coefficients or rate constants). In both cases, the question of coarse-graining is relevant, in the sense that the considered functionals

T. Lelièvre (✉)
Université Paris-Est, CERMICS, École des Ponts ParisTech, 6 et 8 avenue Blaise Pascal, 77455 Marne-La-Vallée Cedex 2, France and INRIA Rocquencourt, MICMAC Team-Project, Domaine de Voluceau, B.P. 105, 78153 Le Chesnay Cedex, France
e-mail: lelievre@cermics.enpc.fr

F. Legoll
Université Paris-Est, Institut Navier, LAMI, École des Ponts ParisTech, 6 et 8 avenue Blaise Pascal, 77455 Marne-La-Vallée Cedex 2, France and INRIA Rocquencourt, MICMAC Team-Project, Domaine de Voluceau, B.P. 105, 78153 Le Chesnay Cedex, France
e-mail: legoll@lami.enpc.fr

B. Engquist et al. (eds.), *Numerical Analysis of Multiscale Computations*, Lecture Notes in Computational Science and Engineering 82, DOI 10.1007/978-3-642-21943-6_13,

typically depend only on a few variables of the system (collective variables, or reaction coordinates). Therefore, it is essential to understand how to obtain coarse-grained models on these variables.

1.1 Coarse-Graining of Thermodynamic Quantities

Computing canonical averages is a standard task in molecular dynamics. For a molecular system whose atom positions are described by a vector $q \in \mathbb{R}^n$, these quantities read

$$\int_{\mathbb{R}^n} \Phi(q)\, d\mu,$$

where $\Phi : \mathbb{R}^n \to \mathbb{R}$ is the observable of interest and μ is the Boltzmann-Gibbs measure,

$$d\mu = Z^{-1} \exp(-\beta V(q))\, dq, \tag{1}$$

where V is the potential energy of the system, β is proportional to the inverse of the system temperature, and

$$Z = \int_{\mathbb{R}^n} \exp(-\beta V(q))\, dq$$

is a normalizing constant. Typically, q represents the position of N particles in dimension d, hence $q \in \mathbb{R}^n$ with $n = dN$.

As mentioned above, observables of interest are often functions of only part of the variable q. For example, q denotes the positions of *all* the atoms of a protein and of the solvent molecules around, and the quantity of interest is only a particular angle between some atoms in the protein, because this angle characterizes the conformation of the protein (and thus the potential energy well in which the system is, is completely determined by the knowledge of this quantity of interest). Another example is the case when $q = (q^1, \dots, q^n)$ denotes the positions of all the atoms of a one-dimensional chain, and quantities of interest are only a function of the total length $q^n - q^1$ of the chain.

We thus introduce the so-called *reaction coordinate*

$$\xi : \mathbb{R}^n \to \mathbb{R},$$

which contains all the information we are interested in. Throughout this article, we assume that it is a smooth function such that $|\nabla \xi|$ is bounded from below by a positive constant, so that the configurational space can be foliated by isosurfaces associated to ξ. A simple case that will be considered below is $\xi(q^1, \dots, q^n) = q^n$.

To this function ξ is naturally associated an effective energy A, called the *free energy*, such that

$$d(\xi \star \mu) = \exp(-\beta A(z))\, dz,$$

where $\xi \star \mu$ denotes the image of the measure μ by ξ. In other words, for any test function $\Phi : \mathbb{R} \to \mathbb{R}$,

$$\int_{\mathbb{R}^n} \Phi(\xi(q))\, Z^{-1} \exp(-\beta V(q))\, dq = \int_{\mathbb{R}} \Phi(z) \exp(-\beta A(z))\, dz. \tag{2}$$

Expressions of A and its derivative are given below (see Sect. 1.4).

The interpretation of (2) is that, when Q is a random variable distributed according to the Boltzmann measure (1), then $\xi(Q)$ is distributed according to the measure $\exp(-\beta A(z))\, dz$. Hence, the free energy A is a relevant quantity for computing thermodynamic quantities, namely canonical averages.

In conclusion, the question of coarse-graining thermodynamic quantities amounts to computing the free energy, and there are several efficient methods to perform such calculations (see for example [6, 20]). In the sequel of this article, we address a particular case, motivated by materials science, where the system under consideration is a one-dimensional chain of atoms, and $\xi(q^1, \dots, q^n) = q^n - q^1$ is the length of the chain (see Fig. 1 below). We are interested in the free energy associated to this reaction coordinate, and its behaviour when the number n of particles goes to $+\infty$. Standard algorithms to compute the free energy then become prohibitively expensive, as the dimension of the system becomes larger and larger. Alternative strategies are needed, and we investigate analytical methods, based on large deviations principles, in Sect. 2.

1.2 Coarse-Graining of Dynamical Quantities

The second topic of this contribution is related to the *dynamics* of the system, and how to coarse-grain it. In short, we will show how to design a dynamics that approximates the path $t \mapsto \xi(Q_t)$, where ξ is the above reaction coordinate.

To make this question precise, we first have to *choose* the full dynamics, which will be the reference one. In the following, we consider the overdamped Langevin dynamics on state space $\mathbb{R}^n$:

$$dQ_t = -\nabla V(Q_t)\, dt + \sqrt{2\beta^{-1}}\, dW_t, \quad Q_{t=0} = Q_0, \tag{3}$$

where W_t is a standard n-dimensional Brownian motion. Under suitable assumptions on V, this dynamics is ergodic with respect to the Boltzmann-Gibbs measure (1) (see [5] and references therein). Hence, for μ-almost all initial conditions Q_0,

$$\lim_{T\to\infty} \frac{1}{T} \int_0^T \Phi(Q_t)\, dt = \int_{\mathbb{R}^n} \Phi(q)\, d\mu \tag{4}$$

almost surely. In practice, this convergence is often very slow, due to some metastabilities in the dynamics: Q_t samples a given well of the potential energy for a long time, before hopping to some other well of V.

An important dynamical quantity we will consider below is the average residence time, that is the mean time that the system spends in a given well, before hopping to another one, when it follows the dynamics (3). Typically, the wells are fully described through ξ (q is in a given well if and only if $\xi(q)$ is in a given interval), so that these times can be obtained from the knowledge of the time evolution of $\xi(Q_t)$, which is expensive to compute since it means simulating the full system.

In Sect. 3 below, we will first present a one-dimensional dynamics of the form

$$d\overline{\eta}_t = b(\overline{\eta}_t)\,dt + \sqrt{2\beta^{-1}}\,\sigma(\overline{\eta}_t)\,dB_t, \tag{5}$$

where B_t is a standard one-dimensional Brownian motion and b and σ are scalar functions, such that $(\overline{\eta}_t)_{0\leq t\leq T}$ is a good approximation (in a sense to be made precise below) of $(\xi(Q_t))_{0\leq t\leq T}$. Hence, the dynamics (5) can be thought of as a coarse-grained, or *effective*, dynamics for the quantity of interest. A natural requirement is that (5) preserves equilibrium quantities, *i.e.* it is ergodic with respect to $\exp(-\beta A(z))\,dz$, the equilibrium measure of $\xi(Q_t)$ when Q_t satisfies (3), but we typically ask for more than that. For example, we would like to be able to recover residence times in the wells from (5), hence bypassing the expensive simulation of $\xi(Q_t)$. We will show below that the effective dynamics we propose indeed fulfills these two requirements.

As a matter of fact, the coarse-grained dynamics

$$d\overline{z}_t = -A'(\overline{z}_t)\,dt + \sqrt{2\beta^{-1}}\,dB_t \tag{6}$$

is a one-dimensional dynamics that is ergodic with respect to $\exp(-\beta A(z))\,dz$. It can thus be thought of as a natural candidate for a dynamics approximating $\xi(Q_t)$, all the more so as practitioners often look at the free energy profile (*i.e.* the function $z \mapsto A(z)$) to get an idea of the dynamics of transition (typically the transition time) between one region indexed by the reaction coordinate (say for example $\{q \in \mathbb{R}^n;\ \xi(q) \leq z_0\}$) and another one (for example $\{q \in \mathbb{R}^n;\ \xi(q) > z_0\}$). If $\xi(Q_t)$ follows a dynamics which is close to (6), then the Transition State Theory says that residence times are a function of the free energy barriers [17, 18], and then it makes sense to look at the free energy to compute some dynamical properties. It is thus often assumed that there is some dynamical information in the free energy A.

In the sequel, we will compare the accuracy (with respect to the original full dynamics) of both coarse-grained dynamics, an effective dynamics, an effective dynamics of type (5) (namely dynamics (67) below) and the dynamics (6) driven by the free energy. Their relation has been investigated from an analytical viewpoint in [19, Sect. 2.3] (see also [11, Sect. 10 and (89)] and [21]).

1.3 Outline of the Article

In this contribution, we mainly review, numerically illustrate and extend results from the two articles [3, 19]. Our aim is to present in a pedagogical and unified manner recent contributions on coarse-graining procedures concerning: (1) a static case inspired by material sciences, namely the computation of stress-strain (namely force-elongation) relation for one-dimensional chains of atoms, in the thermodynamic limit (Sect. 2) and (2) a dynamic case inspired by molecular dynamics computations, namely the derivation of effective dynamics along the reaction coordinate, for overdamped Langevin equations (Sect. 3). Compared to the original articles [3, 19], we propose some extensions of the theoretical results (see *e.g.* Sect. 2.2), some simpler proofs in more restricted settings (in Sect. 3.3) and new numerical experiments (Sects. 2.2.4 and 3.4).

1.4 Notation

We gather here some useful notation and results. Let Σ_z be the submanifold of $\mathbb{R}^n$ of positions at a fixed value of the reaction coordinate:

$$\Sigma_z = \{q \in \mathbb{R}^n; \xi(q) = z\}. \tag{7}$$

Let us introduce μ_{Σ_z}, which is the probability measure μ conditioned at a fixed value of the reaction coordinate:

$$d\mu_{\Sigma_z} = \frac{\exp(-\beta V)\,|\nabla\xi|^{-1}\,d\sigma_{\Sigma_z}}{\displaystyle\int_{\Sigma_z} \exp(-\beta V)\,|\nabla\xi|^{-1}\,d\sigma_{\Sigma_z}}, \tag{8}$$

where the measure σ_{Σ_z} is the Lebesgue measure on Σ_z induced by the Lebesgue measure in the ambient Euclidean space $\mathbb{R}^n \supset \Sigma_z$. By construction, if Q is distributed according to the Gibbs measure (1), then the law of Q conditioned to a fixed value z of $\xi(Q)$ is μ_{Σ_z}. The measure $|\nabla\xi|^{-1}d\sigma_{\Sigma_z}$ is sometimes denoted by $\delta_{\xi(q)-z}(dq)$ in the literature.

We recall the following expressions for the free energy A and its derivative A', also called the *mean force* (see [7]):

$$A(z) = -\beta^{-1}\ln\left(\int_{\Sigma_z} Z^{-1}\exp(-\beta V)\,|\nabla\xi|^{-1}\,d\sigma_{\Sigma_z}\right), \tag{9}$$

$$A'(z) = \int_{\Sigma_z} F\,d\mu_{\Sigma_z}, \tag{10}$$

where F is the so-called *local mean force*:

$$F = \frac{\nabla V\cdot\nabla\xi}{|\nabla\xi|^2} - \beta^{-1}\,div\left(\frac{\nabla\xi}{|\nabla\xi|^2}\right). \tag{11}$$

In the particular case when the reaction coordinate is just one of the cartesian coordinate, say $\xi(q) = q^n$, then

$$A(z) = -\beta^{-1} \ln \left(\int_{\mathbb{R}^{n-1}} Z^{-1} \exp(-\beta V(q^1, \dots, q^{n-1}, z)) \, dq^1 \dots dq^{n-1} \right)$$

and the local mean force is just $F = \partial_{q^n} V$, so that

$$A'(z) = \frac{\int_{\mathbb{R}^{n-1}} \partial_{q^n} V(q^1, \dots, q^{n-1}, z) \exp(-\beta V(q^1, \dots, q^{n-1}, z)) \, dq^1 \dots dq^{n-1}}{\int_{\mathbb{R}^{n-1}} \exp(-\beta V(q^1, \dots, q^{n-1}, z)) \, dq^1 \dots dq^{n-1}}.$$

2 Computing Macroscopic Stress-Strain Relations for One-Dimensional Chains of Atoms

In this section, we wish to compute the stress-strain relation of a one-dimensional chain of atoms, in the thermodynamic limit. More precisely, we consider a chain of $1 + N$ atoms, with its left-end atom fixed, and either submit the right-end atom to a force, and compute the average elongation, or prescribe the elongation, and compute the force. We will show that, in the limit $N \to \infty$, these two relations are identical, and that they can be computed in an extremely efficient manner. In short, passing to the limit $N \to \infty$ makes tractable a computation that is, for finite and large N, very expensive.

The relation between that question and the question of determining the free energy of the system, when the reaction coordinate is the length of the system, will also be discussed.

In the sequel, we first proceed with the nearest neighbour case (see Sect. 2.1). We next address the next-to-nearest neighbour case in Sect. 2.2, which is technically more involved.

2.1 The Nearest Neighbour (NN) Case

We consider a one-dimensional chain of atoms, with positions $q^0, q^1, \dots, q^N$. In this section, we only consider nearest neighbour interaction. In addition to this internal interaction, we assume that the atom at the right boundary of the chain is submitted to an external force f, and that the atom at the left boundary is fixed: $q^0 = 0$. The energy of the chain thus reads

$$\widetilde{E}_f \left(q^1, \dots, q^N \right) = \sum_{i=1}^{N} W \left(q^i - q^{i-1} \right) - f q^N.$$

In the sequel, we will consider the limit when the number N of atoms goes to ∞. We wish to make sure that, even when $N \to \infty$, the system occupies, on average, a finite length. To this aim, we introduce the rescaled positions $u^i = h q^i$, with $h = 1/N$. The energy now reads

$$E_f\left(u^1,\ldots,u^N\right) = \sum_{i=1}^{N} W\left(\frac{u^i - u^{i-1}}{h}\right) - f\frac{u^N}{h}, \tag{12}$$

where again $u^0 = 0$.

For any observable Φ, depending on the variables $u^1,\ldots,u^N$, we define the canonical average of Φ by

$$\langle\Phi\rangle_N^f = Z^{-1}\int_{\mathbb{R}^N} \Phi\left(u^1,\ldots,u^N\right)\exp\left(-\beta E_f\left(u^1,\ldots,u^N\right)\right)du^1\ldots du^N, \tag{13}$$

where the partition function Z reads

$$Z = \int_{\mathbb{R}^N} \exp\left(-\beta E_f\left(u^1,\ldots,u^N\right)\right)du^1\ldots du^N.$$

We assume in the sequel that $W(r)$ grows fast enough to ∞ when $|r| \to \infty$, so that Z is well defined (it is for instance enough that $W(r) \sim_{|r|\to\infty} |r|^\alpha$ with $\alpha > 1$).

We will be interested in the limit of $\langle\Phi\rangle_N^f$, when $N \to \infty$, and when Φ only depends on u^N: $\Phi(u^1,\ldots,u^N) = A(u^N)$ for a given function A.

Remark 1. In (13), we let the variables u^i vary on the whole real line. We do not constrain them to obey $u^{i-1} \le u^i$, which would encode the fact that nearest neighbours remain nearest neighbours. The argument provided here carries through when this constraint is accounted for: we just need to replace the interaction potential W by

$$W_c(y) = \begin{cases} W(y) & \text{when} \quad y \ge 0, \\ +\infty & \text{otherwise.} \end{cases}$$

2.1.1 Computing the Strain for a Given Stress

We first show a simple adaptation of [3, Theorem 1], which is useful to compute averages of general observables, in the thermodynamic limit, for the canonical ensemble at a fixed stress:

Lemma 1. *Assume that $A : \mathbb{R} \to \mathbb{R}$ is continuous, that for some $p \ge 1$, there exists a constant C such that*

$$\forall y \in \mathbb{R}, \quad |A(y)| \le C\left(1 + |y|^p\right),$$

and that

$$\int_{\mathbb{R}} (1 + |y|^p)\exp(-\beta\,[W(y) - fy])\,dy < +\infty.$$

Then

$$\lim_{N\to\infty} \langle A(u^N)\rangle_N^f = A\left(y^\star(f)\right),$$

with

$$y^\star(f) = \frac{\int_{\mathbb{R}} y\,\exp(-\beta\,[W(y) - fy])\,dy}{\int_{\mathbb{R}} \exp(-\beta\,[W(y) - fy])\,dy}. \tag{14}$$

Proof. We observe that

$$\langle A\rangle_N^f = Z^{-1}\int_{\mathbb{R}^N} A\left(u^N\right)\exp\left(-\beta E_f\left(u^1,\dots,u^N\right)\right)du^1\dots du^N$$
$$= Z^{-1}\int_{\mathbb{R}^N} A\left(u^N\right)\exp\left(-\beta\sum_{i=1}^N W_f\left(\frac{u^i-u^{i-1}}{h}\right)\right)du^1\dots du^N,$$

where $W_f(x) = W(x) - fx$. Introducing $y^i = \dfrac{u^i - u^{i-1}}{h}$, a change of variables in the above integral yields

$$\langle A\rangle_N^f = Z^{-1}\int_{\mathbb{R}^N} A\left(\frac{1}{N}\sum_{i=1}^N y^i\right)\exp\left(-\beta\sum_{i=1}^N W_f\left(y^i\right)\right)dy^1\dots dy^N,$$

where, with a slight abuse of notation,

$$Z = \int_{\mathbb{R}^N}\exp\left(-\beta\sum_{i=1}^N W_f\left(y^i\right)\right)dy^1\dots dy^N.$$

Consider now a sequence $\left\{Y^i\right\}_{i=1}^N$ of independent random variables, sharing the same law $z^{-1}\exp\left(-\beta W_f(y)\right)dy$ with

$$z = \int_{\mathbb{R}}\exp(-\beta W_f(y))\,dy.$$

It is clear that

$$\langle A\rangle_N^f = \mathbb{E}\left[A\left(\frac{1}{N}\sum_{i=1}^N Y^i\right)\right].$$

The law of large numbers readily yields that $\frac{1}{N}\sum_{i=1}^N Y^i$ converges almost surely to $y^\star(f)$ defined by (14).

We infer from [3, Theorem 1] that, for any force f, and for any observable A sufficiently smooth, the limit when $N\to\infty$ of $\langle A\rangle_N^f$ is

$$\lim_{N\to\infty}\langle A\rangle_N^f = A(y^\star(f)).$$

Rates of convergence are also provided in the same theorem. □

Numerical simulations illustrating this result are reported in [3, Sect. 2.3].

In the specific case of interest here, namely computing the stress-strain relation, we take $A(u^N) = u^N$, thus $\epsilon_N(f) := \langle A\rangle_N^f$ represents the average length of the chain, for a prescribed force f. We infer from the previous result that

$$\lim_{N\to\infty}\epsilon_N(f) = y^\star(f).$$

We hence have determined the macroscopic elongation, namely $y^\star(f)$, for a prescribed microscopic force f in the chain.

Notice that, in this specific case, A is a linear function, so we actually have $\epsilon_N(f) = y^\star(f)$ for any N. The result of Lemma 1 remains interesting for computing standard deviation of the average length, for example.

Remark 2. The force between atoms j and $j-1$ is $W'\left(\frac{u^j-u^{j-1}}{h}\right)$. Its canonical average, defined by (13), is

$$\begin{aligned}
\sigma_N^j &= Z^{-1}\int_{\mathbb{R}^N} W'\left(\frac{u^j-u^{j-1}}{h}\right)\exp\left(-\beta E_f\left(u^1,\dots,u^N\right)\right) du^1\dots du^N \\
&= Z^{-1}\int_{\mathbb{R}^N} W'(y^j)\exp\left(-\beta\sum_{i=1}^N\left[W(y^i)-fy^i\right]\right) dy^1\dots dy^N \\
&= \frac{\int_{\mathbb{R}} W'(y^j)\exp(-\beta\left[W(y^j)-fy^j\right])\, dy^j}{\int_{\mathbb{R}}\exp(-\beta\left[W(y^j)-fy^j\right])\, dy^j} \\
&= f + \frac{\int_{\mathbb{R}}\left[W'(y^j)-f\right]\exp(-\beta\left[W(y^j)-fy^j\right])\, dy^j}{\int_{\mathbb{R}}\exp(-\beta\left[W(y^j)-fy^j\right])\, dy^j},
\end{aligned}$$

where $y^j = \dfrac{u^j-u^{j-1}}{h}$. Integrating by parts, we see that the second term of the last line vanishes. We hence obtain that the average force between two consecutive atoms is independent of j (the stress is homogeneous in the material), and is equal to its prescribed microscopic value f:

$$\forall j,\ \forall N,\quad \sigma_N^j = f.$$

Imposing a force f on the right boundary atom hence implies that the average force between any two consecutive atoms is equal to f. ◇

2.1.2 Computing the Stress for a Given Strain

In the previous section, we have prescribed a force, and computed an average elongation. We now prescribe the length of the material, by imposing $u^0 = 0$ and $u^N = x$ (see Fig. 1).

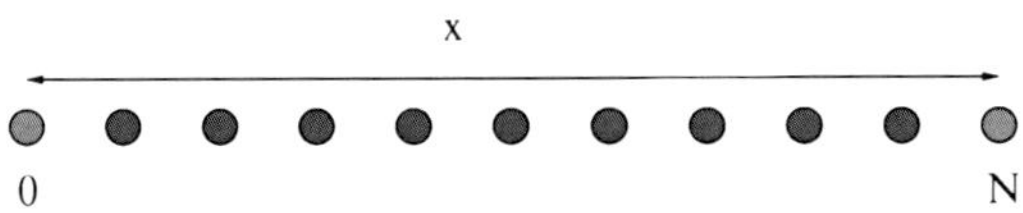

Fig. 1 One-dimensional chain of $1+N$ atoms, where the total length of the system is prescribed at the value x

As we fix the position of atom N, the system is insensitive to any force f imposed on that atom. We hence set $f = 0$. Our aim is to compute the force in the chain,

$$\mathscr{T}_N(x) = \frac{\displaystyle\int_{\mathbb{R}^{N-1}} W'\left(\frac{x-u^{N-1}}{h}\right) \exp\Big(-\beta E_0(u^1,\dots,u^{N-1},x)\Big)\, du^1 \dots du^{N-1}}{\displaystyle\int_{\mathbb{R}^{N-1}} \exp\Big(-\beta E_0(u^1,\dots,u^{N-1},x)\Big)\, du^1 \dots du^{N-1}}, \tag{15}$$

or, more precisely, its limit when $N \to \infty$. Note that, as all the $(u^i - u^{i-1})/h$ play the same role in the above expression, we also have, for any $1 \leq i \leq N-1$,

$$\mathscr{T}_N(x) = \frac{\displaystyle\int_{\mathbb{R}^{N-1}} W'\left(\frac{u^i-u^{i-1}}{h}\right) \exp\Big(-\beta E_0(u^1,\dots,u^{N-1},x)\Big)\, du^1 \dots du^{N-1}}{\displaystyle\int_{\mathbb{R}^{N-1}} \exp\Big(-\beta E_0(u^1,\dots,u^{N-1},x)\Big)\, du^1 \dots du^{N-1}}.$$

The force between atom N and $N-1$ is thus equal to the force between any two consecutive atoms.

We infer from (15) that $\mathscr{T}_N(x) = F'_N(x)$, where

$$F_N(x) = -\frac{1}{\beta N} \ln\left[\int_{\mathbb{R}^{N-1}} \exp\Big(-\beta E_0(u^1,\dots,u^{N-1},x)\Big)\, du^1 \dots du^{N-1}\right].$$

Hence NF_N is the free energy of the material associated to the reaction coordinate $\xi(u^1,\dots,u^N) = u^N$, and F_N is a rescaled free energy (free energy per integrated out particle). Using the variables $y^i = (u^i - u^{i-1})/h$, we also see that $\exp(-\beta N F_N(x))\,dx$ is (up to a normalizing multiplicative constant) the probability distribution of the random variable $\frac{1}{N}\sum_{i=1}^N Y^i$, when $\{Y^i\}_{i=1}^N$ is a sequence of independent random variables, sharing the same law $z^{-1}\exp(-\beta W(y))\,dy$, with

$$z = \int_{\mathbb{R}} \exp(-\beta W(y))\, dy.$$

In the case $W(y) = (y-a)^2/2$, it is possible to analytically compute $F_N(x)$, and to observe that there exists a constant C_N, independent of x, such that $F_N(x) + C_N$ has a finite limit when $N \to \infty$. In the general case, the limit of F_N is given by the following result, which relies on a large deviations result for i.i.d. random variables:

Lemma 2 ([3], Theorem 2). *Assume that the potential W satisfies*

$$\forall \xi \in \mathbb{R}, \quad \int_{\mathbb{R}} \exp(\xi y - \beta W(y))\, dy < +\infty,$$

and $\exp(-\beta W) \in H^1(\mathbb{R})$. *Then*

$$\lim_{N\to+\infty} \left(F_N(x) + \frac{1}{\beta}\ln\frac{z}{N}\right) = F_\infty(x), \tag{16}$$

with

$$F_\infty(x) := \frac{1}{\beta} \sup_{\xi \in \mathbb{R}} \left(\xi x - \ln \left[z^{-1} \int_{\mathbb{R}} \exp(\xi y - \beta W(y)) \, dy \right] \right) \tag{17}$$

and

$$z = \int_{\mathbb{R}} \exp(-\beta W(y)) \, dy.$$

This convergence holds pointwise in x*, and also in* $L^p_{\rm loc}$*, for any* $1 \leq p < \infty$*. As a consequence,* F'_N *converges to* F'_∞ *in* $W^{-1,p}_{\rm loc}$*.*

We hence obtain the macroscopic force $F'_\infty(x)$ for a prescribed elongation x. Numerical simulations that illustrate this result are reported in [3, Sect. 2.3].

Remark 3. The additive term $\beta^{-1} \ln(z/N)$ in (16) can be seen as a normalizing constant. Indeed, as mentioned above, $N F_N$ is a free energy, and the correct normalization for $\exp(-\beta N F_N)$ to be a probability density function is:

$$\int_{\mathbb{R}} \exp \left[-\beta N \left(F_N(x) + \frac{1}{\beta} \ln \frac{z}{N} \right) \right] dx = 1.$$

$\diamond$

Remark 4. F_N is a challenging quantity to compute. One possible method is to compute, for each x, its derivative $F'_N(x)$, and deduce F_N (this is the so-called thermodynamic integration method). Note that $F'_N(x) = \mathscr{T}_N(x)$ is given by (15): it is a canonical average of some observable, in a space of dimension $N - 1 \gg 1$. In contrast, F_∞ is easier to compute, since it only involves one-dimensional integrals or optimization problems. $\diamond$

2.1.3 Equivalence of Stress-Strain Relations in the Thermodynamic Limit

The function we maximize in (17) is concave, so there exists a unique maximizer $\xi(x)$ in (17), that satisfies the Euler-Lagrange equation

$$x = \frac{\int_{\mathbb{R}} y \, \exp(\xi(x) y - \beta W(y)) \, dy}{\int_{\mathbb{R}} \exp(\xi(x) y - \beta W(y)) \, dy}. \tag{18}$$

We observe that

$$F'_\infty(x) = \frac{\xi(x)}{\beta}.$$

On the other hand, recall the definition (14) of $y^\star(f)$:

$$y^\star(f) = \frac{\int_{\mathbb{R}} y \, \exp(-\beta [W(y) - f y]) \, dy}{\int_{\mathbb{R}} \exp(-\beta [W(y) - f y]) \, dy}.$$

Comparing (18) and (14), we see that $y^\star(\beta^{-1}\xi(x)) = y^\star(F'_\infty(x)) = x$. The function $f \mapsto y^\star(f)$ is increasing (because its derivative is positive), thus it is injective, and we also get the converse relation: $F'_\infty(y^\star(f)) = f$.

Otherwise stated, the relation $f \mapsto y^\star(f)$ and $x \mapsto F'_\infty(x)$ are inverse one to each other. So, prescribing a microscopic force f and computing the macroscopic elongation is equivalent to prescribing an elongation and computing the macroscopic force, *in the thermodynamic limit* (namely in the limit $N \to \infty$).

2.2 The Next-to-Nearest Neighbour (NNN) Case

We now consider next-to-nearest neighbour interactions in the chain. Again, the first atom is fixed: $u^0 = 0$, whereas the last one is submitted to an external force f. The (rescaled) energy reads

$$E_f\left(u^1,\dots,u^N\right) = \sum_{i=1}^{N} W_1\left(\frac{u^i - u^{i-1}}{h}\right) + \sum_{i=1}^{N-1} W_2\left(\frac{u^{i+1} - u^{i-1}}{h}\right) - f\frac{u^N}{h}. \quad (19)$$

If $W_2 \equiv 0$, this energy reduces to (12). Averages of observables are again defined by (13).

2.2.1 Computing the Strain for a Given Stress

Our aim, as in Sect. 2.1.1, is to compute the macroscopic strain, which is the average length of the material, that is

$$\epsilon_N(f) = \langle u^N \rangle_N^f,$$

where $\langle \cdot \rangle_N^f$ is the average with respect to the canonical measure associated to E_f. We introduce the notation

$$W_{1f}(x) = W_1(x) - fx,$$

which will be useful in the sequel. A simple adaptation of [3, Theorem 3] yields the following general result:

Lemma 3. *Assume that $A : \mathbb{R} \mapsto \mathbb{R}$ is continuous, and that there exists $p \geq 1$ and $C > 0$ such that*

$$|A(x)| \leq C(1 + |x|^p).$$

Assume also that W_{1f} and W_2 both belong to $L^1_{\rm loc}(\mathbb{R})$, that they are bounded from below, and that, for any $x \in \mathbb{R}$, we have $|W_{1f}(x)| < \infty$ and $|W_2(x)| < \infty$. In addition, we assume that $e^{-\beta W_{1f}}$ and $e^{-\beta W_2}$ both belong to $W^{1,1}_{\rm loc}(\mathbb{R})$, with

$$\int_{\mathbb{R}} (1+|x|^p)\, e^{-\beta W_{1f}(x)} dx < +\infty \quad and \quad \int_{\mathbb{R}} (1+|x|^p)\, e^{-\beta W_2(x)} dx < +\infty.$$

Then

$$\lim_{N\to\infty} \langle A(u^N)\rangle_N^f = A(y^\star(f)), \tag{20}$$

with

$$y^\star(f) = \int_{\mathbb{R}} y\, \psi_f^2(y)\, dy, \tag{21}$$

where ψ_f *solves the variational problem*

$$\lambda_f = \max_{\psi\in L^2(\mathbb{R})} \left\{ \int_{\mathbb{R}^2} \psi(y)\, \psi(z)\, K_f(y,z)\, dy\, dz;\ \int_{\mathbb{R}} \psi^2(y)\, dy = 1 \right\}, \tag{22}$$

with

$$K_f(x,y) := \exp\left[-\beta W_2(x+y) - \frac{\beta}{2} W_{1f}(x) - \frac{\beta}{2} W_{1f}(y) \right]. \tag{23}$$

We only provide here the main arguments to prove this result (see [3, Sec. 3.1.1 and Theorem 3] for details). They will be useful in the sequel. The observable $A(u^N)$ only depends on u^N, thus

$$\begin{aligned}\langle A(u^N)\rangle_N^f &= Z^{-1} \int_{\mathbb{R}^N} A\left(u^N\right) \exp\left(-\beta E_f\left(u^1,\dots,u^N\right)\right) du^1 \dots du^N \\ &= Z^{-1} \int_{\mathbb{R}^N} A\left(u^N\right) \exp\left(-\beta \sum_{i=1}^N W_{1f}\left(\frac{u^i - u^{i-1}}{h}\right) \right. \\ &\qquad \left. -\beta \sum_{i=1}^{N-1} W_2\left(\frac{u^{i+1}-u^{i-1}}{h}\right)\right) du^1 \dots du^N.\end{aligned}$$

Introducing again the variables $y^i = \dfrac{u^i - u^{i-1}}{h}$, we see that

$$\langle A(u^N)\rangle_N^f = Z^{-1} \int_{\mathbb{R}^N} A\left(\frac{1}{N}\sum_{i=1}^N y^i\right) \exp\left(-\beta W_{1f}(y^1)\right) \prod_{i=2}^N k_f(y^{i-1},y^i)\, dy^1 \dots dy^N, \tag{24}$$

with

$$k_f\left(y^{i-1},y^i\right) = \exp\left(-\beta W_{1f}\left(y^i\right) - \beta W_2\left(y^{i-1}+y^i\right)\right).$$

Assume for a moment that $\int_{\mathbb{R}} k_f(a,b)\, db = 1$. Then we see that

$$\langle A(u^N)\rangle_N^f = \mathbb{E}\left[A\left(\frac{1}{N}\sum_{i=1}^N Y^i\right)\right],$$

where $\left\{Y^i\right\}_{i=1}^N$ is a realization of a Markov chain of transition kernel k_f, and where Y^1 has the initial law (up to a normalization constant) $\exp\left(-\beta W_{1f}\left(y^1\right)\right) dy^1$. A law of large numbers argument, now for Markov chains, yields the large N limit of $\langle A(u^N)\rangle_N^f$ (recall that, in the case of the NN model considered in Sect. 2.1.1, this limit is given by a law of large numbers argument for i.i.d. sequences).

In general, of course, $\int_{\mathbb{R}} k_f(a,b)\,db \neq 1$. There is thus a slight technical difficulty in identifying a Markov chain structure in (24). It yet turns out that the above argument can be made rigorous as follows. Consider the variational problem (22), with K_f defined by (23). Under our assumptions, $K_f \in L^2(\mathbb{R}\times\mathbb{R})$. Using standard tools of spectral theory of self-adjoint operators (see *e.g.* [10]), one can prove that this problem has a maximizer (denoted ψ_f), and that, up to changing ψ_f in $-\psi_f$, the maximizer is unique. In addition, one can choose it such that $\psi_f > 0$. We can next define

$$g_f(x,y) := \frac{\psi_f(y)}{\lambda_f \psi_f(x)}\, K_f(x,y), \tag{25}$$

which satisfies

$$\int_{\mathbb{R}} g_f(y,z)\,dz = 1, \quad \int_{\mathbb{R}} \psi_f^2(y)\, g_f(y,z)\,dy = \psi_f^2(z).$$

The average (24) now reads

$$\begin{aligned}\langle A(u^N)\rangle_N^f = Z_g^{-1} \int_{\mathbb{R}^N} & A\left(\frac{1}{N}\sum_{i=1}^N y^i\right) \psi_f(y^1)\, e^{-\frac{\beta}{2}W_{1f}(y^1)} \\ & \times g_f(y^1,y^2)\ldots g_f(y^{N-1},y^N)\, \frac{e^{-\frac{\beta}{2}W_{1f}(y^N)}}{\psi_f(y^N)}\, dy^1\ldots dy^N,\end{aligned} \tag{26}$$

with

$$Z_g = \int_{\mathbb{R}^N} \psi_f(y^1) e^{-\frac{\beta}{2}W_{1f}(y^1)} g_f(y^1,y^2)\ldots g_f(y^{N-1},y^N) \frac{e^{-\frac{\beta}{2}W_{1f}(y^N)}}{\psi_f(y^N)} dy^1\ldots dy^N.$$

Thus

$$\langle A(u^N)\rangle_N^f = \mathbb{E}\left[A\left(\frac{1}{N}\sum_{i=1}^N Y^i\right)\right],$$

where $(Y^1,\ldots,Y^N)$ may now be seen as a realization of a *normalized* Markov chain of kernel g_f, with invariant probability measure ψ_f^2.

Under our assumptions, the Markov chain has a unique invariant measure, and satisfies a law of large numbers with respect to it. This yields the convergence (20). Numerical simulations illustrating this result are reported in [3, Sect. 3.1.3].

In the specific case of interest here, namely computing the stress-strain relation, we take $A(u^N) = u^N$, thus $\epsilon_N(f) := \langle A\rangle_N^f$ represents the average length of the chain, for a prescribed force f. We infer from the previous result that

$$\lim_{N\to\infty} \epsilon_N(f) = y^\star(f).$$

We hence have determined the macroscopic elongation, namely $y^\star(f)$, for a prescribed microscopic force f in the chain.

We conclude this section by showing the following result, which will be useful in the sequel.

Lemma 4. *Under the assumptions of Lemma 3, introduce the asymptotic variance $\sigma^2(f)$ defined by*

$$\sigma^2(f) = \int_{\mathbb{R}} (x - y^\star(f))^2 \, \psi_f^2(x)\, dx + 2 \sum_{i\geq 2} \mathbb{E}\left((\widetilde{Y}_i - y^\star(f))(\widetilde{Y}_1 - y^\star(f))\right), \tag{27}$$

where $\left(\widetilde{Y}_i\right)_{i\geq 1}$ is a Markov chain of transition kernel g_f, and of initial law ψ_f^2, the invariant measure.

Assume that $\sigma^2(f) \neq 0$ almost everywhere. Then the function $f \mapsto y^\star(f)$ is increasing.

Note that the right-hand side of (27) is exactly the variance appearing in the Central Limit Theorem for Markov chains [23, Theorem 17.0.1]. It is thus non-negative. More precisely, we have that

$$\lim_{N\to\infty} N \,\mathrm{Var}\left(\frac{1}{N}\sum_{i=1}^{N} \widetilde{Y}_i\right) = \sigma^2(f),$$

where $\left(\widetilde{Y}_i\right)_{i\geq 1}$ is the Markov chain defined in the above lemma.

Proof. Let $\epsilon_N(f) := \langle u^N \rangle_N^f$. An analytical computation shows that

$$D_N(f) := \frac{d\epsilon_N}{df}(f) = N\beta \left[\langle (u^N)^2 \rangle_N^f - \left(\langle u^N \rangle_N^f\right)^2 \right].$$

Thus the function $f \mapsto \epsilon_N(f)$ is non-decreasing. By Lemma 3, $y^\star(f)$ is the pointwise limit of $\epsilon_N(f)$: it is thus non-decreasing. It remains to prove that it is increasing.

Let us now compute the limit when $N \to \infty$ of $D_N(f)$. Using [3, Theorem 4], we see that

$$\lim_{N\to\infty} D_N(f) = \beta\sigma^2(f),$$

where $\sigma^2(f)$ is defined by (27).

Let us now fix τ and $\overline{\tau} \geq \tau$. Since $D_N(f) \geq 0$, we can use Fatou lemma, which yields that

$$\beta \int_\tau^{\overline{\tau}} \sigma^2(f) df = \int_\tau^{\overline{\tau}} \liminf D_N(f)\, df \leq \liminf \int_\tau^{\overline{\tau}} D_N(f)\, df = y^\star(\overline{\tau}) - y^\star(\tau).$$

As $\sigma^2(f) > 0$ almost everywhere, we thus obtain that $\tau \mapsto y^\star(\tau)$ is an increasing function. □

2.2.2 Computing the Stress for a Given Strain

We now prescribe the length of the material, by imposing $u^0 = 0$ and $u^N = x$. Our aim is to compute the average force in the chain,

$$\mathcal{T}_N(x) = \frac{\displaystyle\int_{\mathbb{R}^{N-1}} A_h\left(u^{N-1}, u^{N-2}; x\right) \exp\left(-\beta E_0\left(u^1, \dots, u^{N-1}, x\right)\right) du^1 \dots du^{N-1}}{\displaystyle\int_{\mathbb{R}^{N-1}} \exp\left(-\beta E_0\left(u^1, \dots, u^{N-1}, x\right)\right) du^1 \dots du^{N-1}}, \tag{28}$$

where E_0 is the energy (19) with $f = 0$, and where the observable A_h is the force at the end of the chain, which reads

$$A_h(u^{N-1}, u^{N-2}; x) = W_1'\left(\frac{x - u^{N-1}}{h}\right) + W_2'\left(\frac{x - u^{N-2}}{h}\right).$$

More precisely, we are interested in $\lim_{N\to\infty} \mathcal{T}_N(x)$.

As in Sect. 2.1.2, we see that $\mathcal{T}_N(x) = F_N'(x)$, with

$$F_N(x) = -\frac{1}{\beta N} \ln\left[\int_{\mathbb{R}^{N-1}} \exp\left(-\beta E_0(u^1, \dots, u^{N-1}, x)\right) du^1 \dots du^{N-1}\right]. \tag{29}$$

Again, NF_N is the free energy associated to the reaction coordinate $\xi(u^1, \dots, u^N) = u^N$, and F_N is a rescaled free energy (free energy per integrated out particle). In the NN case, we have computed the large N limit of $F_N(x)$ using a large deviations result for i.i.d. random variables. Comparing Sects. 2.1.1 and 2.2.1, we also see that moving from a NN setting to a NNN setting implies moving from a framework where random variables are i.i.d. to a framework where they are a realization of a Markov chain. It is hence natural to try and use a large deviations result for Markov chains to compute the large N limit of (29).

We now assume that the underlying Markov chain satisfies the following *pointwise* large deviations result:

Assumption 1 *Consider the Markov chain* $\{Y^i\}_{i\geq 1}$ *of kernel* $k \in L^2(\mathbb{R}\times\mathbb{R})$. *Assume that, for any* $\xi \in \mathbb{R}$, *the function* $\exp(\xi y) k(x,y) \in L^2(\mathbb{R}\times\mathbb{R})$.

Introduce the operator (on $L^2(\mathbb{R})$*)*

$$(Q_\xi \varphi)(y) = \int_{\mathbb{R}} \varphi(x) \exp(\xi y) k(x,y)\, dx$$

and assume that it has a simple and isolated largest eigenvalue $\Lambda(\xi)$, *and that* $\xi \mapsto \ln \Lambda(\xi)$ *is convex.*

Let $\exp(-N\overline{F}_N(x))\,dx$ *be the law of the random variable* $\frac{1}{N}\sum_{i=1}^{N} Y^i$. *We assume the large deviations principle*

$$\lim_{N\to+\infty} \overline{F}_N(x) = \overline{F}_\infty(x), \tag{30}$$

where

$$\overline{F}_\infty(x) := \sup_{\xi\in\mathbb{R}} (\xi x - \ln \Lambda(\xi)). \tag{31}$$

We moreover assume that the convergence (30) *holds pointwise in* x, *and also in* $L^p_{\rm loc}$, *for any* $1 \le p < \infty$. *As a consequence,* $\overline{F}'_N$ *converges to* $\overline{F}'_\infty$ *in* $W^{-1,p}_{\rm loc}$.

Note that similar results in a finite state Markov chain setting are reviewed in [9, pages 60–61] or [8, Sec. 3.1.1] (the continuous state case is addressed in *e.g.* [8, Secs. 6.3 and 6.5]). In the discrete state case, one can prove that $\xi \mapsto \ln \Lambda(\xi)$ is convex (see [9, Exercise V.14]). We will numerically check in the sequel that this assumption is indeed satisfied in the example we consider (see Fig. 2).

Remark 5. We have assumed that the operator Q_ξ has a simple and isolated largest eigenvalue. This can be proved for many kernels k, using for instance Krein-Rutman theorem [28]. In the case of interest in this contribution, we will use the specific expression of the kernel to transform the operator Q_ξ into a self-adjoint Hilbert-Schmidt operator on $L^2(\mathbb{R})$ (see Remark 7 below). We will thus be in position to work with self-adjoint compact operators. ◇

Remark 6. In the NN case, when $k(x,y) = \theta(y) = z^{-1}\exp(-\beta W(y))$, the sequence $\{Y^i\}_{i\ge 1}$ is a sequence of i.i.d. variables sharing the same law $\theta(y)\,dy$. The operator Q_ξ has a unique eigenvalue

$$\Lambda(\xi) = \int_{\mathbb{R}} \exp(\xi y)\,\theta(y)\,dy.$$

We then recover the large deviations result of i.i.d. sequence given in Lemma 2 (see also [12–14, 29]). ◇

We now wish to use Assumption 1 to compute the large N limit of (29). As pointed out in Sect. 2.2.1, there is a slight technical difficulty in identifying a Markov chain structure in the NNN setting, related to the normalization of the Markov chain kernel. We thus cannot readily use Assumption 1. We now detail how to overcome this difficulty.

Consider an observable A that depends only on u_N. In view of (29) and (26), its canonical average reads

$$\begin{aligned}\langle A\rangle_N &= Z^{-1}\int_{\mathbb{R}^N} A\left(u^N\right)\exp\left(-\beta E_0\left(u^1,\dots,u^{N-1},u^N\right)\right)\, du^1\dots du^{N-1}\,du^N \\ &= Z^{-1}\int_{\mathbb{R}} A(x)\exp(-\beta N F_N(x))\,dx \\ &= Z_g^{-1}\int_{\mathbb{R}^N} A\left(\frac{1}{N}\sum_{i=1}^N y^i\right)\psi_0(y^1)\,e^{-\frac{\beta}{2}W_1(y^1)} \\ &\qquad \times g_0(y^1,y^2)\dots g_0(y^{N-1},y^N)\,\frac{e^{-\frac{\beta}{2}W_1(y^N)}}{\psi_0(y^N)}\,dy^1\dots dy^N,\end{aligned}$$

where g_0 is defined by (25) and ψ_0 is the maximizer in (22), when the body force $f = 0$. Let $\mathscr{P}(y^1,\dots,y^N)$ be the probability density of a Markov chain $\{Y^i\}_{i=1}^N$ of kernel g_0, where the law of Y^1 is (up to a normalization constant) $\psi_0(y^1)\exp(-\beta W_1(y^1)/2)dy^1$. Then

$$\int_{\mathbb{R}} A(x)\exp(-\beta N F_N(x))\,dx = C_N\int_{\mathbb{R}^N} A\left(\frac{1}{N}\sum_{i=1}^N y^i\right)\mathscr{P}(y^1,\dots,y^N)\,r(y^N)\,dy^1\dots dy^N, \tag{32}$$

where C_N is a constant that does not depend on the observable A, and

$$r(y^N) = \frac{e^{-\frac{\beta}{2}W_1(y^N)}}{\psi_0(y^N)}.$$

Let now $\alpha_N(x,y^N)\,dx\,dy^N$ be the law of the couple $\left(\frac{1}{N}\sum_{i=1}^N Y^i, Y^N\right)$. We recast (32) as

$$\int_{\mathbb{R}} A(x)\exp(-\beta N F_N(x))\,dx = C_N\int_{\mathbb{R}^2} A(x)\,\alpha_N\left(x,y^N\right)r\left(y^N\right)dx\,dy^N.$$

As this relation holds for any observable A, with a constant C_N independent of A, we obtain

$$\exp(-\beta N F_N(x)) = C_N\int_{\mathbb{R}}\alpha_N\left(x,y^N\right)r\left(y^N\right)dy^N.$$

Assuming that r and $1/r$ are in $L^\infty(\mathbb{R})$, we have

$$C_N\|1/r\|_{L^\infty}^{-1}\int_{\mathbb{R}}\alpha_N\left(x,y^N\right)dy^N \le \exp(-\beta N F_N(x)) \le C_N\|r\|_{L^\infty}\int_{\mathbb{R}}\alpha_N\left(x,y^N\right)dy^N.$$

As a consequence, since the function r is independent of N,

$$\lim_{N\to\infty}(F_N(x)+D_N) = \lim_{N\to\infty}\left[-\frac{1}{\beta N}\ln\int_{\mathbb{R}}\alpha_N\left(x,y^N\right)dy^N\right], \tag{33}$$

where $D_N = \dfrac{1}{\beta N} \ln C_N$. Recall now that

$$\gamma_N(x) = \int_{\mathbb{R}} \alpha_N\left(x, y^N\right) dy^N$$

is the density of $\frac{1}{N}\sum_{i=1}^N Y^i$, where $\left\{Y^i\right\}_{i=1}^N$ is a realization of the Markov chain of kernel g_0. The behaviour of γ_N when $N \to \infty$ is given by Assumption 1:

$$\lim_{N\to+\infty} -\frac{1}{N} \ln \gamma_N(x) = \overline{F}_\infty(x), \tag{34}$$

where $\overline{F}_\infty$ is given by (31). Collecting (33) and (34), we hence obtain that

$$\lim_{N\to\infty} \left(F_N(x) + D_N\right) = \frac{1}{\beta}\overline{F}_\infty(x).$$

We thus have the following result:

Lemma 5. *Assume that W_1 and W_2 both belong to $L^1_{\mathrm{loc}}(\mathbb{R})$, that they are bounded from below, and that, for any $x \in \mathbb{R}$, we have $|W_1(x)| < \infty$ and $|W_2(x)| < \infty$. In addition, we assume that $e^{-\beta W_1}$ and $e^{-\beta W_2}$ both belong to $W^{1,1}_{\mathrm{loc}}(\mathbb{R})$, with*

$$\int_{\mathbb{R}} e^{-\beta W_1(x)} dx < +\infty \quad \textit{and} \quad \int_{\mathbb{R}} e^{-\beta W_2(x)} dx < +\infty,$$

and that, for any $\xi \in \mathbb{R}$, we have $\exp(\xi x - \beta W_1(x)) \in L^1(\mathbb{R})$.

Under Assumption 1 for the kernel g_0 defined by (25), we have that

$$\lim_{N\to+\infty} \left(F_N(x) + C_N\right) = F_\infty(x), \tag{35}$$

where F_N is defined by (29), C_N is a constant that does not depend on x, and F_∞ is given by the Legendre transform

$$F_\infty(x) := \frac{1}{\beta} \sup_{\xi\in\mathbb{R}} \left(\xi x - \ln \Lambda(\xi)\right), \tag{36}$$

where $\Lambda(\xi)$ is the largest eigenvalue of the operator (defined on $L^2(\mathbb{R})$)

$$(Q_\xi \varphi)(y) = \int_{\mathbb{R}} \varphi(x) \exp(\xi y)\, g_0(x, y)\, dx. \tag{37}$$

The convergence (35) holds pointwise in x, and also in L^p_{loc}, for any $1 \le p < \infty$. As a consequence, the macroscopic force in the chain $\mathcal{T}_N(x) = F'_N(x)$ converges to F'_∞ in $W^{-1,p}_{\mathrm{loc}}$.

We hence obtain the macroscopic force $F'_\infty(x)$ for a prescribed elongation x. Note that, under our assumptions, in view of its definition (36), F_∞ is (up to the factor β) the Legendre transform of some function. It is hence always a convex function. Thus, as in the zero temperature case, we observe, in this one-dimensional setting, that the macroscopic constitutive law $x \mapsto F_\infty(x)$ is a convex function.

Remark 7. In view of the definitions (25) of g_0 and (23) of K_0, we see that

$$\frac{(Q_\xi \varphi)(y)}{\psi_0(y)} = \frac{1}{\lambda_0} \int_{\mathbb{R}} \frac{\varphi(x)}{\psi_0(x)} \exp(\xi y) \, K_0(x,y) \, dx.$$

Thus $\Lambda(\xi)$ is also the largest eigenvalue of the operator

$$(\widetilde{Q}_\xi \varphi)(y) = \frac{1}{\lambda_0} \int_{\mathbb{R}} \varphi(x) \exp(\xi y) \, K_0(x,y) \, dx.$$

Furthermore, if λ is an eigenvalue of $\widetilde{Q}_\xi$, then

$$\int_{\mathbb{R}} \varphi(x) \exp(\xi y) \, K_0(x,y) \, dx = \lambda_0 \lambda \varphi(y),$$

where φ is an associated eigenfunction. Thus

$$\int_{\mathbb{R}} \frac{\varphi(x)}{\exp(\xi x/2)} \exp(\xi y/2) \exp(\xi x/2) \, K_0(x,y) \, dx = \lambda_0 \lambda \frac{\varphi(y)}{\exp(\xi y/2)}$$

and $\lambda_0 \lambda$ is an eigenvalue of the operator

$$(\overline{Q}_\xi \varphi)(y) = \int_{\mathbb{R}} \varphi(x) \exp(\xi y/2) \exp(\xi x/2) \, K_0(x,y) \, dx.$$

The converse is also true. As $\Lambda(\xi)$ is the largest eigenvalue of the operator $\widetilde{Q}_\xi$, we have that $\lambda_0 \Lambda(\xi)$ is the largest eigenvalue of the operator $\overline{Q}_\xi$.

As W_2 is bounded from below and $\exp(\xi x - \beta W_1(x)) \in L^1(\mathbb{R})$ for any ξ, we have that $\exp(\xi x/2) \exp(\xi y/2) K_0(x,y) \in L^2(\mathbb{R} \times \mathbb{R})$. Hence $\overline{Q}_\xi$ is a self-adjoint compact operator on $L^2(\mathbb{R})$, which is thus easier to manipulate theoretically and numerically than Q_ξ. In particular, using standard tools of spectral theory of self-adjoint operators (see *e.g.* [10]), one can prove that the largest eigenvalue of $\overline{Q}_\xi$ is simple, and that the associated eigenvector Ψ_ξ (which is unique up to a multiplicative constant) can be chosen such that $\Psi_\xi > 0$. ◇

2.2.3 Equivalence of Stress-Strain Relations in the Thermodynamic Limit

In Sect. 2.2.1, we have identified the function $f \mapsto y^\star(f)$, that associates to a prescribed force f the macroscopic elongation $y^\star(f)$. Next, in Sect. 2.2.2, we have identified the function $x \mapsto F'_\infty(x)$, that associates to a prescribed elongation x the macroscopic force $F'_\infty(x)$. We show now that these functions are reciprocal one to each other.

Consider the optimization problem (36). Since the function $\xi \mapsto \ln \Lambda(\xi)$ is convex (see Assumption 1), there exists a unique maximizer $\xi(x)$ in (36), which satisfies the Euler-Lagrange equation

$$x = \frac{\Lambda'(\xi(x))}{\Lambda(\xi(x))}. \tag{38}$$

We also observe that

$$F'_\infty(x) = \frac{\xi(x)}{\beta}.$$

We see from (38) that we need to compute $\Lambda'(\xi)$. Recall that $\Lambda(\xi)$ is the largest eigenvalue of the operator (37). In view of Remark 7, $\lambda_0 \Lambda(\xi)$ is also the largest eigenvalue of $\overline{Q}_\xi$. Denoting Ψ_ξ the associated eigenfunction satisfying $\|\Psi_\xi\|_{L^2} = 1$ and $\Psi_\xi > 0$, we thus have

$$(\overline{Q}_\xi \Psi_\xi)(y) = \int_{\mathbb{R}} \Psi_\xi(t)\, K_0^\xi(t,y)\, dt = \lambda_0 \Lambda(\xi) \Psi_\xi(y),$$

where

$$K_0^\xi(t,y) = \exp(\xi y/2) \exp(\xi t/2)\, K_0(t,y). \tag{39}$$

Multiplying by $\Psi_\xi(y)$ and integrating, we obtain

$$\int_{\mathbb{R}^2} \Psi_\xi(y)\, \Psi_\xi(t)\, K_0^\xi(t,y)\, dt\, dy = \lambda_0 \Lambda(\xi). \tag{40}$$

We thus have, using that $K_0^\xi(t,y) = K_0^\xi(y,t)$, that

$$\begin{aligned}
\lambda_0 \Lambda'(\xi) &= \int_{\mathbb{R}^2} \frac{d\Psi_\xi}{d\xi}(y)\, \Psi_\xi(t)\, K_0^\xi(t,y)\, dt\, dy + \int_{\mathbb{R}^2} \Psi_\xi(y) \frac{d\Psi_\xi}{d\xi}(t)\, K_0^\xi(t,y)\, dt\, dy \\
&\quad + \int_{\mathbb{R}^2} \Psi_\xi(y)\, \Psi_\xi(t) \frac{dK_0^\xi}{d\xi}(t,y)\, dt\, dy \\
&= 2\lambda_0 \Lambda(\xi) \int_{\mathbb{R}} \frac{d\Psi_\xi}{d\xi}(y)\, \Psi_\xi(y)\, dy + \int_{\mathbb{R}^2} \Psi_\xi(y)\, \Psi_\xi(t) \frac{dK_0^\xi}{d\xi}(t,y)\, dt\, dy.
\end{aligned}$$

In the above expression, the first term vanishes, since, for any ξ, $\displaystyle\int_{\mathbb{R}} \Psi_\xi^2(y)\, dy = 1$. We thus obtain

$$\lambda_0 \Lambda'(\xi) = \int_{\mathbb{R}^2} \Psi_\xi(y)\, \Psi_\xi(t) \frac{t+y}{2}\, K_0^\xi(t,y)\, dt\, dy. \tag{41}$$

Collecting (38), (40) and (41), we see that

$$
\begin{aligned}
x &= \frac{\displaystyle\int_{\mathbb{R}^2} \Psi_{\xi(x)}(y)\, \Psi_{\xi(x)}(t)\, \frac{t+y}{2}\, K_0^{\xi(x)}(t,y)\, dt\, dy}{\displaystyle\int_{\mathbb{R}^2} \Psi_{\xi(x)}(y)\, \Psi_{\xi(x)}(t)\, K_0^{\xi(x)}(t,y)\, dt\, dy} \\
&= \frac{\displaystyle\int_{\mathbb{R}^2} y\; \Psi_{\xi(x)}(y)\, \Psi_{\xi(x)}(t)\, K_0^{\xi(x)}(t,y)\, dt\, dy}{\displaystyle\int_{\mathbb{R}^2} \Psi_{\xi(x)}(y)\, \Psi_{\xi(x)}(t)\, K_0^{\xi(x)}(t,y)\, dt\, dy} \\
&= \frac{\displaystyle\int_{\mathbb{R}} y\; \Psi^2_{\xi(x)}(y)\, dy}{\displaystyle\int_{\mathbb{R}} \Psi^2_{\xi(x)}(y)\, dy} \\
&= \int_{\mathbb{R}} y\; \Psi^2_{\xi(x)}(y)\, dy,
\end{aligned}
\tag{42}
$$

where we have used, at the second line, that $K_0^{\xi(x)}(t,y) = K_0^{\xi(x)}(y,t)$.

On the other hand, we have obtained that the macroscopic elongation $y^\star(f)$, for a prescribed force f, is given by (21), namely

$$
y^\star(f) = \int_{\mathbb{R}} y\; \psi_f^2(y)\, dy,
\tag{43}
$$

where ψ_f is the maximizer of the variational problem (22). As K_f is symmetric, the Euler-Lagrange equation of (22) reads

$$
\begin{aligned}
\lambda_f \psi_f(y) &= \int_{\mathbb{R}} \psi_f(t)\; K_f(t,y)\; dt \\
&= \int_{\mathbb{R}} \psi_f(t)\; K_0(t,y) \exp\left(\beta f \frac{x+y}{2}\right)\; dt \\
&= \int_{\mathbb{R}} \psi_f(t)\; K_0^{\beta f}(t,y)\, dt,
\end{aligned}
$$

where $K_0^{\beta f}$ is defined by (39). Thus ψ_f is an eigenfunction associated to the largest eigenvalue λ_f of the Hilbert-Schmidt operator $\overline{Q}_{\beta f}$ of kernel $K_0^{\beta f}$. By definition of $\Psi_{\beta f}$, and using the fact that the largest eigenvalue of $\overline{Q}_{\beta f}$ is simple, we obtain

$$
\Psi_{\beta f} = \pm \psi_f \quad \text{and} \quad \Lambda(\beta f) = \frac{\lambda_f}{\lambda_0}.
$$

We thus recast (43) as

$$
y^\star(f) = \int_{\mathbb{R}} y\; \Psi_{\beta f}^2(y)\, dy.
\tag{44}
$$

We deduce from the comparison of (42) and (44) that $y^\star(\beta^{-1}\xi(x)) = y^\star(F'_\infty(x)) = x$. Recall now that the function $f \mapsto y^\star(f)$ is increasing, as shown by Lemma 4. It is thus injective, and we also get the converse relation $F'_\infty(y^\star(f)) = f$.

As a consequence, as in the NN setting considered in Sect. 2.1.3, the relation $f \mapsto y^\star(f)$ and $x \mapsto F'_\infty(x)$ are inverse one to each other. Prescribing a microscopic force f and computing the macroscopic elongation is equivalent to prescribing an elongation and computing the macroscopic force, *in the thermodynamic limit.*

2.2.4 Numerical Computation of F'_∞ and Comparison with the Zero Temperature Model

For our numerical tests, we follow the choices made in [3], for the sake of comparison. We thus take the pair interaction potentials

$$W_1(x) = \frac{1}{2}(x-1)^4 + \frac{1}{2}x^2 \quad \text{and} \quad W_2(x) = \frac{1}{4}(x-2.1)^4.$$

Note that these potentials satisfy all the assumptions that we have made above.

We are going to compare the free energy derivative $\mathscr{T}_N(x) = F'_N(x)$ with its thermodynamic limit approximation $F'_\infty(x)$. The reference value $F'_N(x)$ is computed as the ensemble average (28), which is in turn computed as a long-time average along the lines of (3)–(4). To compute $F'_\infty(x)$, we proceed as follows:

(i) We first compute the largest eigenvalue $\Lambda(\xi)$ of the operator (37), for all ξ in some prescribed interval.
(ii) For any fixed x in a prescribed interval, we next consider the variational problem (36), compute its maximizer $\xi(x)$, and obtain $F'_\infty(x)$ using $F'_\infty(x) = \xi(x)/\beta$.

In practice, using Remark 7, we work with the operator $\overline{Q}_\xi$, which is easier to manipulate since it is self-adjoint and we do not need to first solve (22). We thus first compute the largest eigenvalue $\lambda_0\Lambda(\xi)$ of $\overline{Q}_\xi$, and next compute the Legendre transform of the function $\xi \mapsto \ln(\lambda_0\Lambda(\xi))$. The maximizer is the same as that for $F_\infty(x)$. On Fig. 2, we plot the function $\xi \mapsto \ln(\lambda_0\Lambda(\xi))$, and observe that it is convex, in agreement with Assumption 1.

We first study the convergence of $F'_N(x)$ to $F'_\infty(x)$ as N increases, for a fixed chain length $x = 1.4$ and a fixed temperature $1/\beta = 1$. Results are shown on Fig. 3. We indeed observe that $F'_N(x) \to F'_\infty(x)$ when $N \to +\infty$.

We now compare $F'_N(x)$ with its approximation $F'_\infty(x)$, for $N = 100$ and $1/\beta = 1$. Results are shown on Fig. 4. We observe that $F'_\infty(x)$ is a very good approximation of $F'_N(x)$, for any x in the considered interval.

For the sake of comparison, we now identify the zero temperature behaviour of the system, in the thermodynamic limit. At zero temperature, for a finite N, we model the system by minimizing the energy E_0, with prescribed Dirichlet boundary conditions (this corresponds to prescribing the elongation, and computing the force;

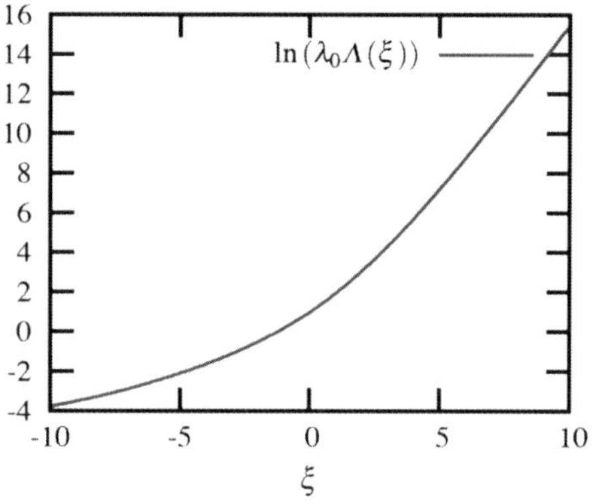

Fig. 2 Plot of $\ln(\lambda_0 \Lambda(\xi))$ as a function of ξ (temperature $1/\beta = 1$)

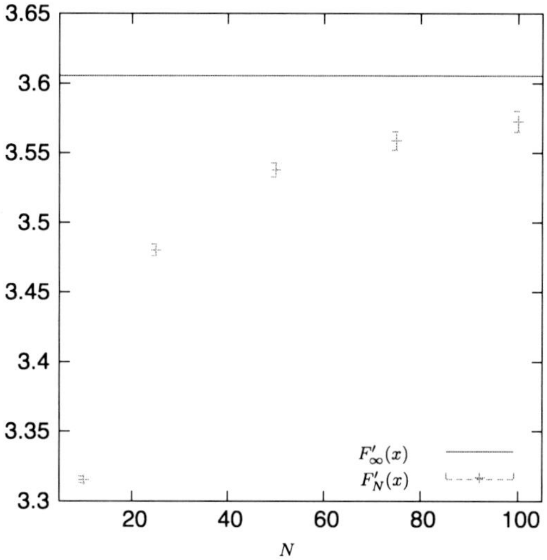

Fig. 3 Convergence of $F'_N(x)$ (shown with error bars computed from 40 independent realizations) to $F'_\infty(x)$ as N increases (temperature $1/\beta = 1$, fixed chain length $x = 1.4$)

alternatively, one could impose Neumann boundary conditions, *i.e.* prescribe a force and compute an elongation):

$$J_N(x) = \frac{1}{N} \inf\left\{E_0\left(u^0, u^1, \dots, u^{N-1}, u^N\right),\ u^0 = 0,\ u^N = x\right\}. \tag{45}$$

We have the following result, which proof will be given below:

Lemma 6. *Let us introduce ϕ defined by*

$$\phi(x) = W_1(x) + W_2(2x). \tag{46}$$

Assume that there exists $\alpha > 0$ such that

$$W_1(x) \geq \alpha x^2, \tag{47}$$

and that W_1 and ϕ are non-negative and strictly convex functions. Then we have the pointwise convergence

$$\lim_{N \to \infty} J_N(x) = \phi(x).$$

Assume in addition that $\phi \in L^p_{\mathrm{loc}}$ for some $1 \leq p < \infty$ and that W_2 is non-negative. Then the above convergence also holds in L^p_{loc}. As a consequence, $J'_N(x)$ converges to $\phi'(x)$ in $W^{-1,p}_{\mathrm{loc}}$.

When the temperature is set to zero, the energy thus converges, in the thermodynamic limit, to $\phi(x)$, and the force (*i.e.* the derivative of the energy with respect to the prescribed Dirichlet boundary condition) converges to $\phi'(x)$. We plot on Fig. 4 the function $x \mapsto \phi'(x)$. We clearly observe the effect of temperature, as $F'_\infty(x)$ for $\beta = 1$ significantly differs from $\phi'(x)$.

Proof (Lemma 6). Let

$$X_N(x) = \left\{\left(u^0, u^1, \dots, u^{N-1}, u^N\right) \in \mathbb{R}^{1+N},\ u^0 = 0,\ u^N = x\right\}$$

be the variational ensemble for the problem (45). The configuration $u^i = ix/N$ clearly belongs to that ensemble. We thus obtain the upper-bound

$$J_N(x) \leq W_1(x) + \frac{N-1}{N} W_2(2x). \tag{48}$$

In the sequel, we first show a lower-bound for $J_N(x)$, and next study its behaviour when $N \to \infty$.

Let us first build a lower bound for $J_N(x)$. Assuming for the sake of simplicity that N is even, and using the short-hand notation $y^i = (u^i - u^{i-1})/h$, we have

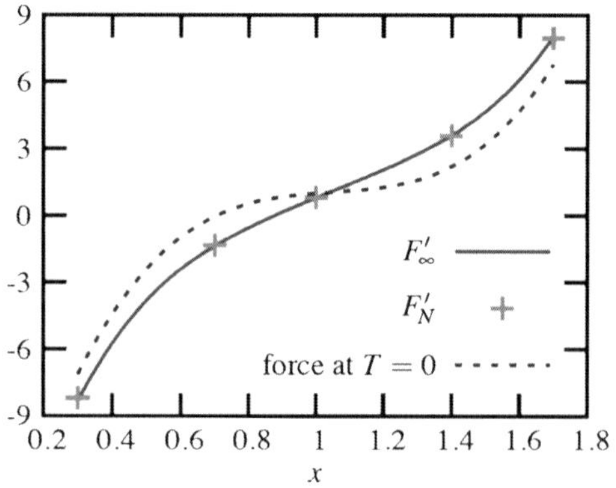

Fig. 4 We plot $F'_N(x)$ and $F'_\infty(x)$ for the temperature $1/\beta = 1$ and $N = 100$. On the scale of the figure, $F'_N(x)$ and $F'_\infty(x)$ are on top of each other. We also plot the zero temperature response $\phi'(x)$

$$\begin{aligned}\frac{1}{N}\sum_{i=1}^{N} W_1\left(\frac{u^i - u^{i-1}}{h}\right) &= \frac{1}{N}\sum_{i=1}^{N} W_1\left(y^i\right)\\ &= \frac{1}{2N} W_1(y^1) + \frac{1}{2N} W_1(y^N)\\ &\quad + \frac{1}{2N}\sum_{i=1}^{N/2}\left[W_1\left(y^{2i-1}\right) + W_1\left(y^{2i}\right)\right]\\ &\quad + \frac{1}{2N}\sum_{i=1}^{N/2-1}\left[W_1\left(y^{2i}\right) + W_1\left(y^{2i+1}\right)\right].\end{aligned}$$

By convexity of W_1, we obtain

$$\begin{aligned}\frac{1}{N}\sum_{i=1}^{N} W_1\left(\frac{u^i - u^{i-1}}{h}\right) &\geq \frac{1}{2N} W_1(y^1) + \frac{1}{2N} W_1(y^N)\\ &\quad + \frac{1}{N}\sum_{i=1}^{N/2} W_1\left[\frac{1}{2}\left(y^{2i-1} + y^{2i}\right)\right]\\ &\quad + \frac{1}{N}\sum_{i=1}^{N/2-1} W_1\left[\frac{1}{2}\left(y^{2i} + y^{2i+1}\right)\right].\end{aligned}$$

Taking into account the next-to-nearest interactions, we thus obtain that, for any $\left(u^0, u^1, \ldots, u^{N-1}, u^N\right) \in \mathbb{R}^{1+N}$,

$$\frac{1}{N}E_0\left(u^0,u^1,\ldots,u^{N-1},u^N\right) \geq \frac{1}{2N}W_1(y^1)+\frac{1}{2N}W_1(y^N) +\frac{1}{N}\sum_{i=1}^{N/2}\phi\left[\frac{1}{2}\left(y^{2i-1}+y^{2i}\right)\right] +\frac{1}{N}\sum_{i=1}^{N/2-1}\phi\left[\frac{1}{2}\left(y^{2i}+y^{2i+1}\right)\right],$$

where ϕ is defined by (46). As ϕ is convex, we deduce that

$$\begin{aligned}\frac{1}{N}E_0(u) \geq{}& \frac{1}{2N}W_1(y^1)+\frac{1}{2N}W_1(y^N)+\frac{1}{2}\phi\left(\frac{1}{N}\sum_{i=1}^{N/2}\left[y^{2i-1}+y^{2i}\right]\right)\\ &+\frac{N-2}{2N}\phi\left(\frac{1}{N-2}\sum_{i=1}^{N/2-1}\left[y^{2i}+y^{2i+1}\right]\right)\\ ={}& \frac{1}{2N}W_1(y^1)+\frac{1}{2N}W_1(y^N)+\frac{1}{2}\phi\left(u^N-u^0\right)\\ &+\frac{N-2}{2N}\phi\left(\frac{N}{N-2}\left(u^{N-1}-u^1\right)\right).\end{aligned}$$

As a consequence, for any configuration $u \in X_N(x)$, we have

$$\frac{1}{N}E_0\left(u^0,u^1,\ldots,u^{N-1},u^N\right) \geq \overline{E}_N(u^1,u^{N-1};x), \tag{49}$$

with

$$\overline{E}_N(u^1,u^{N-1};x) = \frac{1}{2N}W_1(Nu^1)+\frac{1}{2N}W_1(N(x-u^{N-1}))+\frac{1}{2}\phi(x) +\frac{N-2}{2N}\phi\left(\frac{N}{N-2}\left(u^{N-1}-u^1\right)\right). \tag{50}$$

We infer from (49) the lower bound

$$J_N(x) \geq \overline{J}_N(x), \tag{51}$$

with

$$\overline{J}_N(x) = \inf\left\{\overline{E}_N(u^1,u^{N-1};x);\ u^1 \in \mathbb{R},\ u^{N-1} \in \mathbb{R}\right\}. \tag{52}$$

We now study the auxiliary variational problem (52) to determine the limit of $\overline{J}_N(x)$ when $N \to \infty$. Since ϕ is non-negative, and using (47), we have that

$$\overline{E}_N(u^1,u^{N-1};x) \geq \frac{\alpha N}{2}\left[(u^1)^2+(x-u^{N-1})^2\right] \geq 0.$$

As a consequence, $\overline{J}_N(x) \geq 0$, and any minimizing sequence is bounded (by a constant possibly depending on N). Up to extraction, it thus converges to a minimizer, that we denote $\left(\overline{u}^1,\overline{u}^{N-1}\right)$. As W_1 and ϕ are strictly convex, it is easy to see that the hessian matrix of $\overline{E}_N$ is positive definite, hence $\overline{E}_N$ is also strictly convex, hence it has a unique minimizer. The problem (52) is thus well-posed. To underline the dependency of its minimizer with N, we denote it $\left(\overline{u}^1(N),\overline{u}^{N-1}(N)\right)$ in the sequel.

The Euler-Lagrange equation associated to (52) reads

$$W_1'(N\overline{u}^1(N)) = \phi'\left(\frac{N}{N-2}\left(\overline{u}^{N-1}(N)-\overline{u}^1(N)\right)\right) = W_1'(N(x-\overline{u}^{N-1}(N))).$$

As W_1 is strictly convex, this implies that

$$\begin{cases} \overline{u}^1(N) = x-\overline{u}^{N-1}(N), \\ N\overline{u}^1(N) = \chi\left(\frac{N}{N-2}\left(x-2\overline{u}^1(N)\right)\right), \end{cases} \tag{53}$$

where the function $\chi = (W_1')^{-1}\circ\phi'$ is independent of N, and increasing.

Let us now show that $\overline{u}^1(N)$ is bounded with respect to N. If this is not the case, then, without loss of generality, it is possible to find a subsequence $\varphi(N)$ such that $\lim_{N\to\infty}\overline{u}^1(\varphi(N)) = +\infty$. Passing to the limit in the second line of (53), one obtains a contradiction. Thus $\overline{u}^1(N)$ is bounded.

In view of the first line of (53), $\overline{u}^{N-1}(N)$ is also bounded. Up to a subsequence extraction, $(\overline{u}^1(N),\overline{u}^{N-1}(N))$ converges when $N\to\infty$ to some (a,b). We infer from (53) that $a=0$ and $b=x$, thus the limit is unique, and the whole sequence converges:

$$\lim_{N\to\infty}\overline{u}^1(N) = 0, \qquad \lim_{N\to\infty}\overline{u}^{N-1}(N) = x. \tag{54}$$

We next infer from the above limits and (53) that

$$\lim_{N\to\infty} N\overline{u}^1(N) = \lim_{N\to\infty} N(x-\overline{u}^{N-1}(N)) = \chi(x). \tag{55}$$

By definition, we have

$$\overline{J}_N(x) = \inf\left\{\overline{E}_N(u^1,u^{N-1};x);\ u^1\in\mathbb{R},\ u^{N-1}\in\mathbb{R}\right\} = \overline{E}_N(\overline{u}^1(N),\overline{u}^{N-1}(N);x).$$

In view of (50), (54) and (55), we obtain

$$\lim_{N\to\infty}\overline{J}_N(x) = \lim_{N\to\infty}\overline{E}_N(\overline{u}^1(N),\overline{u}^{N-1}(N);x) = \phi(x). \tag{56}$$

Collecting (48), (51) and (56), we obtain the claimed pointwise convergence of $J_N(x)$ to $\phi(x)$.

We now turn to the second assertion of Lemma 6. Under the additional assumption that W_2 is non-negative, we deduce from (48) that, for any N and any x,

$$0 \le J_N(x) \le W_1(x) + W_2(2x) = \phi(x).$$

As $\phi\in L^p_{\rm loc}$, we obtain the convergence of J_N to ϕ in $L^p_{\rm loc}$. □

3 A Coarse-Graining Procedure in the Dynamical Setting

In this section, we present a procedure for coarse-graining a dynamics. More precisely, we consider $Q_t \in \mathbb{R}^n$ solution to the overdamped dynamics (3), and a reaction coordinate $\xi : \mathbb{R}^n \mapsto \mathbb{R}$. Our aim is to find a closed one-dimensional dynamics of type (5) on a process $\overline{\eta}_t$, such that $\overline{\eta}_t$ is a good approximation of $\xi(Q_t)$. In Sects. 3.2 and 3.3, we build such a process (see (67) below), and present an analytical estimation of its accuracy (the obtained estimate is an upper-bound on the "distance" between the laws of $\xi(Q_t)$ and $\overline{\eta}_t$ at any time t). We will next report on some numerical experiments that somewhat check the accuracy of $\overline{\eta}_t$ in a stronger sense (Sect. 3.4).

3.1 Measuring Distances Between Probability Measures

We introduce here some tools that will be useful in the sequel, to measure how close two probability measures are. Consider two probability measures $\nu(dq)$ and $\eta(dq)$. The distance between the two can be measured by the total variation norm $\|\nu - \eta\|_{\mathrm{TV}}$, which amounts to the L^1-norm $\int \left|\psi_\nu(q) - \psi_\eta(q)\right| dq$ in case ν and η have respectively the densities ψ_ν and ψ_η with respect to the Lebesgue measure.

When studying the long-time behaviour of solutions to PDEs (such as long time convergence of the solution of a Fokker-Planck equation to the stationary measure of the corresponding SDE), the notion of relative entropy turns out to be more useful. Under the assumption that ν is absolutely continuous with respect to η (denoted $\nu \ll \eta$ in the sequel), it is defined by

$$H(\nu|\eta) = \int \ln\left(\frac{d\nu}{d\eta}\right) d\nu.$$

The relative entropy provides an upper-bound on the total variation norm, by the Csiszár-Kullback inequality [1]:

$$\|\nu - \eta\|_{\mathrm{TV}} \leq \sqrt{2H(\nu|\eta)}.$$

In the sequel, we will also use the Wasserstein distance with quadratic cost, which is another way to measure distances between probability measures. It is defined, for any two probability measures ν and η with support on a Riemannian manifold Σ, by

$$W(\nu,\eta) = \sqrt{\inf_{\pi \in \Pi(\nu,\eta)} \int_{\Sigma \times \Sigma} d_\Sigma(x,y)^2 \, \pi(dx,dy)}. \tag{57}$$

In the above expression, $d_\Sigma(x,y)$ denotes the geodesic distance between x and y on Σ,

$$d_\Sigma(x,y) = \inf\left\{ \sqrt{\int_0^1 |\dot{\alpha}(t)|^2\,dt};\ \alpha \in C^1([0,1],\Sigma),\ \alpha(0)=x,\ \alpha(1)=y \right\},$$

and $\Pi(\nu,\eta)$ denotes the set of coupling probability measures, that is probability measures π on $\Sigma \times \Sigma$ such that their marginals are ν and η: for any test function Φ,

$$\int_{\Sigma\times\Sigma} \Phi(x)\pi(dx,dy) = \int_\Sigma \Phi(x)\nu(dx) \text{ and } \int_{\Sigma\times\Sigma} \Phi(y)\pi(dx,dy) = \int_\Sigma \Phi(y)\eta(dy).$$

In the sequel, we will need two functional inequalities, that we now recall [1]:

Definition 1. A probability measure η satisfies a logarithmic Sobolev inequality with a constant $\rho > 0$ if, for any probability measure ν such that $\nu \ll \eta$,

$$H(\nu|\eta) \leq \frac{1}{2\rho} I(\nu|\eta),$$

where the Fisher information $I(\nu|\eta)$ is defined by

$$I(\nu|\eta) = \int \left| \nabla \ln\left(\frac{d\nu}{d\eta}\right)\right|^2 d\nu.$$

Definition 2. A probability measure η satisfies a Talagrand inequality with a constant $\rho > 0$ if, for any probability measure ν,

$$W(\nu,\eta) \leq \sqrt{\frac{2}{\rho} H(\nu|\eta)}.$$

We will also need the following important result (see [24, Theorem 1] and [4]):

Lemma 7. *If η satisfies a logarithmic Sobolev inequality with a constant $\rho > 0$, then η satisfies a Talagrand inequality with the same constant $\rho > 0$.*

The following standard result illustrates the usefulness of logarithmic Sobolev inequalities (we refer to [1, 2, 30] for more details on this subject).

Theorem 1. *Consider Q_t solution to the overdamped Langevin equation (3), and assume the stationary measure $\psi_\infty(q)\,dq = Z^{-1}\exp(-\beta V(q))\,dq$ satisfies a logarithmic Sobolev inequality with a constant $\rho > 0$. Then the probability distribution $\psi(t,\cdot)$ of Q_t converges to ψ_∞ exponentially fast, in the sense:*

$$\forall t \geq 0, \quad H(\psi(t,\cdot)|\psi_\infty) \leq H(\psi(0,\cdot)|\psi_\infty)\exp(-2\rho\beta^{-1}t). \tag{58}$$

Conversely, if (58) holds for any initial condition $\psi(0,\cdot)$, then the stationary measure $\psi_\infty(q)\,dq$ satisfies a logarithmic Sobolev inequality with a constant $\rho > 0$.

Proof. The probability distribution function $\psi(t,q)$ of Q_t satisfies the Fokker-Planck equation

$$\partial_t \psi = div(\psi \nabla V) + \beta^{-1} \Delta \psi. \tag{59}$$

As $\nabla \psi_\infty = -\beta \psi_\infty \nabla V$, we recast the above equation as

$$\partial_t \psi = \beta^{-1} div \left[\psi_\infty \nabla \left(\frac{\psi}{\psi_\infty} \right) \right].$$

Note that this equation implies that $\int_{\mathbb{R}^n} \psi(t,q)\, dq$ is a constant. Introduce now the relative entropy

$$\mathscr{E}(t) = H(\psi(t,\cdot)|\psi_\infty) = \int_{\mathbb{R}^n} \ln \left(\frac{\psi(t,q)}{\psi_\infty(q)} \right) \psi(t,q)\, dq.$$

Then

$$\begin{aligned}
\frac{d\mathscr{E}}{dt} &= \int_{\mathbb{R}^n} \ln \left(\frac{\psi}{\psi_\infty} \right) \partial_t \psi + \frac{\psi_\infty}{\psi} \frac{\partial_t \psi}{\psi_\infty} \psi \\
&= \int_{\mathbb{R}^n} \ln \left(\frac{\psi}{\psi_\infty} \right) \beta^{-1} div \left[\psi_\infty \nabla \left(\frac{\psi}{\psi_\infty} \right) \right] \\
&= -\beta^{-1} \int_{\mathbb{R}^n} \nabla \left[\ln \left(\frac{\psi}{\psi_\infty} \right) \right] \psi_\infty \nabla \left(\frac{\psi}{\psi_\infty} \right) \\
&= -\beta^{-1} \int_{\mathbb{R}^n} \left| \nabla \left[\ln \left(\frac{\psi}{\psi_\infty} \right) \right] \right|^2 \psi \\
&= -\beta^{-1} I(\psi(t,\cdot)|\psi_\infty).
\end{aligned} \tag{60}$$

As ψ_∞ satisfies a logarithmic Sobolev inequality with the constant $\rho > 0$, we have that, for any time $t \geq 0$,

$$H(\psi(t,\cdot)|\psi_\infty) \leq (2\rho)^{-1} I(\psi(t,\cdot)|\psi_\infty). \tag{61}$$

We infer from (60) and (61) that

$$\frac{d\mathscr{E}}{dt} \leq -2\rho\beta^{-1} \mathscr{E}.$$

Using the Gronwall lemma, we obtain the claimed result.

Conversely, if

$$\forall t \geq 0, \quad \mathscr{E}(t) \leq \mathscr{E}(0) \exp(-2\rho\beta^{-1} t),$$

we also have

$$\forall t > 0, \quad \frac{\mathscr{E}(t) - \mathscr{E}(0)}{t} \leq \mathscr{E}(0) \frac{\exp(-2\rho\beta^{-1} t) - 1}{t}.$$

By letting t go to 0 and using (60), one obtains the logarithmic Sobolev inequality $I(\psi(0,\cdot)|\psi_\infty) \geq 2\rho H(\psi(0,\cdot)|\psi_\infty)$. □

3.2 Effective Dynamics

Consider Q_t that solves (3). By a simple Itô computation, we have

$$d\xi(Q_t) = \left(-\nabla V \cdot \nabla \xi + \beta^{-1} \Delta \xi\right)(Q_t)\, dt + \sqrt{2\beta^{-1}}\, |\nabla \xi|(Q_t)\, dB_t, \qquad (62)$$

where B_t is the one-dimensional Brownian motion

$$dB_t = \frac{\nabla \xi}{|\nabla \xi|}(Q_t) \cdot dW_t.$$

Of course, (62) is not closed. Following Gyöngy [16], a simple closing procedure is to consider $\widetilde{\eta}_t$ solution to

$$d\widetilde{\eta}_t = \widetilde{b}(t, \widetilde{\eta}_t)\, dt + \sqrt{2\beta^{-1}}\, \widetilde{\sigma}(t, \widetilde{\eta}_t)\, dB_t, \qquad (63)$$

where

$$\widetilde{b}(t,z) = \mathbb{E}\left[\left(-\nabla V \cdot \nabla \xi + \beta^{-1} \Delta \xi\right)(Q_t) \mid \xi(Q_t) = z\right], \qquad (64)$$

$$\widetilde{\sigma}^2(t,z) = \mathbb{E}\left[|\nabla \xi|^2(Q_t) \mid \xi(Q_t) = z\right]. \qquad (65)$$

Note that $\widetilde{b}$ and $\widetilde{\sigma}$ depend on t, since these are expected values conditioned on the fact that $\xi(Q_t) = z$, and the probability distribution function of Q_t of course depends on t.

As shown in [16], this procedure is exact from the point of view of time marginals: at any time t, the random variables $\widetilde{\eta}_t$ and $\xi(Q_t)$ have the same law. This is stated in the following lemma.

Lemma 8 ([19], Lemma 2.3). *The probability distribution function ψ^ξ of $\xi(Q_t)$, where Q_t satisfies (3), satisfies the Fokker-Planck equation associated to (63):*

$$\partial_t \psi^\xi = \partial_z \left(-\widetilde{b}\, \psi^\xi + \beta^{-1} \partial_z (\widetilde{\sigma}^2 \psi^\xi)\right).$$

The problem with equation (63) is that the functions $\widetilde{b}$ and $\widetilde{\sigma}$ are very complicated to compute, since they involve the full knowledge of ψ. Therefore, one cannot consider (63) as a reasonable closure. A natural simplification is to consider a time-independent approximation of the functions $\widetilde{b}$ and $\widetilde{\sigma}$. Considering (64) and (65), we introduce ($\mathbb{E}_\mu$ denoting a mean with respect to the measure μ)

$$\begin{aligned} b(z) &= \mathbb{E}_\mu\left[\left(-\nabla V \cdot \nabla \xi + \beta^{-1} \Delta \xi\right)(Q) \mid \xi(Q) = z\right] \\ &= \int_{\Sigma_z} \left(-\nabla V \cdot \nabla \xi + \beta^{-1} \Delta \xi\right) d\mu_{\Sigma_z}, \end{aligned} \qquad (66)$$

and

$$\sigma^2(z) = \mathbb{E}_\mu\left(|\nabla \xi|^2(Q) \mid \xi(Q) = z\right) = \int_{\Sigma_z} |\nabla \xi|^2\, d\mu_{\Sigma_z},$$

where μ_{Σ_z} is defined by (8). This amounts to replacing the measure $\psi(t,x)$ in (64) (conditioned at the value $\xi(x) = z$) by the equilibrium measure $\psi_\infty(x)$ (conditioned at the value $\xi(x) = z$), and likewise for (65). This simplification especially makes sense if $\xi(Q_t)$ is a slow variable, that is if the characteristic evolution time of $\xi(Q_t)$ is much larger than the characteristic time needed by Q_t to sample the manifold Σ_z. This is quantified in the sequel.

In the spirit of (63), we next introduce the coarse-grained dynamics

$$d\overline{\eta}_t = b(\overline{\eta}_t)\,dt + \sqrt{2\beta^{-1}}\,\sigma(\overline{\eta}_t)\,dB_t, \quad \overline{\eta}_{t=0} = \xi(Q_0). \tag{67}$$

We have proved in [19] that the effective dynamics (67) is ergodic for the equilibrium measure $\xi \star \mu$, that is $\exp(-\beta A(z))\,dz$. In addition, this measure satisfies a detailed balance condition. We have also proved the following error bound, that quantifies the "distance" between the probability distribution function of $\xi(Q_t)$ (at any given time t) and that of $\overline{\eta}_t$.

Proposition 1 ([19], Proposition 3.1). *Assume that ξ is a smooth scalar function such that*

$$\text{for all } q \in \mathbb{R}^n, \quad 0 < m \le |\nabla\xi(q)| \le M < \infty, \tag{68}$$

and that the conditioned probability measures μ_{Σ_z}, defined by (8)*, satisfy a logarithmic Sobolev inequality with a constant ρ uniform in z: for any probability measure ν on Σ_z which is absolutely continuous with respect to the measure μ_{Σ_z}, we have*

$$H(\nu|\mu_{\Sigma_z}) \le \frac{1}{2\rho} I(\nu|\mu_{\Sigma_z}). \tag{69}$$

Let us also assume that the coupling is bounded in the following sense:

$$\kappa = \|\nabla_{\Sigma_z} F\|_{L^\infty} < \infty, \tag{70}$$

where F is the local mean force defined by (11)*.*

Finally, let us assume that $|\nabla\xi|$ is close to a constant on the manifold Σ_z in the following sense:

$$\lambda = \left\| \frac{|\nabla\xi|^2 - \sigma^2\circ\xi}{\sigma^2\circ\xi} \right\|_{L^\infty} < \infty. \tag{71}$$

Assume that, at time $t = 0$, the distribution of the initial conditions of (3) *and* (67) *are consistent one with each other: $\psi^\xi(t = 0,\cdot) = \phi(t = 0,\cdot)$. Then we have the following estimate: for any time $t \ge 0$,*

$$E(t) \le \frac{M^2}{4m^2}\left(\lambda^2 + \frac{m^2\beta^2\kappa^2}{\rho^2}\right)\left(H(\psi(0,\cdot)|\mu) - H(\psi(t,\cdot)|\mu)\right), \tag{72}$$

where $E(t)$ is the relative entropy of the probability distribution function ψ^ξ of $\xi(Q_t)$, where Q_t follows (3)*, with respect to the probability distribution function ϕ of the solution $\overline{\eta}_t$ to* (67)*:*

$$E(t) = H\left(\psi^\xi(t,\cdot)|\phi(t,\cdot)\right) = \int_{\mathbb{R}} \ln\left(\frac{\psi^\xi(t,z)}{\phi(t,z)}\right)\psi^\xi(t,z)\,dz.$$

The above proposition thus yields a uniform-in-time bound on the relative entropy between ψ^ξ and ϕ. In addition, we also know that the effective dynamics is ergodic for $\exp(-\beta A(z))\,dz$, which is the equilibrium measure of $\xi(Q_t)$, in the long-time limit. We thus expect the two probability densities to converge one to each other, in the long-time limit. This is indeed the case, as it is shown in [19, Corollary 3.1]: under some mild assumptions, the L^1 distance between $\psi^\xi(t,\cdot)$ and $\phi(t,\cdot)$ vanishes at an exponential rate in the long-time limit.

The difficulty of the question we address stems from the fact that, in general, $t \to \xi(Q_t)$ is not a Markov process: this is a closure problem. If an appropriate time-scale separation is present in the system (between $\xi(Q_t)$ and the complementary degrees of freedom), then memory effects may be neglected, and $\xi(Q_t)$ be approximated by a Markov process such as (67).

One interest of our approach is to get the error estimate (72), which is not an asymptotic result, and holds for any coarse-grained variable. Of course, this error estimate certainly yields a large error bound in some cases, in particular if ξ is not well-chosen, or when no time-scale separation is present in the dynamics. If bounds (69) and (70) encode a time-scale separation, namely if $\kappa \ll 1$ and $\rho \gg 1$, then the right-hand side of (72) is small, and $\overline{\eta}_t$ solution to (67) is indeed a good approximation of $\xi(Q_t)$.

We would like to emphasize that the effective dynamics (67) may also be obtained using different arguments, such as the Mori-Zwanzig projection approach [15]. In the case when a small parameter is present in the system, one can alternatively use asymptotic expansions of the generator (see [11, 25, 26]).

3.3 The Proof in a Simple Two-Dimensional Case

For the purpose of illustration, we consider in this section an extremely simple case: starting from the overdamped dynamics (3) in *two dimensions* (we write $q = (x,y) \in \mathbb{R}^2$), we want to derive an effective dynamics for the coarse-grained variable $\xi(q) = \xi(x,y) = x$. Although this case is over-simplified, it turns out that the main arguments of our derivation, as well as the proof arguments, can be well understood here.

In that context, the complete dynamics (3) reads

$$\begin{cases} dX_t = -\partial_x V(X_t,Y_t)\,dt + \sqrt{2\beta^{-1}}\,dW_t^x, \\ dY_t = -\partial_y V(X_t,Y_t)\,dt + \sqrt{2\beta^{-1}}\,dW_t^y, \end{cases} \tag{73}$$

with the initial condition $Q_0 = (X_0, Y_0)$. The manifold Σ_z defined by (7) is

$$\Sigma_z = \{(z,y);\ y \in \mathbb{R}\}$$

and the probability measure $d\mu_{\Sigma_z}$ defined by (8) reads

$$d\mu_{\Sigma_z} = \frac{\exp(-\beta V(z,y))dy}{\displaystyle\int_{\mathbb{R}} \exp(-\beta V(z,y))dy} = \frac{\psi_\infty(z,y)dy}{\displaystyle\int_{\mathbb{R}} \psi_\infty(z,y)dy}. \tag{74}$$

We focus on the dynamics of $\xi(X_t, Y_t) = X_t$. In that case, the equation (62) is just the first line of (73), which is obviously not closed in X_t, since Y_t appears. At time t, Q_t is distributed according to the measure $\psi(t,q)$. Hence, the probability distribution function of Y_t, conditioned to the fact that $\xi(Q_t) = X_t = x$, is given by

$$\psi^x_{\text{cond}}(t,y) = \frac{\psi(t,x,y)}{\displaystyle\int_{\mathbb{R}} \psi(t,x,y)\,dy}.$$

Following Gyöngy [16], we introduce the function $\widetilde{b}(t,x)$ defined by (64), which reads in the present context as

$$\widetilde{b}(t,x) = \int_{\mathbb{R}} [-\partial_x V(x,y)]\,\psi^x_{\text{cond}}(t,y)\,dy = -\frac{\displaystyle\int_{\mathbb{R}} \partial_x V(x,y)\,\psi(t,x,y)\,dy}{\displaystyle\int_{\mathbb{R}} \psi(t,x,y)\,dy}. \tag{75}$$

The resulting dynamics (63) reads

$$d\widetilde{X}_t = \widetilde{b}(t,\widetilde{X}_t)\,dt + \sqrt{2\beta^{-1}}\,dW_t^x. \tag{76}$$

We now prove Lemma 8 in that specific context and show that, at any time t, the probability distribution function of $\widetilde{X}_t$ is equal to that of $\xi(Q_t) = X_t$.

Proof (Lemma 8, case $\xi(x,y) = x$). The probability density function $\psi(t,x,y)$ of $Q_t = (X_t, Y_t)$ satisfies the Fokker-Planck (59):

$$\begin{aligned} \partial_t \psi &= div(\psi \nabla V) + \beta^{-1}\Delta\psi \\ &= \partial_x(\psi\,\partial_x V) + \partial_y\left(\psi\,\partial_y V\right) + \beta^{-1}\partial_{xx}\psi + \beta^{-1}\partial_{yy}\psi. \end{aligned} \tag{77}$$

The probability distribution function of $\xi(Q_t) = X_t$ is

$$\psi^\xi(t,x) = \int_{\mathbb{R}} \psi(t,x,y)\,dy.$$

Integrating (77) with respect to y, we obtain

$$\partial_t \psi^\xi = \partial_x \left(\int \psi \partial_x V \, dy \right) + \beta^{-1} \partial_{xx} \psi^\xi$$
$$= -\partial_x \left(\psi^\xi \, \widetilde{b} \right) + \beta^{-1} \partial_{xx} \psi^\xi, \qquad (78)$$

where $\widetilde{b}(t,x)$ is given by (75). We recognize the Fokker-Planck equation associated to the (76). □

As pointed out above, (63) (*i.e.* (76) here) cannot be considered as a reasonable closure, since it involves the function $\widetilde{b}$, which is defined using $\psi(t,x,y)$ (see (75)), which in practice is hardly computable. We thus approximate $\widetilde{b}$ by the function b defined by (66), which amounts to replacing $\psi(t,x,y)$ in (75) by the equilibrium measure $\psi_\infty(x,y)$:

$$b(x) = -\frac{\displaystyle\int_{\mathbb{R}} \partial_x V(x,y) \; \psi_\infty(x,y) \, dy}{\displaystyle\int_{\mathbb{R}} \psi_\infty(x,y) \, dy}.$$

In the spirit of (76), we thus introduce the effective dynamics

$$d\overline{X}_t = b(\overline{X}_t) \, dt + \sqrt{2\beta^{-1}} \, dW_t^x. \qquad (79)$$

We now prove Proposition 1 (error estimator on the effective dynamics), in the specific case at hand here. The assumption (69) means that the measure (74) satisfies, for any z, a logarithmic Sobolev inequality with a constant ρ independent of z. The assumption (70) reads $\kappa = \|\partial_{xy} V\|_{L^\infty} < \infty$, and the assumption (71) is satisfied with $\lambda = 0$ since $\nabla \xi = (1,0)^T$ is a constant vector.

Proof (Proposition 1, case $\xi(x,y) = x$). By definition (see (9)), the free energy A associated to the reaction coordinate ξ satisfies

$$\exp(-\beta A(x)) = \int_{\mathbb{R}} \psi_\infty(x,y) \, dy = Z^{-1} \int_{\mathbb{R}} \exp(-\beta V(x,y)) \, dy,$$

hence

$$A'(x) = \frac{\displaystyle\int \partial_x V(x,y) \psi_\infty(x,y) \, dy}{\displaystyle\int_{\mathbb{R}} \psi_\infty(x,y) \, dy} = -b(x). \qquad (80)$$

The effective dynamics (79) thus reads

$$d\overline{X}_t = -A'(\overline{X}_t) \, dt + \sqrt{2/\beta} \, dW_t^x.$$

Note that, in this specific context, the effective dynamics is of the form (6) (see [19, Sect. 2.3] for a comprehensive discussion of the relation between the effective dynamics and (6)). The probability distribution $\phi(t,x)$ of $\overline{X}_t$ satisfies the

Fokker-Planck equation associated to the above stochastic differential equation, that reads

$$\partial_t \phi = \partial_x \left(\phi \, A'\right) + \beta^{-1} \partial_{xx} \phi. \tag{81}$$

Consider now the relative entropy

$$E(t) = H(\psi^\xi | \phi) = \int_{\mathbb{R}} \ln \left(\frac{\psi^\xi(t,x)}{\phi(t,x)} \right) \psi^\xi(t,x) \, dx.$$

We compute, using (81) and (78), that

$$\begin{aligned}
\frac{dE}{dt} &= \int_{\mathbb{R}} \ln \left(\frac{\psi^\xi}{\phi} \right) \partial_t \psi^\xi - \int_{\mathbb{R}} \frac{\psi^\xi}{\phi} \, \partial_t \phi \\
&= \int_{\mathbb{R}} \ln \left(\frac{\psi^\xi}{\phi} \right) \left[-\partial_x \left(\psi^\xi \, \widetilde{b} \right) + \beta^{-1} \partial_{xx} \psi^\xi \right] - \int_{\mathbb{R}} \frac{\psi^\xi}{\phi} \, \left[\partial_x \left(\phi \, A' \right) + \beta^{-1} \partial_{xx} \phi \right] \\
&= -\beta^{-1} \int_{\mathbb{R}} \partial_x \left[\ln \left(\frac{\psi^\xi}{\phi} \right) \right] \partial_x \psi^\xi + \beta^{-1} \int_{\mathbb{R}} \partial_x \left(\frac{\psi^\xi}{\phi} \right) \partial_x \phi \\
&\qquad + \int_{\mathbb{R}} \psi^\xi \, \partial_x \left(\ln \frac{\psi^\xi}{\phi} \right) \left(\widetilde{b} + A' \right) \\
&= -\beta^{-1} \int_{\mathbb{R}} \partial_x \left[\ln \left(\frac{\psi^\xi}{\phi} \right) \right] \left[\partial_x \psi^\xi - \frac{\psi^\xi \partial_x \phi}{\phi} \right] + \int_{\mathbb{R}} \psi^\xi \, \partial_x \left(\ln \frac{\psi^\xi}{\phi} \right) \left(\widetilde{b} + A' \right) \\
&= -\beta^{-1} \int_{\mathbb{R}} \partial_x \left[\ln \left(\frac{\psi^\xi}{\phi} \right) \right] \phi \, \partial_x \left(\frac{\psi^\xi}{\phi} \right) + \int_{\mathbb{R}} \psi^\xi \, \partial_x \left(\ln \frac{\psi^\xi}{\phi} \right) \left(\widetilde{b} + A' \right) \\
&= -\beta^{-1} I(\psi^\xi | \phi) + \int_{\mathbb{R}} \psi^\xi \, \partial_x \left(\ln \frac{\psi^\xi}{\phi} \right) \left(\widetilde{b} + A' \right).
\end{aligned}$$

Using a Young inequality with a parameter $\alpha > 0$ to be fixed later, we obtain

$$\begin{aligned}
\frac{dE}{dt} &\leq -\beta^{-1} I(\psi^\xi | \phi) + \frac{1}{2\alpha} \int_{\mathbb{R}} \psi^\xi \left(\partial_x \left(\ln \frac{\psi^\xi}{\phi} \right) \right)^2 + \frac{\alpha}{2} \int_{\mathbb{R}} \psi^\xi \left(A' + \widetilde{b} \right)^2 \\
&= \left(\frac{1}{2\alpha} - \beta^{-1} \right) I(\psi^\xi | \phi) + \frac{\alpha}{2} \int_{\mathbb{R}} \psi^\xi \left(A' + \widetilde{b} \right)^2.
\end{aligned} \tag{82}$$

We now observe that, in view of (75) and (80), A' and $-\widetilde{b}$ are averages of the *same* quantity with respect to different probability measures:

$$-\widetilde{b}(t,x) = \int_{\mathbb{R}} \partial_x V(x,y) \, \nu_1^{t,x}(y) \, dy \quad \text{and} \quad A'(x) = \int_{\mathbb{R}} \partial_x V(x,y) \, \nu_2^x(y) \, dy$$

with

$$\nu_1^{t,x}(y) = \frac{\psi(t,x,y)}{\int_{\mathbb{R}} \psi(t,x,y)\,dy} \quad \text{and} \quad \nu_2^x(y) = \frac{\psi_\infty(x,y)}{\int_{\mathbb{R}} \psi_\infty(x,y)\,dy}. \tag{83}$$

We write

$$\begin{aligned} A'(x) + \widetilde{b}(t,x) &= \int_{\mathbb{R}} \partial_x V(x,y)\,\nu_2^x(y)\,dy - \int_{\mathbb{R}} \partial_x V(x,y)\,\nu_1^{t,x}(y)\,dy \\ &= \int_{\mathbb{R}^2} (\partial_x V(x,y_1) - \partial_x V(x,y_2))\,k^{t,x}(y_1,y_2)\,dy_1\,dy_2 \end{aligned}$$

for any probability measure $k^{t,x}$ such that

$$\int_{\mathbb{R}} k^{t,x}(y_1,y_2)\,dy_2 = \nu_2^x(y_1) \quad \text{and} \quad \int_{\mathbb{R}} k^{t,x}(y_1,y_2)\,dy_1 = \nu_1^{t,x}(y_2).$$

Hence,

$$\begin{aligned} \left| A'(x) + \widetilde{b}(t,x) \right| &\leq \|\partial_{xy} V\|_{L^\infty} \int_{\mathbb{R}^2} |y_1 - y_2|\,k^{t,x}(y_1,y_2)\,dy_1\,dy_2 \\ &\leq \|\partial_{xy} V\|_{L^\infty} \left(\int_{\mathbb{R}^2} |y_1 - y_2|^2\,k^{t,x}(y_1,y_2)\,dy_1\,dy_2 \right)^{1/2}. \end{aligned}$$

We now optimize on $k^{t,x}$. Introducing the Wasserstein distance $W(\nu_1^{t,x}, \nu_2^x)$ between $\nu_1^{t,x}$ and ν_2^x (see (57)), we obtain

$$\left| A'(x) + \widetilde{b}(t,x) \right| \leq \|\partial_{xy} V\|_{L^\infty}\; W(\nu_1^{t,x}, \nu_2^x).$$

As recalled above, assumption (69) means that ν_2^x satisfies a logarithmic Sobolev inequality. Thus, it also satisfies a Talagrand inequality (see Lemma 7), hence

$$W(\nu_1^{t,x}, \nu_2^x) \leq \sqrt{\frac{2}{\rho} H(\nu_1^{t,x} | \nu_2^x)} \leq \frac{1}{\rho} \sqrt{I(\nu_1^{t,x} | \nu_2^x)}.$$

As a consequence,

$$\left| A'(x) + \widetilde{b}(t,x) \right| \leq \frac{\|\partial_{xy} V\|_{L^\infty}}{\rho}\; \sqrt{I(\nu_1^{t,x} | \nu_2^x)}.$$

Using (83), we obtain

$$\int_{\mathbb{R}} \psi^{\xi} \left(A' + \widetilde{b}\right)^2 dx \leq \frac{\|\partial_{xy} V\|_{L^\infty}^2}{\rho^2} \int_{\mathbb{R}} \psi^{\xi}(t,x)\, I(\nu_1^{t,x} | \nu_2^x)\, dx$$
$$\leq \frac{\|\partial_{xy} V\|_{L^\infty}^2}{\rho^2} \int_{\mathbb{R}} \psi^{\xi}(t,x) \left[\int_{\mathbb{R}} \left| \partial_y \ln \frac{\psi(t,x,y)}{\psi_\infty(x,y)} \right|^2 \frac{\psi(t,x,y)}{\psi^{\xi}(t,x)}\, dy \right] dx$$
$$\leq \frac{\|\partial_{xy} V\|_{L^\infty}^2}{\rho^2} I(\psi | \psi_\infty).$$

Returning to (82), and using (60), we thus deduce that

$$\frac{dE}{dt} \leq \left(\frac{1}{2\alpha} - \beta^{-1} \right) I(\psi^{\xi} | \phi) + \frac{\alpha}{2} \frac{\|\partial_{xy} V\|_{L^\infty}^2}{\rho^2} I(\psi | \psi_\infty)$$
$$= \left(\frac{1}{2\alpha} - \beta^{-1} \right) I(\psi^{\xi} | \phi) - \frac{\alpha\beta \|\partial_{xy} V\|_{L^\infty}^2}{2\rho^2} \partial_t H(\psi | \psi_\infty).$$

We take $2\alpha = \beta$, so that the first term vanishes, and we are left with

$$\frac{dE}{dt} \leq -\frac{\beta^2 \|\partial_{xy} V\|_{L^\infty}^2}{4\rho^2} \partial_t H(\psi | \psi_\infty).$$

Integrating this inequality between the times 0 and t, and using that $E(0) = 0$, we obtain

$$E(t) \leq \frac{\beta^2 \|\partial_{xy} V\|_{L^\infty}^2}{4\rho^2} \left(H(\psi(t=0) | \psi_\infty) - H(\psi(t,\cdot) | \psi_\infty) \right).$$

As recalled above, assumption (70) reads $\kappa = \|\partial_{xy} V\|_{L^\infty} < \infty$. The above bound is thus exactly the bound (72) in the present context. □

3.4 Numerical Results

In this section, we check the accuracy of the effective dynamics (67) in terms of residence times, and also compare this effective dynamics with the coarse-grained dynamics (6) based on the free energy. We perform such comparison on two test-cases, and evaluate the influence of the temperature on the results. We also provide some analytical explanations for the observed numerical results.

In the following numerical tests, we focus on residence times. We have indeed already underlined that the characteristic behaviour of the dynamics (3) is to sample a given well of the potential energy, then suddenly jump to another basin, and start over. Consequently, an important quantity is the residence time that the system spends in the well, before going to another one.

For all the numerical tests reported in this section, the complete dynamics (3) has been integrated with the Euler-Maruyama scheme

$$Q_{j+1} = Q_j - \Delta t\, \nabla V(Q_j) + \sqrt{2\, \Delta t\, \beta^{-1}}\, G_j,$$

where, for any j, G_j is a n-dimensional vector, whose coordinates are independent and identically distributed (i.i.d.) random variables, distributed according to a normal Gaussian law.

For the simulation of the dynamics (67) and (6), we need to have an expression for the free energy derivative A' and the functions b and σ. These have been computed using the algorithm proposed in [7], on a regular grid of some bounded interval. Values of the functions for points that do not belong to that grid were obtained by linear interpolation. We have again used the Euler-Maruyama scheme to numerically integrate the dynamics (67) and (6).

To compute residence times in a well, we have proceeded as follows (for the sake of clarity, we assume in the following that there are only two wells in the test case at hand). First, the left and the right wells are defined as the sets $\{q \in \mathbb{R}^n;\ \xi(q) \leq \xi^{\rm th}_{\rm left}\}$ and $\left\{q \in \mathbb{R}^n;\ \xi(q) \geq \xi^{\rm th}_{\rm right}\right\}$ respectively, with $\xi^{\rm th}_{\rm right} > \xi^{\rm th}_{\rm left}$. Next, we perform the following computations:

1. We first generate a large number $\mathcal{N}$ of configurations $\{q_i \in \mathbb{R}^n\}_{1 \leq i \leq \mathcal{N}}$, distributed according to the measure μ restricted to the right well: as a consequence, $\xi(q_i) > \xi^{\rm th}_{\rm right}$.
2. We next run the dynamics (3) from the initial condition q_i, and monitor the first time τ_i at which the system reaches a point $q(\tau_i)$ in the left well: $\tau_i = \inf\{t;\ \xi(q_t) < \xi^{\rm th}_{\rm left}\}$.
3. From these $(\tau_i)_{1 \leq i \leq \mathcal{N}}$, we compute an average residence time and a confidence interval. These figures are the reference figures.
4. We next consider the initial conditions $\{\xi(q_i) \in \mathbb{R}\}_{1 \leq i \leq \mathcal{N}}$ for the effective dynamics. By construction, these configurations are distributed according to the equilibrium measure $\xi \star \mu$ (that is $\exp(-\beta A(z))\, dz$) restricted to the right well.
5. From these initial conditions, we run the dynamics (67) or (6) until the left well is reached, and compute, as for the complete description, a residence time and its confidence interval.

3.4.1 A Three Atom Molecule

Our aim in this section is to show that different reaction coordinates, although similar at first sight, can lead to very different results. As explained in [19], the error estimate (72) can then help discriminating between these reaction coordinates.

We consider here a molecule made of three two-dimensional particles, whose positions are q_A, q_B and q_C. The potential energy of the system is

$$V(q) = \frac{1}{2\epsilon}\left(r_{AB} - \ell_{\rm eq}\right)^2 + \frac{1}{2\epsilon}\left(r_{BC} - \ell_{\rm eq}\right)^2 + W_3(\theta_{ABC}), \tag{84}$$

where $r_{AB} = \|q_A - q_B\|$ is the distance between atoms A and B, $\ell_{\rm eq}$ is an equilibrium distance, θ_{ABC} is the angle formed by the three atoms, and $W_3(\theta)$ is a three-body potential, that we choose here to be a double-well potential:

$$W_3(\theta) = \frac{1}{2} k_\theta \left((\theta - \theta_{\text{saddle}})^2 - \delta\theta^2 \right)^2.$$

Wells of W_3 are located at $\theta = \theta_{\text{saddle}} \pm \delta\theta$. The potential (84) represents stiff bonds between particles A and B on the one hand, and B and C on the other hand, with a softer term depending on the angle θ_{ABC}. To remove rigid body motion invariance, we set $q_B = 0$ and $q_A \cdot e_y = 0$. In the following, we work with the parameters $\epsilon = 10^{-3}$, $k_\theta = 208$, $\ell_{\text{eq}} = 1$, $\theta_{\text{saddle}} = \pi/2$ and $\delta\theta = \theta_{\text{saddle}} - 1.187$. All dynamics are integrated with the time step $\Delta t = 10^{-3}$.

We consider two reaction coordinates, that both indicate in which well the system is:

- The angle formed by the three atoms:

$$\xi_1 = \theta_{ABC}.$$

  In that case, wells are defined by $\left\{q \in \mathbb{R}^n;\ \xi_1(q) \le \xi_{\text{left}}^{\text{th}} = \theta_{\text{saddle}} - 0.15\right\}$ and $\left\{q \in \mathbb{R}^n;\ \xi_1(q) \ge \xi_{\text{right}}^{\text{th}} = \theta_{\text{saddle}} + 0.15\right\}$.
- The square of the distance between A and C:

$$\xi_2 = \|q_A - q_C\|^2.$$

  In that case, wells are defined by $\left\{q \in \mathbb{R}^n;\ \xi_2(q) \le \xi_{\text{left}}^{\text{th}} = 1.6\ell_{\text{eq}}^2\right\}$ and $\left\{q \in \mathbb{R}^n;\ \xi_2(q) \ge \xi_{\text{right}}^{\text{th}} = 2.4\ell_{\text{eq}}^2\right\}$.

Note that there is a region of state space that does not belong to any well. This choice allows to circumvent the so-called recrossing problem.

Remark 8. Note that (84) reads

$$V(q) = \frac{1}{2\epsilon} \left(U_{AB}(q)^2 + U_{BC}(q)^2 \right) + W_3(\theta_{ABC})$$

with $U_{AB}(q) = r_{AB} - \ell_{\text{eq}}$ and $U_{BC}(q) = r_{BC} - \ell_{\text{eq}}$. The two first terms in V are much stiffer than the last one. We observe that $\nabla\theta_{ABC} \cdot \nabla U_{AB} = \nabla\theta_{ABC} \cdot \nabla U_{BC} = 0$. Hence, the reaction coordinate ξ_1 is orthogonal to the stiff terms of the potential energy, in contrast to ξ_2.

For potentials of the above type, we have shown in [19, Sect. 3.2] that the coupling constant κ defined by (70) is of the order of ϵ when the reaction coordinate is orthogonal to the stiff terms of the potential energy, and of order 1 otherwise. In turn, the constant ρ defined by (69) typically remains bounded away from 0 when ϵ goes to zero. Ignoring the effect of the constant λ, we hence see that the right-hand side of the error bound (72) is much smaller (and so the effective dynamics is more accurate) when the reaction coordinate is orthogonal to the stiff terms of the potential energy.

Consequently, in the case at hand here, we expect to obtain accurate results with ξ_1, in contrast to ξ_2. This is indeed the case, as shown in the sequel of this section. ◇

We compute the residence time in a given well following the complete description, and compare it with the result given by a reduced description, based either on (67) or (6). Results are gathered in Table 1, for the temperatures $\beta^{-1} = 1$ and $\beta^{-1} = 0.2$. We observe that working with ξ_1 (and either (67) or (6)) leads to very accurate results, independently of the temperature. On the other hand, when working with ξ_2, the reaction coordinate is not orthogonal to the stiff terms of the potential, and both coarse-grained dynamics turn out to be not accurate.

Remark 9. In the case at hand here, $\|\nabla \xi_1\|^2 = \|\nabla \theta_{ABC}\|^2 = r_{BC}^{-2}$. This quantity is almost a constant, since the bond length potential is stiff and the temperature is small. Hence, along the trajectory, we have that $\|\nabla \xi_1\|^2 \approx \ell_{\rm eq}^{-2} = 1$. We pointed out in [19, Sect. 2.3] that, when the reaction coordinate satisfies $\|\nabla \xi\| = 1$, then both coarse-grained dynamics (67) and (6) are identical. This explains why, in the present case, when choosing the reaction coordinate ξ_1, dynamics (67) and (6) give similar results. ◇

Table 1 Three-atom molecule: residence times obtained from the complete description (third column) and from the reduced descriptions (two last columns), for both reaction coordinates (confidence intervals have been computed on the basis of $\mathcal{N} = 15,000$ realizations)

Temperature	Reaction coordinate	Reference residence time	Residence time using (67)	Residence time using (6)
$\beta^{-1} = 1$	$\xi_1 = \theta_{ABC}$	0.700 ± 0.011	0.704 ± 0.011	0.710 ± 0.011
$\beta^{-1} = 1$	$\xi_2 = r_{AC}^2$	0.709 ± 0.015	0.219 ± 0.004	2.744 ± 0.056
$\beta^{-1} = 0.2$	$\xi_1 = \theta_{ABC}$	5784 ± 101	5836 ± 100	5752 ± 101
$\beta^{-1} = 0.2$	$\xi_2 = r_{AC}^2$	5833 ± 88	1373 ± 20	2135 ± 319

We now study how results depend on temperature. Let us first consider the reaction coordinate $\xi_1 = \theta_{ABC}$. Results are shown on Fig. 5. Both coarse-grained dynamics provide extremely accurate results, independently of the temperature. We also observe that we can fit the residence time $\tau_{\rm res}$ according to the relation

$$\tau_{\rm res} \approx \tau_{\rm res}^0 \exp(s\beta) \tag{85}$$

with $\tau_{\rm res}^0 = 0.07521$ and $s = 2.25031$.

By analytical considerations, we now explain why the residence times computed from both coarse-grained dynamics (6) and (67) satisfy the relation (85), with the numerical values of s and $\tau_{\rm res}^0$ reported above.

We first consider the coarse-grained dynamics (6) driven by the free energy. In the case at hand here, it is possible to compute analytically the free energy. Using the internal coordinates r_{AB}, r_{BC} and θ_{ABC}, we indeed infer from (2) that the free energy A_1 does not depend on the temperature and satisfies

$$A_1(\theta_{ABC}) = W_3(\theta_{ABC}).$$

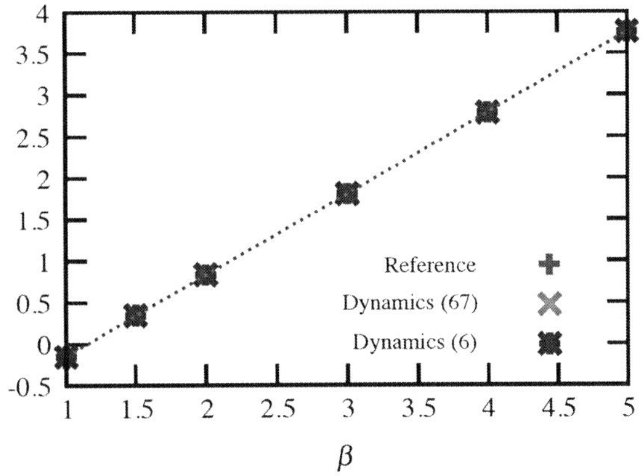

Fig. 5 $\log_{10}$(residence time) as a function of β, for the reaction coordinate $\xi_1 = \theta_{ABC}$

Thus A_1 has two global minimizers, separated by a barrier

$$\Delta A_1 = \frac{1}{2} k_\theta (\delta\theta)^4 \approx 2.25648.$$

The large deviations theory can be used to understand the behaviour of the dynamics (6), in the low temperature regime. It yields the fact that, when $\beta \gg 1$, residence times are given by

$$\tau_{\rm res}^{\rm LD} \approx \tau_{\rm res}^{0,\rm LD} \exp(\beta \Delta A_1) \quad \text{with} \quad \tau_{\rm res}^{0,\rm LD} = \frac{2\pi}{\omega_{\rm SP}\, \omega_{\rm W}}, \tag{86}$$

where $\omega_{\rm SP}^2 = -A_1''(\xi_{\rm SP})$ is the pulsation at the saddle-point $\xi_{\rm SP} = \theta_{\rm saddle}$, and $\omega_{\rm W}^2 = A_1''(\xi_{\rm W})$ is the pulsation at the local minimizer $\xi_{\rm W} = \theta_{\rm saddle} \pm \delta\theta$ (see also [17, (7.9) and (7.10)]). In the present case, we compute that $\omega_{\rm SP} \approx 7.828$ and $\omega_{\rm W} \approx 11.07$, thus $\tau_{\rm res}^{0,\rm LD} \approx 0.0725$, and we find that

$$s \approx \Delta A_1 \quad \text{and} \quad \tau_{\rm res}^0 \approx \tau_{\rm res}^{0,\rm LD}.$$

We thus obtain a good agreement between (85) and (86), as observed on Fig. 5. Note that this agreement holds even up to temperature $\beta^{-1} = 1$.

We now turn to the dynamics (67). We pointed out in Remark 9 that dynamics (67) and (6) are identical in the limit of low temperature, for the reaction coordinate ξ_1. The functions b and σ are plotted for the temperature $\beta^{-1}=1$ on Fig. 6. We observe that, even though the temperature is not very small, we already have $b \approx -W_3' = -A_1'$ and $\sigma \approx 1$. The agreement is even better when the temperature is smaller. This thus explains why results given by both coarse-grained dynamics (67) and (6) can be fitted by the same relation (85), on the whole range of temperature.

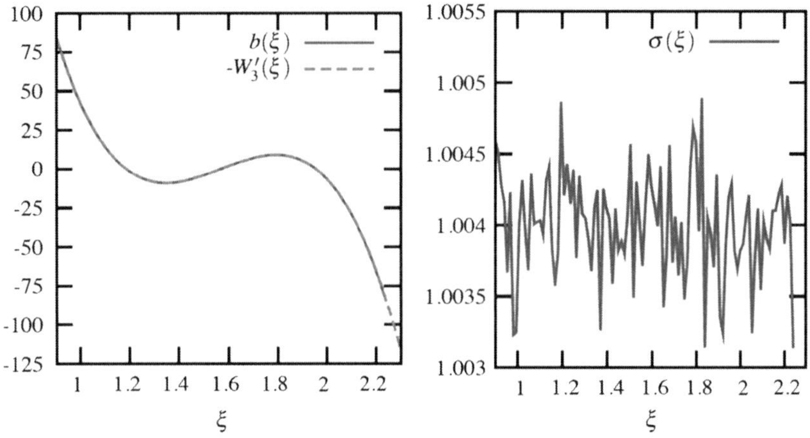

Fig. 6 Plot of the functions b and σ, for the reaction coordinate $\xi_1 = \theta_{ABC}$, at the temperature $\beta^{-1} = 1$

We now consider the reaction coordinate $\xi_2 = r_{AC}^2$. Residence times as a function of the inverse temperature β are shown on Fig. 7. We observe that neither the dynamics (6) nor the dynamics (67) provide accurate results. More precisely, the reference results, the results given by (67) and the results given by (6) can be fitted by

$$\tau_{\rm res}^{\rm ref} \approx \tau_{\rm res}^{0,\rm ref} \exp(s\beta),$$

$$\tau_{\rm res}^{\rm eff} \approx \tau_{\rm res}^{0,\rm eff} \exp(s\beta), \tag{87}$$

$$\tau_{\rm res}^{\rm free} \approx \tau_{\rm res}^{0,\rm free} \exp(s\beta), \tag{88}$$

respectively, with the same parameter $s = 2.21 \pm 0.03$ and

$$\tau_{\rm res}^{0,\rm ref} \approx 0.0768, \quad \tau_{\rm res}^{0,\rm eff} \approx 0.0241, \quad \tau_{\rm res}^{0,\rm free} \approx 0.293.$$

The dependency with respect to the temperature is thus accurately reproduced by both coarse-grained dynamics. The inaccuracy comes from the fact that the prefactor $\tau^{0,\rm ref}$ is ill-approximated.

Again, these numerical observations are in agreement with analytical computations based on the large deviations theory. More precisely, we explain in the sequel why the residence times computed from both coarse-grained dynamics (67) and (6)

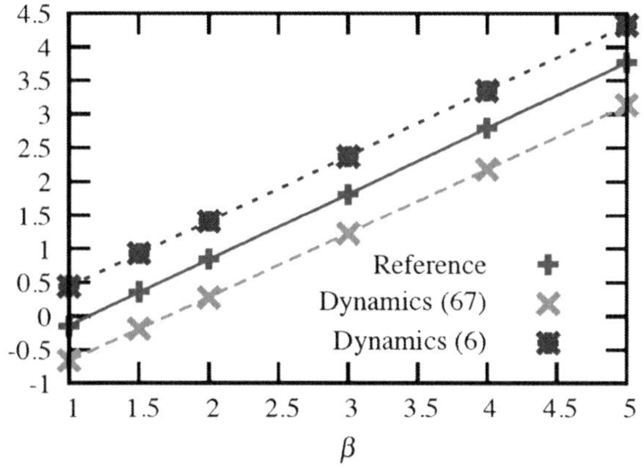

Fig. 7 $\log_{10}$(residence time) as a function of β, for the reaction coordinate $\xi_2 = r_{AC}^2$

satisfy (87) and (88), with the same s, and for the numerical values of s, $\tau_{\rm res}^{0,\rm eff}$ and $\tau_{\rm res}^{0,\rm free}$ reported above.

The functions A_2', b and σ are plotted for two different temperatures on Fig. 8. Although A_2 a priori depends on β (as expected), it turns out this dependency becomes quite weak when $\beta \geq 1$. It turns out that we can fit A_2' by

$$A_2'(\xi) \approx c_5(x-2)^5 + c_4(x-2)^4 + c_3(x-2)^3 + c_2(x-2)^2 + c_1(x-2),$$

with $c_1 = -16.4433$, $c_2 = 3.87398$, $c_3 = 34.2171$, $c_4 = -6.36938$ and $c_5 = -7.89431$. The free energy has thus two local minimizers, $\xi_{\rm W,r} \approx 2.73$ and $\xi_{\rm W,l} \approx 1.25$ and a saddle point, $\xi_{\rm SP} \approx 2$, with

$$A_2(\xi_{\rm SP}) \approx 0, \quad A_2(\xi_{\rm W,r}) \approx -2.1, \quad A_2(\xi_{\rm W,l}) \approx -2.37.$$

We introduce the barriers to go from the right well to the left well (r $\to$ l) and *vice-versa*:

$$\Delta A_2^{\rm r \to l} = A_2(\xi_{\rm SP}) - A_2(\xi_{\rm W,r}) \quad \text{and} \quad \Delta A_2^{\rm l \to r} = A_2(\xi_{\rm SP}) - A_2(\xi_{\rm W,l}).$$

In the case of the dynamics (6) driven by the free energy, and under the assumption that the temperature is low enough so that A_2 becomes independent of β, the large deviations theory can again be used, and yields the fact that residence times are given by

$$\tau_{\rm res,free}^{\rm LD,r \to l} \approx \frac{2\pi}{\omega_{\rm SP}\,\omega_{\rm W,r}} \exp(\beta \Delta A_2^{\rm r \to l}), \quad \tau_{\rm res,free}^{\rm LD,l \to r} \approx \frac{2\pi}{\omega_{\rm SP}\,\omega_{\rm W,l}} \exp(\beta \Delta A_2^{\rm l \to r}),$$

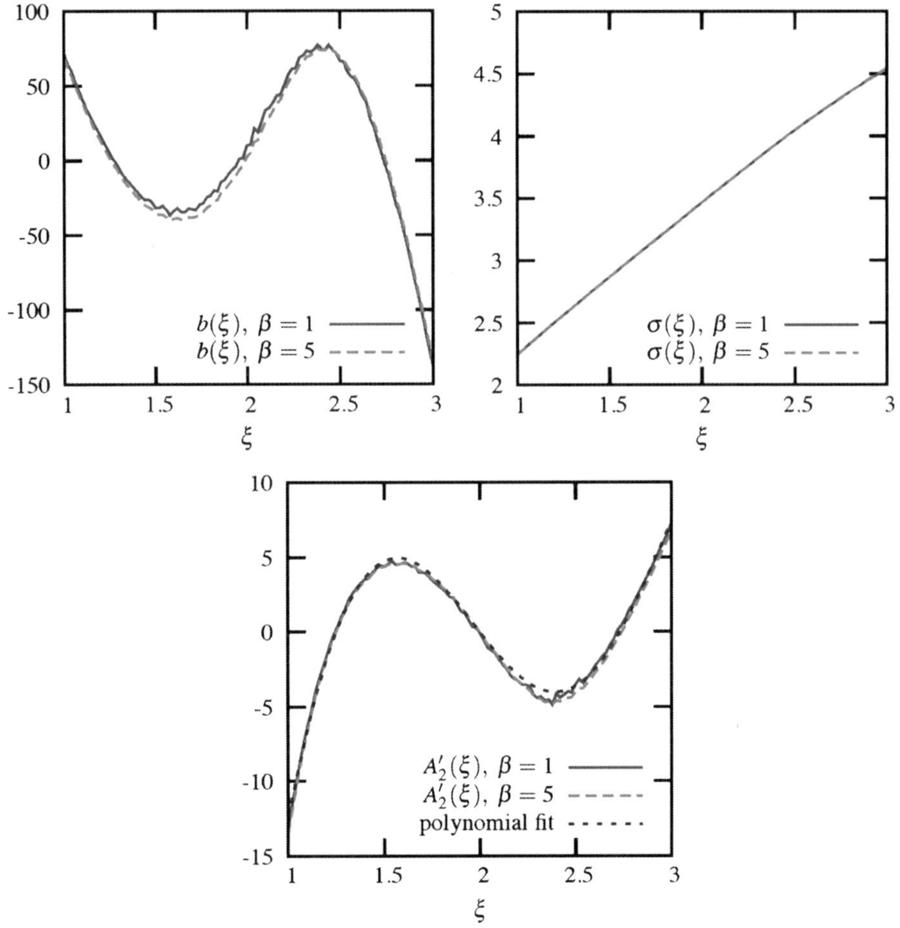

Fig. 8 Plot of the functions b, σ and A_2', for the reaction coordinate $\xi_2 = r_{AC}^2$, at two different temperatures

where $\omega_{\rm SP}^2$, $\omega_{\rm W,l}^2$ and $\omega_{\rm W,r}^2$ are the pulsations at the saddle-point, the left well and the right well, respectively. In the present case, we compute that $\omega_{\rm SP} \approx \sqrt{-c_1} \approx 4.055$, $\omega_{\rm W,l} \approx 5.809$ and $\omega_{\rm W,r} \approx 4.774$.

The left well is deeper than the right well. Hence, in the low temperature limit, the residence time in the left well is much larger than the residence time in the right well, and the probability to be in the left well is higher than the probability to be in the right well. Hence,

$$\tau_{\rm res,free}^{\rm LD} \approx \tau_{\rm res,free}^{\rm LD,l\to r} \approx \tau_{\rm res,free}^{0,\rm LD,l\to r} \exp(\beta \Delta A_2^{\rm l\to r}) \quad \text{with} \quad \tau_{\rm res,free}^{0,\rm LD,l\to r} = \frac{2\pi}{\omega_{\rm SP}\,\omega_{\rm W,l}}. \tag{89}$$

With the parameters that we used, we compute $\tau_{\rm res,free}^{0,\rm LD,l\to r} \approx 0.267$, hence

$$s \approx \Delta A_2^{\mathrm{l}\to\mathrm{r}} \quad \text{and} \quad \tau_{\mathrm{res}}^{0,\mathrm{free}} \approx \tau_{\mathrm{res,free}}^{0,\mathrm{LD,l}\to\mathrm{r}},$$

and we obtain a good agreement between (88) and (89).

We now turn to the dynamics (67). The functions b and σ plotted on Fig. 8 seem to be almost independent of the temperature when $\beta \geq 1$. Following [19, Sect. 2.3] and [11, Sect. 10 and (89)], we introduce the one-to-one function

$$h(\xi) = \int_0^{\xi} \sigma^{-1}(y)\, dy$$

and the coordinate $\zeta = h(\xi_2)$. We next change of variable in the effective dynamics (67) on the reaction coordinate ξ and recast it as

$$d\zeta_t = -\widetilde{A}'(\zeta_t)\, dt + \sqrt{2\beta^{-1}}\, dB_t,$$

where $\widetilde{A}$ turns out to be the free energy associated to the reaction coordinate $\zeta(q) = h(\xi_2(q))$. The residence time to exit the left well is hence given by

$$\tau_{\mathrm{res,eff}}^{\mathrm{LD,l}\to\mathrm{r}} \approx \frac{2\pi}{\widetilde{\omega}_{\mathrm{SP}}\, \widetilde{\omega}_{\mathrm{W,l}}} \exp(\beta \Delta \widetilde{A}^{\mathrm{l}\to\mathrm{r}}).$$

In the regime of low temperature, the second term of (11) is negligible, and we deduce from (10) that $\widetilde{A}(h(\xi)) = A_2(\xi)$, where A_2 is the free energy associated with the reaction coordinate ξ_2. As a consequence,

$$\Delta \widetilde{A}^{\mathrm{l}\to\mathrm{r}} = \Delta A^{\mathrm{l}\to\mathrm{r}}, \quad \widetilde{\omega}_{\mathrm{SP}} = \omega_{\mathrm{SP}}\, \sigma(\xi_{\mathrm{SP}}), \quad \widetilde{\omega}_{\mathrm{W,l}} = \omega_{\mathrm{W,l}}\, \sigma(\xi_{\mathrm{W,l}}).$$

Hence,

$$\tau_{\mathrm{res,eff}}^{\mathrm{LD,l}\to\mathrm{r}} \approx \tau_{\mathrm{res,eff}}^{0,\mathrm{LD,l}\to\mathrm{r}} \exp(\beta \Delta A_2^{\mathrm{l}\to\mathrm{r}}), \tag{90}$$

with

$$\tau_{\mathrm{res,eff}}^{0,\mathrm{LD,l}\to\mathrm{r}} = \frac{2\pi}{\omega_{\mathrm{SP}}\, \omega_{\mathrm{W,l}}\, \sigma(\xi_{\mathrm{SP}})\, \sigma(\xi_{\mathrm{W,l}})}.$$

We thus recover that the dependency of the residence times with temperature is identical between the residence times predicted by the effective dynamics (67) and the residence times predicted by (6): this dependency is exponential, with the same prefactor $\Delta A_2^{\mathrm{l}\to\mathrm{r}}$.

We also compute $\sigma(\xi_{\mathrm{SP}}) \approx 3.465$ and $\sigma(\xi_{\mathrm{W,l}}) \approx 2.563$, so $\tau_{\mathrm{res,eff}}^{0,\mathrm{LD,l}\to\mathrm{r}} \approx 0.03$. Thus the values $\tau_{\mathrm{res}}^{0,\mathrm{eff}}$ and $\tau_{\mathrm{res,eff}}^{0,\mathrm{LD,l}\to\mathrm{r}}$ qualitatively agree, and we obtain a good agreement between (87) and (90).

3.4.2 The Butane Molecule Case

We now consider a system in higher dimension, namely a butane molecule, in the united atom model [22, 27]. We hence only simulate four particles, whose positions are $q^i \in \mathbb{R}^3$, for $1 \leq i \leq 4$. The potential energy reads

$$V(q) = \sum_{i=1}^{3} V_{\text{bond}}\left(\|q^{i+1} - q^i\|\right) + V_{\text{bond-angle}}(\theta_1) + V_{\text{bond-angle}}(\theta_2) + V_{\text{torsion}}(\phi),$$

where θ_1 is the angle formed by the three first particles, θ_2 is the angle formed by the three last particles, and ϕ is the dihedral angle, namely the angle between the plane on which the three first particles lay and the plane on which the three last particles lay, with the convention $\phi \in (-\pi, \pi)$. We work with

$$V_{\text{bond}}(\ell) = \frac{k_2}{2}(\ell - \ell_{eq})^2, \quad V_{\text{bond-angle}}(\theta) = \frac{k_3}{2}(\theta - \theta_{eq})^2$$

and

$$V_{\text{torsion}}(\phi) = c_1(1 - \cos\phi) + 2c_2(1 - \cos^2\phi) + c_3(1 + 3\cos\phi - 4\cos^3\phi).$$

Rigid body motion invariance is removed by setting $q^2 = 0$, $q^1 \cdot e_z = 0$ and $q^3 \cdot e_x = q^3 \cdot e_z = 0$.

In the system of units where the length unit is $\ell_0 = 1.53 \cdot 10^{-10}$ m and the energy unit is such that $k_B T = 1$ at $T = 300$ K, the time unit is $\bar{t} = 364$ fs, and the numerical values of the parameters are $\ell_{eq} = 1$, $k_3 = 208$, $\theta_{eq} = 1.187$, $c_1 = 1.18$, $c_2 = -0.23$, and $c_3 = 2.64$. We will work in the sequel with $k_2 = 1{,}000$. We set the unit of mass such that the mass of each particle is equal to 1.

For these values of the parameters c_i, the function V_{torsion} has a unique global minimum (at $\phi = 0$) and two local non-global minima (see Fig. 9). It is hence a metastable potential. We choose to work with the dihedral angle as reaction coordinate:

$$\xi(q) = \phi.$$

We are interested in the residence time in the main well (around the global minimizer $\phi_0 = 0$) before hopping to any of the two wells around the local minimizers $\phi_{\pm 1} = \pm 2\pi/3$. For each minimizer ϕ_0, ϕ_1 and ϕ_{-1}, the associated well is defined by $\left\{q; |\xi(q) - \phi_i| \leq \xi^{\text{th}}\right\}$, $i = -1, 0, 1$, with $\xi^{\text{th}} = 0.5$.

Remark 10. We observe that

$$\nabla V_{\text{stiff}} \cdot \nabla \xi = 0,$$

where $V_{\text{stiff}}(q) = \sum_{i=1}^{3} V_{\text{bond}}\left(\|q^{i+1} - q^i\|\right) + V_{\text{bond-angle}}(\theta_1) + V_{\text{bond-angle}}(\theta_2)$. In view of [19, Sect. 3.2], we hence expect to obtain accurate results with this choice of reaction coordinate, as it is indeed the case. $\diamond$

As in the previous section, we compute reference residence times by integrating the complete dynamics, and we then consider both coarse-grained dynamics (67) and (6). All computations have been done with the time step $\Delta t = 10^{-3}$. Results are reported in Table 2. We observe that the effective dynamics (67) again yields

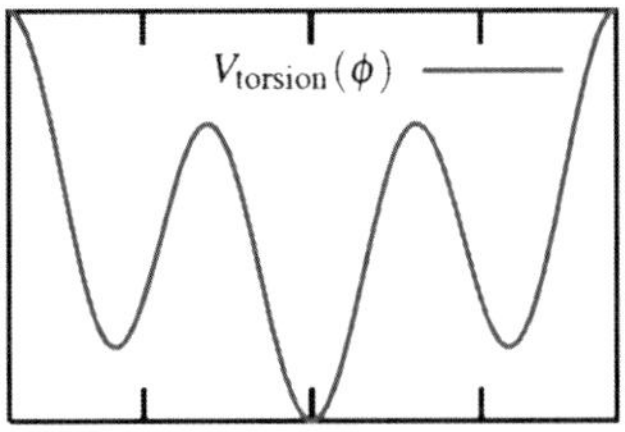

Fig. 9 Torsion angle potential $V_{\text{torsion}}(\phi)$

extremely accurate results. The results obtained by the dynamics (6), although qualitatively correct, are less accurate. This conclusion holds for all the temperatures we considered.

As in the previous section, residence times depend on the temperature following

$$\tau_{\text{res}} \approx \tau_{\text{res}}^0 \exp(s\beta).$$

For both coarse-grained dynamics, the values found for s and τ_{res}^0 agree with predictions based on the large deviations theory. In the case at hand here, it turns out that the free energy associated to the reaction coordinate $\xi(q) = \phi$ is simply $A(\xi) = V_{\text{torsion}}(\xi)$. On Fig. 10, we plot the functions b and σ. We observe that they are almost independent of the temperature (as soon as $\beta \geq 1$), and that σ is almost a constant. Hence, up to the time rescaling $t_{\text{rescale}} = \sigma t$, the effective dynamics reads as the dynamics (6) governed by the free energy. As $\sigma = 1.086 \approx 1$ (see Fig. 10), the dynamics (6) yields qualitatively correct results.

Table 2 Butane molecule: residence times obtained from the complete description (second column) and from the reduced descriptions (two last columns), at different temperatures (confidence intervals have been computed on the basis of $\mathcal{N} = 13{,}000$ realizations)

Temperature	Reference residence time	Residence time using (67)	Residence time using (6)
$\beta^{-1} = 1$	31.9 ± 0.56	32.0 ± 0.56	37.1 ± 0.64
$\beta^{-1} = 0.67$	493 ± 8	490 ± 8	581 ± 9
$\beta^{-1} = 0.5$	7624 ± 113	7794 ± 115	9046 ± 133

Acknowledgements The present contribution is related to a lecture given by TL at a workshop at BIRS on "Numerical analysis of multiscale computations" (December 7-11, 2009). This work is supported in part by the INRIA, under the grant "Action de Recherche Collaborative" HYBRID, and by the Agence Nationale de la Recherche, under grant ANR-09-BLAN-0216-01 (MEGAS).

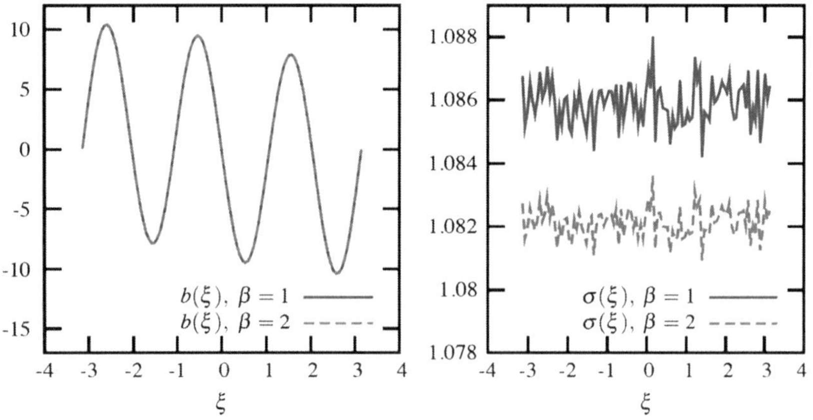

Fig. 10 Plot of the functions b and σ, for the reaction coordinate $\xi = \phi$, at different temperatures

References

1. C. Ané, S. Blachère, D. Chafaï, P. Fougères, I. Gentil, F. Malrieu, C. Roberto, and G. Scheffer. *Sur les inégalités de Sobolev logarithmiques*. Société Mathématique de France, Paris, 2000.
2. A. Arnold, P. Markowich, G. Toscani, and A. Unterreiter. On convex Sobolev inequalities and the rate of convergence to equilibrium for Fokker-Planck type equations. *Comm. Part. Diff. Eq.*, 26:43–100, 2001.
3. X. Blanc, C. Le Bris, F. Legoll, and C. Patz. Finite-temperature coarse-graining of one-dimensional models: mathematical analysis and computational approaches. *Journal of Nonlinear Science*, 20(2):241–275, 2010.
4. S. Bobkov and F. Götze. Exponential integrability and transportation cost related to logarithmic Sobolev inequalities. *J. Funct. Anal.*, 163(1):1–28, 1999.
5. E. Cancès, F. Legoll, and G. Stoltz. Theoretical and numerical comparison of some sampling methods for molecular dynamics. *Math. Mod. Num. Anal. (M2AN)*, 41(2):351–389, 2007.
6. C. Chipot and A. Pohorille, editors. *Free energy calculations*, volume 86 of *Springer Series in Chemical Physics*. Springer, 2007.
7. G. Ciccotti, T. Lelièvre, and E. Vanden-Eijnden. Projection of diffusions on submanifolds: application to mean force computation. *Comm. Pure and Applied Math.*, 61(3):371–408, 2008.
8. A. Dembo and O. Zeitouni. *Large deviations techniques*. Jones and Bartlett Publishers, 1993.
9. F. den Hollander. *Large deviations*, volume 14 of *Fields Institute Monographs*. American Mathematical Society, Providence, RI, 2000.
10. N. Dunford and J.T. Schwartz. *Linear operators. Volume 2: Spectral theory: self adjoint operators in Hilbert space*. Wiley, New York, 1963.
11. W. E and E. Vanden-Eijnden. Metastability, conformation dynamics, and transition pathways in complex systems. In S. Attinger and P. Koumoutsakos, editors, *Multiscale Modelling and Simulation*, pages 35–68. Lect. Notes Comput. Sci. Eng. 39, Springer, 2004.
12. R.S. Ellis. *Entropy, large deviations, and statistical mechanics*, volume 271 of *Grundlehren der Mathematischen Wissenschaften*. Springer-Verlag, New York, 1985.
13. R.S. Ellis. Large deviations and statistical mechanics. In *Particle systems, random media and large deviations, Brunswick, Maine, 1984*, volume 41 of *Contemp. Math.*, pages 101–123. American Mathematical Society, Providence, RI, 1985.
14. R.S. Ellis. An overview of the theory of large deviations and applications to statistical mechanics. *Scand. Actuar. J.*, 1:97–142, 1995. Harald Cramer Symposium, Stockholm, 1993.
15. D. Givon, R. Kupferman, and A.M. Stuart. Extracting macroscopic dynamics: model problems and algorithms. *Nonlinearity*, 17(6):55–127, 2004.

16. I. Gyöngy. Mimicking the one-dimensional marginal distributions of processes having an Itô differential. *Probab. Th. Rel. Fields*, 71:501–516, 1986.
17. P. Hänggi, P. Talkner, and M. Borkovec. Reaction-rate theory: fifty years after Kramers. *Reviews of Modern Physics*, 62(2):251–342, 1990.
18. H.A. Kramers. Brownian motion in a field of force and the diffusion model of chemical reactions. *Physica*, 7(4):284–304, 1940.
19. F. Legoll and T. Lelièvre. Effective dynamics using conditional expectations. *Nonlinearity*, 23(9):2131–2163, 2010.
20. T. Lelièvre, M. Rousset, and G. Stoltz. *Free energy computations: A mathematical perspective*. Imperial College Press, 2010.
21. L. Maragliano, A. Fischer, E. Vanden-Eijnden, and G. Ciccotti. String method in collective variables: minimum free energy paths and isocommittor surfaces. *J. Chem. Phys.*, 125:024106, 2006.
22. M.G. Martin and J.I. Siepmann. Transferable potentials for phase equilibria. I. United-atom description of n-alkanes. *J. Phys. Chem.*, 102:2569–2577, 1998.
23. S.P. Meyn and R.L. Tweedie. *Markov chains and stochastic stability*. Springer, 1993.
24. F. Otto and C. Villani. Generalization of an inequality by Talagrand and links with the logarithmic Sobolev inequality. *J. Funct. Anal.*, 173(2):361–400, 2000.
25. G.C. Papanicolaou. Some probabilistic problems and methods in singular perturbations. *Rocky Mountain J. Math.*, 6(4):653–674, 1976.
26. G.C. Papanicolaou. Introduction to the asymptotic analysis of stochastic equations. In *Modern modeling of continuum phenomena (Ninth Summer Sem. Appl. Math., Rensselaer Polytech. Inst., Troy, N.Y., 1975)*, volume 16 of *Lectures in Appl. Math.*, pages 109–147. Amer. Math. Soc., Providence, R.I., 1977.
27. J.P. Ryckaert and A. Bellemans. Molecular dynamics of liquid alkanes. *Faraday Discuss.*, 66:95–106, 1978.
28. H. Schaefer and M.P. Wolff. *Topological vector spaces*, volume 3 of *Graduate Texts in Mathematics*. Springer-Verlag, New York, 1999. second edition.
29. S.R.S. Varadhan. *Large deviations and applications*. SIAM, Philadelphia, 1984.
30. C. Villani. *Topics in optimal transportation*, volume 58 of *Graduate Studies in Mathematics*. American Mathematical Society, Providence, RI, 2003.

Linear Stationary Iterative Methods for the Force-Based Quasicontinuum Approximation

Mitchell Luskin and Christoph Ortner

Abstract Force-based multiphysics coupling methods have become popular since they provide a simple and efficient coupling mechanism, avoiding the difficulties in formulating and implementing a consistent coupling energy. They are also the only known pointwise consistent methods for coupling a general atomistic model to a finite element continuum model. However, the development of efficient and reliable iterative solution methods for the force-based approximation presents a challenge due to the non-symmetric and indefinite structure of the linearized force-based quasicontinuum approximation, as well as to its unusual stability properties. In this paper, we present rigorous numerical analysis and computational experiments to systematically study the stability and convergence rate for a variety of linear stationary iterative methods.

1 Introduction

Low energy local minima of crystalline atomistic systems are characterized by highly localized defects such as vacancies, interstitials, dislocations, cracks, and grain boundaries separated by large regions where the atoms are slightly deformed from a lattice structure. The goal of atomistic-to-continuum coupling methods [1–4, 15, 16, 22, 26, 28, 32] is to approximate a fully atomistic model by maintaining the accuracy of the atomistic model in a small neighborhood surrounding each localized defect and using the efficiency of continuum coarse-grained models in the vast regions that are only mildly deformed from a lattice structure.

M. Luskin (✉)
School of Mathematics, 206 Church St. SE, University of Minnesota, Minneapolis, MN 55455, USA
e-mail: luskin@umn.edu

C. Ortner
Mathematical Institute, St. Giles' 24–29, Oxford OX1 3LB, UK
e-mail: ortner@maths.ox.ac.uk

B. Engquist et al. (eds.), *Numerical Analysis of Multiscale Computations*, Lecture Notes in Computational Science and Engineering 82, DOI 10.1007/978-3-642-21943-6_14,

Force-based atomistic-to-continuum methods decompose a computational reference lattice into an *atomistic region* $\mathscr{A}$ and a *continuum region* $\mathscr{C}$, and assign forces to representative atoms according to the region they are located in. In the quasicontinuum method, the representative atoms are all atoms in the atomistic region and the nodes of a finite element approximation in the continuum region. The force-based approximation is thus given by [5, 6, 10–12, 32]

$$\mathscr{F}_j^{\mathrm{qcf}}(y) := \begin{cases} \mathscr{F}_j^{\mathrm{a}}(y) & \text{if } j \in \mathscr{A}, \\ \mathscr{F}_j^{\mathrm{c}}(y) & \text{if } j \in \mathscr{C}, \end{cases} \tag{1}$$

where y denotes the positions of the representative atoms which are indexed by j, $\mathscr{F}_j^{\mathrm{a}}(y)$ denotes the atomistic force at representative atom j, and $\mathscr{F}_j^{\mathrm{c}}(y)$ denotes a continuum force at representative atom j.

The force-based quasicontinuum method (QCF) uses a Cauchy-Born strain energy density for the continuum model to achieve a patch test consistent approximation [6, 11, 24]. We recall that a patch test consistent atomistic-to-continuum approximation exactly reproduces the zero net forces of uniformly strained lattices [19, 24, 27]. However, the recently discovered unusual stability properties of the linearized force-based quasicontinuum (QCF) approximation, especially its indefiniteness, present a challenge to the development of efficient and reliable iterative methods [12]. Energy-based quasicontinuum approximations have many attractive features such as more reliable solution methods, but practical patch test consistent, energy-based quasicontinuum approximations have yet to be developed for most problems of physical interest, such as three-dimensional problems with many-body interaction potentials [20, 21, 30].

Rather than attempt an analysis of linear stationary methods for the full nonlinear system, in this paper we restrict our focus to the linearization of a one-dimensional model problem about the uniform deformation, $y_j^F = Fj\epsilon$, where $F > 0$ is a macroscopic strain, $j \in \mathbb{Z}$, and ϵ is the reference interatomic spacing. We then consider linear stationary methods of the form

$$P\left(u^{(n+1)} - u^{(n)}\right) = \alpha r^{(n)}, \tag{2}$$

where P is a nonsingular preconditioning operator, the damping parameter $\alpha > 0$ is fixed throughout the iteration (that is, stationary), and the residual is defined as

$$r^{(n)} := f - L_F^{\mathrm{qcf}} u^{(n)},$$

where f denotes any applied forces and L_F^{qcf} denotes the linearization of the QCF operator (1) about the uniform deformation y^F.

We will see below that our analysis of this simple model problem already allows us to observe many interesting and crucial features of the various methods. For example, we can distinguish which iterative methods converge up to the critical strain F_* (see (9) for a discussion of the critical strain), and we obtain first results on their convergence rates.

We begin in Sects. 2 and 3 by introducing the most important quasicontinuum approximations and outlining their stability properties, which are mostly straightforward generalizations of results from [9–11, 13]. In Sect. 4, we review the basic properties of linear stationary iterative methods.

In Sect. 5, we give an analysis of the Richardson Iteration ($P = I$) and prove a contraction rate of order $1 - O(N^{-2})$ in the ℓ^p_ϵ norm (discrete Sobolev norms are defined in Sect. 2.1), where N is the size of the atomistic system.

In Sect. 6, we consider the iterative solution with preconditioner $P = L_F^{\mathrm{qcl}}$, where L_F^{qcl} is a standard second order elliptic operator, and show that the preconditioned iteration with an appropriately chosen damping parameter α is a contraction up to the critical strain F_* only in $\mathscr{U}^{2,\infty}$ among the common discrete Sobolev spaces. We show, however, that a rate of contraction in $\mathscr{U}^{2,\infty}$ independent of N can be achieved with the elliptic preconditioner L_F^{qcl} and an appropriate choice of the damping parameter α.

In Sect. 7, we consider the popular ghost force correction iteration (GFC) which is given by the preconditioner $P = L_F^{\mathrm{qce}}$, and we show that the GFC iteration ceases to be a contraction for any norm at strains less than the critical strain. This result and others presented in Sect. 7 imply that the GFC iteration might not always reliably reproduce the stability of the atomistic system [9]. We did not find that the GFC method predicted an instability at a reduced strain in our benchmark tests [18] (see also [24]). To explain this, we note that our 1D analysis in this paper can be considered a good model for cleavage fracture, but not for the slip instabilities studied in [18,24]. We are currently attempting to develop a 2D benchmark test for cleavage fracture to study the stability of the GFC method.

2 The QC Approximations and Their Stability

We give a review of the prototype QC approximations and their stability properties in this section. The reader can find more details in [9, 10].

2.1 Function Spaces and Norms

We consider a one-dimensional atomistic chain whose $2N + 1$ atoms have the reference positions $x_j = j\epsilon$ for $\epsilon = 1/N$. The displacement of the boundary atoms will be constrained, so the space of admissible displacements will be given by the *displacement space*

$$\mathscr{U} = \{u \in \mathbb{R}^{2N+1} : u_{-N} = u_N = 0\}.$$

We will use various norms on the space $\mathscr{U}$ which are discrete variants of the usual Sobolev norms that arise naturally in the analysis of elliptic PDEs.

For displacements $v \in \mathscr{U}$ and $1 \le p \le \infty$, we define the ℓ^p_ϵ norms,

$$\|v\|_{\ell^p_\epsilon} := \begin{cases} \left(\epsilon \sum_{\ell=-N+1}^{N} |v_\ell|^p\right)^{1/p}, & 1 \le p < \infty, \\ \max_{\ell=-N+1,\dots,N} |v_\ell|, & p = \infty, \end{cases}$$

and we denote by $\mathscr{U}^{0,p}$ the space $\mathscr{U}$ equipped with the ℓ^p_ϵ norm. The inner product associated with the ℓ^2_ϵ norm is

$$\langle v, w \rangle := \epsilon \sum_{\ell=-N+1}^{N} v_\ell w_\ell \qquad \text{for } v, w \in \mathscr{U}.$$

We will also use $\|f\|_{\ell^p_\epsilon}$ and $\langle f, g\rangle$ to denote the ℓ^p_ϵ-norm and ℓ^2_ϵ-inner product for arbitrary vectors f, g which need not belong to $\mathscr{U}$. In particular, we further define the $\mathscr{U}^{1,p}$ norm

$$\|v\|_{\mathscr{U}^{1,p}} := \|v'\|_{\ell^p_\epsilon},$$

where $(v')_\ell = v'_\ell = \epsilon^{-1}(v_\ell - v_{\ell-1})$, $\ell = -N+1, \dots, N$, and we let $\mathscr{U}^{1,p}$ denote the space $\mathscr{U}$ equipped with the $\mathscr{U}^{1,p}$ norm. Similarly, we define the space $\mathscr{U}^{2,p}$ and its associated $\mathscr{U}^{2,p}$ norm, based on the centered second difference $v''_\ell = \epsilon^{-2}(v_{\ell+1} - 2v_\ell + v_{\ell-1})$ for $\ell = -N+1, \dots, N-1$.

We have that $v' \in \mathbb{R}^{2N}$ for $v \in \mathscr{U}$ has mean zero $\sum_{j=-N+1}^{N} v'_j = 0$. We can thus obtain from [10, (9)] that

$$\max_{\substack{v \in \mathscr{U} \\ \|v'\|_{\ell^q_\epsilon} = 1}} \langle u', v' \rangle \le \max_{\substack{\sigma \in \mathbb{R}^{2N} \\ \|\sigma\|_{\ell^q_\epsilon} = 1}} \langle u', \sigma \rangle = \|u\|_{\mathscr{U}^{1,p}} \le 2 \max_{\substack{v \in \mathscr{U} \\ \|v'\|_{\ell^q_\epsilon} = 1}} \langle u', v' \rangle. \tag{3}$$

We denote the space of linear functionals on $\mathscr{U}$ by $\mathscr{U}^*$. For $g \in \mathscr{U}^*$, $s = 0, 1$, and $1 \le p \le \infty$, we define the negative norms $\|g\|_{\mathscr{U}^{-s,p}}$ by

$$\|g\|_{\mathscr{U}^{-s,p}} := \sup_{\substack{v \in \mathscr{U} \\ \|v\|_{\mathscr{U}^{s,q}} = 1}} \langle g, v \rangle,$$

where $1 \le q \le \infty$ satisfies $\frac{1}{p} + \frac{1}{q} = 1$. We let $\mathscr{U}^{-s,p}$ denote the dual space $\mathscr{U}^*$ equipped with the $\mathscr{U}^{-s,p}$ norm.

For a linear mapping $A : \mathscr{U}_1 \to \mathscr{U}_2$ where $\mathscr{U}_i$ are vector spaces equipped with the norms $\|\cdot\|_{\mathscr{U}_i}$, we denote the operator norm of A by

$$\|A\|_{L(\mathscr{U}_1, \mathscr{U}_2)} := \sup_{v \in \mathscr{U}, v \ne 0} \frac{\|Av\|_{\mathscr{U}_2}}{\|v\|_{\mathscr{U}_1}}.$$

If $\mathscr{U}_1 = \mathscr{U}_2$, then we use the more concise notation

$$\|A\|_{\mathscr{U}_1} := \|A\|_{L(\mathscr{U}_1, \mathscr{U}_1)}.$$

If $A : \mathscr{U}^{0,2} \to \mathscr{U}^{0,2}$ is invertible, then we can define the *condition number* by

$$\text{cond}(A) = \|A\|_{\mathscr{U}^{0,2}} \cdot \|A^{-1}\|_{\mathscr{U}^{0,2}}.$$

When A is symmetric and positive definite, we have that

$$\text{cond}(A) = \lambda_{2N-1}^{A} / \lambda_1^{A}$$

where the eigenvalues of A are $0 < \lambda_1^A \leq \cdots \leq \lambda_{2N-1}^A$. If a linear mapping $A : \mathscr{U} \to \mathscr{U}$ is symmetric and positive definite, then we define the A-inner product and A-norm by

$$\langle v, w \rangle_A := \langle Av, w \rangle, \qquad \|v\|_A^2 = \langle Av, v \rangle.$$

The operator $A : \mathscr{U}_1 \to \mathscr{U}_2$ is *operator stable* if the operator norm $\|A^{-1}\|_{L(\mathscr{U}_2, \mathscr{U}_1)}$ is finite, and a sequence of operators $A_j : \mathscr{U}_{1,j} \to \mathscr{U}_{2,j}$ is *operator stable* if the sequence $\|(A_j)^{-1}\|_{L(\mathscr{U}_{2,j}, \mathscr{U}_{1,j})}$ is uniformly bounded. A symmetric operator $A : \mathscr{U}^{0,2} \to \mathscr{U}^{0,2}$ is called *stable* if it is positive definite, and this implies operator stability. A sequence of positive definite, symmetric operators $A_j : \mathscr{U}^{0,2} \to \mathscr{U}^{0,2}$ is called *stable* if their smallest eigenvalues $\lambda_1^{A_j}$ are uniformly bounded away from zero.

2.2 The Atomistic Model

We now consider a one-dimensional atomistic chain whose $2N + 3$ atoms have the reference positions $x_j = j\epsilon$ for $\epsilon = 1/N$, and interact only with their nearest and next-nearest neighbors.

We denote the deformed positions by y_j, $j = -N - 1, \ldots, N + 1$; and we constrain the boundary atoms and their next-nearest neighbors to match the uniformly deformed state, $y_j^F = Fj\epsilon$, where $F > 0$ is a macroscopic strain, that is,

$$\begin{aligned} y_{-N-1} &= -F(N+1)\epsilon, \qquad & y_{-N} &= -FN\epsilon, \\ y_N &= FN\epsilon, \qquad & y_{N+1} &= F(N+1)\epsilon. \end{aligned} \tag{4}$$

We introduced the two additional atoms with indices $\pm(N+1)$ so that $y = y^F$ is an equilibrium of the atomistic model. The total energy of a deformation $y \in \mathbb{R}^{2N+3}$ is now given by

$$\text{E}^{\text{a}}(y) - \sum_{j=-N}^{N} \epsilon f_j y_j,$$

where

$$\text{E}^{\text{a}}(y) = \sum_{j=-N}^{N+1} \epsilon \phi\left(\frac{y_j - y_{j-1}}{\epsilon}\right) = \sum_{j=-N}^{N+1} \epsilon \phi(y_j') + \sum_{j=-N+1}^{N+1} \epsilon \phi(y_j' + y_{j-1}'). \tag{5}$$

Here, ϕ is a scaled two-body interatomic potential (for example, the normalized Lennard-Jones potential, $\phi(r) = r^{-12} - 2r^{-6}$), and f_j, $j = -N, \dots, N$, are external forces. The equilibrium equations are given by the force balance conditions at the unconstrained atoms,

$$\begin{aligned} -\mathscr{F}_j^{\mathrm{a}}(y^{\mathrm{a}}) &= f_j \qquad \text{for} \quad j = -N+1, \dots, N-1, \\ y_j^{\mathrm{a}} &= Fj\epsilon \qquad \text{for} \quad j = -N-1, -N, N, N+1, \end{aligned} \tag{6}$$

where the atomistic force (per lattice spacing ϵ) is given by

$$\begin{aligned} \mathscr{F}_j^{\mathrm{a}}(y) &:= -\frac{1}{\epsilon}\frac{\partial \mathrm{E}^a(y)}{\partial y_j} \\ &= \frac{1}{\epsilon}\Big\{\big[\phi'(y'_{j+1}) + \phi'(y'_{j+2} + y'_{j+1})\big] - \big[\phi'(y'_j) + \phi'(y'_j + y'_{j-1})\big]\Big\}. \end{aligned} \tag{7}$$

We linearize (7) by letting $u \in \mathbb{R}^{2N+3}$, $u_{\pm N} = u_{\pm(N+1)} = 0$, be a "small" displacement from the uniformly deformed state $y_j^F = Fj\epsilon$; that is, we define

$$u_j = y_j - y_j^F \qquad \text{for } j = -N-1, \dots, N+1.$$

We then linearize the atomistic equilibrium (6) about the uniformly deformed state y^F and obtain a linear system for the displacement u^a,

$$\begin{aligned} (L_F^{\mathrm{a}} u^{\mathrm{a}})_j &= f_j \qquad \text{for} \quad j = -N+1, \dots, N-1, \\ u_j^a &= 0 \qquad \text{for} \quad j = -N-1, -N, N, N+1, \end{aligned}$$

where $(L_F^{\mathrm{a}} v)_j$ is given by

$$(L_F^{\mathrm{a}} v)_j := \phi''_F \left[\frac{-v_{j+1} + 2v_j - v_{j-1}}{\epsilon^2}\right] + \phi''_{2F}\left[\frac{-v_{j+2} + 2v_j - v_{j-2}}{\epsilon^2}\right].$$

Here and throughout we define

$$\phi''_F := \phi''(F) \quad \text{and} \quad \phi''_{2F} := \phi''(2F),$$

where ϕ is the interatomic potential in (5). We will always assume that $\phi''_F > 0$ and $\phi''_{2F} < 0$, which holds for typical pair potentials such as the Lennard-Jones potential under physically realistic deformations.

The stability properties of L_F^{a} can be understood by using a representation derived in [9],

$$\langle L_F^{\mathrm{a}} u, u\rangle = \epsilon A_F \sum_{\ell=-N+1}^{N} |u'_\ell|^2 - \epsilon^3 \phi''_{2F} \sum_{\ell=-N}^{N} |u''_\ell|^2 = A_F \|u'\|_{\ell^2_\epsilon}^2 - \epsilon^2 \phi''_{2F} \|u''\|_{\ell^2_\epsilon}^2, \tag{8}$$

where A_F is the *continuum elastic modulus*

$$A_F = \phi_F'' + 4\phi_{2F}''.$$

We can obtain the following result from the argument in [9, Proposition 1] and [12].

Proposition 1. *If* $\phi_{2F}'' \le 0$, *then*

$$\min_{\substack{u\in\mathbb{R}^{2N+3}\setminus\{0\}\\ u_{\pm N}=u_{\pm(N+1)}=0}} \frac{\langle L_F^{\mathrm{a}} u, u\rangle}{\|u'\|_{\ell_\epsilon^2}^2} = A_F - \epsilon^2 \nu_\epsilon \phi_{2F}'',$$

where

$$\nu_\epsilon := \min_{\substack{u\in\mathbb{R}^{2N+3}\setminus\{0\}\\ u_{\pm N}=u_{\pm(N+1)}=0}} \frac{\|u''\|_{\ell_\epsilon^2}^2}{\|u'\|_{\ell_\epsilon^2}^2}$$

satisfies $0 < \nu_\epsilon \le C$ *for some universal constant* C.

2.2.1 The Critical Strain F_*

The previous result shows that L_F^{a} is positive definite, uniformly as $N \to \infty$, if and only if $A_F > 0$. For realistic interaction potentials, L_F^{a} is positive definite in a ground state $F_0 > 0$. For simplicity, we assume that $F_0 = 1$, and we ask how far the system can be "stretched" by applying increasing macroscopic strains F until it loses its stability. In the limit as $N \to \infty$, this happens at the *critical strain* F_*, which is the smallest number larger than F_0, solving the equation

$$A_{F_*} = \phi''(F_*) + 4\phi''(2F_*) = 0. \tag{9}$$

2.3 The Local QC Approximation (QCL)

The local quasicontinuum (QCL) approximation uses the Cauchy-Born approximation to approximate the nonlocal atomistic model by a local continuum model [5,23, 26]. For next-nearest neighbor interactions, the Cauchy-Born approximation reads

$$\phi\left(\epsilon^{-1}(y_{\ell+1} - y_{\ell-1})\right) \approx \tfrac{1}{2}\left[\phi(2y_\ell') + \phi(2y_{\ell+1}')\right],$$

and results in the QCL energy, for $y \in \mathbb{R}^{2N+3}$ satisfying the boundary conditions (4),

$$\begin{aligned} \mathrm{E}^{\mathrm{qcl}}(y) = &\sum_{j=-N+1}^{N} \epsilon\left[\phi(y_j') + \phi(2y_j')\right] \\ &+ \epsilon\left[\phi(y_{-N}') + \frac{1}{2}\phi(2y_{-N}') + \phi(y_{N+1}') + \frac{1}{2}\phi(2y_{N+1}')\right]. \end{aligned} \tag{10}$$

Imposing the artificial boundary conditions of zero displacement from the uniformly deformed state, $y_j^F = Fj\epsilon$, we obtain the QCL equilibrium equations

$$\begin{aligned} -\mathscr{F}_j^{\mathrm{qcl}}(y^{\mathrm{qcl}}) &= f_j && \text{for} \quad j = -N+1, \dots, N-1, \\ y_j^{\mathrm{qcl}} &= Fj\epsilon && \text{for} \quad j = -N, N, \end{aligned}$$

where

$$\begin{aligned} \mathscr{F}_j^{\mathrm{qcl}}(y) &:= -\frac{1}{\epsilon}\frac{\partial \mathrm{E}^{\mathrm{qcl}}(y)}{\partial y_j} \\ &= \frac{1}{\epsilon}\Big\{\big[\phi'(y'_{j+1}) + 2\phi'(2y'_{j+1})\big] - \big[\phi'(y'_j) + 2\phi'(2y'_j)\big]\Big\}. \end{aligned} \tag{11}$$

We see from (11) that the QCL equilibrium equations are well-defined with only a single constraint at each boundary, and we can restrict our consideration to $y \in \mathbb{R}^{2N+1}$ with $y_{-N} = -F$ and $y_N = F$ as the boundary conditions.

Linearizing the QCL equilibrium (11) about y^F results in the system

$$\begin{aligned} (L_F^{\mathrm{qcl}} u^{\mathrm{qcl}})_j &= f_j && \text{for} \quad j = -N+1, \dots, N-1, \\ u_j^{\mathrm{qcl}} &= 0 && \text{for} \quad j = -N, N, \end{aligned}$$

where

$$L_F^{\mathrm{qcl}} = A_F L$$

and L is the discrete Laplacian, for $v \in \mathscr{U}$, given by

$$(Lv)_j := -v''_j = \left[\frac{-v_{j+1} + 2v_j - v_{j-1}}{\epsilon^2}\right], \quad j = -N+1, \dots, N-1. \tag{12}$$

The QCL operator is a scaled discrete Laplace operator, so

$$\langle L_F^{\mathrm{qcl}} u, u\rangle = A_F \|u'\|_{\ell_\epsilon^2}^2 \qquad \text{for all } u \in \mathscr{U}.$$

In particular, it follows that L_F^{qcl} is stable if and only if $A_F > 0$, that is, if and only if $F < F_*$, where F_* is the critical strain defined in (9).

2.4 The Force-Based QC Approximation (QCF)

The force-based quasicontinuum (QCF) method combines the accuracy of the atomistic model with the efficiency of the QCL approximation by decomposing the computational reference lattice into an *atomistic region* $\mathscr{A}$ and a *continuum region* $\mathscr{C}$, and assigns forces to atoms according to the region they are located in. The QCF operator is given by [5, 6]

$$\mathscr{F}_j^{\text{qcf}}(y) := \begin{cases} \mathscr{F}_j^{\text{a}}(y) & \text{if } j \in \mathscr{A}, \\ \mathscr{F}_j^{\text{qcl}}(y) & \text{if } j \in \mathscr{C}, \end{cases} \tag{13}$$

and the QCF equilibrium equations are given by

$$\begin{aligned} -\mathscr{F}_j^{\text{qcf}}(y^{\text{qcf}}) &= f_j \qquad \text{for} \quad j = -N+1, \dots, N-1, \\ y_j^{\text{qcf}} &= Fj\epsilon \qquad \text{for} \quad j = -N, N. \end{aligned}$$

We note that, since atoms near the boundary belong to $\mathscr{C}$, only one boundary condition is required at each end.

For simplicity, we specify the atomistic and continuum regions as follows. We fix $K \in \mathbb{N}$, $1 \le K \le N-2$, and define

$$\mathscr{A} = \{-K, \dots, K\} \quad \text{and} \quad \mathscr{C} = \{-N+1, \dots, N-1\} \setminus \mathscr{A}.$$

Linearizing (13) about y^F, we obtain

$$\begin{aligned} (L_F^{\text{qcf}} u^{\text{qcf}})_j &= f_j \qquad \text{for} \quad j = -N+1, \dots, N-1, \\ u_j^{\text{qcf}} &= 0 \qquad \text{for} \quad j = -N, N, \end{aligned} \tag{14}$$

where the linearized force-based operator is given explicitly by

$$(L_F^{\text{qcf}} v)_j := \begin{cases} (L_F^{\text{qcl}} v)_j, & \text{for } j \in \mathscr{C}, \\ (L_F^{\text{a}} v)_j, & \text{for } j \in \mathscr{A}. \end{cases}$$

The stability analysis of the QCF operator L_F^{qcf} is less straightforward [10, 11]; we will therefore treat it separately and postpone it to Sect. 3.

2.5 The Original Energy-Based QC Approximation (QCE)

The original energy-based quasicontinuum (QCE) method [26] defines an energy functional by assigning atomistic energy contributions in the atomistic region and continuum energy contributions in the continuum region. For our model problem, we obtain

$$\mathrm{E}^{\text{qce}}(y) = \epsilon \sum_{\ell \in \mathscr{A}} \mathrm{E}_\ell^a(y) + \epsilon \sum_{\ell \in \mathscr{C}} \mathrm{E}_\ell^c(y) \quad \text{for } y \in \mathbb{R}^{2N+1},$$

where

$$\begin{aligned} \mathrm{E}_\ell^c(y) &= \tfrac{1}{2}\big(\phi(2y'_\ell) + \phi(y'_\ell) + \phi(y'_{\ell+1}) + \phi(2y'_{\ell+1})\big), \quad \text{and} \\ \mathrm{E}_\ell^a(y) &= \tfrac{1}{2}\big(\phi(y'_{\ell-1} + y'_\ell) + \phi(y'_\ell) + \phi(y'_{\ell+1}) + \phi(y'_{\ell+1} + y'_{\ell+2})\big). \end{aligned}$$

The QCE method is patch tests inconsistent [7,8,25,31], which can be seen from the existence of "ghost forces" at the interface, that is, $\nabla \mathrm{E}^{\mathrm{qce}}(y^F) = g^F \neq 0$. Hence, the linearization of the QCE equilibrium equations about y^F takes the form (see [8, Sect. 2.4] and [7, Sect. 2.4] for more detail)

$$\begin{aligned} (L_F^{\mathrm{qce}} u^{\mathrm{qce}})_j - g_j^F &= f_j \qquad \text{for} \quad j = -N+1, \ldots, N-1, \\ u_j^{\mathrm{qce}} &= 0 \qquad \text{for} \quad j = -N, N, \end{aligned} \tag{15}$$

where, for $0 \le j \le N-1$, we have

$$(L_F^{\mathrm{qce}} v)_j = \phi_F'' \frac{-v_{j+1} + 2v_j - v_{j-1}}{\epsilon^2}$$
$$+ \phi_{2F}'' \begin{cases} 4\dfrac{-v_{j+2} + 2v_j - v_{j-2}}{4\epsilon^2}, & 0 \le j \le K-2, \\ 4\dfrac{-v_{j+2} + 2v_j - v_{j-2}}{4\epsilon^2} + \dfrac{1}{\epsilon}\dfrac{v_{j+2} - v_j}{2\epsilon}, & j = K-1, \\ 4\dfrac{-v_{j+2} + 2v_j - v_{j-2}}{4\epsilon^2} - \dfrac{2}{\epsilon}\dfrac{v_{j+1} - v_j}{\epsilon} + \dfrac{1}{\epsilon}\dfrac{v_{j+2} - v_j}{2\epsilon}, & j = K, \\ 4\dfrac{-v_{j+1} + 2v_j - v_{j-1}}{\epsilon^2} - \dfrac{2}{\epsilon}\dfrac{v_j - v_{j-1}}{\epsilon} + \dfrac{1}{\epsilon}\dfrac{v_j - v_{j-2}}{2\epsilon}, & j = K+1, \\ 4\dfrac{-v_{j+1} + 2v_j - v_{j-1}}{\epsilon^2} + \dfrac{1}{\epsilon}\dfrac{v_j - v_{j-2}}{2\epsilon}, & j = K+2, \\ 4\dfrac{-v_{j+1} + 2v_j - v_{j-1}}{\epsilon^2}, & K+3 \le j \le N-1, \end{cases}$$

and where the vector of "ghost forces," g, is defined by

$$g_j^F = \begin{cases} 0, & 0 \le j \le K-2, \\ -\frac{1}{2\epsilon}\phi_{2F}', & j = K-1, \\ \frac{1}{2\epsilon}\phi_{2F}', & j = K, \\ \frac{1}{2\epsilon}\phi_{2F}', & j = K+1, \\ -\frac{1}{2\epsilon}\phi_{2F}', & j = K+2, \\ 0, & K+3 \le j \le N-1. \end{cases}$$

The equations for $j = -N+1, \ldots, -1$ follow from symmetry.

The following result is a new sharp stability estimate for the QCE operator L_F^{qce}. Its somewhat technical proof is given in Appendix 7.

Theorem 1. *If $K \ge 1$, $N \ge K+2$, and $\phi_{2F}'' \le 0$, then*

$$\inf_{\substack{u \in \mathscr{U} \\ \|u'\|_{\ell_\epsilon^2} = 1}} \langle L_F^{\mathrm{qce}} u, u \rangle = A_F + \lambda_K \phi_{2F}'',$$

where $\frac{1}{2} \le \lambda_K \le 1$. Asymptotically, as $K \to \infty$, we have

$$\lambda_K \sim \lambda_* + O(e^{-cK}) \quad \text{where } \lambda_* \approx 0.6595 \text{ and } c \approx 1.5826.$$

2.6 The Quasi-Nonlocal QC Approximation (QNL)

The QCF method is the simplest idea to circumvent the interface inconsistency of the QCE method, but gives non-conservative equilibrium equations [5]. An alternative energy-based approach was suggested in [14, 33], which is based on a modification of the energy at the interface. The quasi-nonlocal approximation (QNL) is given by the energy functional

$$\mathrm{E}^{\mathrm{qnl}}(y) := \epsilon \sum_{\ell=-N+1}^{N} \phi(y'_\ell) + \epsilon \sum_{\ell \in \mathscr{A}} \phi(y'_\ell + y'_{\ell+1}) + \epsilon \sum_{\ell \in \mathscr{C}} \tfrac{1}{2}\left[\phi(2y'_\ell) + \phi(2y'_{\ell+1})\right],$$

where we set $\phi(y'_{-N}) = \phi(y'_{N+1}) = 0$. The QNL approximation is patch test consistent; that is, $y = y^F$ is an equilibrium of the QNL energy functional.

The linearization of the QNL equilibrium equations about y^F is

$$\begin{aligned} (L_F^{\mathrm{qnl}} u^{\mathrm{qnl}})_j &= f_j \qquad \text{for} \quad j = -N+1, \ldots, N-1, \\ u_j^{\mathrm{qnl}} &= 0 \qquad \text{for} \quad j = -N, N, \end{aligned}$$

where

$$(L_F^{\mathrm{qnl}} v)_j = \phi''_F \frac{-v_{j+1} + 2v_j - v_{j-1}}{\epsilon^2} + \phi''_{2F} \begin{cases} 4\dfrac{-v_{j+2} + 2v_j - v_{j-2}}{4\epsilon^2}, & 0 \le j \le K-1, \\ 4\dfrac{-v_{j+2} + 2v_j - v_{j-2}}{4\epsilon^2} - \dfrac{-v_{j+2} + 2v_{j+1} - v_j}{\epsilon^2}, & j = K, \\ 4\dfrac{-v_{j+1} + 2v_j - v_{j-1}}{\epsilon^2} + \dfrac{-v_j + 2v_{j-1} - v_{j-2}}{\epsilon^2}, & j = K+1, \\ 4\dfrac{-v_{j+1} + 2v_j - v_{j-1}}{\epsilon^2}, & K+2 \le j \le N-1. \end{cases} \tag{16}$$

We can repeat our stability analysis for the periodic QNL operator in [9, Sec. 3.3] verbatim to obtain the following result.

Proposition 2. *If $K < N-1$, and $\phi_{2F} \le 0$, then*

$$\inf_{\substack{u \in \mathscr{U} \\ \|u'\|_{\ell^2_\epsilon} = 1}} \langle L_F^{\mathrm{qnl}} u, u \rangle = A_F.$$

Remark 1. Since $\phi''_{2F} = (A_F - \phi''_F)/4$, the linearized operators $(\phi''_F)^{-1} L_F^{\mathrm{a}}$, $(\phi''_F)^{-1} L_F^{\mathrm{qcl}}$, $(\phi''_F)^{-1} L_F^{\mathrm{qcf}}$, $(\phi''_F)^{-1} L_F^{\mathrm{qce}}$, and $(\phi''_F)^{-1} L_F^{\mathrm{qnl}}$ depend only on A_F/ϕ''_F, N and K.

3 Stability and Spectrum of the QCF operator

In this section, we give various properties of the linearized QCF operator, most of which are variants of our results in [10, 11]. We first give a result for the non-coercivity of the QCF operator which lies at the heart of many of the difficulties one encounters in analyzing the QCF method.

Theorem 2 (Theorem 1, [11]). *If $\phi''_F > 0$ and $\phi''_{2F} \in \mathbb{R} \setminus \{0\}$ then, for sufficiently large N, the operator L_F^{qcf} is* not *positive-definite. More precisely, there exist $N_0 \in \mathbb{N}$ and $C_1 \geq C_2 > 0$ such that, for all $N \geq N_0$ and $2 \leq K \leq N/2$,*

$$-C_1 N^{1/2} \leq \inf_{\substack{v \in \mathscr{U} \\ \|v'\|_{\ell_\epsilon^2}=1}} \langle L_F^{\mathrm{qcf}} v, v \rangle \leq -C_2 N^{1/2}.$$

The proof of Theorem 2 yields also the following asymptotic result on the operator norm of L_F^{qcf}. Its proof is a straightforward extension of [11, Lemma 2], which covers the case $p = 2$, and we therefore omit it.

Lemma 1. *Let $\phi''_{2F} \neq 0$, then there exists a constant $C_3 > 0$ such that for sufficiently large N, and for $2 \leq K \leq N/2$,*

$$C_3^{-1} N^{1/p} \leq \left\| L_F^{\mathrm{qcf}} \right\|_{L(\mathscr{U}^{1,p},\, \mathscr{U}^{-1,p})} \leq C_3 N^{1/p}.$$

As a consequence of Theorem 2 and Lemma 1, we analyzed the stability of L_F^{qcf} in alternative norms. By following the proof of [10, Theorem 3] verbatim (see also [10, Remark 3]), we can obtain the following sharp stability result.

Proposition 3. *If $A_F > 0$ and $\phi''_{2F} \leq 0$, then L_F^{qcf} is invertible with*

$$\left\| (L_F^{\mathrm{qcf}})^{-1} \right\|_{L(\mathscr{U}^{0,\infty},\, \mathscr{U}^{2,\infty})} \leq 1/A_F.$$

If $A_F = 0$, then L_F^{qcf} is singular.

This result shows that L_F^{qcf} is operator stable up to the critical strain F_* at which the atomistic model loses its stability as well (cf. Sect. 2.2).

3.1 Spectral Properties of L_F^{qcf} in $\mathscr{U}^{0,2} = \ell_\epsilon^2$

The spectral properties of the L_F^{qcf} operator are fundamental for the analysis of the performance of iterative methods in Hilbert spaces. The basis of our analysis of L_F^{qcf} in the Hilbert space $\mathscr{U}^{0,2}$ is the surprising observation that, even though L_F^{qcf} is non-normal, it is nevertheless diagonalizable and its spectrum is identical to that of

L_F^{qnl}. We first observed this numerically in [10, Sect. 4.4] for the case of periodic boundary conditions. A proof has since been given in [13, Sect. 3], which translates verbatim to the case of Dirichlet boundary conditions and yields the following result.

Lemma 2. *For all $N \geq 4$, $1 \leq K \leq N-2$, we have the identity*

$$L_F^{\mathrm{qcf}} = L^{-1} L_F^{\mathrm{qnl}} L. \tag{17}$$

In particular, the operator L_F^{qcf} is diagonalizable and its spectrum is identical to the spectrum of L_F^{qnl}.

We denote the eigenvalues of L_F^{qnl} (and L_F^{qcf}) by

$$0 < \lambda_1^{\mathrm{qnl}} \leq \cdots \leq \lambda_\ell^{\mathrm{qnl}} \leq \cdots \leq \lambda_{2N-1}^{\mathrm{qnl}}.$$

The following lemma gives a lower bound for λ_1^{qnl}, an upper bound for $\lambda_{2N-1}^{\mathrm{qnl}}$, and consequently an upper bound for $\mathrm{cond}(L_F^{\mathrm{qnl}}) = \lambda_{2N-1}^{\mathrm{qnl}}/\lambda_1^{\mathrm{qnl}}$.

Lemma 3. *If $K < N-1$ and $\phi_{2F}'' \leq 0$, then*

$$\lambda_1^{\mathrm{qnl}} \geq 2A_F, \qquad \lambda_{2N-1}^{\mathrm{qnl}} \leq \big(A_F - 4\phi_{2F}''\big)\epsilon^{-2} = \phi_F''\epsilon^{-2}, \quad \textit{and}$$
$$\mathrm{cond}(L_F^{\mathrm{qnl}}) = \frac{\lambda_{2N-1}^{\mathrm{qnl}}}{\lambda_1^{\mathrm{qnl}}} \leq \left(\frac{\phi_F''}{2A_F}\right)\epsilon^{-2}.$$

For the analysis of iterative methods, we are also interested in the condition number of a basis of eigenvectors of L_F^{qcf} as N tends to infinity. Employing Lemma 2, we can write $L_F^{\mathrm{qcf}} = L^{-1}\Lambda^{\mathrm{qcf}}L$ where L is the discrete Laplacian operator and Λ^{qcf} is diagonal. The columns of L^{-1} are poorly scaled; however, a simple rescaling was found in [13, Theorem 3.3] for periodic boundary conditions. The construction and proof translate again verbatim to the case of Dirichlet boundary conditions and yield the following result (note, in particular, that the main technical step, [13, Lemma 4.6] can be applied directly).

Lemma 4. *Let $A_F > 0$, then there exists a matrix V of eigenvectors for the force-based QC operator L_F^{qcf} such that* $\mathrm{cond}(V)$ *is bounded above by a constant that is independent of N.*

3.2 Spectral Properties of L_F^{qcf} in $\mathscr{U}^{1,2}$

In our analysis below, particularly in Sects. 6.1 and 6.2, we will see that the preconditioner $L_F^{\mathrm{qcl}} = A_F L$ is a promising candidate for the efficient solution of the QCF system. The operator $L^{1/2}$ can be understood as a basis transformation to an orthonormal basis in $\mathscr{U}^{1,2}$. Hence, it will be useful to study the spectral properties of L_F^{qcf} in that space. The relevant (generalized) eigenvalue problem is

$$L_F^{\mathrm{qcf}} v = \lambda L v, \qquad v \in \mathscr{U}, \tag{18}$$

which can, equivalently, be written as

$$L^{-1} L_F^{\mathrm{qcf}} v = \lambda v, \qquad v \in \mathscr{U}, \tag{19}$$

or as

$$L^{-1/2} L_F^{\mathrm{qcf}} L^{-1/2} w = \lambda w, \qquad w \in \mathscr{U}, \tag{20}$$

with the basis transform $w = L^{1/2} v$, in either case reducing it to a standard eigenvalue problem in ℓ_ϵ^2. Since L and $L^{1/2}$ commute, Lemma 2 immediately yields the following result.

Lemma 5. *For all $N \geq 4$, $1 \leq K \leq N-2$ the operator $L^{-1} L_F^{\mathrm{qcf}}$ is diagonalizable and its spectrum is identical to the spectrum of $L^{-1} L_F^{\mathrm{qnl}}$.*

We gave a proof in [12] of the following lemma, which completely characterizes the spectrum of $L^{-1} L_F^{\mathrm{qnl}}$, and thereby also the spectrum of $L^{-1} L_F^{\mathrm{qcf}}$. We denote the spectrum of $L^{-1} L_F^{\mathrm{qnl}}$ (and $L^{-1} L_F^{\mathrm{qcf}}$) by $\{\mu_j^{\mathrm{qnl}} : j = 1, \ldots, 2N-1\}$.

Lemma 6. *Let $K \leq N-2$ and $A_F > 0$, then the (unordered) spectrum of $L^{-1} L_F^{\mathrm{qnl}}$ (that is, the $\mathscr{U}^{1,2}$-spectrum) is given by*

$$\mu_j^{\mathrm{qnl}} = \begin{cases} A_F - 4\phi_{2F}'' \sin^2\left(\frac{j\pi}{4K+4}\right), & j = 1, \ldots, 2K+1, \\ A_F, & j = 2K+2, \ldots, 2N-1. \end{cases}$$

In particular, if $\phi_{2F}'' \leq 0$, then

$$\frac{\max_j \mu_j^{\mathrm{qnl}}}{\min_j \mu_j^{\mathrm{qnl}}} = 1 - \frac{4\phi_{2F}''}{A_F} \sin^2\left(\frac{(2K+1)\pi}{4K+4}\right) = \frac{\phi_F''}{A_F} + \frac{4\phi_{2F}''}{A_F} \sin^2\left(\frac{\pi}{4K+4}\right) = \frac{\phi_F''}{A_F} + O(K^{-2}).$$

We conclude this study by stating a result on the condition number of the matrix of eigenvectors for the eigenvalue problem (20). Letting $\tilde{V}$ be an orthogonal matrix of eigenvectors of $L^{-1/2} L_F^{\mathrm{qnl}} L^{-1/2}$ and $\tilde{\Lambda}$ the corresponding diagonal matrix, then Lemma 2 yields

$$\begin{aligned} L^{-1/2} L_F^{\mathrm{qcf}} L^{-1/2} &= L^{-1}\big[L^{-1/2} L_F^{\mathrm{qnl}} L^{-1/2}\big] L \\ &= (\tilde{V}^T L)^{-1} \tilde{\Lambda} (\tilde{V}^T L). \end{aligned}$$

Clearly, $\mathrm{cond}(\tilde{V}^T L) = O(N^2)$, which gives the following result.

Lemma 7. *If $A_F > 0$, then there exists a matrix $\widetilde{W}$ of eigenvectors for the preconditioned force-based QC operator $L^{-1/2} L_F^{\mathrm{qcf}} L^{-1/2}$, such that $\mathrm{cond}(\widetilde{W}) = O(N^2)$ as $N \to \infty$.*

4 Linear Stationary Iterative Methods

In this section, we investigate linear stationary iterative methods for solving the linearized QCF (14). These are iterations of the form

$$P\left(u^{(n)} - u^{(n-1)}\right) = \alpha r^{(n-1)}, \tag{21}$$

where P is a nonsingular preconditioner, the step size parameter $\alpha > 0$ is constant (that is, stationary), and the residual is defined as

$$r^{(n)} := f - L_F^{\mathrm{qcf}} u^{(n)}.$$

The iteration error

$$e^{(n)} := u^{\mathrm{qcf}} - u^{(n)}$$

satisfies the recursion

$$P e^{(n)} = \left(P - \alpha L_F^{\mathrm{qcf}}\right) e^{(n-1)},$$

or equivalently,

$$e^{(n)} = \left(I - \alpha P^{-1} L_F^{\mathrm{qcf}}\right) e^{(n-1)} =: G e^{(n-1)}, \tag{22}$$

where the operator $G = I - \alpha P^{-1} L_F^{\mathrm{qcf}} : \mathscr{U} \to \mathscr{U}$ is called the *iteration matrix*. By iterating (22), we obtain that

$$e^{(n)} = \left(I - \alpha P^{-1} L_F^{\mathrm{qcf}}\right)^n e^{(0)} = G^n e^{(0)}. \tag{23}$$

Before we investigate various preconditioners, we briefly review the classical theory of linear stationary iterative methods [29]. We see from (23) that the iterative method (21) converges for every initial guess $u^{(0)} \in \mathscr{U}$ if and only if $G^n \to 0$ as $n \to \infty$. For a given norm $\|v\|$, for $v \in \mathscr{U}$, we can see from (23) that the reduction in the error after n iterations is bounded above by

$$\|G^n\| = \sup_{e^{(0)} \in \mathscr{U}} \frac{\|e^{(n)}\|}{\|e^{(0)}\|}.$$

It can be shown [29] that the convergence of the iteration for every initial guess $u^{(0)} \in \mathscr{U}$ is equivalent to the condition $\rho(G) < 1$, where $\rho(G)$ is the *spectral radius* of G,

$$\rho(G) = \max\{|\lambda_i| : \lambda_i \text{ is an eigenvalue of } G\}.$$

In fact, the Spectral Radius Theorem [29] states that

$$\lim_{n\to\infty} \|G^n\|^{1/n} = \rho(G)$$

for any vector norm on $\mathscr{U}$. However, if $\rho(G) < 1$ and $\|G\| \geq 1$, the Spectral Radius Theorem does not give any information about how large n must be to obtain

$\|G^n\| \leq 1$. On the other hand, if $\rho(G) < 1$, then there exists a norm $\|\cdot\|$ such that $\|G\| < 1$, so that G itself is a contraction [17]. In this case, we have the stronger contraction property that

$$\|e^{(n)}\| \leq \|G\|\|e^{(n-1)}\| \leq \|G\|^n\|e^{(0)}\|.$$

In the remainder of this section, we will analyze the norm of the iteration matrix, $\|G\|$, for several preconditioners P, using appropriate norms in each case.

5 The Richardson Iteration ($P = I$)

The simplest example of a linear iterative method is the Richardson iteration, where $P = I$. If follows from Lemma 4 that there exists a similarity transform S such that

$$L_F^{\mathrm{qcf}} = S^{-1}\Lambda^{\mathrm{qnl}}S, \tag{24}$$

where $\mathrm{cond}(S) \leq C$ (where C is independent of N), and Λ^{qnl} is the diagonal matrix of $\mathscr{U}^{0,2}$-eigenvalues $(\lambda_j^{\mathrm{qnl}})_{j=1}^{2N-1}$ of L_F^{qcf}. As an immediate consequence, we obtain the identity

$$G_{\mathrm{id}}(\alpha) = I - \alpha L_F^{\mathrm{qcf}} = S^{-1}\big(I - \alpha\Lambda^{\mathrm{qnl}}\big)S,$$

which yields

$$\|G_{\mathrm{id}}(\alpha)\|_{\ell_\epsilon^2} \leq \mathrm{cond}(S)\|I - \alpha\Lambda^{\mathrm{qnl}}\|_{\ell_\epsilon^2} \leq C \max_{j=1,\dots,2N-1}\big|1 - \alpha\lambda_j^{\mathrm{qnl}}\big|. \tag{25}$$

If $A_F > 0$, then it follows from Proposition 2 that $\lambda_j^{\mathrm{qnl}} > 0$ for all j, and hence that the iteration matrix $G_{\mathrm{id}}(\alpha) := I - \alpha L_F^{\mathrm{qcf}}$ is a contraction in the $\|\cdot\|_{\ell_\epsilon^2}$ norm if and only if $0 < \alpha < \alpha_{\max}^{\mathrm{id}} := 2/\lambda_{2N-1}^{\mathrm{qnl}}$. It follows from Lemma 3 that $\alpha_{\max}^{\mathrm{id}} \leq (2\epsilon^2)/\phi_F''$.

We can minimize the contraction constant for $G_{\mathrm{id}}(\alpha)$ in the $\|v\|_{S^TS}$ norm by choosing $\alpha = \alpha_{\mathrm{opt}}^{\mathrm{id}} := 2/(\lambda_1^{\mathrm{qnl}} + \lambda_{2N-1}^{\mathrm{qnl}})$, and in this case we obtain from Lemma 3 that

$$\big\|G_{\mathrm{id}}\big(\alpha_{\mathrm{opt}}^{\mathrm{id}}\big)\big\|_{\ell_\epsilon^2} \leq C\,\frac{\lambda_{2N-1}^{\mathrm{qnl}} - \lambda_1^{\mathrm{qnl}}}{\lambda_{2N-1}^{\mathrm{qnl}} + \lambda_1^{\mathrm{qnl}}} \leq C\left(1 - \frac{2A_F\epsilon^2}{\phi_F''}\right).$$

It thus follows that the contraction constant for $G_{\mathrm{id}}(\alpha)$ in the $\|\cdot\|_{\ell_\epsilon^2}$ norm is only of the order $1 - \mathrm{O}(\epsilon^2)$, even with an optimal choice of α. This is the same generic behavior that is typically observed for Richardson iterations for discretized second-order elliptic differential operators.

5.1 Numerical Example for the Richardson Iteration

In Fig. 1, we plot the error in the Richardson iteration against the iteration number. As a typical example, we use the right-hand side

$$f(x) = h(x)\cos(3\pi x), \quad \text{where} \quad h(x) = \begin{cases} 1, & x \geq 0, \\ -1, & x < 0, \end{cases} \tag{26}$$

which is smooth in the continuum region but has a discontinuity in the atomistic region. We choose $\phi_F'' = 1$, $A_F = 0.5$, and the optimal $\alpha = \alpha_{\text{opt}}^{\text{id}}$ discussed above (we note that $G_{\text{id}}(\alpha_{\text{opt}}^{\text{id}})$ depends only on A_F/ϕ_F'' and N, but $e^{(0)}$ depends on A_F and ϕ_F'' independently) . We observe initially a much faster convergence rate than the one predicted because the initial residual for (26) has a large component in the eigenspaces corresponding to the intermediate eigenvalues λ_j^{qnl} for $1 < j < 2N - 1$. However, after a few iterations the convergence behavior approximates the predicted rate.

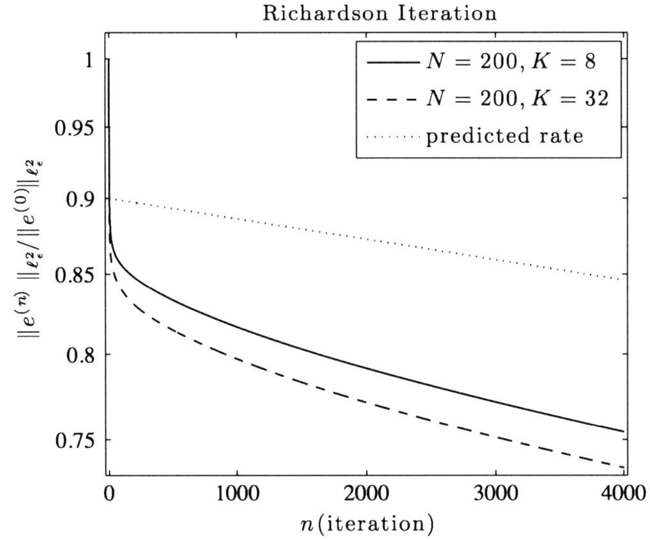

Fig. 1 Normalized ℓ^2_ϵ-error of successive Richardson iterations for the linear QCF system with $N = 200, K = 8, 32, \phi_F'' = 1, A_F = 0.5$, optimal $\alpha = \alpha_{\text{opt}}^{\text{id}}$, right-hand side (26), and starting guess $u^{(0)} = 0$

6 Preconditioning with QCL ($P = L_F^{\rm qcl} = A_F L$)

We have seen in Sect. 5 that the Richardson iteration with the trivial preconditioner $P = I$ converges slowly, and with a contraction rate of the order $1 - O(\epsilon^2)$. The goal of a (quasi-)optimal preconditioner for large systems is to obtain a performance that is independent of the system size. We will show in the present section that the preconditioner $P = A_F L$ (the system matrix for the QCL method) has this desirable quality.

Of course, preconditioning with $P = A_F L$ comes at the cost of solving a large linear system at each iteration. However, the QCL operator is a standard elliptic operator for which efficient solution methods exist. For example, the preconditioner $P = A_F L$ could be replaced by a small number of multigrid iterations, which would lead to a solver with optimal complexity. Here, we will ignore these additional complications and assume that P is inverted exactly.

Throughout the present section, the iteration matrix is given by

$$G_{\rm qcl}(\alpha) := I - \alpha (L_F^{\rm qcl})^{-1} L_F^{\rm qcf} = I - \alpha (A_F L)^{-1} L_F^{\rm qcf}, \tag{27}$$

where $\alpha > 0$ and $A_F = \phi_F'' + 4\phi_{2F}'' > 0$. We will investigate whether, if $\mathscr{U}$ is equipped with a suitable topology, $G_{\rm qcl}(\alpha)$ becomes a contraction. To demonstrate that this is a non-trivial question, we first show that in the spaces $\mathscr{U}^{1,p}$, $1 \le p < \infty$, which are natural choices for elliptic operators, this result does not hold.

Proposition 4. *If* $2 \le K \le N/2$, $\phi_{2F}'' \ne 0$, *and* $p \in [1, \infty)$, *then for any* $\alpha > 0$ *we have*

$$\| G_{\rm qcl}(\alpha) \|_{\mathscr{U}^{1,p}} \sim N^{1/p} \qquad \text{as } N \to \infty.$$

Proof. We have from (3) and $q = p/(p-1)$ the inequality

$$\begin{aligned}
\| L^{-1} L_F^{\rm qcf} \|_{\mathscr{U}^{1,p}} &= \max_{\substack{u \in \mathscr{U} \\ \|u'\|_{\ell_\epsilon^p} = 1}} \| (L^{-1} L_F^{\rm qcf} u)' \|_{\ell_\epsilon^p} \\
&\le 2 \max_{\substack{u,v \in \mathscr{U} \\ \|u'\|_{\ell_\epsilon^p} = 1,\ \|v'\|_{\ell_\epsilon^q} = 1}} \left\langle (L^{-1} L_F^{\rm qcf} u)', v' \right\rangle \\
&= 2 \max_{\substack{u,v \in \mathscr{U} \\ \|u'\|_{\ell_\epsilon^p} = 1,\ \|v'\|_{\ell_\epsilon^q} = 1}} \left\langle L (L^{-1} L_F^{\rm qcf} u), v \right\rangle \\
&= 2 \max_{\substack{u,v \in \mathscr{U} \\ \|u'\|_{\ell_\epsilon^p} = 1,\ \|v'\|_{\ell_\epsilon^q} = 1}} \left\langle L_F^{\rm qcf} u, v \right\rangle \\
&= 2 \| L_F^{\rm qcf} \|_{L(\mathscr{U}^{1,p},\ \mathscr{U}^{-1,p})},
\end{aligned}$$

as well as the reverse inequality

$$\| L_F^{\rm qcf} \|_{L(\mathscr{U}^{1,p},\ \mathscr{U}^{-1,p})} \le \| L^{-1} L_F^{\rm qcf} \|_{\mathscr{U}^{1,p}}.$$

The result now follows from the definition of $G_{\rm qcl}(\alpha)$ in (27), Lemma 1, and the fact that $\alpha > 0$ and $A_F > 0$. □

We will return to an analysis of the QCL preconditioner in the space $\mathscr{U}^{1,2}$ in Sect. 6.3, but will first attempt to prove convergence results in alternative norms.

6.1 Analysis of the QCL Preconditioner in $\mathscr{U}^{2,\infty}$

We have found in our previous analyses of the QCF method [10, 11] that it has superior properties in the function spaces $\mathscr{U}^{1,\infty}$ and $\mathscr{U}^{2,\infty}$. Hence, we will now investigate whether α can be chosen such that $G_{\rm qcl}(\alpha)$ is a contraction, uniformly as $N \to \infty$. In [10], we have found that the analysis is easiest with the somewhat unusual choice $\mathscr{U}^{2,\infty}$. Hence we begin by analyzing $G_{\rm qcl}(\alpha)$ in this space.

To begin, we formulate a lemma in which we compute the operator norm of $G_{\rm qcl}(\alpha)$ explicitly. Its proof is slightly technical and is therefore postponed to Appendix 7.

Lemma 8. *If $N \geq 4$, then*

$$\big\| G_{\rm qcl}(\alpha) \big\|_{\mathscr{U}^{2,\infty}} = \Big| 1 - \alpha\big(1 - \tfrac{2\phi''_{2F}}{A_F}\big) \Big| + \alpha \Big| \tfrac{2\phi''_{2F}}{A_F} \Big|.$$

What is remarkable (though not necessarily surprising) about this result is that the operator norm of $G_{\rm qcl}(\alpha)$ is independent of N and K. This immediately puts us into a position where we can obtain contraction properties of the iteration matrix $G_{\rm qcl}(\alpha)$, that are uniform in N and K. It is worth noting, though, that the optimal contraction rate is not uniform as A_F approaches zero; that is, the preconditioner does not give uniform efficiency as the system approaches its stability limit.

Theorem 3. *Suppose that $N \geq 4$, $A_F > 0$, and $\phi''_{2F} \leq 0$, and define*

$$\alpha_{\rm opt}^{\rm qcl,2,\infty} := \frac{A_F}{A_F + 2|\phi''_{2F}|} = \frac{2A_F}{\phi''_F + A_F} \quad \textit{and} \quad \alpha_{\max}^{\rm qcl,2,\infty} := \frac{2A_F}{\phi''_F}.$$

Then $G_{\rm qcl}(\alpha)$ is a contraction of $\mathscr{U}^{2,\infty}$ if and only if $0 < \alpha < \alpha_{\max}^{\rm qcl,2,\infty}$, and for any such choice the contraction rate is independent of N and K. The optimal choice is $\alpha = \alpha_{\rm opt}^{\rm qcl,2,\infty}$, which gives the contraction rate

$$\big\| G_{\rm qcl}(\alpha_{\rm opt}^{\rm qcl,2,\infty}) \big\|_{\mathscr{U}^{2,\infty}} = \frac{1 - \frac{A_F}{\phi''_F}}{1 + \frac{A_F}{\phi''_F}} < 1.$$

Proof. Note that $\alpha_{\rm opt}^{\rm qcl,2,\infty} = 1/\big(1 - \frac{2\phi''_{2F}}{A_F}\big)$. Hence, if we assume, first, that $0 < \alpha \leq \alpha_{\rm opt}^{\rm qcl,2,\infty}$, then

$$\|G_{\mathrm{qcl}}(\alpha)\|_{\mathscr{U}^{2,\infty}} = 1-\alpha\big(1-2\tfrac{\phi''_{2F}}{A_F}\big)-2\alpha\tfrac{\phi''_{2F}}{A_F} = 1-\alpha =: m_1(\alpha).$$

The optimal choice is clearly $\alpha = \alpha_{\mathrm{opt}}^{\mathrm{qcl},2,\infty}$ which gives the contraction rate

$$\big\|G_{\mathrm{qcl}}\big(\alpha_{\mathrm{opt}}^{\mathrm{qcl},2,\infty}\big)\big\|_{\mathscr{U}^{2,\infty}} = \alpha_{\mathrm{opt}}^{\mathrm{qcl},2,\infty}\left|\frac{2\phi''_{2F}}{A_F}\right| = \frac{2|\phi''_{2F}|}{\phi''_F+2\phi''_{2F}} = \frac{1-\frac{A_F}{\phi''_F}}{1+\frac{A_F}{\phi''_F}}.$$

Alternatively, if $\alpha \ge \alpha_{\mathrm{opt}}^{\mathrm{qcl},2,\infty}$, then

$$\big\|G_{\mathrm{qcl}}(\alpha)\big\|_{\mathscr{U}^{2,\infty}} = \alpha\big(1-\tfrac{4\phi''_{2F}}{A_F}\big)-1 = \alpha\frac{\phi''_F}{A_F}-1 =: m_2(\alpha).$$

This value is strictly increasing with α, hence the optimal choice is again $\alpha = \alpha_{\mathrm{opt}}^{\mathrm{qcl},2,\infty}$.

Moreover, we have $m_2(\alpha) < 1$ if and only if

$$\alpha < \frac{2A_F}{\phi''_F} = \alpha_{\max}^{\mathrm{qcl},2,\infty}.$$

Since, for $\alpha = \alpha_{\mathrm{opt}}^{\mathrm{qcl},2,\infty}$ we have $m_1(\alpha) = m_2(\alpha) < 1$, it follows that $\alpha_{\max}^{\mathrm{qcl},2,\infty} > \alpha_{\mathrm{opt}}^{\mathrm{qcl},2,\infty}$ (as a matter of fact, the condition $\alpha_{\max}^{\mathrm{qcl},2,\infty} > \alpha_{\mathrm{opt}}^{\mathrm{qcl},2,\infty}$ is equivalent to $A_F > 0$). In conclusion, we have shown that $\|G_{\mathrm{qcl}}(\alpha)\|_{\mathscr{U}^{2,\infty}}$ is independent of N and K and that it is strictly less than one if and only if $\alpha < \alpha_{\max}^{\mathrm{qcl},2,\infty}$, with optimal value $\alpha = \alpha_{\mathrm{opt}}^{\mathrm{qcl},2,\infty}$. □

As an immediate corollary, we obtain the following general convergence result.

Corollary 1. *Suppose that* $N \ge 4$, $A_F > 0$, $\phi''_{2F} \le 0$, *and suppose that* $\|\cdot\|_X$ *is a norm defined on* $\mathscr{U}$ *such that*

$$\|u\|_X \le C\,\|u\|_{\mathscr{U}^{2,\infty}} \qquad \forall u \in \mathscr{U}.$$

Moreover, suppose that $0 < \alpha < \alpha_{\max}^{\mathrm{qcl},2,\infty}$. *Then, for any* $u \in \mathscr{U}$,

$$\big\|G_{\mathrm{qcl}}(\alpha)^n u\big\|_X \le \hat{q}^n C\,\|u\|_{\mathscr{U}^{2,\infty}} \to 0 \quad \text{as } n \to \infty,$$

where $\hat{q} := \|G_{\mathrm{qcl}}(\alpha)\|_{\mathscr{U}^{2,\infty}} < 1$.

In particular, the convergence is uniform among all N, K *and all possible initial values* $u \in \mathscr{U}$ *for which a uniform bound on* $\|u\|_{\mathscr{U}^{2,\infty}}$ *holds.*

Proof. We simply note that, according to Theorem 3, for $0 < \alpha < \alpha_{\max}^{\mathrm{qcl},2,\infty}$, we have

$$\big\|G_{\mathrm{qcl}}(\alpha)^n\big\|_{\mathscr{U}^{2,\infty}} \le \hat{q}^n,$$

where $\hat{q} := \|G_{\mathrm{qcl}}(\alpha)\|_{\mathscr{U}^{2,\infty}} < 1$ is a number that is independent of N and K. Hence, we have

$$\left\| G_{\rm qcl}(\alpha)^n u \right\|_X \le C \left\| G_{\rm qcl}(\alpha)^n u \right\|_{\mathscr{U}^{2,\infty}} \le C \hat{q}^n \|u\|_{\mathscr{U}^{2,\infty}}. \qquad \square$$

Remark 2. Although we have seen in Theorem 3 and Corollary 1 that the linear stationary method with preconditioner $A_F L$ and with sufficiently small step size α is convergent, this convergence may still be quite slow if the initial data is "rough." Particularly in the context of defects, we may, for example, be interested in the convergence properties of this iteration when the initial residual is small or moderate in $\mathscr{U}^{1,p}$, for some $p \in [1, \infty]$, but possibly of order $O(N)$ in the $\mathscr{U}^{2,\infty}$-norm. We can see from the following Poincaré and inverse inequalities

$$\|u\|_{\mathscr{U}^{1,\infty}} \le \frac{1}{2} \|u\|_{\mathscr{U}^{2,\infty}} \qquad \text{and} \qquad \|u\|_{\mathscr{U}^{2,\infty}} \le 2N \|u\|_{\mathscr{U}^{1,\infty}} \qquad \text{for all } u \in \mathscr{U};$$

that the application of Corollary 1 to the case $X = \mathscr{U}^{1,\infty}$ gives the estimate

$$\left\| G_{\rm qcl}(\alpha)^n u \right\|_{\mathscr{U}^{1,\infty}} \le \hat{q}^n N \|u\|_{\mathscr{U}^{1,\infty}} \qquad \text{for all } u \in \mathscr{U}.$$

Similarly, with $X = \mathscr{U}^{1,2}$, we obtain

$$\left\| G_{\rm qcl}(\alpha)^n u \right\|_{\mathscr{U}^{1,2}} \le \hat{q}^n N^{3/2} \|u\|_{\mathscr{U}^{1,2}} \qquad \text{for all } u \in \mathscr{U}. \tag{28}$$

We have seen in Proposition 4 that a direct convergence analysis in $\mathscr{U}^{1,p}$, $p < \infty$, may be difficult with analytical methods, hence we focus in the next section on the case $\mathscr{U}^{1,\infty}$.

6.2 Analysis of the QCL Preconditioner in $\mathscr{U}^{1,\infty}$

As before, we first compute the operator norm of the iteration matrix explicitly. The proof of the following lemma is again postponed to the Appendix 7.

Lemma 9. *If $K \ge 3$, $N \ge \max(9, K+3)$, and $\phi''_{2F} \le 0$, then*

$$\left\| G_{\rm qcl}(\alpha) \right\|_{\mathscr{U}^{1,\infty}} = \begin{cases} \left|1-\alpha\right| + \alpha 4 \left|\frac{\phi''_{2F}}{A_F}\right| & \text{for } 0 \le \alpha \le \alpha_{\rm opt}^{\rm qcl,1,\infty}, \\ \left|1-\alpha\left(1-2\frac{\phi''_{2F}}{A_F}\right)\right| + \alpha(6+2\epsilon-4\epsilon K)\left|\frac{\phi''_{2F}}{A_F}\right| & \text{for } \alpha_{\rm opt}^{\rm qcl,1,\infty} \le \alpha, \end{cases}$$

where

$$\alpha_{\rm opt}^{\rm qcl,1,\infty} := \left[1 + (2+\epsilon-2\epsilon K)\left|\frac{\phi''_{2F}}{A_F}\right|\right]^{-1}$$

satisfies $\alpha_{\rm opt}^{\rm qcl,2,\infty} \le \alpha_{\rm opt}^{\rm qcl,1,\infty} \le 1$.

Again we note that the operator norm is independent, but now up to terms of order $O(\epsilon K)$, of the system size.

Theorem 4. *Suppose that $K \ge 3$, $N \ge \max(9, K+3)$, and $\phi''_{2F} < 0$, then the following statements are true:*

(i) If $\phi''_F + 8\phi''_{2F} \le 0$*, then* $G_{\rm qcl}(\alpha)$ *is* not *a contraction of* $\mathscr{U}^{1,\infty}$*, for any value of* α*.*

(ii) If $\phi''_F + 8\phi''_{2F} > 0$*, then* $G_{\rm qcl}(\alpha)$ *is a contraction for sufficiently small* α*. More precisely, setting*

$$\alpha_{\rm max}^{{\rm qcl},1,\infty} := \frac{2A_F}{A_F + (8+2\epsilon-4\epsilon K)|\phi''_{2F}|},$$

we have that $G_{\rm qcl}(\alpha)$ *is a contraction of* $\mathscr{U}^{1,\infty}$ *if and only if* $0<\alpha<\alpha_{\rm max}^{{\rm qcl},1,\infty}$*. The operator norm* $\|G_{\rm qcl}(\alpha)\|_{\mathscr{U}^{1,\infty}}$ *is minimized by choosing* $\alpha = \alpha_{\rm opt}^{{\rm qcl},1,\infty}$ *(cf. Lemma 9) and in this case*

$$\left\|G_{\rm qcl}\big(\alpha_{\rm opt}^{{\rm qcl},1,\infty}\big)\right\|_{\mathscr{U}^{1,\infty}} = 1 - \frac{\phi''_F + 8\phi''_{2F}}{\phi''_F + (2-\epsilon+2\epsilon K)\phi''_{2F}} < 1.$$

Proof. Suppose, first, that $0<\alpha\le\alpha_{\rm opt}^{{\rm qcl},1,\infty}$. Since $\alpha_{\rm opt}^{{\rm qcl},1,\infty} \le 1$ it follows that

$$\left\|G_{\rm qcl}(\alpha)\right\|_{\mathscr{U}^{1,\infty}} = 1 - \alpha\frac{\phi''_F + 8\phi''_{2F}}{A_F},$$

and hence $\|G_{\rm qcl}(\alpha)\|_{\mathscr{U}^{1,\infty}} < 1$ if and only if $\phi''_F + 8\phi''_{2F} > 0$. In that case $\|G_{\rm qcl}(\alpha)\|_{\mathscr{U}^{1,\infty}}$ is strictly decreasing in $(0,\alpha_{\rm opt}^{{\rm qcl},1,\infty}]$.

Since $\alpha_{\rm opt}^{{\rm qcl},1,\infty} \ge \alpha_{\rm opt}^{{\rm qcl},2,\infty} = (1-2\frac{\phi''_{2F}}{A_F})^{-1}$ we can see that $\|G_{\rm qcl}(\alpha)\|_{\mathscr{U}^{1,\infty}}$ is always strictly increasing in $[\alpha_{\rm opt}^{{\rm qcl},1,\infty},+\infty)$ and hence if $\phi''_F + 8\phi''_{2F} > 0$, then $\alpha = \alpha_{\rm opt}^{{\rm qcl},1,\infty}$ minimizes the operator norm $\|G_{\rm qcl}(\alpha)\|_{\mathscr{U}^{1,\infty}}$. Moreover, straightforward computations show that $\alpha_{\rm max}^{{\rm qcl},1,\infty} > \alpha_{\rm opt}^{{\rm qcl},1,\infty}$ and that $\|G_{\rm qcl}(\alpha)\|_{\mathscr{U}^{1,\infty}} < 1$ if and only if $0<\alpha<\alpha_{\rm max}^{{\rm qcl},1,\infty}$.

We remark that the optimal value of α in $\mathscr{U}^{1,\infty}$, that is $\alpha = \alpha_{\rm opt}^{{\rm qcl},1,\infty}$, is not the same as the optimal value, $\alpha_{\rm opt}^{{\rm qcl},2,\infty}$, in $\mathscr{U}^{2,\infty}$. However, it is easy to see that $\alpha_{\rm opt}^{{\rm qcl},1,\infty} = \alpha_{\rm opt}^{{\rm qcl},2,\infty} + O(\epsilon K)$, and hence, even though $\alpha_{\rm opt}^{{\rm qcl},2,\infty}$ is not optimal in $\mathscr{U}^{1,\infty}$ it is still close to the optimal value. On the other hand, $\alpha_{\rm max}^{{\rm qcl},1,\infty}$ and $\alpha_{\rm max}^{{\rm qcl},2,\infty}$ are not close, since, if $4\epsilon K - 2\epsilon < 1$, then

$$\alpha_{\rm max}^{{\rm qcl},1,\infty} \le \frac{2A_F}{\phi''_F + 3|\phi''_{2F}|} < \frac{2A_F}{\phi''_F} = \alpha_{\rm max}^{{\rm qcl},2,\infty}. \qquad \square$$

In summary, we have seen that the contraction property of $G_{\rm qcl}(\alpha)$ in $\mathscr{U}^{1,\infty}$ is significantly more complicated than in $\mathscr{U}^{2,\infty}$, and that, in fact, $G_{\rm qcl}(\alpha)$ is *not* a contraction for all macroscopic strains F up to the critical strain F_*.

6.3 Analysis of the QCL Preconditioner in $\mathscr{U}^{1,2}$

Even though we were able to prove uniform contraction properties for the QCL-preconditioned iterative method in $\mathscr{U}^{2,\infty}$, we have argued above that these are not entirely satisfactory in the presence of irregular solutions containing defects. Hence we analyzed the iteration matrix $G_{\rm qcl}(\alpha) = I - \alpha(A_F L)^{-1} L_F^{\rm qcf}$ in $\mathscr{U}^{1,\infty}$, but there we showed that it is not a contraction up to the critical load F_*. To conclude our results for the QCL preconditioner, we present a discussion of $G_{\rm qcl}(\alpha)$ in the space $\mathscr{U}^{1,2}$.

We begin by noting that it follows from (22) that

$$\begin{aligned} P^{1/2}e^{(n)} &= P^{1/2}G_{\rm qcl}(\alpha)e^{(n-1)} = P^{1/2}\left(I - \alpha P^{-1}L_F^{\rm qcf}\right)P^{-1/2}\left(P^{1/2}e^{(n-1)}\right) \\ &= \left(I - \alpha P^{-1/2}L_F^{\rm qcf}P^{-1/2}\right)\left(P^{1/2}e^{(n-1)}\right) =: \widetilde{G}_{\rm qcl}(\alpha)\left(P^{1/2}e^{(n-1)}\right). \end{aligned}$$

Since $\|P^{1/2}v\|_{\ell^2_\epsilon} = A_F^{1/2}\|v\|_{\mathscr{U}^{1,2}}$ for $v \in \mathscr{U}$, it follows that $G_{\rm qcl}(\alpha)$ is a contraction in $\mathscr{U}^{1,2}$ if and only if $\widetilde{G}_{\rm qcl}(\alpha)$ is a contraction in ℓ^2_ϵ. Unfortunately, we have shown in Proposition 4 that $\|G_{\rm qcl}(\alpha)\|_{\mathscr{U}^{1,2}} \sim N^{1/2}$ as $N \to \infty$. Hence, we will follow the idea used in Sect. 5 and try to find an alternative norm with respect to which $\widetilde{G}_{\rm qcl}(\alpha)$ is a contraction.

From Lemma 5 we deduce that there exists a similarity transform $\tilde{S}$ such that $\mathrm{cond}(\tilde{S}) \le N^2$, and such that

$$L^{-1/2}L_F^{\rm qcf}L^{-1/2} = \tilde{S}^{-1}\widetilde{\Lambda}^{\rm qnl}\tilde{S},$$

where $\widetilde{\Lambda}^{\rm qnl}$ is the diagonal matrix of $\mathscr{U}^{1,2}$-eigenvalues $(\mu_j^{\rm qnl})_{j=1}^{2N-1}$ of $L_F^{\rm qnl}$. As an immediate consequence we obtain

$$\widetilde{G}_{\rm qcl}(\alpha) = \tilde{S}^{-1}\big(I - \tfrac{\alpha}{A_F}\widetilde{\Lambda}^{\rm qnl}\big)\tilde{S}.$$

Proceeding as in Sect. 5, we would obtain that $\|\widetilde{G}_{\rm qcl}(\alpha)\|_{\ell^2_\epsilon} \le O(N^2)$. Instead, we observe that

$$\begin{aligned} \big\|G_{\rm qcl}(\alpha)u\big\|_{\tilde{S}^T\tilde{S}} &= \big\|\tilde{S}\widetilde{G}_{\rm qcl}(\alpha)u\big\|_{\ell^2_\epsilon} = \big\|(I - \tfrac{\alpha}{A_F}\widetilde{\Lambda}^{\rm qnl})\tilde{S}u\big\|_{\ell^2_\epsilon} \\ &\le \big\|I - \tfrac{\alpha}{A_F}\widetilde{\Lambda}^{\rm qnl}\big\|_{\ell^2_\epsilon}\|\tilde{S}u\|_{\ell^2_\epsilon} = \max_{j=1,\dots,2N-1}\big|1 - \tfrac{\alpha}{A_F}\mu_j^{\rm qnl}\big|\|u\|_{\tilde{S}^T\tilde{S}}, \end{aligned}$$

that is,

$$\big\|\widetilde{G}_{\rm qcl}(\alpha)\big\|_{\tilde{S}^T\tilde{S}} \le \max_{j=1,\dots,2N-1}\big|1 - \tfrac{\alpha}{A_F}\mu_j^{\rm qnl}\big|. \tag{29}$$

Thus, we can conclude that $\widetilde{G}_{\rm qcl}(\alpha)$ is a contraction in the $\|\cdot\|_{\tilde{S}^T\tilde{S}}$-norm if and only if $0 < \alpha < \alpha_{\max}^{\rm qcl,1,2} := 2A_F/\mu_{2N-1}^{\rm qnl}$. Moreover, we obtain the error bound

$$\|e^{(n)}\|_{\mathscr{U}^{1,2}} \le \mathrm{cond}(\tilde{S})\,\tilde{q}^n\,\|e^{(0)}\|_{\mathscr{U}^{1,2}} \le N^2\tilde{q}^n\|e^{(0)}\|_{\mathscr{U}^{1,2}},$$

where $\tilde{q} := \left\| \widetilde{G}_{\mathrm{qcl}}(\alpha) \right\|_{\tilde{S}^T \tilde{S}}$. This is slightly worse in fact, than (28), however, we note that this large prefactor cannot be seen in the following numerical experiment.

Moreover, optimizing the contraction rate with respect to α leads to the choice $\alpha_{\mathrm{opt}}^{\mathrm{qcl},1,2} := 2A_F/(\mu_1^{\mathrm{qnl}} + \mu_{2N-1}^{\mathrm{qnl}})$, and in this case we obtain from Lemma 6 that

$$\tilde{q} = \tilde{q}_{\mathrm{opt}} := \left\| \widetilde{G}_{\mathrm{qcl}}\big(\alpha_{\mathrm{opt}}^{\mathrm{qcl},1,2}\big) \right\|_{\tilde{S}^T \tilde{S}} = \frac{\mu_{2N-1}^{\mathrm{qnl}} - \mu_1^{\mathrm{qnl}}}{\mu_{2N-1}^{\mathrm{qnl}} + \mu_1^{\mathrm{qnl}}} \le \frac{1 - \frac{A_F}{\phi_F''}}{1 + \frac{A_F}{\phi_F''}},$$

where the upper bound is sharp in the limit $K \to \infty$. It is particularly interesting to note that the contraction rate obtained here is precisely the same as the one in $\mathscr{U}^{2,\infty}$ (cf. Theorem 3). Moreover, it can be easily seen from Lemma 6 that $\alpha_{\mathrm{opt}}^{\mathrm{qcl},1,2} \to \alpha_{\mathrm{opt}}^{\mathrm{qcl},2,\infty}$ as $K \to \infty$, which is the optimal stepsize according to Theorem 3. We further have that $\alpha_{\max}^{\mathrm{qcl},1,2} \to \alpha_{\max}^{\mathrm{qcl},2,\infty}$ as $K \to \infty$.

6.4 Numerical Example for QCL-Preconditioning

We now apply the QCL-preconditioned stationary iterative method to the QCF system with right-hand side (26), $\phi_F'' = 1$, $A_F = 0.2$, and the optimal value $\alpha = \alpha_{\mathrm{opt}}^{\mathrm{qcl},2,\infty}$ (we note that $G_{\mathrm{id}}(\alpha_{\mathrm{opt}}^{\mathrm{qcl},2,\infty})$ depends only on A_F/ϕ_F'' and N, but $e^{(0)}$ depends on A_F and ϕ_F'' independently). The error for successive iterations in the $\mathscr{U}^{1,2}$, $\mathscr{U}^{1,\infty}$ and $\mathscr{U}^{2,\infty}$-norms are displayed in Fig. 2. Even though our theory, in this case, predicts a perfect contractive behavior only in $\mathscr{U}^{2,\infty}$ and (partially) in $\mathscr{U}^{1,2}$, we nevertheless observe perfect agreement with the optimal predicted rate also in the $\mathscr{U}^{1,\infty}$-norms. As a matter of fact, the parameters are chosen so that case (1) of Theorem 4 holds, that is, $G_{\mathrm{qcl}}(\alpha)$ is *not* a contraction of $\mathscr{U}^{1,\infty}$. A possible explanation why we still observe this perfect asymptotic behavior is that the norm of $G_{\mathrm{qcl}}(\alpha)$ is attained in a subspace that is never entered in this iterative process. This is also supported by the fact that the exact solution is uniformly bounded in $\mathscr{U}^{2,\infty}$ as $N, K \to \infty$, which is a simple consequence of Proposition 3.

7 Preconditioning with QCE ($P = L_F^{\mathrm{qce}}$): Ghost-Force Correction

We have shown in [5, 12] that the popular *ghost force correction method (GFC)* is equivalent to preconditioning the QCF equilibrium equations by the QCE equilibrium equations. The ghost force correction method in a quasi-static loading can thus be reduced to the question whether the iteration matrix

$$G_{\mathrm{qce}} := I - (L_F^{\mathrm{qce}})^{-1} L_F^{\mathrm{qcf}}$$

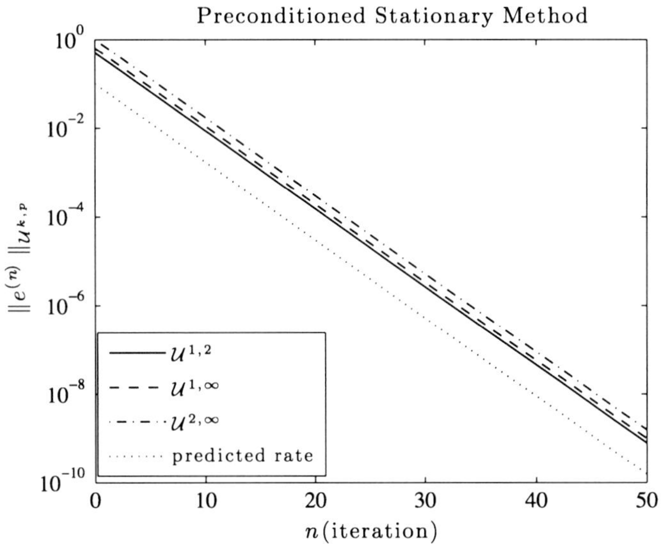

Fig. 2 Error of the QCL-preconditioned linear stationary iterative method for the QCF system with $N = 800$, $K = 32$, $\phi''_F = 1$, $A_F = 0.2$, optimal value $\alpha = \alpha_{\rm opt}^{\rm qcl,2,\infty}$, and right-hand side (26). In this case, the iteration matrix $G_{\rm qcl}(\alpha)$ is *not* a contraction of $\mathcal{U}^{1,\infty}$. Even though our theory predicts a perfect contractive behavior only in $\mathcal{U}^{2,\infty}$, we observe perfect agreement with the optimal predicted rate also in the $\mathcal{U}^{1,2}$ and $\mathcal{U}^{1,\infty}$-norms

is a contraction. Due to the typical usage of the preconditioner $L_F^{\rm qce}$ in this case, we do not consider a step size α in this section. The purpose of the present section is (1) to investigate whether there exist function spaces in which $G_{\rm qce}$ is a contraction; and (2) to identify the range of the macroscopic strains F where $G_{\rm qce}$ is a contraction.

We begin by recalling the fundamental stability result for the $L_F^{\rm qce}$ operator, Theorem 1:

$$\inf_{\substack{u\in\mathcal{U}\\ \|u'\|_{\ell^2_\epsilon}=1}} \langle L_F^{\rm qce} u, u\rangle = A_F + \lambda_K \phi''_{2F},$$

where $\lambda_K \sim \lambda_* + O(e^{-cK})$ with $\lambda_* \approx 0.6595$. This result shows that the GFC iteration must necessarily run into instabilities before the deformation reaches the critical strain F_c^*. This is made precise in the following corollary which states that there is no norm with respect to which $G_{\rm qce}$ is a contraction up to the critical strain F_*.

Corollary 2. *Fix N and K, and let $\|\cdot\|_X$ be an arbitrary norm on the space $\mathcal{U}$, then, upon understanding $G_{\rm qce}$ as dependent on ϕ''_F and ϕ''_{2F}, we have*

$$\|G_{\rm qce}\|_X \to +\infty \quad as \quad A_F + \lambda_K \phi''_{2F} \to 0.$$

Despite this negative result, we may still be interested in the question of whether the GFC iteration is a contraction in "very stable regimes," that is, for macroscopic strains which are far away from the critical strain F_*. Naturally, we are particularly interested in the behavior as $N \to \infty$, that is, we will investigate in which function spaces the operator norm of $G_{\rm qce}$ remains bounded away from one as $N \to \infty$. Theorem 2 on the unboundedness of $L_F^{\rm qcf}$ immediately provides us with the following negative answer.

Proposition 5. *If* $2 \le K \le N/2$, $\phi''_{2F} \ne 0$, *and* $A_F + \lambda_K \phi''_{2F} > 0$, *then*

$$\|G_{\rm qce}\|_{\mathscr{U}^{1,2}} \sim N^{1/2}, \qquad \text{as } N \to \infty.$$

Proof. It is an easy exercise to show that, if $A_F + \lambda_K \phi''_{2F} > 0$, then the $\mathscr{U}^{1,2}$-norm is equivalent to the norm induced by $L_F^{\rm qce}$, that is,

$$C^{-1}\|u\|_{\mathscr{U}^{1,2}} \le \|u\|_{L_F^{\rm qce}} \le C\|u\|_{\mathscr{U}^{1,2}}.$$

Hence, we have $\|G_{\rm qce}\|_{\mathscr{U}^{1,2}} \approx \|G_{\rm qce}\|_{L_F^{\rm qce}}$ and by the same argument as in the proof of Proposition 4, and using again the uniform norm-equivalence, we can deduce that

$$\big\|G_{\rm qce}\big\|_{\mathscr{U}^{1,2}} \approx \big\|L_F^{\rm qcf}\big\|_{L(\mathscr{U}^{1,2},\,\mathscr{U}^{-1,2})} \pm 1 \sim N^{1/2}, \quad \text{as } N \to \infty. \qquad \square$$

Since the operator $(L_F^{\rm qce})^{-1} L_F^{\rm qcf}$ is more complicated than that of $(A_F L)^{-1} L_F^{\rm qcf}$, which we analyzed in the previous section, we continue to investigate the contraction properties of $G_{\rm qce}$ in various different norms in numerical experiments. In Fig. 3, we plot the operator norm of $G_{\rm qce}$, in the function spaces

$$\mathscr{U}^{k,p}, \quad k = 0,1,2, \quad p = 1,2,\infty,$$

against the system size N (see Appendix 7 for a description of how we compute $\|G_{\rm qce}\|_{\mathscr{U}^{k,p}}$). This experiment is performed for $A_F/\phi''_F = 0.8$ which is at some distance from the singularity of $L_F^{\rm qce}$ (we note that $G_{\rm qce}$ depends only on A_F/ϕ''_F and N since both $(\phi''_F)^{-1} L_F^{\rm qcf}$ and $(\phi''_F)^{-1} L_F^{\rm qce}$ depend only on A_F/ϕ''_F and N). The experiments suggests clearly that $\|G_{\rm qce}\|_{\mathscr{U}^{k,p}} \to \infty$ as $N \to \infty$ for all norms except for $\mathscr{U}^{1,\infty}$ and $\mathscr{U}^{2,1}$.

Hence, in a second experiment, we investigate how $\|G_{\rm qce}\|_{\mathscr{U}^{1,\infty}}$ and $\|G_{\rm qce}\|_{\mathscr{U}^{2,1}}$ behave, for fixed N and K, as $A_F + \lambda_K \phi''_{2F}$ approaches zero. The results of this experiment, which are displayed in Fig. 4, confirm the prediction of Corollary 2 that $\|G_{\rm qce}\|_{\mathscr{U}^{k,p}} \to \infty$ as $A_F + \lambda_K \phi''_{2F}$ approaches zero. Indeed, they show that $\|G_{\rm qce}\|_{\mathscr{U}^{k,p}} > 1$ already much earlier, namely around a strain F where $A_F \approx 0.52$ and $A_F + \lambda_K \phi''_{2F} \approx 0.44$.

Our conclusion based on these analytical results and numerical experiments is that the GFC method is not universally reliable near the limit strain F_*, that is, under conditions near the formation or movement of a defect it can fail to converge to a stable solution of the QCF equilibrium equations as the quasi-static loading step tends to zero or the number of GFC iterations tends to infinity. Even though the simple model problem that we investigated here cannot, of course, provide a

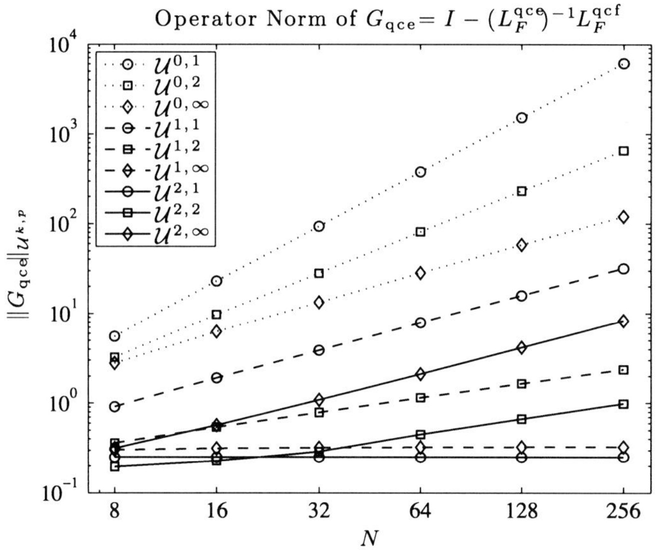

Fig. 3 Graphs of the operator norm $\|G_{\text{qce}}\|_{\mathcal{U}^{k,p}}$, $k = 0, 1, 2$, $p = 1, 2, \infty$, plotted against the number of atoms, N, with atomistic region size $K = \lceil \sqrt{N} \rceil - 1$, and $A_F/\phi''_F = 0.8$. (The graph for the $\mathcal{U}^{1,p}$-norms, $p = 1, \infty$, are only estimates up to a factor of $1/2$; cf. Appendix 7.) The graphs clearly indicate that $\|G_{\text{qce}}\|_{\mathcal{U}^{k,p}} \to \infty$ as $N \to \infty$ in all spaces except for $\mathcal{U}^{1,\infty}$ and $\mathcal{U}^{2,1}$

definite statement, it shows at the very least that further investigations for more realistic model problems are required.

Conclusion

We proposed and studied linear stationary iterative solution methods for the QCF method with the goal of identifying iterative schemes that are efficient and reliable for all applied loads. We showed that, if the local QC operator is taken as the preconditioner, then the iteration is guaranteed to converge to the solution of the QCF system, up to the critical strain. What is interesting is that the choice of function space plays a crucial role in the efficiency of the iterative method. In $\mathcal{U}^{2,\infty}$, the convergence is always uniform in N and K, however, in $\mathcal{U}^{1,\infty}$ this is only true if the macroscopic strain is at some distance from the critical strain. This indicates that, in the presence of defects (that is, non-smooth solutions), the efficiency of a QCL-preconditioned method may be reduced. Further investigations for more realistic model problems are required to shed light on this issue.

We also showed that the popular GFC iteration must necessarily run into instabilities before the deformation reaches the critical strain F_c^*. Even for macroscopic strains that are far lower than the critical strain F_*, we showed that $\|G_{\text{qce}}\|_{\mathcal{U}^{1,2}}$

$\sim N^{1/2}$. We then gave numerical experiments that suggest that $\|G_{\rm qce}\|_{\mathscr{U}^{k,p}} \to \infty$ as $N \to \infty$ for all tested norms except for $\mathscr{U}^{1,\infty}$ and $\mathscr{U}^{2,1}$.

The results presented in this paper demonstrate the challenge for the development of reliable and efficient iterative methods for force-based approximation methods. Further analysis and numerical experiments for two and three dimensional problems are needed to more fully assess the implications of the results in this paper for realistic materials applications.

Acknowledgements This work was supported in part by the National Science Foundation under DMS-0757355, DMS-0811039, the PIRE Grant OISE-0967140, the Institute for Mathematics and Its Applications, and the University of Minnesota Supercomputing Institute. This work was also supported by the Department of Energy under Award Number DE-SC0002085. CO was supported by the EPSRC grant EP/H003096/1 "Analysis of Atomistic-to-Continuum Coupling Methods."

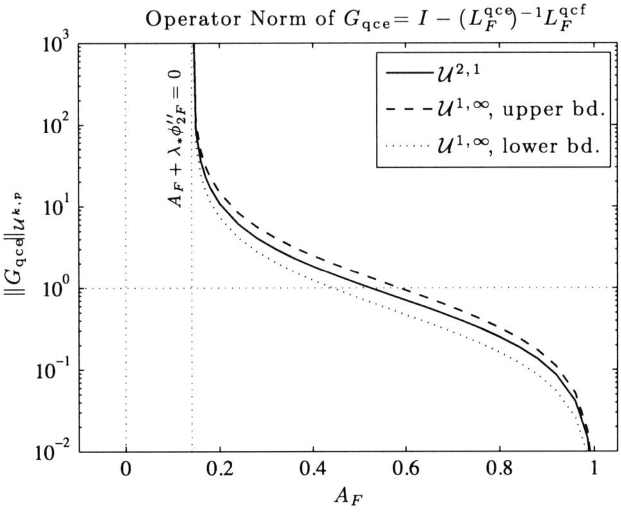

Fig. 4 Graphs of the operator norm $\|G_{\rm qce}\|_{\mathscr{U}^{k,p}}$, $(k,p) \in \{(1,\infty),(2,1)\}$, for fixed $N = 256, K = 15$, $\phi''_F = 1$, plotted against A_F. For the case $\mathscr{U}^{1,\infty}$ only estimates are available and upper and lower bounds are shown instead (cf. Appendix 7). The graphs confirm the result of Corollary 2 that $\|G_{\rm qce}\|_{\mathscr{U}^{k,p}} \to \infty$ as $A_F + \lambda_K \phi''_{2F} \to 0$. Moreover, they clearly indicate that $\|G_{\rm qce}\|_{\mathscr{U}^{k,p}} > 1$ already for strains F in the region $A_F \approx 0.5$, which are much lower than the critical strain at which $L_F^{\rm qce}$ becomes singular

Appendix

Proof of Theorem 1

The purpose of this appendix is to prove the sharp stability result for the operator $L_F^{\rm qce}$, formulated in Theorem 1. Using Formula (23) in [9] we obtain the following representation of $L_F^{\rm qce}$,

$$\begin{aligned}\langle L_F^{\rm qce} u, u\rangle = &\left\{ \sum_{\ell=-N+1}^{-K-2} \epsilon A_F |u'_\ell|^2 + \sum_{\ell=K+3}^{N} \epsilon A_F |u'_\ell|^2 \right\} \\ &+ \left\{ \sum_{\ell=-K+2}^{K-1} \epsilon\Big(A_F |u'_\ell|^2 - \epsilon^2 \phi''_{2F} |u''_\ell|^2 \Big) \right\} \\ &+ \epsilon \Big\{ (A_F - \phi''_{2F})(|u'_{-K+1}|^2 + |u'_K|^2) + A_F(|u'_{-K}|^2 + |u'_{K+1}|^2) \\ &\qquad + (A_F + \phi''_{2F})(|u'_{-K-1}|^2 + |u'_{K+2}|^2) \\ &\quad - \tfrac{1}{2}\epsilon^2 \phi''_{2F}(|u''_{-K}|^2 + |u''_{-K-1}|^2 + |u''_K|^2 + |u''_{K+1}|^2) \Big\}. \end{aligned} \tag{30}$$

If $\phi''_{2F} < 0$, then we can see from this decomposition that there is a loss of stability at the interaction between atoms $-K-2$ and $-K-1$ as well as between atoms $K+1$ and $K+2$. It is therefore natural to test this expression with a displacement $\hat{u}$ defined by

$$\hat{u}'_\ell = \begin{cases} 1, & \ell = -K-1, \\ -1, & \ell = K+2, \\ 0, & \text{otherwise.} \end{cases}$$

From (30), we easily obtain

$$\langle L_F^{\rm qce} \hat{u}, \hat{u}\rangle = A_F + \tfrac{1}{2}\phi''_{2F}.$$

In particular, we see that, if $A_F + \frac{1}{2}\phi''_{2F} < 0$, then $L_F^{\rm qce}$ is indefinite. On the other hand, it was shown in [8] that $L_F^{\rm qce}$ is positive definite provided $A_F + \phi''_{2F} > 0$. (As a matter of fact, the analysis in [8] is for periodic boundary conditions, however, since the Dirichlet displacement space is contained in the periodic displacement space the result is also valid for the present case.)

Thus, we have shown that

$$\inf_{\substack{u \in \mathscr{U} \\ \|u'\|_{\ell^2_\epsilon} = 1}} \langle L_F^{\rm qce} u, u\rangle = A_F + \mu \phi''_{2F}, \quad \text{where } \tfrac{1}{2} \le \mu \le 1.$$

To conclude the proof of Theorem 1, we need to show that μ depends only on K and that the stated asymptotic result holds.

From (30) it follows that $L_F^{\rm qce}$ can be written in the form

$$\langle L_F^{\mathrm{qce}} u, u \rangle = (u')^T \mathscr{H} u',$$

where we identify u' with the vector $u' = (u'_\ell)_{\ell=-N+1}^{N}$ and where $\mathscr{H} \in \mathbb{R}^{2N \times 2N}$. Writing $\mathscr{H} = \phi''_F \mathscr{H}_1 + \phi''_{2F} \mathscr{H}_2$, we can see that $\mathscr{H}_1 = \mathrm{Id}$ and that $\mathscr{H}_2$ has the entries

$$\mathscr{H}_2 = \begin{pmatrix} \ddots & \ddots & \ddots & & & & & & \\ & 1 & 2 & 1 & & & & & \\ & & 1 & 2 & 1 & & & & \\ & & & 1 & 3/2 & 1/2 & & & \\ & & & & 1/2 & 3 & 1/2 & & \\ & & & & & 1/2 & 9/2 & 0 & \\ & & & & & & 0 & 4 & 0 \\ & & & & & & & 0 & 4 & 0 \\ & & & & & & & & \ddots & \ddots & \ddots \end{pmatrix}.$$

Here, the row with entries $[1, 3/2, 1/2]$ denotes the Kth row (in the coordinates u'_k). This form can be verified, for example, by appealing to (30). Let $\sigma(A)$ denote the spectrum of a matrix A. Since, by assumption, $\phi''_{2F} \le 0$, the smallest eigenvalue of $\mathscr{H}$ is given by

$$\min \sigma(\mathscr{H}) = \phi''_F + \phi''_{2F} \max \sigma(\mathscr{H}_2),$$

that is, we need to compute the largest eigenvalue $\bar{\lambda}$ of $\mathscr{H}_2$. Since $\mathscr{H}_2 e_k = 4 e_k$ for $k = K+3, K+4, \dots$ and for $K = -K-2, -K-3, \dots$, and since eigenvectors are orthogonal, we conclude that all other eigenvectors depend only on the submatrix describing the atomistic region and the interface. In particular, $\bar{\lambda}$ depends only on K but not on N. This proves the claim of Theorem 1 that λ_K depends indeed only on K.

We thus consider the $\{-K-1, \dots, K+2\}$-submatrix $\bar{\mathscr{H}}_2$, which has the form

$$\bar{\mathscr{H}}_2 = \begin{pmatrix} 9/2 & 1/2 & & & & & & & \\ 1/2 & 3 & 1/2 & & & & & & \\ & 1/2 & 3/2 & 1 & & & & & \\ & & 1 & 2 & 1 & & & & \\ & & & \ddots & \ddots & \ddots & & & \\ & & & & 1 & 2 & 1 & & \\ & & & & & 1 & 3/2 & 1/2 & \\ & & & & & & 1/2 & 3 & 1/2 \\ & & & & & & & 1/2 & 9/2 \end{pmatrix}.$$

Letting $\bar{\mathscr{H}}_2 \psi = \lambda \psi$, then for $\ell = -K+2, \dots, K-1$,

$$\psi_{\ell-1} + 2\psi_\ell + \psi_{\ell+1} = \lambda \psi_\ell,$$

and hence, ψ has the general form

$$\psi_\ell = a z^\ell + b z^{-\ell}, \qquad \ell = -K+1, \dots, K,$$

leaving ψ_ℓ undefined for $\ell \in \{-K, -K-1, K+1, K+2\}$ for now, and where $z, 1/z$ are the two roots of the polynomial

$$z^2 + (2-\lambda)z + 1 = 0.$$

In particular, we have

$$z = (\tfrac{1}{2}\lambda - 1) + \sqrt{(\tfrac{1}{2}\lambda - 1)^2 - 1} > 1. \tag{31}$$

To determine the remaining degrees of freedom, we could now insert this general form into the eigenvalue equation and attempt to solve the resulting problem. This leads to a complicated system which we will try to simplify.

We first note that, for any eigenvector ψ, the vector $(\psi_{K-\ell})$ is also an eigenvector, and hence we can assume without loss of generality that ψ is skew-symmetric about $\ell = 1/2$. This implies that $a = -b$. Since the scaling is irrelevant for the eigenvalue problem, we therefore make the *ansatz* $\psi_\ell = z^\ell - z^{-\ell}$. Next, we notice that for K sufficiently large the term $z^{-\ell}$ is exponentially small and therefore does not contribute to the eigenvalue equation near the right interface. We may safely ignore it if we are only interested in the asymptotics of the eigenvalue $\bar{\lambda}$ as $K \to \infty$. Thus, letting $\hat{\psi}_\ell = z^\ell$, $\ell = 1, \ldots, K$ and $\hat{\psi}_\ell$ unknown, $\ell = K+1, K+2$, we obtain the system

$$\begin{aligned} z^{K-1} + \tfrac{3}{2}z^K + \tfrac{1}{2}\hat{\psi}_{K+1} &= \hat{\lambda} z^K, \\ \tfrac{1}{2}z^K + 3\hat{\psi}_{K+1} + \tfrac{1}{2}\hat{\psi}_{K+2} &= \hat{\lambda}\hat{\psi}_{K+1}, \\ \tfrac{1}{2}\hat{\psi}_{K+1} + \tfrac{9}{2}\hat{\psi}_{K+2} &= \hat{\lambda}\hat{\psi}_{K+2}. \end{aligned}$$

The free parameters $\hat{\psi}_{K+1}, \hat{\psi}_{K+2}$ can be easily determined from the first two equations. From the final equation we can then compute $\hat{\lambda}$. Upon recalling from (31) that $\hat{z}$ can be expressed in terms of $\hat{\lambda}$, and conversely that $\hat{\lambda} = (\hat{z}^2+1)/\hat{z}+2$, we obtain a polynomial equation of degree five for $\hat{z}$,

$$q(\hat{z}) := 4\hat{z}^5 - 12\hat{z}^4 + 9\hat{z}^3 - 3\hat{z}^2 - 4\hat{z} + 2 = 0.$$

Mathematica was unable to factorize q symbolically, hence we computed its roots numerically to twenty digits precision. It turns out that q has three real roots and two complex roots. The largest real root is at $\hat{z} \approx 2.206272296$ which gives the value

$$\hat{\lambda} = \frac{\hat{z}^2 + 1}{\hat{z}} + 2 \approx 4.659525505897.$$

The relative errors that we had previously neglected are in fact of order $\hat{z}^{-2K}$, and hence we obtain

$$\lambda_K = \lambda_* + O(e^{-cK}), \qquad \text{where} \quad \lambda_* \approx 0.6595 \quad \text{and} \quad c \approx 1.5826.$$

This concludes the proof of Theorem 1. □

Proofs of Lemmas 8 and 9

In this appendix, we prove two technical lemmas from Sect. 6.1. Throughout, the iteration matrix $G_{\rm qcl}(\alpha)$ is given by

$$G_{\rm qcl}(\alpha) := I - \alpha (A_F L)^{-1} L_F^{\rm qcf},$$

where $\alpha > 0$ and $A_F = \phi_F'' + 4\phi_{2F}'' > 0$. We begin with the proof of Lemma 8, which is more straightforward.

Proof (Proof of Lemma 8). Using the basic definition of the operator norm, and the fact that $Lz = -z''$, we obtain

$$\|G_{\rm qcl}(\alpha)\|_{\mathscr{U}^{2,\infty}} = \max_{\substack{u \in \mathscr{U} \\ \|u''\|_{\ell_\epsilon^\infty}=1}} \|(G_{\rm qcl}(\alpha)u)''\|_{\ell_\epsilon^\infty} = \max_{\substack{u \in \mathscr{U} \\ \|u''\|_{\ell_\epsilon^\infty}=1}} \|-LG_{\rm qcl}(\alpha)u\|_{\ell_\epsilon^\infty}.$$

We write the operator $-LG_{\rm qcl}(\alpha) = -L + \frac{\alpha}{A_F} L_F^{\rm qcf}$ as follows:

$$\big[-LG_{\rm qcl}(\alpha)u\big]_\ell = \begin{cases} u_\ell'' - \frac{\alpha}{A_F}(A_F u_\ell''), & \text{if } \ell \in \mathscr{C}, \\ u_\ell'' - \frac{\alpha}{A_F}\big(\phi_F'' u_\ell'' + \phi_{2F}''(u_{\ell-1}'' + 2u_\ell'' + u_{\ell+1}'')\big), & \text{if } \ell \in \mathscr{A}. \end{cases} \tag{32}$$

In the continuum region, we simply obtain

$$\big[-LG_{\rm qcl}(\alpha)u\big]_\ell = (1-\alpha)u_\ell'' \qquad \text{for } \ell \in \mathscr{C}.$$

If $\ell \in \mathscr{A}$, we manipulate (32), using the definition of $A_F = \phi_F'' + 4\phi_{2F}''$, which yields

$$\begin{aligned}\big[-LG_{\rm qcl}(\alpha)u\big]_\ell &= \Big[1 - \tfrac{\alpha}{A_F}\big(\phi_F'' + 2\phi_{2F}''\big)\Big]u_\ell'' + \Big[-\tfrac{\alpha}{A_F}\phi_{2F}''\Big](u_{\ell-1}'' + u_{\ell+1}'') \\ &= \Big[1 - \alpha\big(1 - \tfrac{2\phi_{2F}''}{A_F}\big)\Big]u_\ell'' + \Big[-\alpha\tfrac{\phi_{2F}''}{A_F}\Big](u_{\ell-1}'' + u_{\ell+1}'').\end{aligned}$$

In summary, we have obtained

$$\big[-LG_{\rm qcl}(\alpha)u\big]_\ell = \begin{cases} [1-\alpha]u_\ell'', & \text{if } \ell \in \mathscr{C}, \\ \Big[1 - \alpha\big(1 - \frac{2\phi_{2F}''}{A_F}\big)\Big]u_\ell'' + \Big[-\alpha\frac{\phi_{2F}''}{A_F}\Big](u_{\ell-1}'' + u_{\ell+1}'') & \text{if } \ell \in \mathscr{A}. \end{cases}$$

It is now easy to see that

$$\|G_{\rm qcl}(\alpha)\|_{L(\mathscr{U}^{2,\infty},\, \mathscr{U}^{2,\infty})} \le \max\Big\{|1-\alpha|, \big|1 - \alpha\big(1 - \tfrac{2\phi_{2F}''}{A_F}\big)\big| + \alpha\big|\tfrac{2\phi_{2F}''}{A_F}\big|\Big\}.$$

As a matter of fact, in view of the estimate

$$\big|1 - \alpha\big(1 - \tfrac{2\phi_{2F}''}{A_F}\big)\big| + \alpha\big|\tfrac{2\phi_{2F}''}{A_F}\big| \ge |1-\alpha| - \alpha\big|\tfrac{2\phi_{2F}''}{A_F}\big| + \alpha\big|\tfrac{2\phi_{2F}''}{A_F}\big| = |1-\alpha|,$$

the upper bound can be reduced to

$$\|G_{\mathrm{qcl}}(\alpha)\|_{L(\mathscr{U}^{2,\infty},\,\mathscr{U}^{2,\infty})} \le \big|1-\alpha\big(1-\tfrac{2\phi''_{2F}}{A_F}\big)\big| + \alpha\tfrac{2|\phi''_{2F}|}{A_F}. \tag{33}$$

To show that the bound is attained, we construct a suitable test function. We define $u \in \mathscr{U}$ via

$$u''_{-1} = u''_1 = \mathrm{sign}\Big[-\alpha\tfrac{2\phi''_{2F}}{A_F}\Big], \quad u''_0 = \mathrm{sign}\Big[1-\alpha\big(1-\tfrac{2\phi''_{2F}}{A_F}\big)\Big],$$

(note that $0 \in \mathscr{A}$ for any $K \ge 0$) and the remaining values of u''_ℓ in such a way that $\sum_{\ell=-N+1}^{N} u''_\ell = 0$. If $N \ge 4$, then there exists at least one function $u \in \mathscr{U}$ with these properties and it attains the bound (33). Thus, the bound in (33) is an equality, which concludes the proof of the lemma. □

Before we prove Lemma 9, we recall an explicit representation of $L^{-1}L_F^{\mathrm{qcf}}$ that was useful in our analysis in [10]. The proof of the following result is completely analogous to that of [10, Lemma 14] and is therefore sketched only briefly. It is also convenient for the remainder of the section to define the following atomistic and continuum regions for the strains:

$$\mathscr{A}' = \{-K+1,\dots,K\} \quad \text{and} \quad \mathscr{C}' = \{-N+1,\dots,N\}\setminus\mathscr{A}'.$$

Lemma 10. *Let $u \in \mathscr{U}$ and $z = L^{-1}L_F^{\mathrm{qcf}}u$, then*

$$z'_\ell = \sigma(u')_\ell - \overline{\sigma(u')} + \phi''_{2F}\big(\tilde\alpha_{-K}(u')h_{-K,\ell} - \tilde\alpha_K(u')h_{K,\ell}\big),$$

where $\sigma(u')$, $h_{\pm K} \in \mathbb{R}^{2N}$ and $\overline{\sigma(u')}$, $\tilde\alpha_{\pm K}(u') \in \mathbb{R}$ are defined as follows:

$$\sigma(u')_\ell = \begin{cases} \phi''_F u'_\ell + \phi''_{2F}(u'_{\ell-1} + 2u'_\ell + u'_{\ell+1}), & \ell \in \mathscr{A}', \\ (\phi''_F + 4\phi''_{2F})u'_\ell, & \ell \in \mathscr{C}', \end{cases}$$

$$\overline{\sigma(u')} = \frac{1}{2N}\sum_{\ell=-N+1}^{N} \sigma(u')_\ell = \tfrac{\epsilon}{2}\phi''_{2F}\big[u'_{K+1} - u'_K - u'_{-K+1} + u'_{-K}\big],$$

$$\tilde\alpha_{-K}(u') = u'_{-K+1} - 2u'_{-K} + u'_{-K-1}, \quad \tilde\alpha_K(u') = u'_{K+2} - 2u'_{K+1} + u'_K, \quad \textit{and}$$

$$h_{\pm K,\ell} = \begin{cases} \tfrac12(1 \mp \epsilon K), & \ell = -N+1,\dots,\pm K, \\ \tfrac12(-1 \mp \epsilon K), & \ell = \pm K+1,\dots,N. \end{cases}$$

Proof. In the notation introduced above, the variational representation of L_F^{qcf} from [10, Sec. 3] reads

$$\langle L_F^{\mathrm{qcf}}u, v\rangle = \langle \sigma(u'), v'\rangle + \phi''_{2F}\big[\tilde\alpha_{-K}(u')v_{-K} - \tilde\alpha_K(u')v_K\big] \qquad \forall u, v \in \mathscr{U}.$$

Using the fact that $v_{\pm N} = 0$ and $\sum_\ell v'_\ell = 0$, it is easy to see that the discrete delta-functions appearing in this representation can be rewritten as

$$v_{\pm K} = \langle h_{\pm K}, v' \rangle.$$

Hence, we deduce that the function $z = L^{-1} L_F^{\rm qcf}$ is given by

$$\langle z', v' \rangle = \langle L_F^{\rm qcf} u, v \rangle = \big\langle \sigma(u') + \phi''_{2F} [\tilde{\alpha}_{-K}(u') h_{-K} - \tilde{\alpha}_K(u') h_K], v' \big\rangle \quad \forall v \in \mathscr{U}.$$

In particular, it follows that

$$z' = \sigma(u') + \phi''_{2F} [\tilde{\alpha}_{-K}(u') h_{-K} - \tilde{\alpha}_K(u') h_K] + C,$$

where C is chosen so that $\sum_\ell z'_\ell = 0$. Since $h_{\pm K}$ are constructed so that $\sum_\ell h_{\pm K,\ell} = 0$, we only subtract the mean of $\sigma(u')$. Hence, $C = -\overline{\sigma(u')}$, for which the stated formula is quickly verified. □

Proof (Proof of Lemma 9). Let $u \in \mathscr{U}$ with $\|u'\|_{\ell^\infty_\epsilon} \le 1$. Setting $z = G_{\rm qcl}(\alpha) u$, and employing Lemma 10, we obtain

$$\begin{aligned} z'_\ell &= u'_\ell - \tfrac{\alpha}{A_F} \Big[\sigma_\ell(u') - \overline{\sigma(u')} + \phi''_{2F} (\tilde{\alpha}_{-K}(u') h_{-K,\ell} - \tilde{\alpha}_K(u') h_{K,\ell}) \Big] \\ &= \Big[u'_\ell - \tfrac{\alpha}{A_F} \sigma_\ell(u') \Big] + \alpha \tfrac{\phi''_{2F}}{A_F} \Big[\tfrac{\epsilon}{2} (u'_{K+1} - u'_K - u'_{-K+1} + u'_{-K}) \\ &\qquad\qquad - \tilde{\alpha}_{-K}(u') h_{-K,\ell} + \tilde{\alpha}_K(u') h_{K,\ell} \Big] \\ &:= R_\ell + S_\ell. \end{aligned}$$

We will estimate the terms R_ℓ and S_ℓ separately.

To estimate the first term, we distinguish whether $\ell \in \mathscr{C}'$ or $\ell \in \mathscr{A}'$. A quick computation shows that $R_\ell = (1-\alpha) u'_\ell$ for $\ell \in \mathscr{C}'$. On the other hand, for $\ell \in \mathscr{A}'$ we have

$$\begin{aligned} R_\ell &= \Big[1 - \tfrac{\alpha}{A_F} (\phi''_F + 2\phi''_{2F}) \Big] u'_\ell - \alpha \tfrac{\phi''_{2F}}{A_F} (u'_{\ell-1} + u'_{\ell+1}) \\ &= \Big[1 - \alpha \big(1 - \tfrac{2\phi''_{2F}}{A_F}\big) \Big] u'_\ell - \alpha \tfrac{\phi''_{2F}}{A_F} (u'_{\ell-1} + u'_{\ell+1}), \qquad \forall \ell \in \mathscr{A}'. \end{aligned}$$

Since $\|u'\|_{\ell^\infty_\epsilon} \le 1$, we can thus obtain

$$|R_\ell| \le \begin{cases} |1-\alpha|, & \ell \in \mathscr{C}', \\ \Big| 1 - \alpha \big(1 - \tfrac{2\phi''_{2F}}{A_F}\big) \Big| + \alpha \Big| \tfrac{2\phi''_{2F}}{A_F} \Big|, & \ell \in \mathscr{A}'. \end{cases} \tag{34}$$

As a matter of fact, these bounds can be attained for certain ℓ, by choosing suitable test functions. For example, by choosing $u \in \mathscr{U}$ with $u'_N = \mathrm{sign}(1-\alpha)$ we obtain $R_N = |1-\alpha|$, that is, R_N attains the bound (34). By choosing $u \in \mathscr{U}$ such that

$$u'_0 = u'_2 = \mathrm{sign}\Big(-\tfrac{\phi''_{2F}}{A_F} \Big) = 1 \quad \text{and} \quad u'_1 = \mathrm{sign}\Big(1 - \alpha \big(1 - \tfrac{2\phi''_{2F}}{A_F} \big) \Big),$$

we obtain that R_1 attains the bound (34). In both cases one needs to choose the remaining free u'_ℓ so that $|u'_\ell| \le 1$ and $\sum_\ell u'_\ell = 0$, which guarantees that such functions $u \in \mathscr{U}$ really exist. This can be done under the conditions imposed on N and K.

To estimate S_ℓ, we note that this term depends only on a small number of strains around the interface. We can therefore expand it in terms of these strains and their coefficients and then maximize over all possible interface contributions. Thus, we rewrite S_ℓ as follows:

$$S_\ell = \alpha \tfrac{\phi''_{2F}}{A_F} \Big\{ u'_{-K-1}[-h_{-K,\ell}] + u'_{-K}[2h_{-K,\ell} + \tfrac{\epsilon}{2}] + u'_{-K+1}[-h_{-K,\ell} - \tfrac{\epsilon}{2}] \\ u'_K[h_{K,\ell} - \tfrac{\epsilon}{2}] + u'_{K+1}[-2h_{K,\ell} + \tfrac{\epsilon}{2}] + u'_{K+2}[h_{K,\ell}]\Big\}.$$

This expression is maximized by taking u'_ℓ to be the sign of the respective coefficient (taking into account also the outer coefficient $\alpha \frac{\phi''_{2F}}{A_F}$), which yields

$$\begin{aligned} |S_\ell| \le\ & \alpha \big|\tfrac{\phi''_{2F}}{A_F}\big| \Big\{ |h_{-K,\ell}| + |2h_{-K,\ell} + \tfrac{\epsilon}{2}| + |h_{-K,\ell} + \tfrac{\epsilon}{2}| + |h_{K,\ell} - \tfrac{\epsilon}{2}| \\ & + |2h_{K,\ell} - \tfrac{\epsilon}{2}| + |h_{K,\ell}| \Big\} \\ =\ & \alpha \big|\tfrac{\phi''_{2F}}{A_F}\big| \Big\{ |4h_{-K,\ell} + \epsilon| + |4h_{K,\ell} - \epsilon| \Big\}. \end{aligned}$$

The equality of the first and second line holds because the terms $\pm\frac{\epsilon}{2}$ do not change the signs of the terms inside the bars. Inserting the values for $h_{\pm K,\ell}$, we obtain the bound

$$|S_\ell| \le \begin{cases} \alpha 4 \big|\frac{\phi''_{2F}}{A_F}\big|, & \ell \in \mathscr{C}', \\ \alpha(4 + 2\epsilon - 4\epsilon K)\big|\frac{\phi''_{2F}}{A_F}\big|, & \ell \in \mathscr{A}', \end{cases}$$

and we note that this bound is attained if the values for u'_ℓ, $\ell = -K-1, -K, -K+1, K, K+1, K+2$, are chosen as described above.

Combining the analyses of the terms R_ℓ and S_ℓ, it follows that

$$\|z'\|_{\ell^\infty_\epsilon} \le \max\Big\{ |1-\alpha| + \alpha 4 \big|\tfrac{\phi''_{2F}}{A_F}\big|, \\ \big|1 - \alpha\big(1 - \tfrac{2\phi''_{2F}}{A_F}\big)\big| + \alpha(6 + 2\epsilon - 4\epsilon K)\big|\tfrac{\phi''_{2F}}{A_F}\big| \Big\}.$$

To see that this bound is attained, we note that, under the condition that $K \ge 3$ and $N \ge K+3$, the constructions at the interface to maximize S_ℓ and the constructions to maximize R_ℓ do not interfere. Moreover, under the additional condition $N \ge \max(9, K+3)$, sufficiently many free strains u'_ℓ remain to ensure that $\sum_\ell u'_\ell = 0$ for a test function $u \in \mathscr{U}$, $\|u'\|_{\ell^\infty_\epsilon} = 1$, for which both R_ℓ and S_ℓ attain the stated bound. That is, we have shown that

$$\begin{aligned} \big\|G_{\rm qcl}(\alpha)\big\|_{\mathscr{U}^{1,\infty}} &= \max\Big\{ |1-\alpha| + \alpha 4 \tfrac{|\phi''_{2F}|}{A_F}, \\ &\qquad \big|1 - \alpha\big(1 - \tfrac{2\phi''_{2F}}{A_F}\big)\big| + \alpha(6 + 2\epsilon - 4\epsilon K)\tfrac{|\phi''_{2F}|}{A_F} \Big\} \\ &=: \max\{m_{\mathscr{C}}(\alpha), m_{\mathscr{A}}(\alpha)\}. \end{aligned}$$

To conclude the proof, we need to evaluate this maximum explicitly. To this end we first define $\alpha_1 = (1 - \frac{2\phi''_{2F}}{A_F})^{-1} < 1$. For $0 \le \alpha \le \alpha_1$, we have

$$\begin{aligned} m_{\mathscr{A}}(\alpha) &= 1 - \alpha + \alpha(4 + 2\epsilon - 4\epsilon K)\big|\tfrac{\phi''_{2F}}{A_F}\big| \\ &\le 1 - \alpha + \alpha 4\big|\tfrac{\phi''_{2F}}{A_F}\big| = m_{\mathscr{C}}(\alpha), \end{aligned}$$

that is, $\|G_{\rm qcl}(\alpha)\|_{\mathscr{U}^{1,\infty}} = m_{\mathscr{C}}(\alpha)$. Conversely, for $\alpha \ge 1$, we have

$$\begin{aligned} m_{\mathscr{A}}(\alpha) &= \alpha\Big(1 + (8 + 2\epsilon - 4\epsilon K)\tfrac{|\phi''_{2F}|}{A_F}\Big) - 1 \\ &= m_{\mathscr{C}}(\alpha) + \alpha\Big(4 + 2\epsilon - 4\epsilon K)\tfrac{|\phi''_{2F}|}{A_F}\Big) \ge m_{\mathscr{C}}(\alpha), \end{aligned}$$

that is, $\|G_{\rm qcl}(\alpha)\|_{\mathscr{U}^{1,\infty}} = m_{\mathscr{A}}(\alpha)$. Since, in $[\alpha_1, 1]$, $m_{\mathscr{C}}$ is strictly decreasing and $m_{\mathscr{A}}$ is strictly increasing, there exists a unique $\alpha_2 \in [\alpha_1, 1]$ such that $m_{\mathscr{C}}(\alpha_2) = m_{\mathscr{A}}(\alpha_2)$ and such that the stated formula for $\|G_{\rm qcl}(\alpha)\|_{\mathscr{U}^{1,\infty}}$ holds. A straightforward computation yields the value for $\alpha_2 = \alpha_{\rm opt}^{\rm qcl,1,\infty}$ stated in the lemma. □

Computation of $\|G_{\rm qce}\|_{\mathscr{U}^{k,p}}$

We have computed $\|G_{\rm qce}\|_{\mathscr{U}^{k,p}}$ for $k = 0, 2, p = 1, 2, \infty$, from the standard formulas for the operator norm [17, 29] of the matrix $G_{\rm qce}$ and $LG_{\rm qce}L^{-1}$ with respect to ℓ^p_ϵ. For $k = 1$ and $p = 2$, the norm is also easy to obtain by solving a generalized eigenvalue problem.

The cases $k = 1$ and $p = 1, \infty$ are more difficult. In these cases, the operator norm of $G_{\rm qce}$ in $\mathscr{U}^{1,p}$ can be estimated in terms of the ℓ^p_ϵ-operator norm of the conjugate operator $\widehat{G} = I - (\widehat{L}_F^{\rm qce})^{-1}\widehat{L}_F^{\rm qcf} : \mathbb{R}^{2N} \to \mathbb{R}^{2N}$ (see Lemma 6 for an analogous definition of the conjugate operator $\widehat{L}_F^{\rm qnl} : \mathbb{R}^{2N} \to \mathbb{R}^{2N}$). It is not difficult to see that $\|G_{\rm qce}\|_{\mathscr{U}^{1,p}} = \|\widetilde{G}\|_{\ell^p_\epsilon, \mathbb{R}^{2N}_*}$ for $\widetilde{G} = I - (\widetilde{L}_F^{\rm qce})^{-1}\widetilde{L}_F^{\rm qcf} : \mathbb{R}^{2N}_* \to \mathbb{R}^{2N}_*$ where we recall that $\mathbb{R}^{2N}_* = \{\varphi \in \mathbb{R}^{2N} : \sum_\ell \varphi_\ell = 0\}$ (see Lemma 6 similarly for an analogous definition of the restricted conjugate operator $\widetilde{L}_F^{\rm qnl} : \mathbb{R}^{2N}_* \to \mathbb{R}^{2N}_*$), it follows from (3) that we have only computed $\|G_{\rm qce}\|_{\mathscr{U}^{1,p}}$ for $p = 1, \infty$ up to a factor of $1/2$. More precisely,

$$\|G_{\rm qce}\|_{\mathscr{U}^{1,p}} \le \|\widehat{G}\|_{\ell^p_\epsilon} \le 2\|G_{\rm qce}\|_{\mathscr{U}^{1,p}}$$

Finally We note that we can obtain $\widehat{L}_F^{\rm qcf}$ from the representation given in Lemma 10 and that $\widehat{L}_F^{\rm qce}$ can be directly obtained from (15).

References

1. P. Bauman, H. B. Dhia, N. Elkhodja, J. Oden, and S. Prudhomme. On the application of the Arlequin method to the coupling of particle and continuum models. *Computational Mechanics*, 42:511–530, 2008.
2. T. Belytschko and S. P. Xiao. A bridging domain method for coupling continua with molecular dynamics. *Computer Methods in Applied Mechanics and Engineering*, 193:1645–1669, 2004.
3. N. Bernstein, J. R. Kermode, and G. Csányi. Hybrid atomistic simulation methods for materials systems. *Reports on Progress in Physics*, 72:pp. 026501, 2009.
4. X. Blanc, C. Le Bris, and F. Legoll. Analysis of a prototypical multiscale method coupling atomistic and continuum mechanics. *M2AN Math. Model. Numer. Anal.*, 39(4):797–826, 2005.
5. M. Dobson and M. Luskin. Analysis of a force-based quasicontinuum approximation. *M2AN Math. Model. Numer. Anal.*, 42(1):113–139, 2008.
6. M. Dobson and M. Luskin. Iterative solution of the quasicontinuum equilibrium equations with continuation. *Journal of Scientific Computing*, 37:19–41, 2008.
7. M. Dobson and M. Luskin. An analysis of the effect of ghost force oscillation on the quasicontinuum error. *Mathematical Modelling and Numerical Analysis*, 43:591–604, 2009.
8. M. Dobson and M. Luskin. An optimal order error analysis of the one-dimensional quasicontinuum approximation. *SIAM. J. Numer. Anal.*, 47:2455–2475, 2009.
9. M. Dobson, M. Luskin, and C. Ortner. Accuracy of quasicontinuum approximations near instabilities. *Journal of the Mechanics and Physics of Solids*, 58:1741–1757, 2010. arXiv:0905.2914v2.
10. M. Dobson, M. Luskin, and C. Ortner. Sharp stability estimates for force-based quasicontinuum methods. *SIAM J. Multiscale Modeling & Simulation*, 8:782–802, 2010. arXiv:0907.3861.
11. M. Dobson, M. Luskin, and C. Ortner. Stability, instability, and error of the force-based quasicontinuum approximation. *Archive for Rational Mechanics and Analysis*, 197:179–202, 2010. arXiv:0903.0610.
12. M. Dobson, M. Luskin, and C. Ortner. Iterative methods for the force-based quasicontinuum approximation. *Computer Methods in Applied Mechanics and Engineering*, to appear. arXiv:0910.2013v3.
13. M. Dobson, C. Ortner, and A. Shapeev. The spectrum of the force-based quasicontinuum operator for a homogeneous periodic chain. arXiv:1004.3435.
14. W. E, J. Lu, and J. Yang. Uniform accuracy of the quasicontinuum method. *Phys. Rev. B*, 74(21):214115, 2004.
15. V. Gavini, K. Bhattacharya, and M. Ortiz. Quasi-continuum orbital-free density-functional theory: A route to multi-million atom non-periodic DFT calculation. *J. Mech. Phys. Solids*, 55:697–718, 2007.
16. M. Gunzburger and Y. Zhang. A quadrature-rule type approximation for the quasicontinuum method. *Multiscale Modeling and Simulation*, 8:571–590, 2010.
17. E. Isaacson and H. Keller. *Analysis of Numerical Methods*. Wiler, New York, 1966.
18. B. V. Koten, X. H. Li, M. Luskin, and C. Ortner. A computational and theoretical investigation of the accuracy of quasicontinuum methods. In I. Graham, T. Hou, O. Lakkis, and R. Scheichl, editors, *Numerical Analysis of Multiscale Problems*. Springer, to appear. arXiv:1012.6031.
19. B. V. Koten and M. Luskin. Development and analysis of blended quasicontinuum approximations. arXiv:1008.2138v2, 2010.
20. X. H. Li and M. Luskin. An analysis of the quasi-nonlocal quasicontinuum approximation of the embedded atom model. *International Journal for Multiscale Computational Engineering*, to appear. arXiv:1008.3628v4.
21. X. H. Li and M. Luskin. A generalized quasi-nonlocal atomistic-to-continuum coupling method with finite range interaction. *IMA Journal of Numerical Analysis*, to appear. arXiv:1007.2336.
22. P. Lin. Convergence analysis of a quasi-continuum approximation for a two-dimensional material without defects. *SIAM J. Numer. Anal.*, 45(1):313–332 (electronic), 2007.

23. R. Miller and E. Tadmor. The Quasicontinuum Method: Overview, Applications and Current Directions. *Journal of Computer-Aided Materials Design*, 9:203–239, 2003.
24. R. Miller and E. Tadmor. Benchmarking multiscale methods. *Modelling and Simulation in Materials Science and Engineering*, 17:053001 (51pp), 2009.
25. P. Ming and J. Z. Yang. Analysis of a one-dimensional nonlocal quasicontinuum method. *Multiscale Modeling and Simulation*, 7:1838–1875, 2009.
26. M. Ortiz, R. Phillips, and E. B. Tadmor. Quasicontinuum Analysis of Defects in Solids. *Philosophical Magazine A*, 73(6):1529–1563, 1996.
27. C. Ortner. The role of the patch test in 2D atomistic-to-continuum coupling methods. arXiv:1101.5256, 2011.
28. C. Ortner and E. Süli. Analysis of a quasicontinuum method in one dimension. *M2AN Math. Model. Numer. Anal.*, 42(1):57–91, 2008.
29. Y. Saad. *Iterative Methods for Sparse Linear Systems*, volume 2. Society for Industrial and Applied Mathematics (SIAM), 2003.
30. A. V. Shapeev. Consistent energy-based atomistic/continuum coupling for two-body potential: 1D and 2D case. arXiv:1010.0512, 2010.
31. V. B. Shenoy, R. Miller, E. B. Tadmor, D. Rodney, R. Phillips, and M. Ortiz. An adaptive finite element approach to atomic-scale mechanics–the quasicontinuum method. *J. Mech. Phys. Solids*, 47(3):611–642, 1999.
32. L. E. Shilkrot, R. E. Miller, and W. A. Curtin. Coupled atomistic and discrete dislocation plasticity. *Phys. Rev. Lett.*, 89(2):025501, 2002.
33. T. Shimokawa, J. Mortensen, J. Schiotz, and K. Jacobsen. Matching conditions in the quasicontinuum method: Removal of the error introduced at the interface between the coarse-grained and fully atomistic region. *Phys. Rev. B*, 69(21):214104, 2004.

Analysis of an Averaging Operator for Atomic-to-Continuum Coupling Methods by the Arlequin Approach

Serge Prudhomme, Robin Bouclier, Ludovic Chamoin, Hachmi Ben Dhia, and J. Tinsley Oden

Abstract A new coupling term for blending particle and continuum models with the Arlequin framework is investigated in this work. The coupling term is based on an integral operator defined on the overlap region that matches the continuum and particle solutions in an average sense. The present exposition is essentially the continuation of a previous work (Bauman et al., On the application of the Arlequin method to the coupling of particle and continuum models, *Computational Mechanics*, 42, 511–530, 2008) in which coupling was performed in terms of an H^1-type norm. In that case, it was shown that the solution of the coupled problem was mesh-dependent or, said in another way, that the solution of the continuous coupled problem was not the intended solution. This new formulation is now consistent with the problem of interest and is virtually mesh-independent when considering a particle model consisting of a distribution of heterogeneous bonds. The mathematical properties of the formulation are studied for a one-dimensional model of harmonic springs, with varying stiffness parameters, coupled with a linear elastic bar, whose modulus is determined by classical homogenization. Numerical examples are presented for one-dimensional and two-dimensional model problems

S. Prudhomme (✉) · J.T. Oden
Institute for Computational Engineering and Sciences (ICES), The University of Texas at Austin, 1 University Station C0200, Austin, TX 78712, USA
e-mail: serge@ices.utexas.edu; oden@ices.utexas.edu

R. Bouclier · L. Chamoin
Laboratoire de Mécanique et Technologie (LMT-Cachan), Ecole Normale Supérieure de Cachan, 61 Avenue du Président Wilson, 94235 Cachan Cedex, France
e-mail: bouclier@ices.utexas.edu; chamoin@lmt.ens-cachan.fr

H.B. Dhia
Laboratoire de Mécanique des Sols, Structures et Matériaux (MSSMat), Ecole Centrale Paris, 1 Grande Voie des Vignes, 92290 Châtenay-Malabry, France
e-mail: hachmi.ben-dhia@ecp.fr

B. Engquist et al. (eds.), *Numerical Analysis of Multiscale Computations*, Lecture Notes in Computational Science and Engineering 82, DOI 10.1007/978-3-642-21943-6_15,

that illustrate the approximation properties of the new coupling term and the effect of mesh size.

1 Introduction

Development of multiscale methods for the simulation of material responses is an important research area in which one of the objectives is to combine models so as to capture only the relevant scales in the prediction of complex phenomena. The goal in this work is to develop a new multiscale method to predict the static response of materials that can be described by particle models based on harmonic potentials. Multiscale modeling is commonly classified into information passing modeling, in which information computed at small scales is used in large-scale models, such as in the Heterogeneous Multiscale method [20, 21], and concurrent modeling, in which two or more models are concurrently used to capture the various scales inherent in a given physical phenomenon, see e.g. [22, 23]. We are interested here in concurrent modeling for the simulation of problems that involve both a particle model and a continuum model. The major difficulty in this case is to consistently blend the two models so as to provide accurate approximations of the solution to the full particle model, viewed as the base model but often intractable for large simulation domains. Several methods have been proposed over the years, such as the quasi-continuum method [17–19, 25, 28], the handshake method [14], or the bridging scale approach [29], to name a few. An alternative approach based on the Arlequin framework [8, 9, 11–13] has recently been proposed in [5, 10, 26]. The Arlequin framework involves an overlap region in which the energies of the two models are combined by a partition of unity and where the two solutions are matched by introducing Lagrange multipliers. The bridging domain method of Belytschko and Xiao [7] is in many ways similar to the Arlequin method and was numerically investigated in [30]. A related methodology has also been proposed in [2, 3, 24] in which forces, rather than energies, are blended together. The method proposed in [5] was further employed to develop an adaptive procedure based on goal-oriented error estimates (see [4, 6, 27]) to control the position of the overlap region so as to deliver estimates of quantities of interest within prescribed tolerances.

Well-posedness of the Arlequin problems for the continuous and finite element formulations was investigated in detail in [5] in the case of a one-dimensional model of harmonic springs, with periodically varying stiffness coefficients, coupled with a linear elastic bar. Couplings of the displacement fields obtained from the particle and continuum models were defined based on an L^2-norm or an H^1-norm. It was then proved that the continuous formulation and corresponding discretization of the continuous formulation, by the finite element method for instance, yield well-posed problems only in the H^1-norm case. However, it was recognized at that time that the solution of the coupled problem was mesh-dependent in the sense that the finite element approximation of the continuum model would lock on the particle solution on the overlap region when elements for the Lagrange multiplier were

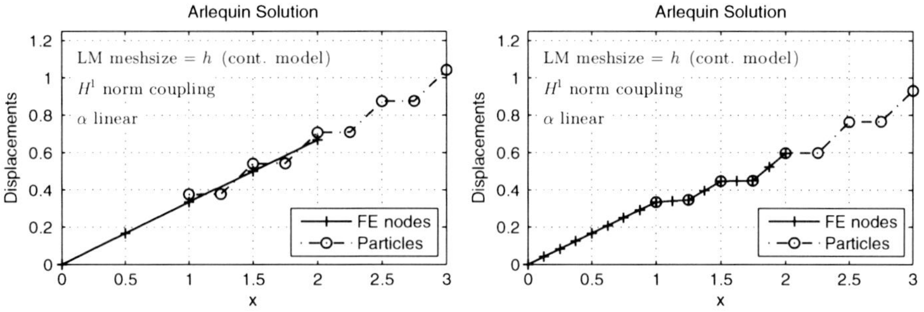

Fig. 1 Solutions of the coupled problem based on the Arlequin framework as proposed in [5] using either a coarse (*left*) or fine (*right*) finite element discretization of the continuum model. "Coarse" and "fine" here are defined with respect to the equilibrium length between particles. The coupling term is based on an H^1-type norm. One observes that the continuum solution on the overlap region locks onto the particle solution in the case of the fine mesh for the Lagrange multiplier and FE solution and thus fails to reflect the large-scale behavior of the displacement field

chosen equal to or smaller than the distance between particles. This issue could be circumscribed by selecting the mesh size for the Lagrange multiplier to be at least larger than the size or a multiple of the size of the representative cell defined to calibrate the parameter(s) of the continuum model, in which case the method would produce satisfactory results. If elements were set too small for the Lagrange multiplier, the continuum solution would fail to reproduce the large-scale behavior of the displacement fields and would pollute the whole solution of the coupled problem. These effects are illustrated in Fig. 1. We propose here a new formulation of the coupling term based on an integral operator that matches the continuum and particle solutions in an average sense. The advantage of this new formulation is that it yields a mesh-independent displacement field. We show in this paper that this new Arlequin formulation yields a well-posed coupled problem and illustrate its efficiency via simple one-dimensional and two-dimensional problems.

The paper is organized as follows: in Sect. 2, we present the particle model and the continuum model and show how the latter is derived from the former by simple homogenization. We introduce the averaging operator and describe the new coupling formulation based on the Arlequin framework in Sect. 3. We show that the coupled problem is well-posed in Sect. 4 and describe the corresponding finite element formulation in Sect. 5. One-dimensional and two-dimensional numerical experiments are presented in Sect. 6 and are followed by conclusions in Sect. 7.

2 Particle and Continuum Model Problems

2.1 Particle Model

We consider here a system of $n + 1$ particles assembled in a one-dimensional chain and connected by n covalent bonds modeled in terms of harmonic springs with stiffness $k_i > 0$ and equilibrium length l_i, $i = 1, \ldots, n$. The initial positions of the particles are given by x_i and the system undergoes displacements y_i when subjected to force f applied at x_n (see Fig. 2). We also suppose that the particle on the left end is fixed, i.e. $y_0 = 0$. The potential energy of such a system is given by

$$\mathscr{E}_d(y) = \frac{1}{2}\sum_{i=1}^{n} k_i\,(y_i - y_{i-1})^2 - f y_n. \tag{1}$$

Introducing the vector space $W_0 = \{z \in \mathbb{R}^{n+1} : \; z_0 = 0\}$ of vectors $z = [z_0, z_1, \ldots, z_n]^T$, the equilibrium state $y \in W_0$ of such a system is obtained as a minimizer of the potential energy, i.e.

$$y = \mathrm{argmin}_{z \in W_0}\, \mathscr{E}_d(z). \tag{2}$$

In other words, the solution w of above minimization problem is a stationary point of $\mathscr{E}_d(z)$ and satisfies

$$\lim_{\theta \to 0} \frac{1}{\theta}\left(\mathscr{E}_d(y + \theta z) - \mathscr{E}_d(y)\right) = 0, \qquad \forall z \in W_0.$$

It follows that Problem (2) can be recast in variational form as

$$\boxed{\text{Find } y \in W_0 \text{ such that} \quad B(y,z) = F(z), \quad \forall z \in W_0,} \tag{3}$$

where the bilinear form $B(\cdot,\cdot)$ and linear form $F(\cdot)$ are defined as:

$$\begin{cases} B(y,z) = \displaystyle\sum_{i=1}^{n} k_i\,(y_i - y_{i-1})\,(z_i - z_{i-1}), \\ F(z) = f z_n. \end{cases} \tag{4}$$

In this paper, we are interested in materials in which the stiffness k_i may vary from one bond to the other. Nevertheless we suppose that the distribution of the bonds are such that the large scales of the material response could be accurately described by a continuum model over representative volume elements (RVE). For instance, in the case of periodic distributions of the bond stiffness k_i, the representative volume element is simply chosen of the same length as one period of the distribution. More complex distributions, for example random, could also be considered (see for example [15]) but the size of the RVE would be unknown a priori.

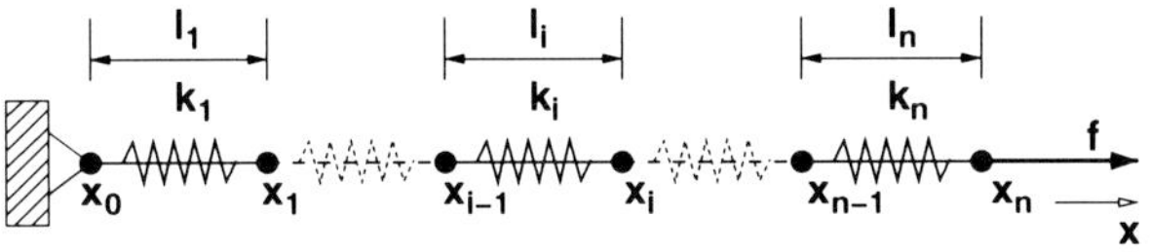

Fig. 2 System of $n + 1$ particles connected with n harmonic springs

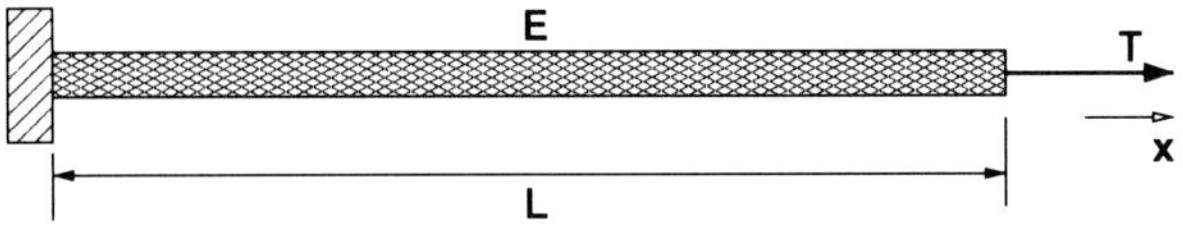

Fig. 3 Elastic bar of length L with modulus of elasticity E and subjected to traction T

For simplicity in the presentation, we will not present here cases where the energy potentials involve next-nearest neighbors. This has been partially treated in [16].

2.2 Continuum Model

If one is interested in large-scale features of the response (in the sense that the scale of those features would be much larger than the representative length-scale of the particle system, e.g. $\max_i(l_i)$), a possible approximation of the particle model can be obtained by employing a linearly elastic continuum model. In this case, the system of springs is replaced by an elastic bar with modulus E and of length L; see Fig. 3. Moreover, the bar is subjected to traction $T = f/A$ at the right end, A being the cross-sectional area of the bar, and is kept fixed at $x = 0$. Displacement in the bar is denoted by the field u. The total energy of the system is then given by

$$\mathscr{E}_c = \int_0^L \frac{A}{2}\sigma(u)\epsilon(u)\,dx - AT(L)u(L), \tag{5}$$

where $\sigma(u)$ and ϵ denote the stress and strain in the bar. Here the material is supposed to obey Hooke's law, $\sigma = E\epsilon$, with E constant. Using $\epsilon = u'$, we have

$$\mathscr{E}_c = \int_0^L \frac{AE}{2}\left(u'\right)^2\,dx - AT(L)u(L). \tag{6}$$

As with the spring model, the equilibrium state for the continuum model is found by minimizing the energy (6). This minimization yields the following problem:

$$\boxed{\text{Find } u \in V \text{ such that: } \quad \int_0^L Eu'v'dx = T(L)v(L) \qquad \forall v \in V,} \tag{7}$$

where V is the space of trial and test functions, i.e. $V = \{v \in H^1(0,L) : v(0) = 0\}$.

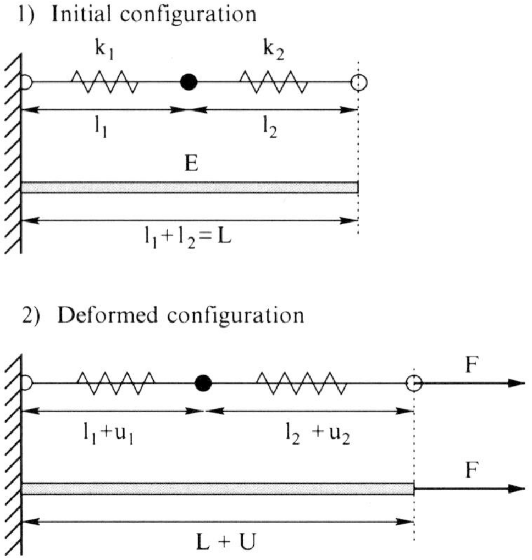

Fig. 4 Homogenization of spring model on a representative cell

2.3 Calibration of Continuum Model

Starting with the original particle model, it is possible to determine a compatible continuum model by properly calibrating the elastic modulus. Following classical homogenization approaches, the main idea here is to introduce a representative volume element, that, if subjected to a given loading, should provide the same global response at equilibrium, i.e. the same global displacement, when using either the particle or continuum model.

To illustrate the concept, we consider here the simple case of a representative cell consisting of a pair of springs with properties (k_1, l_1) and (k_2, l_2), as shown in Fig. 4. We assume that the system is held fixed on the left-hand side and is subjected to the force F to the right, such that the displacement in the first and second springs are u_1 and u_2, respectively. Suppose now that we can replace the system of two springs by a unique spring with properties (K, L) such that $L = l_1 + l_2$. If subjected to the same loading conditions, we would observe the global displacement $U = u_1 + u_2$. From constitutive laws, we also have the relations:

$$F = KU = k_1 u_1 = k_2 u_2, \tag{8}$$

so that

$$\frac{F}{K} = U = u_1 + u_2 = \frac{F}{k_1} + \frac{F}{k_2}, \tag{9}$$

which implies that:

$$\frac{1}{K} = \frac{1}{k_1} + \frac{1}{k_2}, \qquad \text{i.e.} \quad K = \frac{k_1 k_2}{k_1 + k_2}. \tag{10}$$

Finally, replacing the spring model by linear elasticity, we would obtain the following Young's modulus:

$$\boxed{EA = KL = \frac{k_1 k_2}{k_1 + k_2}(l_1 + l_2),} \tag{11}$$

where A is the cross-sectional area of the equivalent bar. For simplicity, we take A equal to unity.

Remark 1. The above relation can naturally be extended to the case of one RVE made of N springs. In this case, we would have:

$$EA = \left[\sum_{j=1}^{N} 1/k_j\right]^{-1} \sum_{i=1}^{N} l_i. \tag{12}$$

It is then straightforward to show that:

$$EA = \sum_{i=1}^{N}\left[\sum_{j=1}^{N} k_i/k_j\right]^{-1} k_i l_i \geq \min_{1\leq i\leq N}(k_i l_i)\left[\sum_{i=1}^{N} 1/k_i\right]\left[\sum_{j=1}^{N} 1/k_j\right]^{-1} = \min_{1\leq i\leq N}(k_i l_i). \tag{13}$$

In the same manner, we have:

$$EA = \sum_{i=1}^{N}\left[\sum_{j=1}^{N} k_i/k_j\right]^{-1} k_i l_i \leq \max_{1\leq i\leq N}(k_i l_i)\left[\sum_{i=1}^{N} 1/k_i\right]\left[\sum_{j=1}^{N} 1/k_j\right]^{-1} = \max_{1\leq i\leq N}(k_i l_i). \tag{14}$$

In other words, with $A = 1$, one gets:

$$\min_{1\leq i\leq N}(k_i l_i) \leq E \leq \max_{1\leq i\leq N}(k_i l_i), \tag{15}$$

i.e. the value of E is necessarily larger than the minimal value of $k_i l_i$ and smaller than the maximal value of $k_i l_i$.

Remark 2. Starting from the relation $U = u_1 + u_2$, we can write:

$$\frac{U}{L} L = \frac{u_1}{l_1} l_1 + \frac{u_2}{l_2} l_2. \tag{16}$$

We recognize in above equation the strains $\bar{\epsilon} = lU/L, \epsilon_1 = u_1/l_1, \epsilon_2 = u_2/l_2$, which are constant in each spring. Therefore, we can derive the following relationship:

$$\int_{\text{RVE}} \bar{\epsilon} dx = \int_{\text{RVE}} \epsilon dx, \tag{17}$$

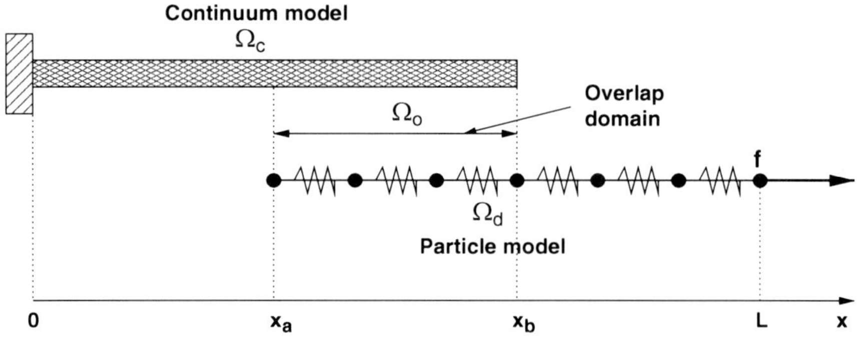

Fig. 5 Arlequin model that replaces the particle model with a combined particle and spring model

where $\epsilon = \epsilon_1$ in the first spring and $\epsilon = \epsilon_2$ in the second spring. This relation shows that the averaged strain over the representative volume element is the same whether it is computed from the particle model or the continuum model. This relationship will motivate our new formulation of the coupling method based on an averaging operator.

3 Coupling Method with Averaging Operator

We recall that our objective is to develop a coupling method to blend the particle model with the continuum model in $\Omega = (0, L)$. We assume that the continuum model is selected in region $\Omega_c = (0, x_b)$ while the particle model is chosen in domain $\Omega_d = (x_a, L)$ such that $\Omega = \Omega_c \bigcup \Omega_d$ and $\Omega_o = \Omega_c \bigcap \Omega_d = (x_a, x_b)$, $|\Omega_o| \neq 0$. We will refer to Ω_o as the overlap region. We denote by $|\Omega_c|$, $|\Omega_d|$, and $|\Omega_o|$, the length of domains Ω_c, Ω_d, and Ω_o, respectively. In doing so, the particle model is reduced from $n + 1$ to $m + 1$ particles, supposedly with $m \ll n$.

Remark 3. We assume in this work that there are $m_o + 1$ particles lying in the overlap region and that there is one particle located at x_a and one at x_b as shown in Fig. 5. The restrictive assumption that is made here is that the overlap region exactly coincides with a given set of complete springs. In other words, the domain Ω_o is not allowed to only cover part of a spring. However, the domain Ω_o can be made of one or several RVE's.

3.1 Energy of the Coupled System

The Arlequin method is an energy-based method in which the energy contributions from two models are blended together via the partition of unity:

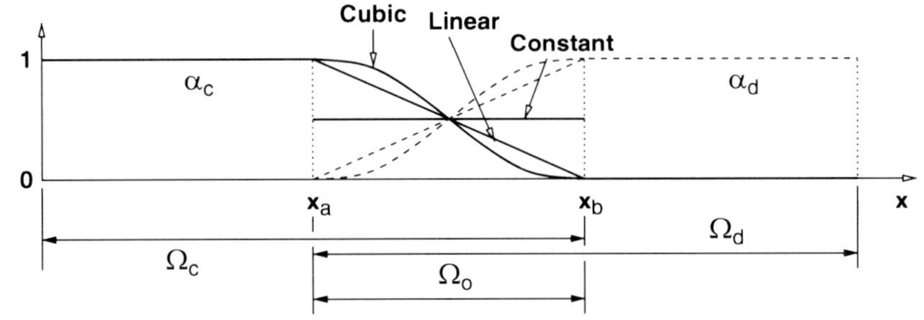

Fig. 6 Plot of different functions used for α_c and α_d

$$\alpha_c(x) + \alpha_d(x) = 1, \quad \forall x \in \Omega,$$

with

$$\alpha_c(x) = \begin{cases} 1, & \forall x \in \Omega_c \setminus \Omega_o, \\ 0, & \forall x \in \Omega_d \setminus \Omega_o, \end{cases} \qquad \alpha_d(x) = \begin{cases} 0, & \forall x \in \Omega_c \setminus \Omega_o, \\ 1, & \forall x \in \Omega_d \setminus \Omega_o. \end{cases}$$

Weight coefficients with respect to each bond are also introduced as:

$$\alpha_i = \frac{1}{l_i} \int_{x_{i-1}}^{x_i} \alpha_d(x)\, dx = 1 - \frac{1}{l_i} \int_{x_{i-1}}^{x_i} \alpha_c(x)\, dx, \quad i = 1, \ldots, m. \tag{18}$$

In the overlap region Ω_o, the coefficient α_c (and thus α_d) can be chosen in different ways. Some intuitive and apparently attractive candidates are for example, the constant, linear, or cubic functions, as shown in Fig. 6. For example, the cubic function can be explicitly written as:

$$\alpha_c(x) = \left[\frac{x_b - x}{x_b - x_a} \right]^2 \left[1 + 2\frac{(x - x_a)}{(x_b - x_a)} \right], \qquad \forall x \in \Omega_o. \tag{19}$$

The total energy of the molecular system can now be replaced by:

$$\hat{\mathscr{E}}(u, w) = \hat{\mathscr{E}}_c(u) + \hat{\mathscr{E}}_d(w),$$

where

$$\begin{aligned} \hat{\mathscr{E}}_c(u) &= \frac{1}{2} \int_{\Omega_c} \alpha_c(x) E \left(u'\right)^2 dx, \\ \hat{\mathscr{E}}_d(w) &= \frac{1}{2} \sum_{i=1}^{m} \alpha_i k_i (w_i - w_{i-1})^2 - f w_m, \end{aligned} \tag{20}$$

with f, once again, being the external force applied at L, i.e. to the particle indexed by m.

3.2 Averaging Coupling Operator

The objective being to properly couple the two models, the displacements u and w need to be matched with respect to some appropriate measure. In order to be able to compare u and w on Ω_o, the displacement vector w needs first to be converted into a function in $H^1(\Omega_o)$. A possible approach is to introduce an interpolation operator $\Pi_o : \mathbb{R}^{m_o+1} \to H^1(\Omega_o)$, which associates with each displacement vector w (restricted to the particles in Ω_o) the piecewise linear interpolant $\Pi_o w$ on Ω_o. Other interpolation schemes are imaginable, but for the sake of simplicity, we shall only consider the linear interpolant in the present work. We also introduce the restriction operator $\mathrm{R}_o : H^1(\Omega_c) \to H^1(\Omega_o)$ that restricts continuum displacements u to Ω_o.

In our previous work [5], we realized that, when using the finite element method for the discretization of the continuum model, matching the displacements $\mathrm{R}_o u$ and $\Pi_o w$ or/and the associated strains $(\mathrm{R}_o u)'$ and $(\Pi_o w)'$ at every point on the overlap region yielded erroneous results as soon as the mesh size was chosen smaller than the size of the representative volume element. In that case, the solution of the continuum model would indeed lock itself to the solution of the particle model. Our objective in this work is to define a formulation that is independent of the finite element mesh size.

In view of homogenization, the continuum model is derived by matching strain averages computed from the two models. An obvious choice is then to match the average of $(\mathrm{R}_o u)'$ with the average of $(\Pi_o w)'$ over a representative volume element, and in order to constrain rigid body motions, to match the average of the displacements $\mathrm{R}_o u$ and $\Pi_o w$ over the overlap Ω_o. Definition of these averages is straightforward except at the boundaries of Ω_o. We thus propose to define the averaging operators as follows, where the size of the RVE is denoted by ξ (see Fig. 7). Let $v \in H^1(\Omega_o)$, then

$$v^*(x) = \begin{cases} \dfrac{1}{\xi}\displaystyle\int_{x_a}^{x_a+\xi} v'dy = \dfrac{v(x_a+\xi)-v(x_a)}{\xi}, & \forall x \in [x_a, x_a+\xi/2], \\ \dfrac{1}{\xi}\displaystyle\int_{x-\xi/2}^{x+\xi/2} v'dy = \dfrac{v(x+\xi/2)-v(x-\xi/2)}{\xi}, & \forall x \in (x_a+\xi/2, x_b-\xi/2), \\ \dfrac{1}{\xi}\displaystyle\int_{x_b-\xi}^{x_b} v'dy = \dfrac{v(x_b)-v(x_b-\xi)}{\xi}, & \forall x \in [x_b-\xi/2, x_b]. \end{cases} \tag{21}$$

We also introduce the average[1] of a function $v \in H^1(\Omega_o)$ on Ω_o as:

$$\overline{v} = \frac{1}{|\Omega_o|}\int_{\Omega_o} v dx. \tag{22}$$

[1] In what follows, averages on Ω_o will always be denoted by a bar over the corresponding quantity.

Notice that the averaging operators $(\cdot)^*$ and $\overline{(\cdot)}$ are linear operators. As a result, the mismatch on overlap Ω_o between the solutions of the continuum and particle models can be measured as:

$$\begin{aligned}\mathscr{M}(\mathrm{R}_o u - \Pi_o w) &= \beta_0 \left|\overline{\mathrm{R}_o u} - \overline{\Pi_o w}\right|^2 + \beta_1 \int_{\Omega_o} \left|(\mathrm{R}_o u)^* - (\Pi_o w)^*\right|^2 dx \\ &= \beta_0 \left|\overline{(\mathrm{R}_o u - \Pi_o w)}\right|^2 + \beta_1 \int_{\Omega_o} \left|(\mathrm{R}_o u - \Pi_o w)^*\right|^2 dx,\end{aligned} \tag{23}$$

where (β_0, β_1) are non-negative weight parameters chosen such that the terms in above expression are of the same unit or dimensionless.

Remark 4. We readily observe that $\mathscr{M}$ defines a seminorm on $H^1(\Omega_o)$ as it is positive but not necessarily definite. Indeed, there exist non-vanishing functions $\mu \in H^1(\Omega_o)$ such that $\mathscr{M}(\mu) = 0$. Such functions are simply those that satisfy $\overline{\mu} = 0$ and $\mu^*(x) = 0$, $\forall x \in \Omega_o$. Let us introduce the subspace M_0 of $H^1(\Omega_o)$ as:

$$M_0 = \{\mu \in H^1(\Omega_o) : \overline{\mu} = 0 \text{ and } \mu^*(x) = 0,\ \forall x \in \Omega_o\}. \tag{24}$$

Functions in M_0 are those that are continuous with zero-mean and that are ξ-periodic on Ω_o. Let us restrict ourselves to the case where Ω_o exactly covers one RVE. Functions in $H^1(\Omega_o)$ can be represented in terms of Fourier Series as:

$$\mu(x) = a_0 + a_1 x + \sum_{k=1}^{\infty} b_k \sin k\pi \frac{x - x_a}{\xi}, \tag{25}$$

where a_0, a_1, and b_k are real numbers. Note that the family of functions $\sin k\pi(x - x_a)/\xi$ is linearly independent and complete in $H_0^1(\Omega_o)$ [1]. We then have two cases:

1. For k even, we observe that the functions $\mu(x) = \sin k\pi(x - x_a)/\xi$ have all zero mean, are ξ-periodic, and satisfy $\mu(x_b) = \mu(x_a) = 0$.
2. For k odd, we can show that the functions:

$$\mu(x) = \sin\left(k\pi \frac{x - x_a}{\xi}\right) - \frac{2}{k\pi} \tag{26}$$

   have zero mean and are ξ-periodic. However, these functions do not necessarily vanish at the endpoints of Ω_o.

Therefore, the functions μ_0 in M_0 can be represented by linear combinations in the form:

$$\mu_0(x) = \sum_{k=1}^{\infty} b_k \left[\sin\left(2k\pi \frac{x - x_a}{\xi}\right)\right] + c_k \left[\sin\left((2k-1)\pi \frac{x - x_a}{\xi}\right) - \frac{2/\pi}{(2k-1)}\right]. \tag{27}$$

It follows that any function in $H^1(\Omega_o)$ can be expanded as:

$$\mu(x) = a_0 + a_1 x + \mu_0(x), \tag{28}$$

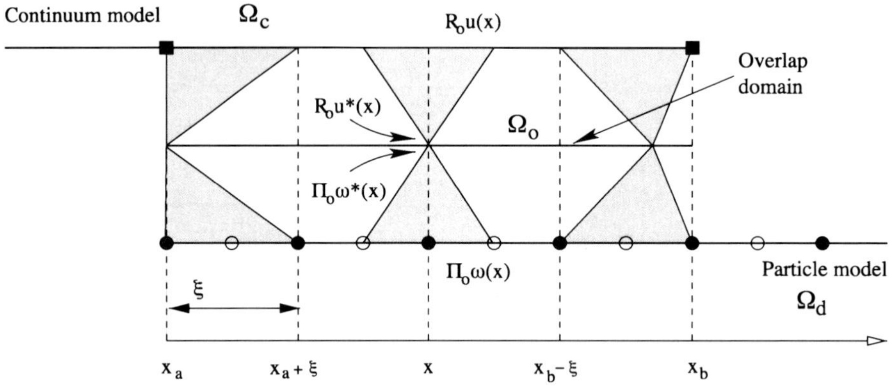

Fig. 7 Domain for the definition of the averaging operator

where a_0 and a_1 are real numbers (that may take different values than those in (25)) and μ_0 is given by (27). Note that $\mathscr{M}$ now defines a norm on the quotient space $H^1(\Omega_o)/M_0$.

3.3 Formulation of the Coupled Problem

Let $V_c = \{v \in H^1(\Omega_c) : v(0) = 0\}$ and $V_d = \{z \in \mathbb{R}^{m+1}\}$ be the vector spaces of test functions for the continuum and discrete models, respectively. The norms on V_c and V_d are chosen as:

$$\|v\|_{V_c} = \sqrt{\int_{\Omega_c} E|v'|^2 dx} \qquad \text{and} \qquad \|z\|_{V_d} = \sqrt{|z|^2_{V_d} + \delta|\overline{z}|^2}, \tag{29}$$

where we have introduced the seminorm $|\cdot|_{V_d}$ on V_d and average of z on Ω_o as:

$$|z|_{V_d} = \sqrt{\sum_{i=1}^{m} k_i (z_i - z_{i-1})^2} \qquad \text{and} \qquad \overline{z} = \frac{1}{|\Omega_o|} \sum_{i=1}^{m_o} l_i \frac{z_i + z_{i-1}}{2} = \overline{\Pi_o z}, \tag{30}$$

with δ a dimensionally consistent weighting constant that we define below. The vector space for the Lagrange multipliers and associated norm are given as $M = H^1(\Omega_o)/M_0$ and:

$$\|\mu\|_M = \sqrt{\beta_0|\overline{\mu}|^2 + \beta_1 \int_{\Omega_o} |\mu^*|^2 dx} = \sqrt{\beta_0|\overline{\mu}|^2 + \beta_1 \|\mu^*\|^2_{L^2(\Omega_o)}}, \tag{31}$$

with associated inner product:

$$(\lambda, \mu)_M = \beta_0 \overline{\lambda} \overline{\mu} + \beta_1 \int_{\Omega_o} \lambda^* \mu^* dx. \tag{32}$$

We also define the bilinear form $b(\cdot, \cdot)$ on $M \times X$ such that:

$$b(\mu, V) = (\mu, \mathrm{R}_o v - \Pi_o z)_M, \tag{33}$$

where, for the sake of simplicity in the notation, we have introduced the product space $X = V_c \times V_d$ with pairs of X denoted, for example, as $U = (u, w)$, $V = (v, z)$, and with norm:

$$\|V\|_X = \sqrt{\|v\|_{V_c}^2 + \|z\|_{V_d}^2}. \tag{34}$$

We now define the kernel space of $b(\cdot, \cdot)$ as the subspace of X such that:

$$X_0 = \{V \in X : b(\mu, V) = 0, \ \forall \mu \in M\}. \tag{35}$$

The coupled problem consists in finding $U \in X$ such that U minimizes the total energy and satisfies the constraint $\|\mathrm{R}_o u - \Pi_o w\|_M = 0$, i.e.

$$\hat{\mathscr{E}}(U) = \hat{\mathscr{E}}_c(u) + \hat{\mathscr{E}}_d(w) = \min_{\substack{V \in X \\ \|\mathrm{R}_o v - \Pi_o z\|_M = 0}} \left(\hat{\mathscr{E}}_c(v) + \hat{\mathscr{E}}_d(z) \right). \tag{36}$$

The minimization problem (36) can be recast into the following saddle point problem:

$$\boxed{\text{Find } U \in X, \lambda \in M \text{ such that} \quad L(U, \lambda) = \inf_{V \in X} \sup_{\mu \in M} L(V, \mu),} \tag{37}$$

where the Lagrangian reads:

$$L(V, \mu) = \hat{\mathscr{E}}_c(v) + \hat{\mathscr{E}}_d(z) + (\mu, \mathrm{R}_o v - \Pi_o z)_M = \frac{1}{2} a(V, V) - l(V) + b(\mu, V), \tag{38}$$

with

$$\begin{aligned} a(U, V) &= \int_{\Omega_c} \alpha_c E u' v' dx + \sum_{i=1}^{m} \alpha_i k_i (w_i - w_{i-1})(z_i - z_{i-1}), \\ l(V) &= f z_m. \end{aligned} \tag{39}$$

The coupled problem can then be recast in mixed form as:

$$\boxed{\begin{aligned} &\text{Find } U \in X, \lambda \in M \text{ such that:} \\ &\qquad a(U, V) + b(\lambda, V) = l(V), \quad \forall V \in X, \\ &\qquad b(\mu, U) = 0, \quad \forall \mu \in M. \end{aligned}} \tag{40}$$

We analyze below the mathematical properties of this coupled problem.

4 Mathematical Analysis of the Coupling Method

The main objective of this section is to show that Problem (40) is well-posed for $\beta_0 > 0$ and $\beta_1 > 0$. We present here a detailed proof and explicitly derive the bounding constants associated with the problem. Proofs of continuity of the forms $a(\cdot,\cdot)$ and $l(\cdot)$ were shown in [5]. We show below that the coupling term $b(\cdot,\cdot)$ is continuous and satisfies the Babuška-Brezzi condition and that form $a(\cdot,\cdot)$ is coercive. For simplicity of the proofs, we shall consider in this section that the overlap region Ω_o exactly coincides with one RVE.

Lemma 1 (Continuity of b). *Let $b(\cdot,\cdot)$ be as defined in* (32)*. Then, for all $\mu \in M$, $V = (v,z) \in X$, there exists a constant $M_b > 0$ such that:*

$$|b(\mu, V)| \leq M_b \|\mu\|_M \|V\|_X,$$

with

$$M_b = \sqrt{\beta_0\left(\frac{|\Omega_c|^2}{2E|\Omega_o|} + \frac{1}{\delta}\right) + \beta_1\left(\frac{1}{E} + \frac{1}{\min_i k_i l_i}\right)}, \tag{41}$$

where $\min_i$ *means the minimum over all values indexed by* $i = 1,2,\ldots,m_o$.

Proof. Let $\mu \in M$ and $V \in X$, such that $\mathrm{R}_o v \in M$ and $\Pi_o z \in M$. From the definition of the bilinear form $b(\cdot,\cdot)$ (32) and by using Cauchy-Schwarz, we have:

$$|b(\mu,V)| = (\mu, \mathrm{R}_o v - \Pi_o z)_M \leq \|\mu\|_M \|\mathrm{R}_o v - \Pi_o z\|_M \leq \|\mu\|_M (\|\mathrm{R}_o v\|_M + \|\Pi_o z\|_M).$$

Now, by definition of the norm, we have

$$\|\mathrm{R}_o v\|_M^2 = \beta_0 \overline{\mathrm{R}_o v}^2 + \beta_1 \|(\mathrm{R}_o v)^*\|_{L^2(\Omega_o)}^2. \tag{42}$$

Then, using Lemma A-2 in [5], the fact that $|\Omega_c| \geq |\Omega_o|$, and Poincaré inequality, we get:

$$\overline{\mathrm{R}_o v}^2 \leq \frac{1}{|\Omega_o|}\|\mathrm{R}_o v\|_{L^2(\Omega_o)}^2 \leq \frac{1}{|\Omega_o|}\|v\|_{L^2(\Omega_c)}^2 \leq \frac{|\Omega_c|^2}{2E|\Omega_o|}\|v\|_{V_c}^2. \tag{43}$$

For the other term, since $\mathrm{R}_o v \in M$, $\mathrm{R}_o v$ is linear on the RVE, and by assumption, on Ω_o. Then $(\mathrm{R}_o v)'$ is constant on Ω_o and it implies that $(\mathrm{R}_o v)^* = (\mathrm{R}_o v)'$, $\forall x \in \Omega_o$. It follows that:

$$\|(\mathrm{R}_o v)^*\|_{L^2(\Omega_o)}^2 = \|(\mathrm{R}_o v)'\|_{L^2(\Omega_o)}^2 = |(\mathrm{R}_o v)|_{H^1(\Omega_o)}^2 \leq |v|_{H^1(\Omega_c)}^2 = \frac{1}{E}\|v\|_{V_c}^2. \tag{44}$$

Then,

$$\|\mathrm{R}_o v\|_M \leq \|v\|_{V_c} \sqrt{\frac{\beta_0|\Omega_c|^2}{2E|\Omega_o|} + \frac{\beta_1}{E}}. \tag{45}$$

In the same way, since $\Pi_o z$ is linear, we have

$$\|\Pi_o z\|_M^2 = \beta_0 \overline{\Pi_o z}^2 + \beta_1 \|(\Pi_o z)^*\|^2_{L^2(\Omega_o)} = \beta_0 \overline{z}^2 + \beta_1 \|(\Pi_o z)'\|^2_{L^2(\Omega_o)}, \quad (46)$$

and

$$\begin{aligned}\|(\Pi_o z)'\|^2_{L^2(\Omega_o)} &= \int_{\Omega_o} (\Pi_o z)'^2 dx = \sum_{i=1}^{m_o} l_i \left(\frac{z_i - z_{i-1}}{l_i}\right)^2 \\ &= \sum_{i=1}^{m_o} \frac{1}{k_i l_i} k_i (z_i - z_{i-1})^2 \leq \left(\min_i k_i l_i\right)^{-1} |z|^2_{V_d}.\end{aligned} \quad (47)$$

Therefore,

$$\|\Pi_o z\|_M^2 \leq \frac{\beta_0}{\delta} \delta \overline{z}^2 + \frac{\beta_1}{\min_i k_i l_i} |z|^2_{V_d} \leq \left(\frac{\beta_0}{\delta} + \frac{\beta_1}{\min_i k_i l_i}\right) \|z\|^2_{V_d}. \quad (48)$$

We combine above results and find

$$M_b = \sqrt{\beta_0 \left(\frac{|\Omega_c|^2}{2E|\Omega_o|} + \frac{1}{\delta}\right) + \beta_1 \left(\frac{1}{E} + \frac{1}{\min_i k_i l_i}\right)}, \quad (49)$$

which completes the proof. □

Lemma 2 (Inf-sup condition for b**).** *Let* $\beta_1 > 0$. *Then, with above notation and definitions, there exists a constant* $\gamma_b > 0$ *such that:*

$$\inf_{\mu \in M} \sup_{V \in X} \frac{|b(\mu, V)|}{\|\mu\|_M \|V\|_X} \geq \gamma_b,$$

with

$$\gamma_b = \min\left(\sqrt{\frac{\beta_0}{2\delta}}, \sqrt{\frac{2\beta_1}{2E + \delta|\Omega_o|}}\right).$$

Proof. Let $\mu \in M \subset H^1(\Omega_o)$. It is sufficient to construct a pair $\hat{V} \in X$ such that

$$\sup_{V \in X} \frac{|b(\mu, V)|}{\|V\|_X} \geq \frac{|b(\mu, \hat{V})|}{\|\hat{V}\|_X} \geq \gamma_b \|\mu\|_M. \quad (50)$$

Since $M \subset H^1(\Omega_o)$, $\mu(x_a)$ is well defined and denoted by μ_a. We introduce the function $\hat{\mu}(x) = \mu(x) - \mu_a$ on $H^1(\Omega_o)$ and observe that $\hat{\mu}(x_a) = \mu(x_a) - \mu_a = 0$. Let $\hat{v} \in V_c$ such that $\hat{v} = \hat{\mu}$ on Ω_o and $\hat{v} = 0$ on $\Omega_c \backslash \Omega_o$ and let $\hat{z} \in V_d$ such that $\hat{z}_i = -\mu_a$, $\forall i = 1, \ldots, m$. Thus, taking $\hat{V} = (\hat{v}, \hat{z})$, we have:

$$\frac{|b(\mu, \hat{V})|}{\|\hat{V}\|_X} = \frac{|(\mu, \mathrm{R}_o \hat{v} - \Pi_o \hat{z})_M|}{\|(\hat{v}, \hat{z})\|_X} = \frac{|(\mu, \mu - \mu_a + \mu_a)_M|}{\|(\hat{v}, \hat{z})\|_X} = \frac{\|\mu\|_M^2}{\|(\hat{v}, \hat{z})\|_X}. \quad (51)$$

It suffices to show that $\|\mu\|_M / \|(\hat{v}, \hat{z})\|_X$ is greater than a positive constant independent of μ. We have

$$\begin{aligned}\|(\hat{v},\hat{z})\|_X^2 &= \|\hat{v}\|_{V_c}^2 + \|\hat{z}\|_{V_d}^2 = \int_{\Omega_o} E|\hat{v}'|^2 dx + |\hat{z}|_{V_d}^2 + \delta|\bar{\hat{z}}|^2 \\ &= \delta\mu_a^2 + \int_{\Omega_o} E|\mu'|^2 dx = \delta\mu_a^2 + \int_{\Omega_o} E|\mu^*|^2 dx,\end{aligned} \tag{52}$$

where we have used the fact that μ is linear on Ω_o, i.e. μ' is constant and $\mu' = \mu^*$. Then, rewriting $\mu = \mu'(x - x_a) + \mu_a$ and taking the average, we also have:

$$\mu_a = \bar{\mu} - \frac{1}{2}|\Omega_o|\mu', \tag{53}$$

and

$$\mu_a^2 \le 2\overline{\mu}^2 + \frac{1}{2}|\Omega_o|^2(\mu')^2 = 2\overline{\mu}^2 + \frac{1}{2}|\Omega_o| \int_{\Omega_o} |\mu^*|^2 dx. \tag{54}$$

It follows that:

$$\begin{aligned}\|(\hat{v},\hat{z})\|_X^2 &\le 2\delta\overline{\mu}^2 + \left(\frac{2E + \delta|\Omega_o|}{2}\right) \int_{\Omega_o} |\mu^*|^2 dx \\ &\le \max\left(\frac{2\delta}{\beta_0}, \left(\frac{2E + \delta|\Omega_o|}{2\beta_1}\right)\right)\|\mu\|_M^2,\end{aligned} \tag{55}$$

and we conclude that

$$\frac{\|\mu\|_M}{\|(\hat{v},\hat{z})\|_X} \ge \min\left(\sqrt{\frac{\beta_0}{2\delta}}, \sqrt{\frac{2\beta_1}{2E + \delta|\Omega_o|}}\right), \tag{56}$$

which completes the proof. □

We now show the coercivity of a in the case where $\alpha_c = \alpha_d = 1/2$ on Ω_o. We believe that the result also holds when α_c is a continuous piecewise linear function but are not able to provide here a rigorous proof.

Lemma 3 (Coercivity of a). *Let $\alpha_c = \alpha_d = 1/2$. Then, with above notation and definitions, there exists a constant $\gamma_a > 0$ such that:*

$$\begin{cases} \inf\limits_{U \in X_0} \sup\limits_{V \in X_0} \dfrac{|a(U,V))|}{\|U\|_X \|V\|_X} > \gamma_a, \\ \sup\limits_{U \in X_0} a(U,V) > 0, \qquad \forall V \in X_0, V \neq 0, \end{cases} \tag{57}$$

with

$$\gamma_a = \frac{1}{2}\min_i\left(\frac{E}{k_i l_i}\right)\min_i\left(\frac{k_i l_i}{E}\right)\min\left(\frac{1}{2}, \frac{E}{\delta}\frac{|\Omega_o|}{|\Omega_c|^2}\right), \tag{58}$$

where $\min\limits_i$ means the minimum over all values indexed by $i = 1, 2, \ldots, m_o$.

Proof. It suffices to show that $a(\cdot,\cdot)$ is coercive on X_0. Let $V = (v, z) \in X_0$. By definition of the bilinear form, and the fact that $\alpha_c = 1$ on $\Omega_c \backslash \Omega_o$ and $\alpha_d = 1$ on

$\Omega_d \backslash \Omega_o$, we have

$$\begin{aligned} a(V,V) &= \int_{\Omega_c} \alpha_c E|v'|^2 \, dx + \sum_{i=1}^{m} \alpha_i k_i (z_i - z_{i-1})^2 \\ &= \int_{\Omega_c \backslash \Omega_o} E|v'|^2 \, dx + \sum_{i=m_o+1}^{m} k_i (z_i - z_{i-1})^2 \\ &\quad + \int_{\Omega_o} \alpha_c E|(\mathrm{R}_o v)'|^2 \, dx + \sum_{i=1}^{m_o} \alpha_i k_i (z_i - z_{i-1})^2. \end{aligned} \tag{59}$$

We provide here a general approach to show the coercivity. We first decompose the overlap terms in above equation into the following contributions:

$$\begin{aligned} &\int_{\Omega_o} \alpha_c E|(\mathrm{R}_o v)'|^2 \, dx + \sum_{i=1}^{m_o} \alpha_i k_i (z_i - z_{i-1})^2 \\ &\quad = \frac{1}{2}\left(\int_{\Omega_o} \alpha_c E|(\mathrm{R}_o v)'|^2 \, dx + \sum_{i=1}^{m_o} \alpha_i k_i (z_i - z_{i-1})^2 \right) \\ &\quad + \frac{1}{2}\left(\int_{\Omega_o} \alpha_c E|(\mathrm{R}_o v)'|^2 \, dx + \sum_{i=1}^{m_o} \alpha_i k_i (z_i - z_{i-1})^2 \right). \end{aligned} \tag{60}$$

Since $V \in X_0$, the functions v and vectors z satisfy:

$$b(\mu, V) = (\mu, \mathrm{R}_o v - \Pi_o z)_M = 0, \qquad \forall \mu \in M. \tag{61}$$

In other words, $\mathrm{R}_o v - \Pi_o z \in M_o$, meaning that $\overline{\mathrm{R}_o v} = \overline{\Pi_o z}$ and that $v(x_b) - v(x_a) = z_{m_o} - z_o$ (where we appeal again to the fact that Ω_o consists of just one representative volume element). Let $\mu_o = \mathrm{R}_o v - \Pi_o z$ with $\overline{\mu_o} = 0$ and $\mu_o(x_a) = \mu_o(x_b)$. We also introduce the parameter $\kappa = \min_i (k_i l_i)/E$ and recall from Remark 1 that $\kappa \leq 1$. We have:

$$\begin{aligned} \frac{1}{2}\int_{\Omega_o} \alpha_c E|(\mathrm{R}_o v)'|^2 dx &\geq \frac{\kappa}{2}\int_{\Omega_o} \alpha_c E|(\mathrm{R}_o v)'|^2 dx \\ &\geq \frac{\kappa}{2}\int_{\Omega_o} \alpha_c E|(\Pi_o z)' + \mu_o'|^2 dx \\ &\geq \frac{\kappa}{2}\int_{\Omega_o} \alpha_c E|(\Pi_o z)'|^2 dx + \frac{\kappa}{2}\int_{\Omega_o} \alpha_c E|\mu_o'|^2 dx + \kappa \int_{\Omega_o} \alpha_c E(\Pi_o z)'\mu_o' dx. \end{aligned} \tag{62}$$

Using the fact that:

$$(\Pi_o z)' = \frac{z_i - z_{i-1}}{l_i}, \qquad \forall x \in (x_{i-1}, x_i), \tag{63}$$

the first integral can be rewritten as:

$$\begin{aligned}\int_{\Omega_o} \alpha_c E|(\Pi_o z)'|^2 dx &= \sum_{i=1}^{m_o} \int_{x_i-1}^{x_i} \alpha_c E\left[\frac{z_i - z_{i-1}}{l_i}\right]^2 dx \\ &= \sum_{i=1}^{m_o} \frac{E}{k_i l_i} k_i (z_i - z_{i-1})^2 \left[\frac{1}{l_i}\int_{x_i-1}^{x_i} \alpha_c dx\right],\end{aligned} \tag{64}$$

and, using the definition of α_i, we get:

$$\int_{\Omega_o} \alpha_c E|(\Pi_o z)'|^2 dx \geq \min_i \left(\frac{E}{k_i l_i}\right) \sum_{i=1}^{m_o} (1-\alpha_i) k_i (z_i - z_{i-1})^2. \tag{65}$$

It follows that

$$\begin{aligned}\frac{1}{2}\int_{\Omega_o} \alpha_c E|(\mathrm{R}_o v)'|^2 dx \geq & \frac{\kappa}{2} \min_i \left(\frac{E}{k_i l_i}\right) \sum_{i=1}^{m_o} (1-\alpha_i) k_i (z_i - z_{i-1})^2 \\ & + \frac{\kappa}{2}\int_{\Omega_o} \alpha_c E|\mu_o'|^2 dx + \kappa \int_{\Omega_o} \alpha_c E(\Pi_o z)' \mu_o' dx.\end{aligned} \tag{66}$$

In the same way, we have:

$$\begin{aligned}\frac{1}{2}\sum_{i=1}^{m_0} \alpha_i k_i (z_i - z_{i-1})^2 &\geq \frac{\kappa}{2}\int_{\Omega_o} (1-\alpha_c) E|(\Pi_o z)'|^2 dx \\ &\geq \frac{\kappa}{2}\int_{\Omega_o} (1-\alpha_c) E|(\mathrm{R}_o v)' - \mu_o'|^2 dx \\ &\geq \frac{\kappa}{2}\int_{\Omega_o} (1-\alpha_c) E|(\mathrm{R}_o v)'|^2 dx \\ &\quad + \frac{\kappa}{2}\int_{\Omega_o} (1-\alpha_c) E|\mu_o'|^2 dx - \kappa \int_{\Omega_o} (1-\alpha_c) E(\mathrm{R}_o v)' \mu_o' dx.\end{aligned} \tag{67}$$

Using (66) and (67) in (60), we obtain:

$$\begin{aligned}&\int_{\Omega_o} \alpha_c E|(\mathrm{R}_o v)'|^2 \, dx + \sum_{i=1}^{m_o} \alpha_i k_i (z_i - z_{i-1})^2 \\ &\quad \geq \frac{\kappa}{2}\left[\int_{\Omega_o} \alpha_c E|(\mathrm{R}_o v)'|^2 dx + \int_{\Omega_o} (1-\alpha_c) E|(\mathrm{R}_o v)'|^2 dx\right] \\ &\quad + \frac{\kappa}{2}\int_{\Omega_o} (1-\alpha_c) E|\mu_o'|^2 dx - \kappa \int_{\Omega_o} (1-\alpha_c) E(\mathrm{R}_o v)' \mu_o' dx \\ &\quad + \frac{\kappa}{2}\min_i \left(\frac{E}{k_i l_i}\right)\left[\sum_{i=1}^{m_o} \alpha_i k_i (z_i - z_{i-1})^2 + \sum_{i=1}^{m_o} (1-\alpha_i) k_i (z_i - z_{i-1})^2\right] \\ &\quad + \frac{\kappa}{2}\int_{\Omega_o} \alpha_c E|\mu_o'|^2 dx + \kappa \int_{\Omega_o} \alpha_c E(\Pi_o z)' \mu_o' dx.\end{aligned} \tag{68}$$

Simplifying, we get:

$$\begin{aligned}
&\int_{\Omega_o} \alpha_c E|(\mathrm{R}_o v)'|^2\, dx + \sum_{i=1}^{m_o} \alpha_i k_i (z_i - z_{i-1})^2 \\
&\quad \geq \frac{1}{2} \min_i \left(\frac{E}{k_i l_i}\right) \min_i \left(\frac{k_i l_i}{E}\right) \left[\int_{\Omega_o} E|(\mathrm{R}_o v)'|^2\, dx + \sum_{i=1}^{m_o} k_i (z_i - z_{i-1})^2\right] \\
&\qquad + \frac{\kappa}{2} \int_{\Omega_o} E|\mu_o'|^2 dx - \kappa \int_{\Omega_o} (1-\alpha_c) E (\mathrm{R}_o v)' \mu_o' dx + \kappa \int_{\Omega_o} \alpha_c E (\Pi_o z)' \mu_o' dx.
\end{aligned} \tag{69}$$

We note that the last three terms, denoted by $\mathscr{K}$, can be combined as follows:

$$\begin{aligned}
\mathscr{K} &= \frac{\kappa}{2} \int_{\Omega_o} (2\alpha_c - 1)\, E \left[2(\mathrm{R}_o v)' - \mu_o'\right] \mu_o' dx \\
&= \frac{\kappa}{2} \int_{\Omega_o} (2\alpha_c - 1)\, E \left[|(\mathrm{R}_o v)'|^2 - |(\Pi_o z)'|^2\right] dx.
\end{aligned} \tag{70}$$

The goal would be to show that $\mathscr{K} \geq 0$ for all $(v, z) \in X_0$ for any admissible profile of α_c on Ω_o. Unfortunately, we are only able to date to prove that $\mathscr{K} = 0$ if $\alpha_c = 1/2$. It is not clear at this point whether the result would hold in the case where α_c is continuous piecewise linear.

Finally, setting $\alpha_c = 1/2$, we may proceed as follows:

$$\begin{aligned}
a(V,V) &\geq \int_{\Omega_c \setminus \Omega_o} E|v'|^2\, dx + \sum_{i=m_o+1}^{m} k_i (z_i - z_{i-1})^2 \\
&\quad + \frac{1}{2} \min_i \left(\frac{E}{k_i l_i}\right) \min_i \left(\frac{k_i l_i}{E}\right) \left[\int_{\Omega_o} E|v'|^2\, dx + \sum_{i=1}^{m_o} k_i (z_i - z_{i-1})^2\right] \\
&\geq \frac{1}{2} \min_i \left(\frac{E}{k_i l_i}\right) \min_i \left(\frac{k_i l_i}{E}\right) \left[\int_{\Omega_c} E|v'|^2\, dx + \sum_{i=1}^{m} k_i (z_i - z_{i-1})^2\right] \\
&\geq \gamma \left(\|v\|_{V_c}^2 + |z|_{V_d}^2\right),
\end{aligned} \tag{71}$$

where we have introduced the constant γ as:

$$\gamma = \frac{1}{2} \min_i \left(\frac{E}{k_i l_i}\right) \min_i \left(\frac{k_i l_i}{E}\right). \tag{72}$$

Applying first the Poincaré inequality, i.e.

$$a(V,V) \geq \gamma \left(\frac{1}{2} \|v\|_{V_c}^2 + \frac{E}{|\Omega_c|^2} \|v\|_{L^2(\Omega_c)}^2 + |z|_{V_d}^2\right), \tag{73}$$

and then Lemma A-2 in [5], as well as the fact that X_0 consists of those functions v and vectors z such that $\overline{v} = \overline{\Pi_o z} = \overline{z}$ on Ω_o, i.e.

$$\|v\|_{L^2(\Omega_c)}^2 \geq \|v\|_{L^2(\Omega_o)}^2 \geq \overline{v}^2 |\Omega_o| = \overline{z}^2 |\Omega_o|, \tag{74}$$

we finally obtain:

$$a(V,V) \geq \gamma \left(\frac{1}{2} \|v\|_{V_c}^2 + |z|_{V_d}^2 + \frac{E}{\delta} \frac{|\Omega_o|}{|\Omega_c|^2} \delta \bar{z}^2 \right) \geq \gamma \min \left(\frac{1}{2}, \frac{E}{\delta} \frac{|\Omega_o|}{|\Omega_c|^2} \right) \|V\|_X^2, \tag{75}$$

which completes the proof. □

From above lemmas, we may conclude that the Arlequin problem is well-posed as long as $\beta_0 > 0$ and $\beta_1 > 0$ (and restriction that $\alpha_c = 1/2$ on overlap domain).

5 Finite Element Formulation

We introduce in this section the finite element formulation of the coupled problem. Let V_c^h and M^h be finite element subspaces of the vector spaces V_c and M, respectively, and let X^h be the product space $X^h = V_c^h \times V_d$. The subspace V_c^h can be constructed as the space spanned by the piecewise linear continuous functions defined with respect to the set of nodes $x_i = ih$, $i = 0, \ldots, N^e$, where N^e denotes the number of elements in the mesh. In the case of M^h, we are clearly faced with several choices since the elements associated with V_c^h and M^h do not have to match. However, for the sake of simplicity, one possibility is to restrict ourselves to cases where each node of the mesh associated with M^h coincides either with a particle or with a node of V_c^h or both. However, M^h needs to be constructed in such a way that the condition $M^h \subset M = H^1(\Omega_o) \backslash M_0$ be satisfied, that is, we need to make sure that functions of M_0 are excluded from M^h. Let $\widetilde{M}^h$ be the vector space spanned by continuous piecewise linear functions defined on Ω_o and let h_M be the mesh size associated with $\widetilde{M}^h$ (assume a uniform grid here). If the overlap region consists of one RVE and if n_s denotes the number of springs in one RVE, we have been able to observe numerically that the number of modes in $\widetilde{M}^h \cap M_0$ is given by:

$$n_0 = \begin{cases} \dfrac{n_s l_i}{h_M} - 1, & \text{if } h_M < \xi, \\ 0, & \text{otherwise,} \end{cases} \tag{76}$$

where ξ is the size of the RVE and l_i is the equilibrium length of each bond (assumed constant here). It follows that a convenient way to construct the finite element space M^h is to consider continuous piecewise linear functions defined with respect to elements of size $h_M = \xi$ (or a multiple of ξ).

Finally, we introduce the notation $U_h = (u_h, w_h)$ and $V_h = (v_h, z)$. Then, Problem (40) is approximated as follows:

$$\boxed{\begin{array}{l} \text{Find } U_h \in X^h,\ \lambda_h \in M^h \text{ such that:} \\ \qquad a(U_h, V_h) + b(\lambda_h, V_h) = l(V_h), \quad \forall V_h \in X^h, \\ \qquad b(\mu_h, U_h) = 0, \quad \forall \mu_h \in M^h. \end{array}} \tag{77}$$

We note that although V_d is a finite-dimensional space and, consequently does not need to be discretized using finite elements, we will use the notation w_h to denote the solution of the particle model in (77) to emphasize that w_h directly depends on the choice of V_h and M_h. We can show that above problem is also well-posed when $\beta_0 > 0$, $\beta_1 > 0$, and $\alpha_c = 1/2$.

6 Numerical Results

6.1 One-Dimensional Numerical Results

In this section, we present some numerical experiments to illustrate our theoretical study of a one-dimensional coupled problem, i.e. a model of harmonic springs, with varying coefficients, coupled with a linear elastic bar, whose Young's modulus is determined by simple homogenization. Unless otherwise stated, we consider in the following experiments the domain $\Omega = (0,3)$. The continuum model is used in the subdomain $\Omega_c = (0,2)$ while the particle model is used in $\Omega_d = (1,3)$ and the weight coefficients α_c and α_d are chosen linear in the overlap domain. Moreover, the force f applied at $x = 3$ is chosen in such a way that the displacement at the right end of the domain, when using the continuum model everywhere in Ω, is equal to unity. We also restrict ourselves to the cases where the equilibrium lengths of the springs are all equal. We also recall that the discrete problem is well-posed if the mesh size used to discretize the Lagrange multiplier space is at least larger than (a multiple of) the size of the repesentative cell. Hence, in the following, the size of the elements used to define M^h is always taken equal to the size of the representative volume element.

6.1.1 Overlap Region Composed of One RVE

Let us start by studying the very simple case of an overlap domain composed of only one RVE. As the objective is to propose a method that is well suited to solve problems dealing with highly heterogeneous particle models, we study here the particle case of a periodic distribution of springs with two spring stiffness parameters for which it is straightforward to derive an equivalent continuum model. Thus, the particle model is chosen to be composed of $m = 4$ springs in Ω_d, i.e. five particles, and that the values of the spring stiffness are $k_1 = 100$ and $k_2 = 1$. The particle structure is then constructed, for m even, as:

$$k_{2j-1} = k_1, \qquad k_{2j} = k_2, \qquad \forall j = 1, \ldots, m/2. \tag{78}$$

The equilibrium length of each spring is chosen constant as $l = l_i = 0.5$ and the corresponding Young's modulus E is then given by:

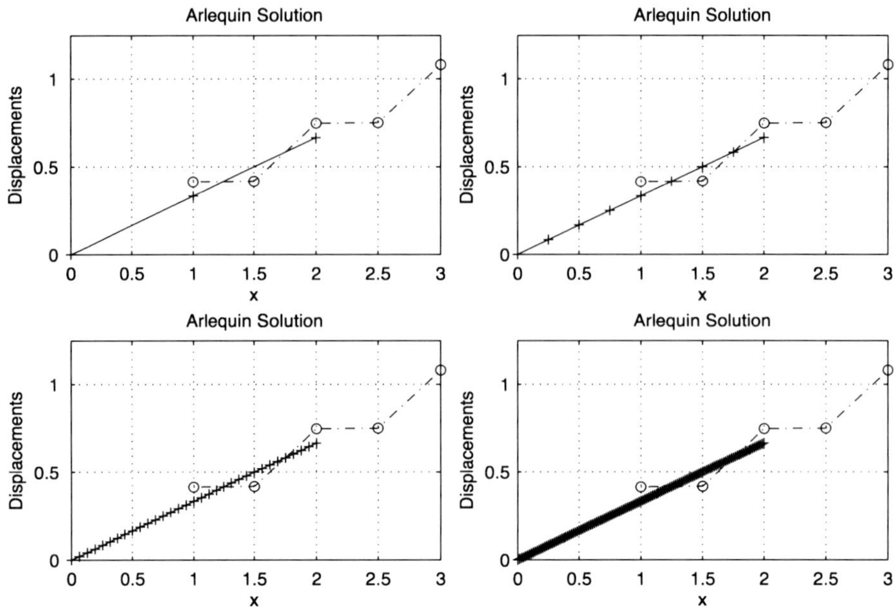

Fig. 8 Arlequin solution in case of one RVE in the overlap region for several mesh sizes

$$E = \frac{k_1 k_2}{k_1 + k_2} 2l = \frac{100}{101} \times 2 \times 0.5 = 0.99010, \tag{79}$$

using the expression derived in (11). In the following set of experiments, we study the effect of the mesh size on the Arlequin solution. The Arlequin solutions for four different mesh sizes, namely $h = 2l$, $h = l/2$, $h = l/4$, and $h = l/32$, where h is the size of the elements for V_c^h, are shown in Fig. 8. As expected, the coupled solution is independent of the mesh size as the displacement z_m of the right end particle is equal for all cases to 1.08168. Notice however that z_m is different from unity, as one might have expected from the choice of the loading force f applied to particle m. This is simply due to the fact that the displacement of the particles is averaged around the continuum solution on the overlap region. If we average the particle solution in Ω_d, we would then obtain a displacement equal to unity since the slope of the continuum solution and that of the averaged particle solution are identical.

6.1.2 Overlap Region Composed of Several RVE's

We now repeat the same experiments in the case where the size of the overlap region is equal to the size of several RVEs. We keep the same periodic distribution as before, that is, the RVE is made of two springs with stiffness coefficients $k_1 = 100$ and $k_2 = 1$. We consider the case where the overlap region is composed of

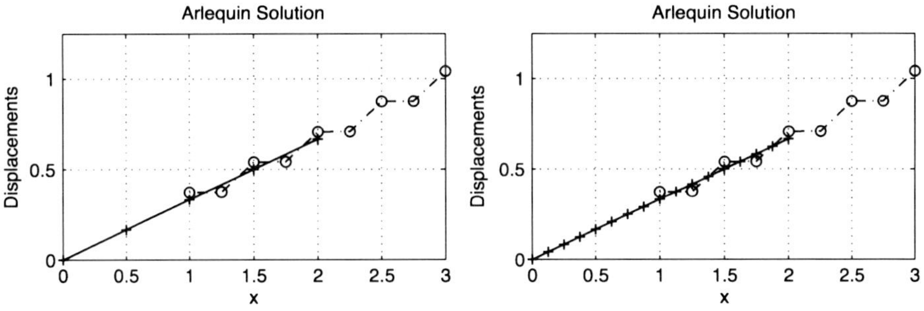

Fig. 9 Arlequin solution in case of two RVEs in the overlap region for two different mesh sizes

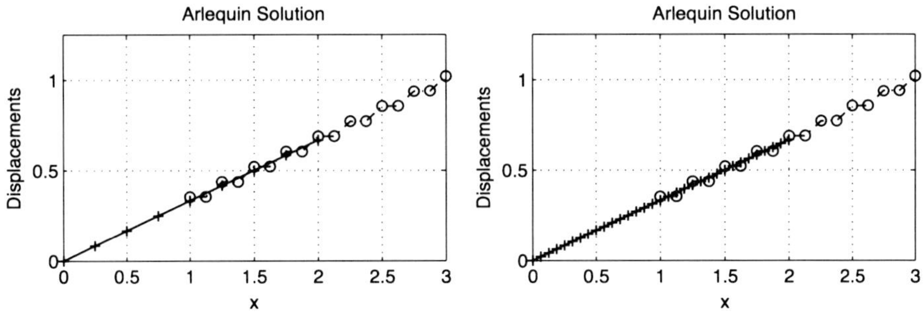

Fig. 10 Arlequin solution in case of four RVEs in the overlap region for two different mesh sizes

two RVEs and the particle structure is made of $m = 8$ springs ($l = 0.25$), and the case of four RVEs in the overlap region and a particle model composed of $m = 16$ springs ($l = 0.125$). For both cases, we compute the solutions on two different mesh sizes, namely $h = 2l$ and $h = l/2$, as shown in Figs. 9 and 10. We can see that the method produces the correct results as expected. The displacements at the right end particle are $z_m = 1.04084$ and $z_m = 1.02042$ in the case of two RVEs and four RVEs, respectively. These displacements get actually closer to unity since the smaller the equilibrium lengths are, the closer the particle solution gets to the continuum solution.

6.1.3 An Example with a Large Number of Particles

In more practical cases, we are interested in systems that are composed of many particles. The objective is to use the particle model around a tiny zone to model the small scale behavior of the material, and in the remaining zone, to use the continuum model in order to reduce the cost of the simulation. We consider here the case of the structure made of a chain of 1001 particles connected by 1000 springs in the domain $\Omega = (0, 1)$, as shown in Fig. 11 . We define $\Omega_c = (0, 0.8)$ and $\Omega_d = (0.796, 1)$.

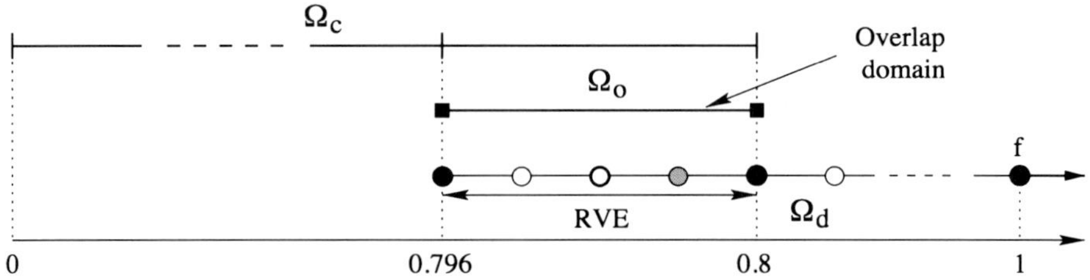

Fig. 11 Implementation of the Arlequin method on a system of 1,001 particles

We assume for the particle model a periodic distribution of four springs with spring constants $k_1 = 100, k_2 = 1, k_3 = 50$, and $k_4 = 10$, and equilibrium length $l = 0.001$, for which we get the equivalent Young's modulus E as:

$$E = \left[\frac{1}{k_1^{-1} + k_2^{-1} + k_3^{-1} + k_4^{-1}}\right] 4l = \left[\frac{1}{0.01 + 1.00 + 0.02 + 0.10}\right] \times 4 \times 0.001, \tag{80}$$

that is, $E = 3.539823 \times 10^{-3}$. Notice that the definition of the geometry implies that the overlap domain Ω_o is made of just one representative cell. With the idea of considering a critical and practical experiment, we discretize Ω_c with a mesh made of two elements. The first element covers the continuum region $\Omega_c \backslash \Omega_o$ while the second element covers the whole overlap region Ω_o. The Arlequin solution is shown in Fig. 12. We observe that the large-scale displacement in the whole structure is perfectly linear and that the displacement at $x = 1$ is again closer to unity ($z_m = 0.99969$) than in the previous results since the equilibrium length of the springs is here reduced to $l = 0.001$. These results clearly demonstrate that we can consider an extreme configuration of a continuum model discretized with only two elements (one for the whole continuum region and one for the coupling zone) to deliver accurate simulations. In other words, only one element is sufficient to model the behavior of the material in $\Omega_c \backslash \Omega_o$ (since the model is linear) and one element to discretize the overlap region (composed of one RVE) is enough to couple the two models.

6.1.4 Simulation of a Defect

The goal in using the proposed coupling method is to replace the particle model by a continuum model in the region where only the large-scale contributions to the values of quantities of interest are significant and where the continuum model remains compatible with the particle model. The hope then is that the particle model would only be required in a small region of the whole domain, around a defect or a geometrical singularity for instance. We propose here to consider a one-dimensional structure, fixed at both extremities and subjected to a point force applied at the center particle (see Fig. 13), in which the stiffness coefficients in the middle bonds are purposely weakened as follows:

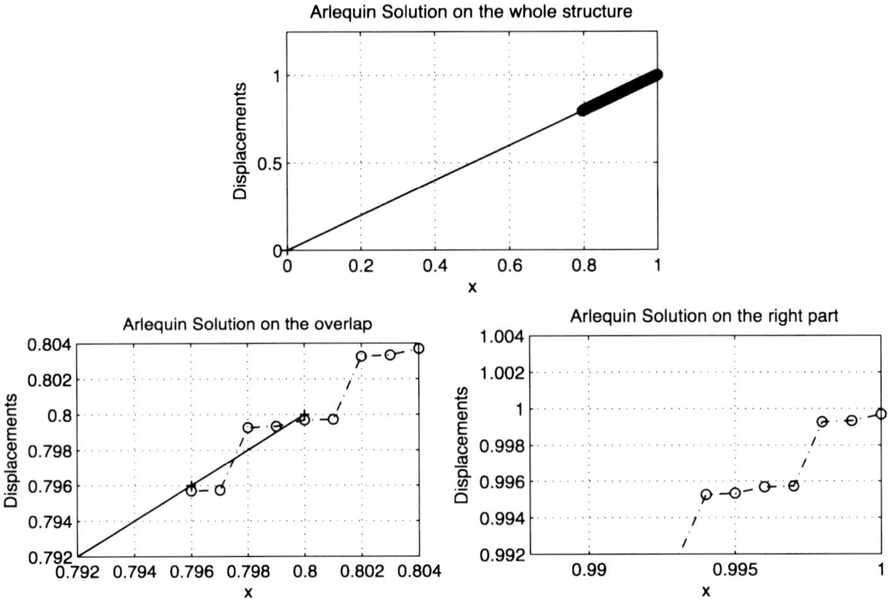

Fig. 12 Arlequin solution in case of a system with many particles using a mesh only composed of two elements

$$k_i^* = k_i \left[\frac{1}{1 + 20e^{-5(x-L/2)^2}} \right], \tag{81}$$

where L is the length of the structure. The main objective here is to model a pseudo-defect in the chain of particles around which the continuum model is no longer compatible with the particle model. The domain is given by $\Omega = (0.0, 5.2)$, i.e. $L = 5.2$, and the particle model is kept only in the subdomain $\Omega_d = (1.4, 3.8)$. The equilibrium length of the bonds is set to $l = 0.1$. Furthermore, we assume that the particle model is defined as a periodic distribution of two spring stiffness parameters $k_1 = 100$ and $k_2 = 30$ along which the proposed defect is superimposed. The Young's modulus of the continuum model is computed by ignoring the defect in the particle model, i.e. by considering the stiffness coefficients k_i rather than k_i^*. Using (79), its value is found to be $E = 4.61538$. In order to study the influence of the position and size of the overlap region onto the Arlequin solution, we consider four different configurations of the coupling zones defined by the overlap regions $\Omega_{o,1} = (1.4, 1.4 + 0.2j)$ and $\Omega_{o,2} = (3.8 - 0.2j, 3.8)$, on the left and on the right of the particle model, respectively, with $j = 1, \dots, 4$ (see Fig. 13). In other words, the size of the region in which the particle model is used is enlarged as the overlap regions are made of 4, 3, 2, and 1 RVE's by varying j from 4 to 1. Finally, the length of the elements is set to $h = 2l$ in $\Omega_c = \Omega_{c,1} \cup \Omega_{c,2}$.

The results are shown in Fig. 14. The first solution is obtained using $j = 4$ and the last one using $j = 1$. The maximum displacement, which corresponds to the displacement of the particle at the center, is reported for each configuration

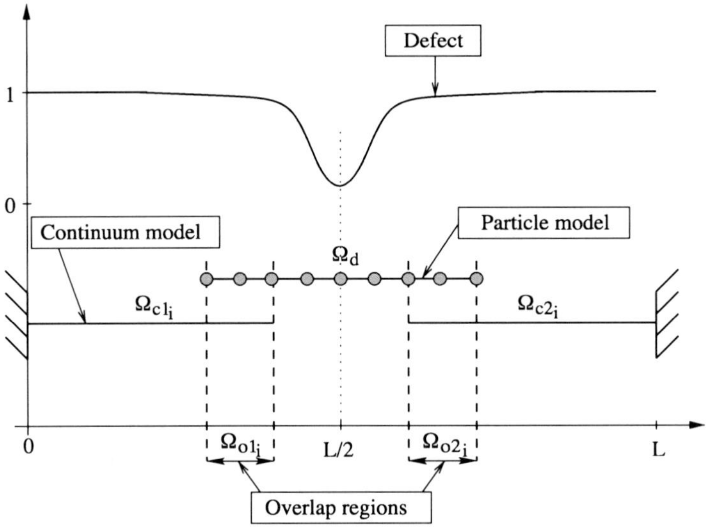

Fig. 13 Definition of the coupled model for the simulation of a defect

in Table 1. We observe that the approximations of the displacement become more accurate when the overlap regions are positioned away from the defect. This is due to the fact that the continuum model is not compatible with the particle model in the vicinity of the defect since the former is calibrated from the latter without taking the defect into account. However, in the case of the configuration with $j = 1$, the models become compatible with each other and the proposed coupling term provides an accurate solution around the defect with respect to the solution of the full particle model (not shown here).

Table 1 Maximum displacement for various values of the number of RVE's, n_{RVE}, in each overlap region

n_{RVE}	Maximum displacement
4	1.05137
3	1.12014
2	1.13450
1	1.13670

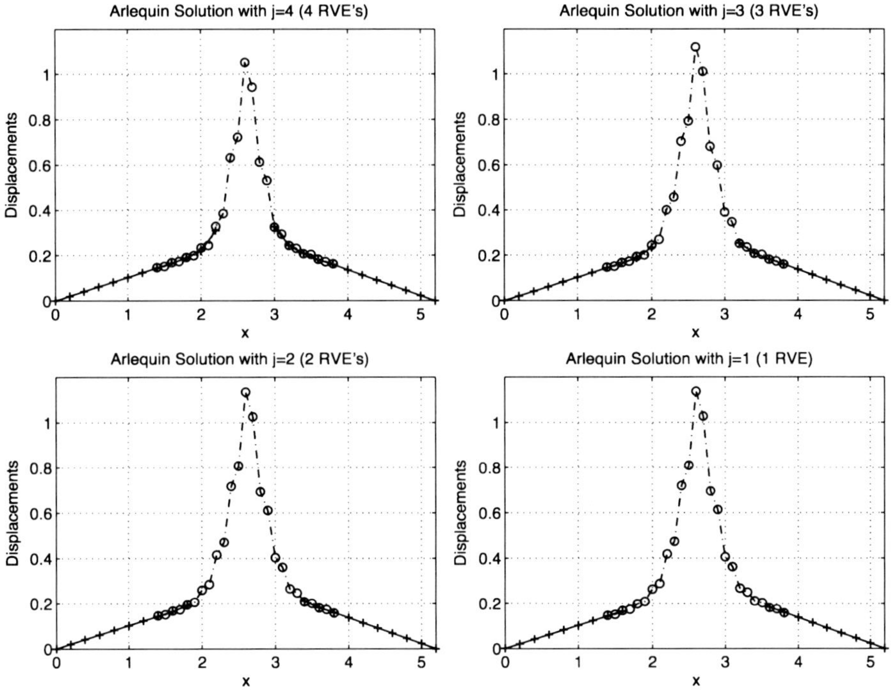

Fig. 14 Arlequin solutions obtained for different configurations of the coupling regions defined by $\Omega_{o,1} = (1.4, 1.4 + 0.2j)$ and $\Omega_{o,2} = (3.8 - 0.2j, 3.8)$, on the left and on the right of the particle model, respectively, with $j = 1, \ldots, 4$

6.2 Two-Dimensional Numerical Results

In this section, we apply the Arlequin formulation using the new coupling operator to the case of two-dimensional problems. In particular, we consider a uniform lattice in which the interactions between particles are modeled in terms of harmonic springs. The particles are supposed to interact only with their nearest neighbors: in the x- and y-directions, the stiffness parameter for each bond is given by k while in the diagonal direction, the stiffness coefficient is set to k_d. The Representative Volume Element is easily identified here as the cell defined by four lattice sites since it represents the smallest substructure within the periodic structure. The RVE is utilized to compute the material coefficients (Young's modulus and Poisson's ratio) of the compatible linear elasticity model.

The system of interest is made of 11×11 particles and is subjected to a point force applied to the particle located at the center of the domain. For large values of the force, displacements in the vicinity of the centered particle are expected to vary rapidly, implying that the linear elasticity model would incorrectly predict the large associated strains. In this simple example, we choose to employ the particle model in the subdomain at the center of the domain, of size corresponding to four RVE's, and

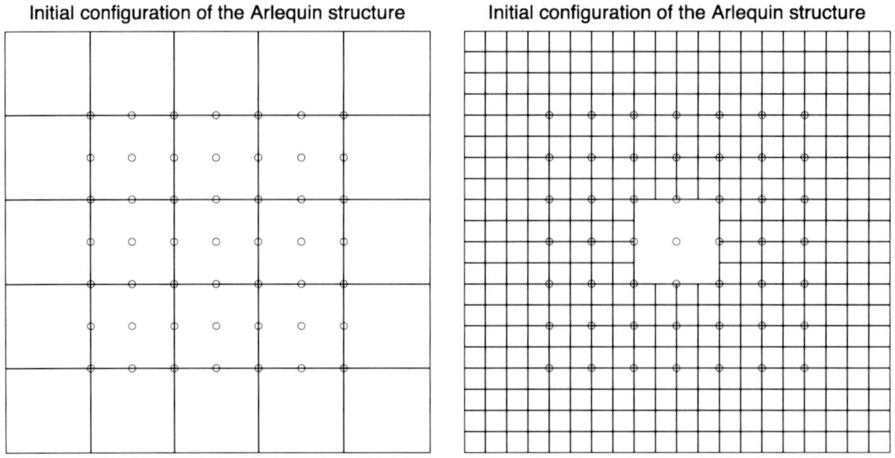

Fig. 15 Arlequin configuration of the coupled problem using a coarse mesh (*left*) and a fine mesh (*right*) for the discretization of the continuum model. The particle model is reduced to the subdomain in the center and the overlap region consists of a layer around the particle region

to construct the overlap region as the layer around the particle region, of thickness corresponding to the size of two RVE's. The continuum model is selected in the remainder of the domain and is discretized using quadrilateral bilinear elements (see Fig. 15). Finally, the system is subjected to homogeneous Dirichlet boundary conditions along the boundary $\partial\Omega$.

In order to test the method, we consider in what follows a coarse mesh and a fine mesh: on the coarse mesh, the finite elements have a mesh size equal to the size of two RVE's for the discretization of the continuum solution and of the Lagrange multiplier as shown on the left of Fig. 15; on the fine mesh, the elements are half the size of one RVE for the continuum solution and twice the size of one RVE for the Lagrange multiplier as shown on the right of Fig. 15.

Finally, we compute two Arlequin solutions on each of the two meshes: in the first Arlequin formulation, the coupling term is defined in terms of the H^1 norm as described in [5] while in the second formulation, the coupling term is defined using the proposed averaging operator. In both formulations, the weighting coefficients are chosen constant on the overlap region and equal to one half. On the coarse mesh, the two solutions are identical as expected (see Fig. 16). The fact that large elements are used in the formulation is equivalent to an averaging over a representative volume element. However, the two solutions are different on the fine mesh (see Fig. 17). This is due to the locking phenomenon in the case of the H^1 norm coupling, that is, the displacements of the continuum solution lock themselves to the displacements of the particle solution on the overlap region. A better approximation, better in the sense that the solution is closer to that of the full particle model, is therefore obtained using the formulation involving the averaging operator.

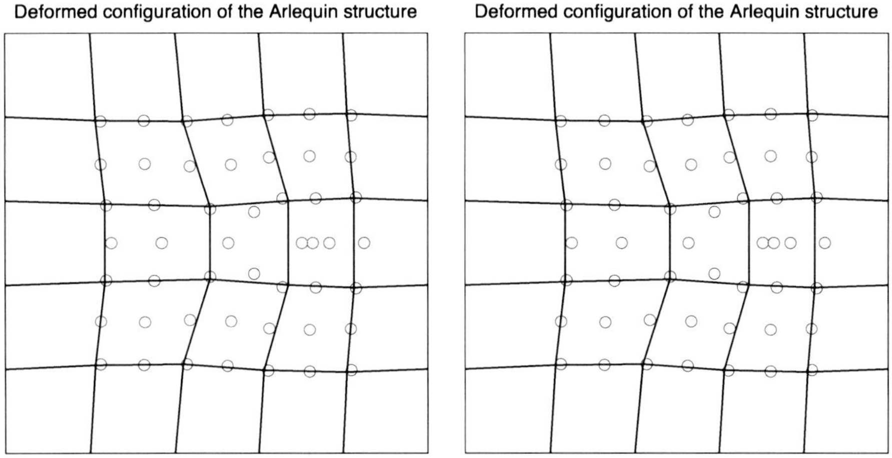

Fig. 16 Deformed configuration on coarse mesh using the Arlequin framework: (*Left*) the coupling term is defined in terms of the H^1 norm; (*Right*) the coupling term is defined in terms of the proposed averaging operator. The two solutions are identical as expected

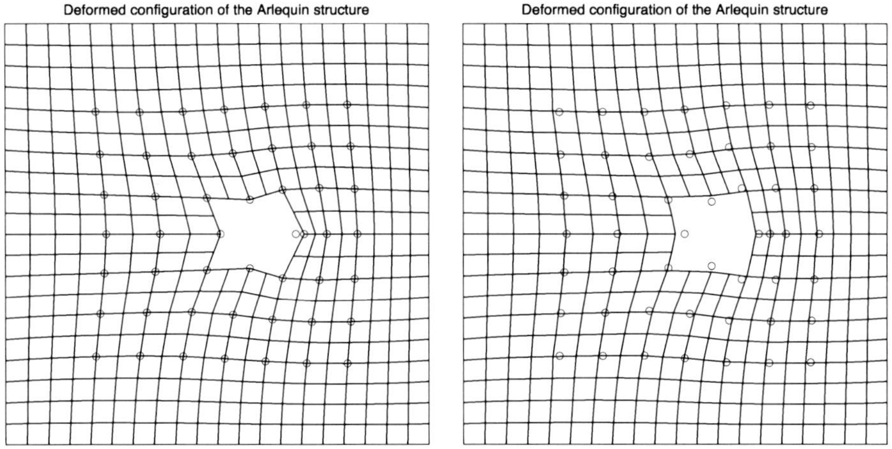

Fig. 17 Deformed configuration on fine mesh using the Arlequin framework: (*Left*) the coupling term is defined in terms of the H^1 norm; (*Right*) the coupling term is defined in terms of the proposed averaging operator. The two solutions are now different

7 Conclusion

We have presented in this paper a new expression for the coupling term when blending a particle model with a continuum model using the Arlequin framework. The coupling method belongs to the family of concurrent methods for solving multiscale problems. It constitutes an improved version of a previously proposed coupling method, described in [5], which had the major drawback of being mesh-dependent in the sense that meshes had to be carefully selected in order to obtain the intended solution of the problem. In particular, it was shown that the method produced satisfactory results as long as the mesh size of the finite elements used to discretize the Lagrange multiplier space was at least larger than (a multiple of) the size of the representative cell defined to calibrate the parameter(s) of the continuum model. In the new coupling method, the selection of meshes used to discretize the continuum solution and the Lagrange multiplier is immediately determined from the formulation of the continuous problem.

The new coupling term is constructed in terms of an averaging operator defined on a representative cell. The cell size determines in some sense the scale at which the continuum model and particle model can exchange information. Indeed, the parameters of the continuum model are usually identified through homogenization from the solution of the particle model computed on one representative cell. We have shown here that the resulting coupled problem is mathematically well-posed and that its discretization by the finite element method provides approximations that converge to the exact solution of the problem as the mesh size goes to zero. We have illustrated on one- and two-dimensional examples that the proposed approach is well suited for problems in which the bonds between particles are heterogeneously distributed. Systems in the present study were considered periodic as compatible continuum models can straightforwardly be derived through classical homogenization techniques.

The study of coupling methods based on the Arlequin framework for blending particle and continuum models is by no means complete. This work only represents one step in the development of general coupling methods. In particular, it would be interesting to investigate the extension of this formulation to stochastic systems for which the notion of representative volume element is not well defined. A preliminary study on this subject is described in [15] based on the coupling method proposed in [5]. Our objective in the near future would be to reconsider stochastic particle systems using the new averaging operator for coupling the continuum and particle models.

Acknowledgements Support of this work by DOE under contract DE-FG02-05ER25701 is gratefully acknowledged. The second author, Robin Bouclier, would like also to thank ICES for hosting him for his internship during the Summer of 2009.

References

1. Babuška, I., Banerjee, U., Osborn, J.E.: On principles for the selection of shape functions for the generalized finite element method. Computer Methods in Applied Mechanics and Engineering **191**(49), 5595–5629 (2002)
2. Badia, S., Bochev, P., Fish, J., Gunzburger, M., Lehoucq, R., Nuggehally, M., Parks, M.: A force-based blending model for atomistic-to-continuum coupling. International Journal for Multiscale Computational Engineering **5**, 387–406 (2007)
3. Badia, S., Parks, M., Bochev, P., Gunzburger, M., Lehoucq, R.: On atomistic-to-continuum (atc) coupling by blending. Multiscale Modeling and Simulation **7**, 381–406 (2008)

4. Bauman, P.T.: Adaptive multiscale modeling of polymeric materials using goal-oriented error estimation, arlequin coupling, and goals algorithms. Ph.D. thesis, The University of Texas at Austin (2008)
5. Bauman, P.T., Ben Dhia, H., Elkhodja, N., Oden, J.T., Prudhomme, S.: On the application of the Arlequin method to the coupling of particle and continuum models. Computational Mechanics **42**, 511–530 (2008)
6. Bauman, P.T., Oden, J.T., Prudhomme, S.: Adaptive multiscale modeling of polymeric materials: Arlequin coupling and goals algorithms. Computer Methods in Applied Mechanics and Engineering **198**, 799–818 (2008)
7. Belytschko, T., Xiao, S.P.: Coupling methods for continuum model with molecular model. International Journal for Multiscale Computational Engineering **1**(1), 115–126 (2003)
8. Ben Dhia, H.: Multiscale mechanical problems: the Arlequin method. Comptes Rendus de l'Académie des Sciences Paris Série IIB **326**(12), 899–904 (1998)
9. Ben Dhia, H.: Global local approaches: the Arlequin framework. European Journal of Computational Mechanics **15**(1–3), 67–80 (2006)
10. Ben Dhia, H., Elkhodja, N.: Coupling of atomistic and continuum models in the Arlequin framework. In: Proceedings of the 8eme Congrès de Mécanique, pp. 133–135. El Jadida, Maroc (2007)
11. Ben Dhia, H., Rateau, G.: Mathematical analysis of the mixed Arlequin method. Comptes Rendus de l'Académie des Sciences Paris Série I **332**, 649–654 (2001)
12. Ben Dhia, H., Rateau, G.: Application of the Arlequin method to some structures with defects. Revue Européenne des Eléments Finis **332**, 649–654 (2002)
13. Ben Dhia, H., Rateau, G.: The Arlequin method as a flexible engineering design tool. Int. J. Numer. Meth. Engng. **62**(11), 1442–1462 (2005)
14. Broughton, J.Q., Abraham, F.F., Bernstein, N., Kaxiras, E.: Concurrent coupling of length scales: Methodology and application. Phys. Rev. B **60**(4), 2391–2403 (1999)
15. Chamoin, L., Oden, J.T., Prudhomme, S.: A stochastic coupling method for atomic-to-continuum Monte Carlo simulations. Comput. Methods Appl. Mech. Engng. **197**, 3530–3546 (2008)
16. Chamoin, L., Prudhomme, S., Ben Dhia, H., Oden, J.T.: Ghost forces and spurious effects in atomic-to-continuum coupling methods by the Arlequin approach. Int. J. Numer. Meth. Engng. **83**(8-9), 1081–1113 (2010)
17. Curtin, W., Miller, R.: Atomistic/continuum coupling in computational material science. Modeling and Simulation in Materials Science and Engineering **11**, R33–R68 (2003)
18. Dobson, M., Luskin, M.: Analysis of a force-based quasicontinuum approximation. Mathematical Modelling and Numerical Analysis **42**(1), 113–139 (2009)
19. Dobson, M., Luskin, M.: An optimal order error analysis of the one-dimensional quasicontinuum approximation. SIAM Journal on Numerical Analysis **47**(4), 2455–2475 (2009)
20. E, W., Engquist, B., Huang, Z.: Heterogeneous multiscale method: A general methodology for multiscale modeling. Physical Review B **67**, 092,101 (2003)
21. E, W., Engquist, B., Li, X., Ren, W., Vanden-Eijnden, E.: Heterogeneous multiscale methods: A review. Communications in Computational Physics **2**(3), 367–450 (2007)
22. Fish, J.: Bridging the scales in nano engineering and science. Journal of Nanoparticle Research **8**(6), 577–594 (2006)
23. Fish, J.: Multiscale Methods: Bridging the Scales in Science and Engineering. Oxford University Press (2009)
24. Fish, J., Nuggehally, M., Shephard, M., Picu, C., Badia, S., Parks, M., Gunzburger, M.: Concurrent atc coupling based on a blend of the continuum stress and the atomic force. Computer Methods in Applied Mechanics and Engineering **196**, 4548–4560 (2007)
25. Miller, R.E., Tadmor, E.B.: The quasicontinuum method: overview, applications and current directions. Journal of Computer-Aided Materials Design **9**, 203–239 (2002)
26. Prudhomme, S., Ben Dhia, H., Bauman, P.T., Elkhodja, N., Oden, J.T.: Computational analysis of modeling error for the coupling of particle and continuum models by the Arlequin method. Computer Methods in Applied Mechanics and Engineering **197**, 3399–3409 (2008)

27. Prudhomme, S., Chamoin, L., Ben Dhia, H., Bauman, P.T.: An adaptive strategy for the control of modeling error in two-dimensional atomic-to-continuum coupling simulations. Computer Methods in Applied Mechanics and Engineering **198**, 1887–1901 (2009)
28. Tadmor, E.B., Ortiz, M., Phillips, R.: Quasicontinuum analysis of defects in solids. Philosophical Magazine **A73**, 1529–1563 (1996)
29. Wagner, G.J., Liu, W.K.: Coupling of atomistic and continuum simulations using a bridging scale decomposition. Journal of Computational Physics **190**, 249–274 (2003)
30. Xiao, S.P., T. Belytschko, T.: A bridging domain method for coupling continua with molecular dynamics. Computer Methods in Applied Mechanics and Engineering **193**, 1645–1669 (2004)

A Coupled Finite Difference – Gaussian Beam Method for High Frequency Wave Propagation

Nicolay M. Tanushev, Yen-Hsi Richard Tsai, and Björn Engquist

Abstract Approximations of geometric optics type are commonly used in simulations of high frequency wave propagation. This form of technique fails when there is strong local variation in the wave speed on the scale of the wavelength or smaller. We propose a domain decomposition approach, coupling Gaussian beam methods where the wave speed is smooth with finite difference methods for the wave equations in domains with strong wave speed variation. In contrast to the standard domain decomposition algorithms, our finite difference domains follow the energy of the wave and change in time. A typical application in seismology presents a great simulation challenge involving the presence of irregularly located sharp inclusions on top of a smoothly varying background wave speed. These sharp inclusions are small compared to the domain size. Due to the scattering nature of the problem, these small inclusions will have a significant effect on the wave field. We present examples in two dimensions, but extensions to higher dimensions are straightforward.

N.M. Tanushev (✉) · Y.-H. R. Tsai · B. Engquist
Department of Mathematics and Institute for Computational Engineering and Sciences (ICES), The University of Texas at Austin, 1 University Station, C1200, Austin, TX 78712, USA
e-mail: nicktan@math.utexas.edu; ytsai@ices.utexas.edu; engquist@ices.utexas.edu

B. Engquist et al. (eds.), *Numerical Analysis of Multiscale Computations*, Lecture Notes in Computational Science and Engineering 82, DOI 10.1007/978-3-642-21943-6_16,

1 Introduction

In this paper, we consider the scalar wave equation,

$$\begin{aligned} \Box u = u_{tt} - c^2(x)\Delta u &= 0, \qquad (t,x) \in [0,T]\times\mathbb{R}^d, \\ u(0,x) &= f(x), \\ u_t(0,x) &= g(x)\ , \end{aligned} \tag{1}$$

where d is the number of space dimensions. We will mainly focus on $d = 2$, though the extension of the methods presented here to three or more spatial dimensions is straight forward. The wave (1) is well-posed in the energy norm,

$$\|u(t,\cdot)\|_E^2 = \int_{\mathbb{R}^d}\left[\frac{|u_t(t,x)|^2}{c^2(x)} + |\nabla u(t,x)|^2\right]dx, \tag{2}$$

and it is often useful to define the point-wise energy function,

$$E[u](t,x) = \frac{|u_t(t,x)|^2}{c^2(x)} + |\nabla u(t,x)|^2, \tag{3}$$

and the energy inner product,

$$< u, v >_E = \int_{\mathbb{R}^d}\left[\frac{u_t(t,x)\bar{v}_t(t,x)}{c^2(x)} + \nabla u(t,x)\cdot\nabla\bar{v}(t,x)\right]dx.$$

High frequency solutions to the wave (1) are necessary in many scientific applications. While the equation has no scale, "high frequency" in this case means that there are many wave oscillations in the domain of interest and these oscillations are introduced into the wave field from the initial conditions. In simulations of high frequency wave propagation, direct discretization methods are notoriously computationally costly and typically asymptotic methods such as geometric optics [4], geometrical theory of diffraction [8], and Gaussian beams [2, 5–7] are used to approximate the wave field. All of these methods rely on the underlying assumption that the wave speed $c(x)$ does not significantly vary on the scale of the wave oscillations. While there are many interesting examples in scientific applications that satisfy this assumption, there are also many cases in which it is violated, for example in seismic exploration, where inclusions in the subsurface composition of the earth can cause the wave speed to vary smoothly on the scale of seismic wavelengths or even smaller scales. In this paper, we are interested in designing coupled simulation methods that are both fast and accurate for domains in which the wave speed is rapidly varying in some subregions of the domain and slowly varying in the rest.

In typical domain decomposition algorithms, the given initial-boundary value problem (IBVP) is solved using numerical solutions of many similar IBVPs on smaller subdomains with fixed dimensions. The union of these smaller domains constitutes the entire simulation domain. In our settings, there are two major differences to the case above. First, the equations and numerical methods in the subdomains are different: we have subdomains in which the wave equation is solved by a finite difference method while in other subdomains the ODEs defined by the Gaussian beam method are solved. Second, we consider situations in which the wave energy concentrates on small subregions of the given domain, so our domain decomposition method requires subdomains which follow the wave energy propagation and thus change size and location as a function of time. Since our method couples two different models of wave propagation, we will refer to it as the hybrid method. These types of methods are also often called heterogeneous domain decomposition [11]. We will describe how information is exchanged among the subdomains as well as how to change the subdomain size without creating instability and undesired numerical effects.

Our strategy will be to use an asymptotic method in subregions of the domain that satisfy the slowly varying sound speed assumption and a local direct method based on standard centered differences in subregions that do not. This hybrid domain decomposition approach includes three steps. The first is to translate a Gaussian beam representation of the high frequency wave field to data for a full wave equation finite difference simulation. Since a finite difference method needs the values of the solution on two time levels, this coupling can be accomplished by simply evaluating the Gaussian beam solution on the finite difference grid. The next step is to perform the finite difference simulation of the wave equation in an efficient manner. For this, we design a local finite difference method that simulates the wave equation in a localized domain, which moves with the location of a wave energy. Since this is a major issue, we have devote a section of this paper to its description and provide some examples. The last step is to translate a general wave field from a finite difference simulation to a superposition of Gaussian beams. To accomplish this, we use the method described in [14] for decomposing a general high frequency wave field $(u, u_t) = (f, g)$ into a sum of Gaussian beams. The decomposition algorithm is a greedy iterative method. At the $(N + 1)$ decomposition step, a set of initial values for the Gaussian beam ODE system is found such that the Gaussian beam wave field given by these initial values will approximates the residual between the wave field (f, g) and the wave field generated by previous (N) Gaussian beams at a fixed time. These new initial values are directly estimated from the residual wave field and are then locally optimized in the energy norm using the Nelder-Mead method [10]. The procedure is repeated until a desired tolerance or maximum number of beams is reached.

Since Gaussian beam methods are not widely known, we begin with a condensed description of Gaussian beams. After this presentation, we give two examples that show the strengths and weaknesses of using Gaussian beams. We develop the local finite difference method as a stand alone method for wave propagation. Finally, we combine Gaussian beams and the local finite difference method to form the hybrid domain decomposition method. We present two examples to show the strength of the hybrid method.

2 Gaussian Beams

Since Gaussian beams play a central role in the hybrid domain decomposition method, we will briefly describe their construction. For a general construction and analysis of Gaussian beams, we refer the reader to [9, 12, 13].

Gaussian beams are approximate high frequency solutions to linear PDEs which are concentrated on a single ray through space–time. They are closely related to geometric optics. In both approaches, the solution of the PDE is assumed to be of the form $a(t,x)e^{ik\phi(t,x)}$, where k is the large high frequency parameter, a is the amplitude of the solution, and ϕ is the phase. Upon substituting this ansatz into the PDE, we find the eikonal and transport equations that the phase and amplitude functions have to satisfy, respectively. In geometric optics ϕ is real valued, while in Gaussian beams ϕ is complex valued. To form a Gaussian beam solution, we first pick a characteristic ray for the eikonal equation and solve a system of ODEs in t along it to find the values of the phase, its first and second order derivatives and amplitude on the ray. To define the phase and amplitude away from this ray to all of space–time, we extend them using a Taylor polynomial. Heuristically speaking, along each ray we propagate information about the phase and amplitude that allows us to reconstruct them locally in a Gaussian envelope.

For the wave equation, the system of ODEs that define a Gaussian beam are

$$
\begin{aligned}
\dot{\phi}_0(t) &= 0\,,\\
\dot{y}(t) &= -c(y(t))p(t)/|p(t)|\,,\\
\dot{p}(t) &= |p(t)|\nabla c(y(t))\,,\\
\dot{M}(t) &= -A(t) - M(t)B(t) - B^{\top}(t)M(t) - M(t)C(t)M(t)\,,\\
\dot{a}_0(t) &= a_0(t)\left(-\frac{p(t)}{2|p(t)|}\cdot\frac{\partial c}{\partial x}(y(t)) - \frac{p(t)\cdot M(t)p(t)}{2|p(t)|^3} + \frac{c(y(t))\mathrm{Tr}[M(t)]}{2|p(t)|}\right),
\end{aligned}
$$

where

$$
\begin{aligned}
A(t) &= -|p(t)|\frac{\partial^2 c}{\partial x^2}(y(t))\,,\\
B(t) &= -\frac{p(t)}{|p(t)|}\otimes\frac{\partial c}{\partial x}(y(t))\,,\\
C(t) &= -\frac{c(y(t))}{|p(t)|}\left(Id_{d\times d} - \frac{p(t)\otimes p(t)}{|p(t)|^2}\right).
\end{aligned}
$$

The quantities $\phi_0(t)$ and $a_0(t)$ are scalar valued, $y(t)$ and $p(t)$ are in $\mathbb{R}^d$, and $M(t)$, $A(t)$, $B(t)$, and $C(t)$ are $d\times d$ matrices. Given initial values, the solution to this system of ODEs will exists for $t\in[0,T]$, provided that $M(0)$ is symmetric and its imaginary part is positive definite. Furthermore, $M(t)$ will remain symmetric with a positive definite imaginary part for $t\in[0,T]$. For a proof, we refer the reader to [12]. Under the restriction on $M(0)$, the ODEs allow us to define the phase and amplitude for the Gaussian beam using:

$$\phi(t,x) = \phi_0(t) + p(t)\cdot[x - y(t)] + \frac{1}{2}[x - y(t)]\cdot M(t)[x - y(t)],$$
$$a(t,x) = a_0(t)\ . \tag{4}$$

Furthermore, since $\dot{\phi}_0(t) = 0$, for fixed k, we can absorb this constant phase shift into the amplitude and take $\phi_0(t) = 0$. Thus, the Gaussian beam solution is given by

$$v(t,x) = a(t,x)e^{ik\phi(t,x)}\ . \tag{5}$$

We will assume that the initial values for these ODEs are given and that they satisfy the conditions on $M(0)$. The initial values for the ODEs are tied directly to the Gaussian beam wave field at $t = 0$, $v(0,x)$ and $v_t(0,x)$. As can be easily seen, the initial conditions for the Gaussian beam will not be of the general form of the conditions for the wave equation given in (1). However, using a decomposition method such as the methods described in [14] or [1], we can approximate the general high frequency initial conditions for (1) as a superposition of individual Gaussian beams. Thus, for the duration of this paper, we will assume that the initial conditions for the wave (1) are the same as those for a single Gaussian beam:

$$u(0,x) = a(0,x)e^{ik\phi(0,x)},$$
$$u_t(0,x) = [a_t(0,x) + ik\phi_t(0,x)a(0,x)]\, e^{ik\phi(0,x)}\ . \tag{6}$$

Note that $a_t(0,x)$ and $\phi_t(0,x)$ are directly determined by the Taylor polynomials (4) and the ODEs above.

3 Motivating Examples

We begin with an example that shows the strengths of using Gaussian beams and, with a small modification, the shortcomings. Suppose that we consider the wave (1) in two dimension for $(t,x_1,x_2) \in [0,2.5]\times[-1.5,1.5]\times[-3,0.5]$, sound speed $c(x) = \sqrt{1-0.05x_2}$, and the Gaussian beam initial conditions given in (6) with,

$$\phi(0,x) = (x_2 - 1) + i(x_1 - 0.45)^2/2 + i(x_2 - 1)^2/2\ ,$$
$$a(0,x) = 1.$$

We take the high frequency parameter $k = 100$. To obtain a numerical solution to the wave (1), we can use either a direct method or the Gaussian beam method. As the direct method, we use the standard second order finite difference method based on the centered difference formulas for both space and time:

$$\frac{u_{\ell,m}^{n+1} - 2u_{\ell,m}^{n} + u_{\ell,m}^{n-1}}{\Delta t^2} \\ = c_{\ell,m}^2 \left[\frac{u_{\ell+1,m}^{n} - 2u_{\ell,m}^{n} + u_{\ell-1,m}^{n}}{\Delta x^2} + \frac{u_{\ell,m+1}^{n} - 2u_{\ell,m}^{n} + u_{\ell,m-1}^{n}}{\Delta y^2} \right], \tag{7}$$

where n is the time level index, ℓ and m are the x and y spatial indices respectively.

Since we need to impose artificial boundaries for the numerical simulation domain, we use first order absorbing boundary conditions (ABC) [3]. The first order ABC amount to using the appropriate one-way wave equation,

$$u_t = \pm c(x,y)u_x \quad \text{or} \quad u_t = \pm c(x,y)u_y, \tag{8}$$

on each of the boundaries, so that waves are propagated out of the simulation domain and not into it. For example, on the left boundary, $x = -1.5$, we use $u_t = cu_x$ with upwind discretization,

$$\frac{u_{\ell,m}^{n+1} - u_{\ell,m}^{n}}{\Delta t} = c_{\ell,m} \left[\frac{u_{\ell+1,m}^{n} - u_{\ell,m}^{n}}{\Delta x} \right], \tag{9}$$

for ℓ equal to its lowest value.

To resolve the oscillations, using 10 points per wavelength, for this particular domain size and value for k, we need roughly 500 points in both the x_1 and x_2 directions. However, to maintain low numerical dispersion for the finite difference solution, we need to use a finer the grid. The grid refinement will the given in terms of the coarse, 10 points per wavelength, grid. For example, a grid with a refinement factor of 3 will have 30 points per wavelength. Note that such grid refinement is not necessary for the Gaussian beam solution. Thus, while we compute the finite difference solution on the refined grid, we only use the refined solution values on the coarser grid for comparisons. For determining the errors in each solution, we compare with the "exact" solution computed using the finite difference method with a high refinement factor of 10.

For this particular example, the sound speed, the finite difference solution and Gaussian beam solution at the final time are shown in Fig. 1. In order to have a meaningful comparison, the grid refinement for the finite difference solution was chosen so that the errors in the finite difference solution are comparable to the ones in the Gaussian beam solution. Both the accuracy and computation times are shown in Table 1. The Gaussian beam solution was computed more than 3,500 times faster than the finite difference solution and the total error for both the Gaussian beam and the finite difference solution is $\approx 10\%$. Near the center of the beam, where the Gaussian beam envelope is greater than 0.25, the Gaussian beam solution is slightly more accurate with a local error of $\approx 7\%$. The Gaussian beam solution is an asymptotic solution, thus its error decreases for larger values of k. In terms of complexity analysis, as we are using a fixed number of points per wavelength to represent the wave field, the Gaussian beam solution is computed in $\mathcal{O}(1)$ steps and evaluated on the grid in $\mathcal{O}(k^2)$. The finite difference solution is computed in

$\mathcal{O}(k^3)$ steps. Additionally, for larger values of k, we would need to increase the grid refinement for the finite difference solution in order to maintain the same level of accuracy as in the Gaussian beam solution. Therefore, it is clear why the Gaussian beam solution method is advantageous for high frequency wave propagation.

Table 1 Comparisons of the finite difference (FD) method and Gaussian beam (GB) method with sound speed with no inclusion. Shown are the total error for each method in the energy norm as a percent of the total energy at each time, the local error (Loc Err) as a percent of the local energy at $t = 2.5$, and the total computational time (C Time) for obtaining the solution at each time. The local error is computed near the beam center, where the Gaussian envelope is greater than 0.25. The finite difference solution is computed with a refinement factor of 6

	t = 0.625	t = 1.25	t = 1.875	t = 2.5,	Loc Err	C Time
FD	1.9%	3.8%	5.6%	7.3%	7.4%	7773.1
GB	2.4%	4.8%	7.2%	9.7%	7.0%	1.6

Now, suppose that we modify the sound speed to have an inclusion, so that the sound speed changes on the same scale as the wave oscillations as shown in Fig. 1 and that we use the same initial conditions as before. The inclusion is positioned in such a way, so that the ray mostly avoids the inclusion, while the wave field on the left side of the ray interacts with the inclusion. Since all of the quantities that define the Gaussian beam are computed on the ray, the Gaussian beam coefficients are similar to the coefficients in the example without the inclusion. However, as can be seen from the full finite difference calculation in Fig. 1, the wave field at $t = 2.5$ is very different from the wave field at $t = 2.5$ for the sound speed with no inclusion shown in the same figure. The solution errors shown in Table 2 demonstrate that, while the Gaussian beam computation time is again more than 3,500 times faster, the error renders the solution essentially useless. Thus, the Gaussian beam solution is not a good approximation of the exact solution in this case. This, of course, is due to the fact that the asymptotic assumption, that the sound speed is slowly varying, is violated. Therefore, for a sound speed with an inclusion of this form, the Gaussian beam method cannot be used and we have to compute the wave field using a method that does not rely on this asymptotic assumption.

4 Local Finite Difference Method

By examining the example in the previous section, it is clear that a large portion of the computational time for the finite difference solution is spent simulating the wave equation where the solution is nearly zero. To exploit this property of the solution, we propose to use finite differences to compute the solution only locally where the wave energy is concentrated. Since the wave energy propagates in the domain, the

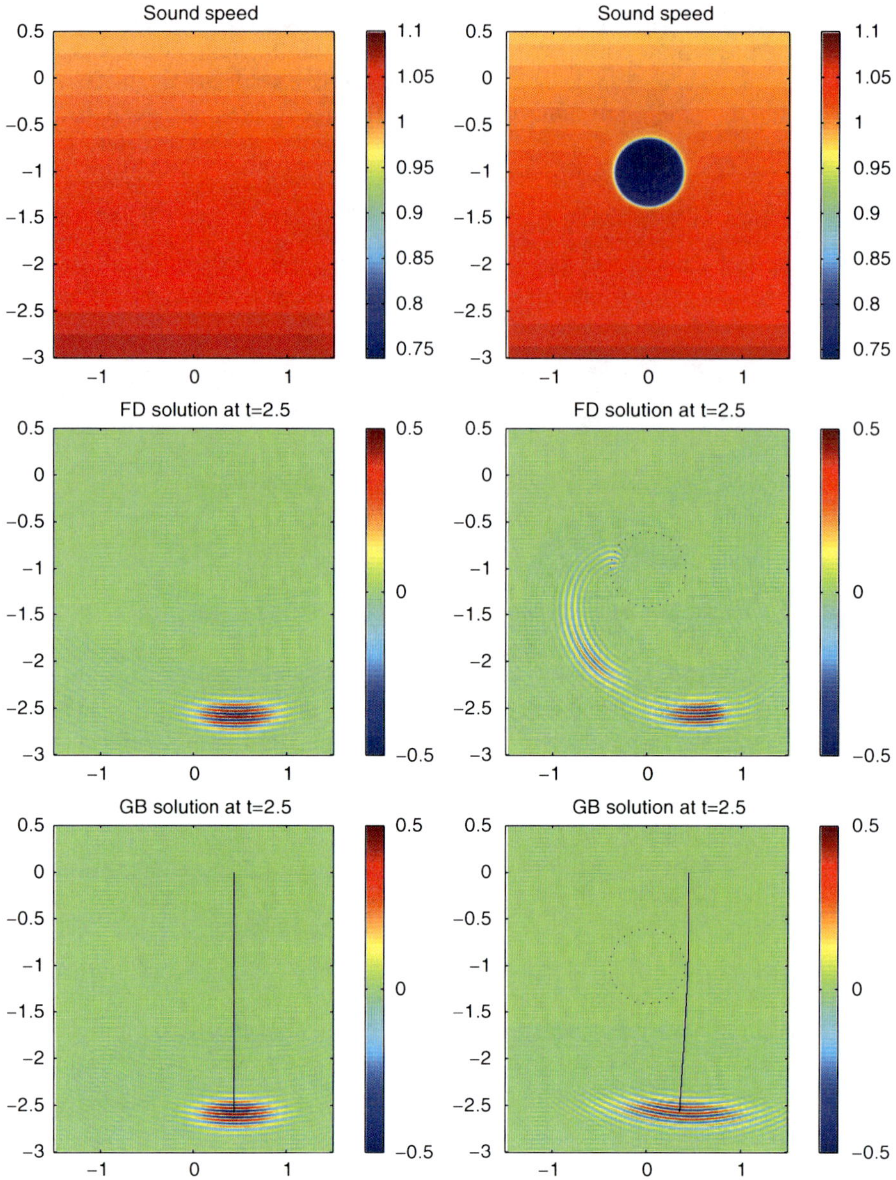

Fig. 1 The first column shows the wave field for simulations with sound speed without an inclusion: sound speed, the finite difference (FD) solution at the final time, and the Gaussian beam (GB) solution at the final time. The second column shows the same graphs for simulations with sound speed containing an inclusion. The line shows the ray for the Gaussian beam. At $t = 0$, the Gaussian beam is centered at the beginning of the line and at $t = 2.5$, it is centered at the end of the line. The dotted circle outlines the location of the inclusion in the sound speed. For each of the wave fields, only the real part is shown

Table 2 Comparisons of the finite difference (FD) method and Gaussian beam (GB) method for a sound speed with inclusion. Shown are the total error for each method in the energy norm as a percent of the total energy at each time, the local error (Loc Err) as a percent of the local energy at $t = 2.5$, and the total computational time (C Time) for obtaining the solution at each time. The local error is computed near the beam center, where the Gaussian envelope is greater than 0.25. The finite difference solution is computed with a refinement factor of 6

	t = 0.625	t = 1.25	t = 1.875	t = 2.5	Loc Err	C Time
FD	1.9%	3.9%	5.6%	7.0%	7.3%	7717.8
GB	6.1%	94.5%	91.2%	90.9%	43.9%	1.5

region in which we carry out the local wave equation simulation must also move with the waves. We emphasize that we are not using Gaussian beams at this stage.

To be more precise, we propose to simulate the wave equation in a domain $\Omega(t)$, that is a function of time and at every t, $\Omega(t)$ contains most of the wave energy. For computational ease, we select $\Omega(t)$ to be a rectangular region.The initial simulation domain $\Omega(0)$ is selected from the initial data by thresholding the energy function (3) to contain most of the wave energy. Since solutions of the wave (1) have finite speed of propagation, the energy moves at the speed of wave propagation and thus the boundaries of $\Omega(t)$ do not move too rapidly. In terms of finite difference methods, this means that if we ensure that the Courant-Friedrichs-Lewy (CFL) condition is met, the boundaries of $\Omega(t)$ will not move by more than a spatial grid point between discrete time levels t and $t + \Delta t$. Whether $\Omega(t)$ increases or decreases by one grid point (or stays the same) at time level $t + \Delta t$ is determined by thresholding the energy function (3) of u at time level t near the boundary of $\Omega(t)$.

Using the standard second order finite difference method, we discretize the wave (1) using a centered in time, centered in space finite difference approximation (7). Since the solution is small near the boundary of $\Omega(t)$, there are several different boundary conditions that we could implement to obtain a solution. The easiest and most straightforward approach is to simply use Dirichlet boundary conditions with $u = 0$. Another approach is to use absorbing boundary conditions. We investigate the case where absorbing boundary conditions are applied to a single layer of grid nodes immediately neighboring the outer most grid nodes of $\Omega(t)$ (single layer ABC) and absorbing boundary conditions are applied again to the layer of grid nodes immediately neighboring the first ABC layer (double layer ABC). For example, for the depicted grid nodes in Fig. 2, $u^{n+1}_{\ell+1,m}$ and $u^{n+1}_{\ell,m}$ are computed by

$$u^{n+1}_{\ell+1,m} = u^{n}_{\ell+1,m} + c_{\ell+1,m}\frac{\Delta t}{\Delta x}\left[u^{n}_{\ell+2,m} - u^{n}_{\ell+1,m}\right],$$

$$u^{n+1}_{\ell,m} = u^{n}_{\ell,m} + c_{\ell,m}\frac{\Delta t}{\Delta x}\left[u^{n}_{\ell+1,m} - u^{n}_{\ell,m}\right].$$

For both Dirichlet and absorbing boundary conditions, when the domain $\Omega(t)$ is expanding, the finite difference stencils will need to use grid nodes that are outside of $\Omega(t)$ and the boundary layers. We artificially set the wave field to be equal to zero at such grid nodes and we will refer to them as "reclaimed grid nodes". Figure 2 shows the domain of influence of the reclaimed nodes for the Dirichlet boundary conditions and the double layer ABC. In this figure, solid lines connect the reclaimed nodes with nodes whose values are computed directly using the reclaimed nodes. Dashed lines connect the reclaimed nodes with nodes whose values are computed using the reclaimed nodes, but through the values of another node. Finally, dotted lines indicate one more level in the effect of the reclaimed nodes. The point of using double layer ABC is to minimize the influence of the reclaimed nodes, as can be seen in Fig. 2. Note that there are no solid line connections between the reclaimed nodes and the nodes in $\Omega(t)$ for double layer ABC. Furthermore, the artificial Dirichlet boundary conditions reflect energy back into the computational domain $\Omega(t)$ which may make it larger compared to $\Omega(t)$ for the solution obtained by double layer ABC as shown in Fig. 3.

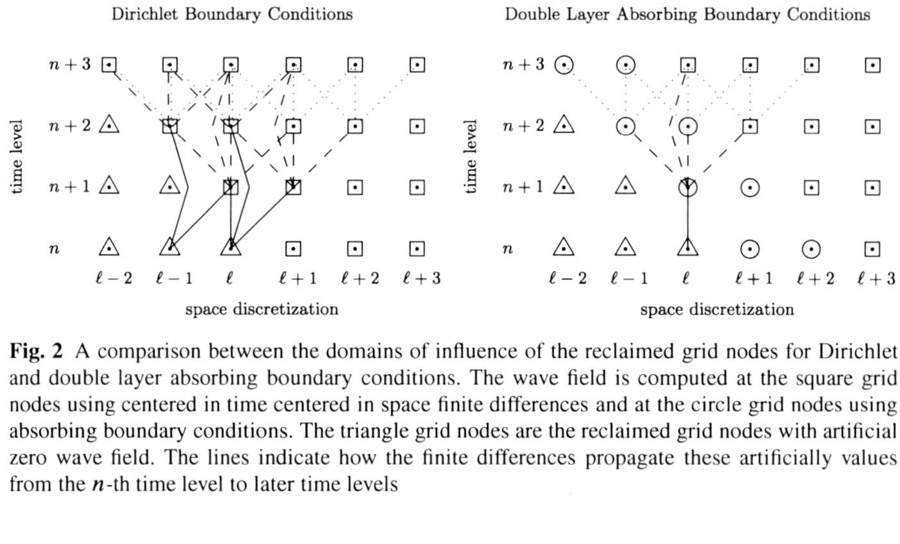

Fig. 2 A comparison between the domains of influence of the reclaimed grid nodes for Dirichlet and double layer absorbing boundary conditions. The wave field is computed at the square grid nodes using centered in time centered in space finite differences and at the circle grid nodes using absorbing boundary conditions. The triangle grid nodes are the reclaimed grid nodes with artificial zero wave field. The lines indicate how the finite differences propagate these artificially values from the n-th time level to later time levels

Finally, we note that due to finite speed of wave propagation, we can design boundary conditions that will not need reclaimed grid nodes. However, these boundary conditions may have a finite difference stencil that spans many time levels and this stencil may need to change depending on how $\Omega(t)$ changes in time. Numerically, we observed a large improvement when using double layer ABC instead of Dirichlet boundary conditions. However, using triple or quadruple layer ABC did not give a significant improvement over the double layer ABC. Thus, for computational simplicity, we use the above double layer absorbing boundary conditions for the simulations that follow.

Using the local finite difference method, we compute the solution to the wave (1) as in the previous section for the example with a sound speed with inclusion, using a refinement factor of 6. To determine $\Omega(0)$, we threshold the energy function (3) at $1/100$ of its maximum. For computational time comparison, we also compute the full finite difference solution, also with a refinement factor of 6. These parameters

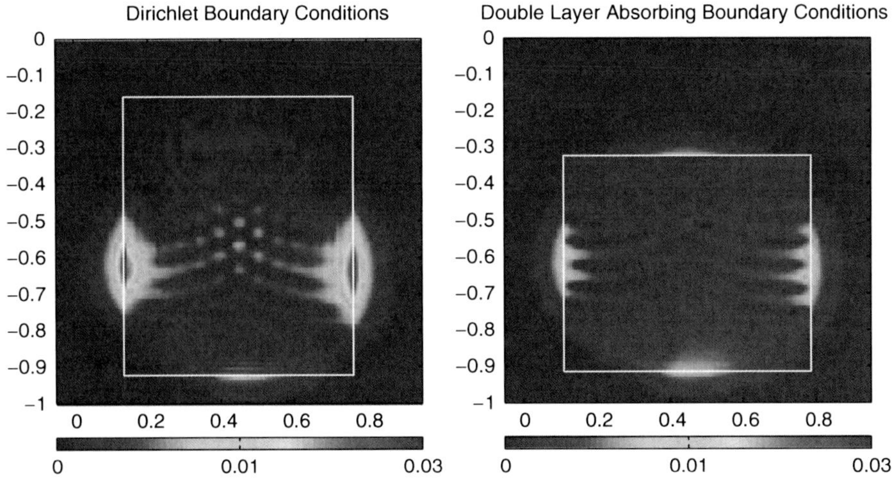

Fig. 3 A comparison between Dirichlet boundary conditions and double layer absorbing boundary conditions for the local finite difference method. The absolute value of the difference between each solution and the finite difference solution for the full domain is plotted at time $t = 0.625$. The domain $\Omega(0.625)$ is outlined in white. Note that overall the double layer absorbing boundary conditions solution is more accurate than the Dirichlet boundary condition solution. Also, note that $\Omega(0.625)$ is smaller for the double layer absorbing boundary conditions

were chosen so that the final error is $\approx 7\%$ and comparable for both solutions. The wave field, along with $\Omega(t)$, are shown in Fig. 4 at $t = \{0, 0.625, 1.25, 1.875, 2.5\}$. The comparisons of accuracy and computation time between the local and full finite difference solutions are shown in Table 3. The error in both solutions is equivalent, but the local finite difference solution in computed 5 times faster. Furthermore, if the local finite difference method is used to simulate the wave field from a Gaussian beam, we need $\mathcal{O}(k)$ steps in time as in the full finite difference method, but the local finite difference method requires $\mathcal{O}(k)$ grid points in space as opposed to $\mathcal{O}(k^2)$ grid points that the full finite difference method requires. This is because the energy from a Gaussian beam is concentrated in a $k^{-1/2}$ neighborhood of its center and this is a two dimensional example.

Table 3 Comparisons of the full finite difference (FD) method and the local finite difference (LFD) method with sound speed with inclusion. Shown are the total error for each method in the energy norm in as a percent of the total energy and the total computational time (C Time) for obtaining the solution at $t = \{0.625, 1.25, 1.875, 2.5\}$

	t = 0.625	t = 1.25	t = 1.875	t = 2.5	C Time
FD	1.9%	3.9%	5.6%	7.0%	7717.8
LFD	2.4%	4.4%	6.0%	7.3%	1535.5

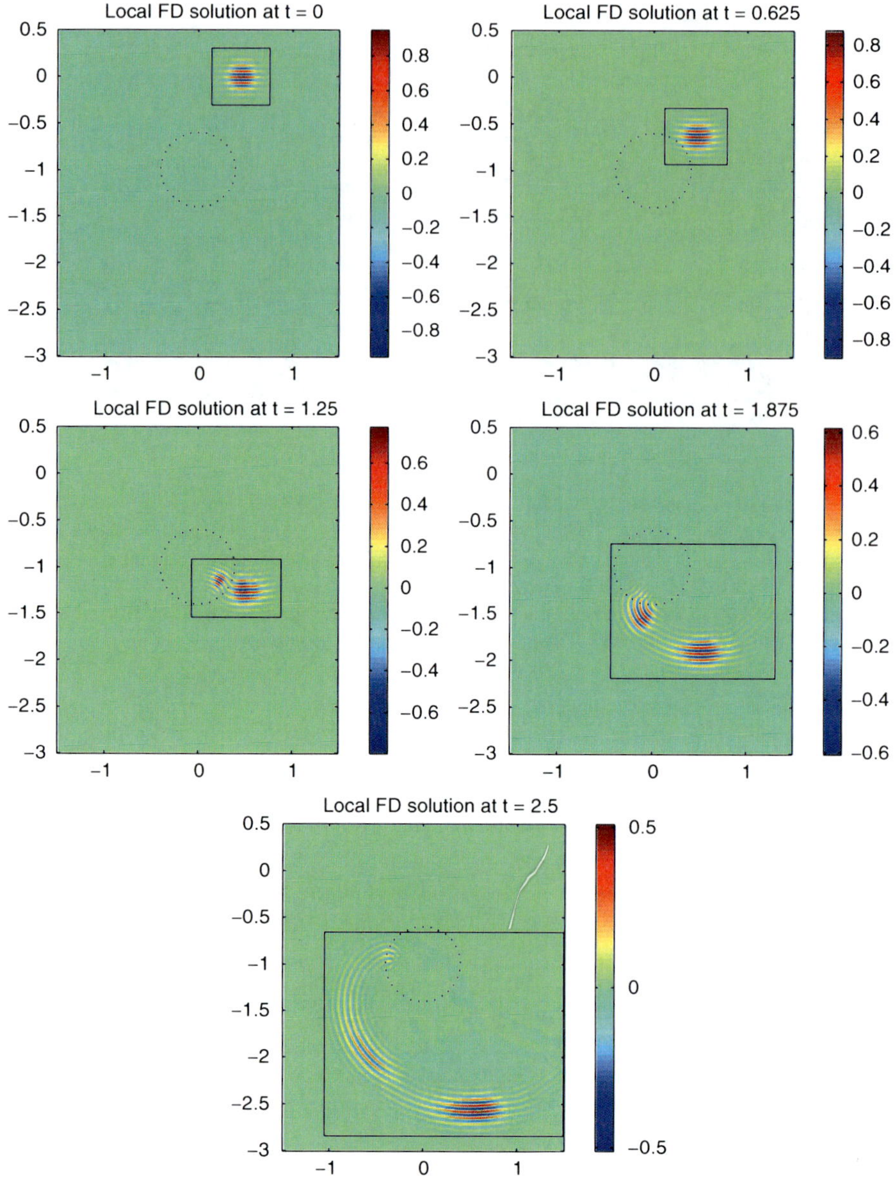

Fig. 4 This figure shows the wave field computed using the local finite difference method for the sound speed with inclusion. The black rectangle outlines the local computational domain, $\Omega(t)$, and the dotted circle outlines the location of the inclusion in the sound speed. Only the real part of the wave fields is shown

Finally, we remark that if instead of finding one rectangle that contains the bulk of the energy we found several, the solution in each of these rectangles can be computed independently. On a parallel computer, this would give another advantage over full finite difference simulations, as there is no need for information exchange between the computations on each rectangle, even if these rectangles overlap. The linear nature of the wave equation allows for the global solution to be obtained by simply adding the solutions from each of the separate local finite difference simulations. Furthermore, the generalization to more than two dimensions is straight forward and the computational gain is even greater in higher dimensions.

5 Hybrid Method

Upon further examination of the inclusion example in Sect. 3 and the wave field simulations in Sect. 4, we note that the Gaussian beam solution has small error for some time initially (see Table 2) and that after the wave energy has interacted with the inclusion in the sound speed, it again appears to have Gaussian beam like characteristics (see Fig. 4, $t > 2$). We can immediately see the effect of the large variation of the sound speed on the wave field. The large gradient roughly splits the wave field into two components, one that continues on nearly the same path as before and one that is redirected to the side. This also shows why the Gaussian beam solution is not a very good approximation. For a single Gaussian beam to represent a wave field accurately, the wave field has to stay coherent; it cannot split into two or more separate components. However, once the wave field has been split into several components by the inclusion, it will propagate coherently until it reaches another region of large sound speed variation. By following the propagation of wave energy in time, while it is near a region of high sound speed variation, we employ the local finite different method and the Gaussian beam method otherwise.

To be able to use such a hybrid method, we need to be able to couple the two different simulation methods. Switching from a Gaussian beam description to a local finite difference description is straightforward. The local finite difference requires the wave field at a time t and $t + \Delta t$, which can be obtained simply by evaluating the Gaussian beam solution on the finite difference grid. The opposite, moving from a local finite difference to a Gaussian beam description, is more difficult to accomplish. For this step we use the decomposition algorithm given in [14]. As discussed in the introduction, this decomposition method is a greedy iterative method. At each iteration the parameters for a single Gaussian beam are estimated and then locally optimized using the Nelder-Mead algorithm [10]. The method is then iterated over the residual wave field. The decomposition is complete when a certain tolerance is met or a maximum number of Gaussian beams is reached. For completeness, we give the algorithm of [14] below:

1. With $n = 1$, let (u^n, u_t^n) be the wave field at a fixed t and suppress t to simplify the notation.
2. Find a candidate Gaussian beam:

   - Estimate Gaussian beam center:
     $\rightarrow$ Let $\tilde{y}^n = \arg\max\{E[u^n](y)\}$ (see equation (3)).
   - Estimate propagation direction:
     $\rightarrow$ Let $G(x) = \exp(-k|x - \tilde{y}^n|^2/2)$.
     $\rightarrow$ Let $p^n = \arg\max\{|\mathscr{F}[u^n(x)G(x)]| + |\mathscr{F}[u_t^n(x)G(x)/k]|\}$, with $\mathscr{F}$ the scaled Fourier transform, $\{x \rightarrow kp\}$.
     $\rightarrow$ Let $\tilde{\phi}_t^n = c(y^n)|\tilde{p}^n|$.
   - Let $\tilde{M}^n = iI$, with I the identity matrix.

3. Minimize the difference between the Gaussian beam and u^n in the energy norm using the Nelder–Mead method with $(\tilde{y}^n, \tilde{\phi}_t^n, \tilde{p}^n, \tilde{M}^n)$ as the initial Gaussian beam parameters:

   - Subject to the constraints, $\mathrm{Im}\{M\}$ is positive definite, entries of M are less than $\sqrt{k}$ in magnitude, $1/\sqrt{k} \leq |p| \leq \sqrt{k}$, and $|\phi_t|^2 = c^2(y)|p|^2$, let

   $$(y^n, \phi_t^n, p^n, M^n) = \arg\ \min \left\{ \left\| u^n - \frac{< u^n, B >_E}{||B||_E^2} B \right\|_E^2 \right\}$$

     where B be the Gaussian beam defined by the parameters $(y^n, \phi_t^n, p^n, M^n)$ and amplitude 1 (see equations (4) and (5)).
   - Let $B^n(x,t)$ be the Gaussian beam defined by the parameters $(y^n, \phi_t^n, p^n, M^n)$ and amplitude 1.
   - Let $a^n = \frac{<u^n, B^n>_E}{||B^n||_E^2}$.

4. The n-th Gaussian beam is given by the parameters $(y^n, \phi_t^n, p^n, M^n, a^n)$. Subtract its wave field:

   $$u^{n+1} = u^n - a^n B^n \text{ and } u_t^{n+1} = u_t^n - a^n B_t^n .$$

5. Re-adjust the previous $n - 1$ beams:

   - For the j-th beam, let $w = u^{n+1} + a^j B^j$ and repeat step 3 with $u^n = w$, $n = j$, and $(y^j, \phi_t^j, p^j, M^j)$ as the Gaussian beam parameters.
   - Let $u^{n+1} = w - a^j B^j$.

6. Re-adjust all beam amplitudes together

   - Let Λ be the matrix of inner products $\Lambda_{j\ell} =< B^\ell, B^j >_E$, and $b^j =< u^1, B^j >_E$.
   - Solve $\Lambda a = b$ and let $u^{n+1} = u^1 - \sum_{j=1}^n a^j B^j$.

7. Repeat steps starting with step 2, until $||u^{n+1}||_E$ is small or until a prescribed number of Gaussian beams is reached.

The final step in designing the hybrid method is deciding when and where to use which method. By looking at the magnitude of the gradient of the sound speed and the value of k, we can decompose the simulation domain into two subdomains D_G and D_L, which represent the Gaussian beam, small sound speed gradient, subdomain and the local finite difference, large gradient, subdomain respectively. When the Gaussian beam ray enters D_L, we switch from the Gaussian beam method to the local finite difference method. Deciding when to switch back to a Gaussian beam description is again more complicated. One way is to monitor the energy function (3) and when a substantial portion of it is supported in D_G, we use the decomposition method to convert that part of the energy into a superposition of a few Gaussian beams. Since calculating the energy function is computationally expensive, it should not be done at every time level of the local finite difference simulation. From the sound speed and size of D_L, we can estimate a maximum speed of propagation for the wave energy, thus a minimum time to exit D_L, and use that as a guide for evaluating the energy function. Additionally, we can look at the overlap between D_G and the local finite difference simulation domain $\Omega(t)$ as a guide for checking the energy function. A more crude, but faster, approach is to use the original ray to estimate the time that it takes for the wave energy to pass through D_L. We use this approach in the examples below. Furthermore, we note that the linearity property of the wave equation allows us to have a joint Gaussian beam and local finite difference description of the wave field. We can take the part of the local finite difference wave field in D_G and represent it as Gaussian beams. If there is a significant amount of energy left in D_L, we propagate the two wave fields concurrently one using Gaussian beams and the other using the local finite difference method. The total wave field is then the sum of the Gaussian beam and local finite difference wave fields.

There are two advantages of the hybrid method over the full and local finite difference methods. One is a decrease in simulation time. The other is due to the particular application to seismic exploration. For seismic wave fields, the ray based nature of Gaussian beams provides a connection between the energy on the initial surface and its location at the final time. Furthermore, this energy is supported in a tube in space–time and thus it only interacts with the sound speed inside this tube. Unfortunately, for finite difference based methods there is only the domain of dependence and this set can be quite large compared to the Gaussian beam space–time tube. For example, if the sound speed model is modified locally, only Gaussian beams that have space–time tubes that pass through the local sound speed modifications will need to be re-computed to obtain the total wave field. In contrast, a local sound speed modification requires that the entire finite difference solution be re-computed. For the hybrid method, if we decompose the wave field in single beam whenever we switch back to the Gaussian beam description then at any given time, we will either have a Gaussian beam wave field or a local finite difference wave field. After the simulation is complete we can interpolate the Gaussian beam coefficients to times for which the wave field is given by the local finite difference. Note that the resulting interpolated wave field will not satisfy the wave equation, however we will once again have a space–time tube that follows the energy propagation. Thus, we are interested in using the hybrid method to obtain a one beam solution that approximates the wave field better than the Gaussian beam method.

5.1 Example: Double Slit Experiment

In the simplest version of the Hybrid method, we consider an example in which we first use the local finite difference method to solve the wave equation for a given amount of time, then we switch to a Gaussian beam representation of the field. For this example we are interested in simulating the wave field in a double slit experiment, where coherent waves pass through two slits that are spaced closely together and their width is $\mathscr{O}(k^{-1})$, with $k = 50$. In the finite difference method, the slits are implemented as Dirichlet boundary conditions. It is clear that due to the diffraction phenomenon near the two slits, the Gaussian beam method alone will not give an accurate representation of the wave field. The wave field simulated using the hybrid method is shown in Fig. 5 and the error and computational time are shown in Table 4. Note that with 14 Gaussian beams, the computational time for the hybrid solution is still a factor of 3 faster than the full finite difference solution and a factor of 2 faster than the local finite difference solution.

Table 4 Comparisons of the full finite difference (FD), the local finite difference (LFD) and the hybrid (H) methods for the double slit experiment. Shown for each method are the total error in the energy norm in terms of percent of total energy and the total computational time (C Time) for obtaining the solution at $t = \{1.25, 2.5, 3.75, 5\}$. The norms are computed only on $y < 0$, since we are only interested in the wave field that propagates through the two slits

	t = 1.25	t = 2.5	t = 3.75	t = 5	C Time
FD	5.91%	10.6%	14.8%	19.1%	470
LFD	6.13%	11%	15.8%	19.7%	270
H	6.13%	12.7%	24.2%	33.9%	150

5.2 Example: Sound Speed with Inclusion

Finally, to demonstrate the hybrid method, we apply it to computing the wave field for the sound speed with inclusion and compare it to the previously discussed methods. For these experiments $k = 100$. The wave field is first computed using Gaussian beams until the beam is close to the inclusion at $t = 0.5$. Then, the solution is propagated with the local finite difference method until most of the wave energy has moved past the inclusion at $t = 2$. The resulting field is then decomposed into

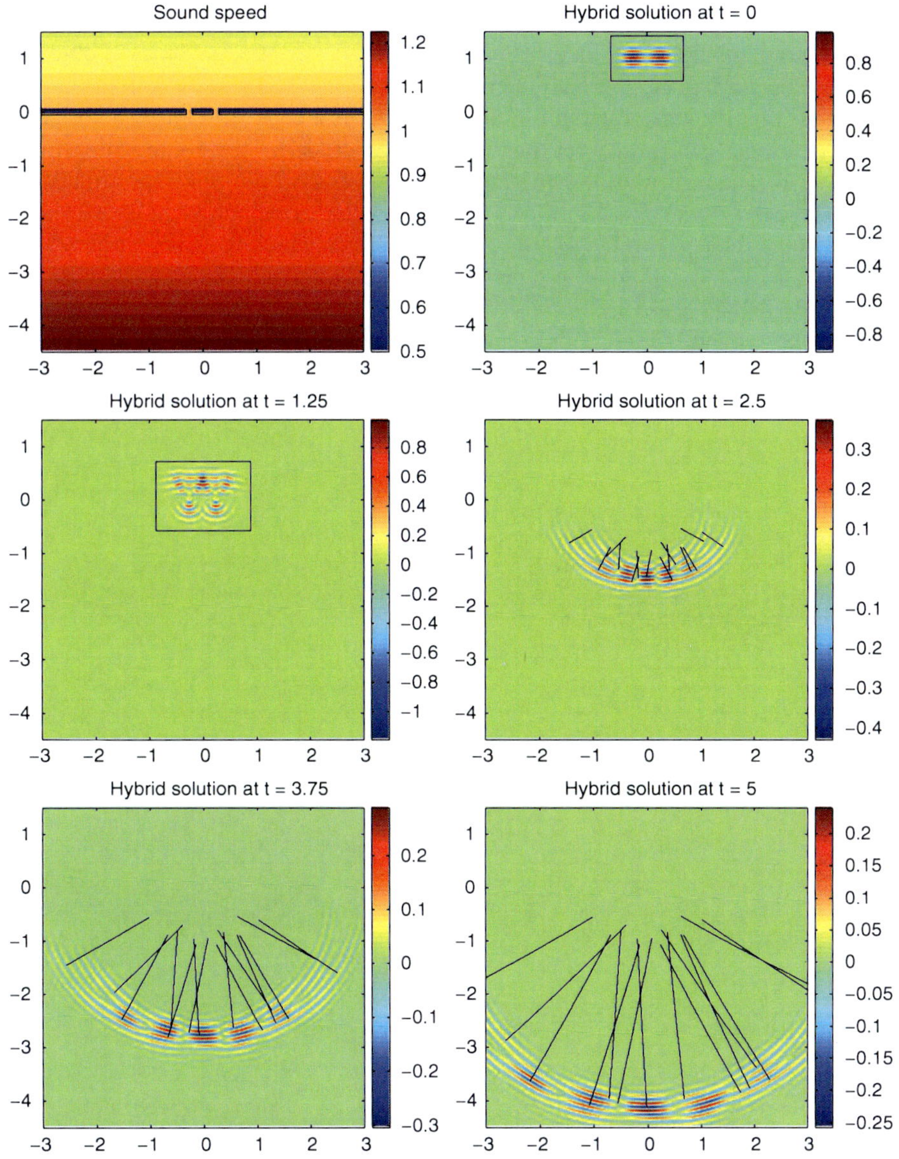

Fig. 5 The wave field obtained by the hybrid method for the double slit experiment. The first panel shows the sound speed and the double slit Dirichlet boundary condition region. The local finite difference domain is outlined by the black rectangle at $t = \{0, 1.25\}$. At $t = \{2.5, 3.75, 5\}$, the black lines indicate the ray for each of the Gaussian beams

one beam (the hybrid one-beam solution) or into two beams (the hybrid two-beam solution) using the decomposition algorithm of [14]. The wave fields for the one and two beam hybrid solutions are shown in Fig. 6. The errors and computation times for the methods discussed in this paper are shown in Table 5. The local finite difference calculations are done with a refinement factor of 5 and $\Omega(t)$ is obtained by thresholding the energy function at $1/10$ of its maximum. This thresholding was chosen so that the final errors in the local finite difference solution are similar to the error in the hybrid solution making the comparison of the computation times meaningful. The errors for the one and two beam hybrid solutions are $\approx$62% and $\approx$37% respectively at $t = 2.5$. This may seem rather large, but we note that this is a large improvement over the Gaussian beam solution which has an error of $\approx$91%. Furthermore, this is a single Gaussian beam approximation of the wave field locally and this wave field is not necessarily of Gaussian beam form. Locally, near the beam centers, the H1 and H2 solutions are more accurate. The computational time for the H1 and H2 hybrid solutions is 2 times faster compared to the local finite difference solution and 10 times faster than the full finite difference solution.

Table 5 Comparisons of the methods for a sound speed with inclusion. Shown for each method are the total error in the energy norm in terms of percent of total energy at each time, the local errors as a percent of the local energy near the beam center for the first beam (Loc Err 1) and near the second beam center (Loc Err 2), and the total computational time (C Time) for obtaining the solution at each time. The local error is computed near the beam center, where the Gaussian envelope is greater than 0.25. Legend: GB – Gaussian beam, LFD – Local finite difference, H1 – Hybrid method with one beam, H2 – hybrid method with two beams

	t = 0.675	t = 1.25	t = 1.875	t = 2.5	Loc Err 1	Loc Err 2	C Time
FD	3.3%	6.6%	9.4%	11.8%	12.3%	10.8%	4446.1
GB	6.1%	94.5%	91.2%	90.9%	42.2%	99.9%	1.5
LFD	6.6%	9.6%	11.9%	14.4%	12.4%	10.8%	781.0
H1	3.9%	7.4%	10.2%	62.0%	12.7%	100.0%	401.5
H2	3.9%	7.4%	10.2%	36.7%	12.7%	25.9%	417.9

6 Conclusion

In this paper, we develop a new hybrid method for high frequency wave propagation. We couple a Gaussian beam approximation of high frequency wave propagation to a local finite difference method in parts of the domains that contain strong variations in the wave speed. The coupling is accomplished either by translating the Gaussian beam representation into a wave field representation on a finite difference grid or by approximating the finite difference solution with a superposition of Gaussian beams. The local finite difference computations are performed on a moving computational domain with absorbing boundary conditions. This direct method is only used at times when a significant portion of the wave field energy is traveling through parts of the domain that contain large variations in the wave speed. The rest of the high frequency wave propagation is accomplished by the Gaussian beam method.

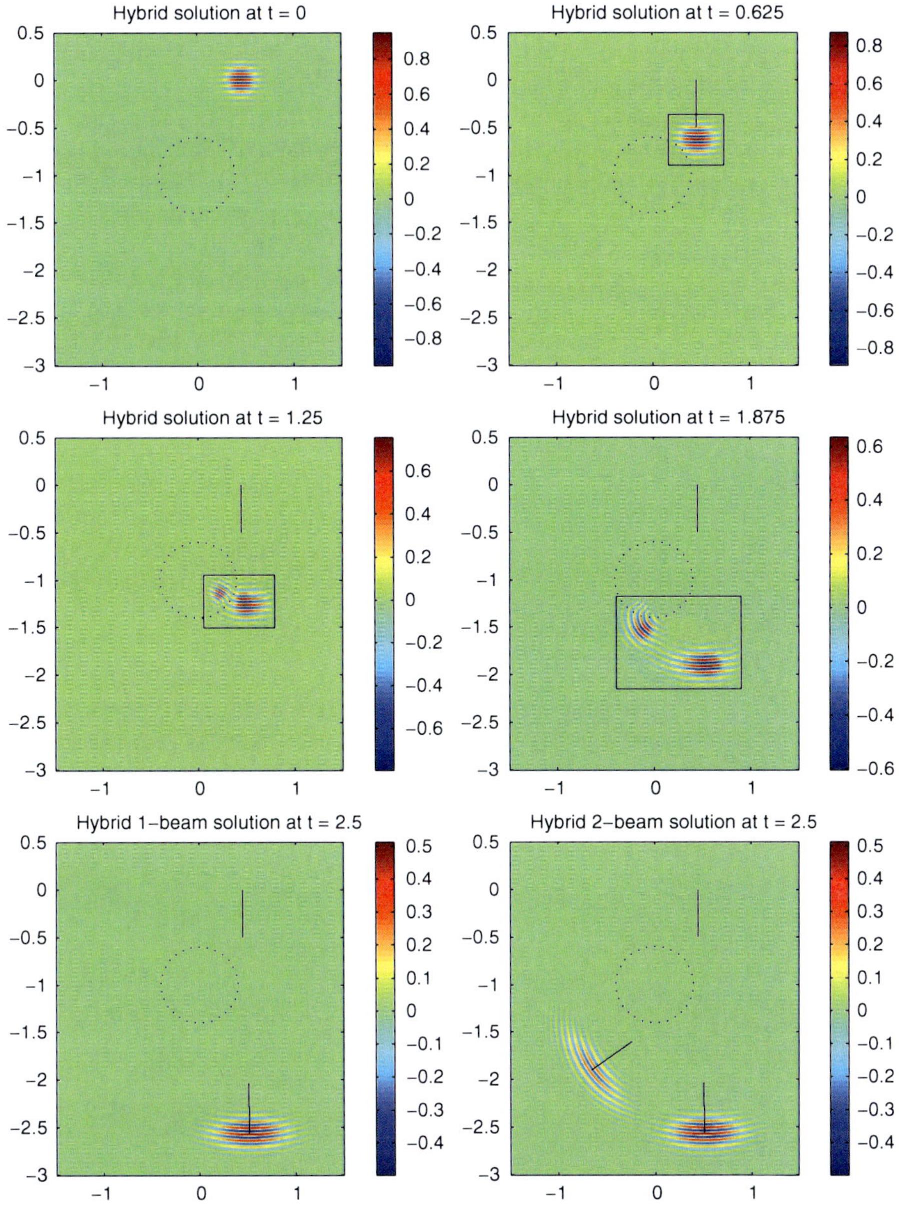

Fig. 6 The wave field for the hybrid H1 and H2 solution. The top two rows show the real part of the wave field which is the same for both the 1–beam and 2–beam hybrid solutions at $t = \{0, 0.625, 1.25, 1.875\}$. Times $t = \{.625, 1.25, 1.875\}$ are during the local finite difference calculation and the black rectangle outline the local finite difference domain $\Omega(t)$. The real part of the wave field for the 1–beam and 2-beam hybrid solutions are shown in the last row at $t = 2.5$. In each panel, the black lines indicate the ray for each of the Gaussian beams

Two numerical test examples show that the hybrid technique can retain the overall computational efficiency of the Gaussian beam method. At the same time the accuracy of the Gaussian beam methods in domains with smooth wave speed field is kept and the accuracy of the finite difference method in domains with strong variation in the wave speed is achieved. Furthermore, the hybrid method maintains the ability to follow the wave energy as it propagates from the initial surface through the domain as in traditional Gaussian beam and other ray based methods.

Acknowledgements The authors would like to thank Sergey Fomel, Ross Hill and Olof Runborg for helpful discussions and acknowledge the financial support of the NSF. The authors were partially supported under NSF grant No. DMS-0714612. NT was also supported under NSF grant No. DMS-0636586 (UT Austin RTG).

References

1. G. Ariel, B. Engquist, N. Tanushev, and R. Tsai. Gaussian beam decomposition of high frequency wave fields using expectation-maximization. *J. Comput. Phys.*, 230(6):2303–2321, 2011.
2. V. Červený, M. Popov, and I. Pšenčík. Computation of wave fields in inhomogeneous media - Gaussian beam approach. *Geophys. J. R. Astr. Soc.*, 70:109–128, 1982.
3. B. Engquist and A. Majda. Absorbing boundary conditions for the numerical simulation of waves. *Mathematics of Computation*, 31(139):629–651, 1977.
4. B. Engquist and O. Runborg. Computational high frequency wave propagation. *Acta Numer.*, 12:181–266, 2003.
5. S. Gray, Y. Xie, C. Notfors, T. Zhu, D. Wang, and C. Ting. Taking apart beam migration. *The Leading Edge*, Special Section:1098–1108, 2009.
6. R. Hill. Gaussian beam migration. *Geophysics*, 55:1416–1428, 1990.
7. R. Hill. Prestack Gaussian-beam depth migration. *Geophysics*, 66(4):1240–1250, 2001.
8. J. Keller. Geometrical theory of diffraction. *Journal of Optical Society of America*, 52:116–130, 1962.
9. H. Liu and J. Ralston. Recovery of high frequency wave fields for the acoustic wave equation. *Multiscale Modeling & Simulation*, 8(2):428–444, 2009.
10. J. Nelder and R. Mead. A simplex method for function minimization. *The Computer Journal*, 7(4):308–313, 1965.
11. A. Quarteroni, F. Pasquarelli, and A. Valli. Heterogeneous domain decomposition: principles, algorithms, applications. In *Fifth International Symposium on Domain Decomposition Methods for Partial Differential Equations (Norfolk, VA, 1991)*, pages 129–150. SIAM, Philadelphia, PA, 1991.
12. J. Ralston. Gaussian beams and the propagation of singularities. In *Studies in partial differential equations*, volume 23 of *MAA Stud. Math.*, pages 206–248. Math. Assoc. America, Washington, DC, 1982.
13. N. Tanushev. Superpositions and higher order Gaussian beams. *Commun. Math. Sci.*, 6(2):449–475, 2008.
14. N. Tanushev, B. Engquist, and R. Tsai. Gaussian beam decomposition of high frequency wave fields. *J. Comput. Phys.*, 228(23):8856–8871, 2009.

Editorial Policy

1. Volumes in the following three categories will be published in LNCSE:

i) Research monographs
ii) Tutorials
iii) Conference proceedings

Those considering a book which might be suitable for the series are strongly advised to contact the publisher or the series editors at an early stage.

2. Categories i) and ii). Tutorials are lecture notes typically arising via summer schools or similar events, which are used to teach graduate students. These categories will be emphasized by Lecture Notes in Computational Science and Engineering. **Submissions by interdisciplinary teams of authors are encouraged.** The goal is to report new developments – quickly, informally, and in a way that will make them accessible to non-specialists. In the evaluation of submissions timeliness of the work is an important criterion. Texts should be well-rounded, well-written and reasonably self-contained. In most cases the work will contain results of others as well as those of the author(s). In each case the author(s) should provide sufficient motivation, examples, and applications. In this respect, Ph.D. theses will usually be deemed unsuitable for the Lecture Notes series. Proposals for volumes in these categories should be submitted either to one of the series editors or to Springer-Verlag, Heidelberg, and will be refereed. A provisional judgement on the acceptability of a project can be based on partial information about the work: a detailed outline describing the contents of each chapter, the estimated length, a bibliography, and one or two sample chapters – or a first draft. A final decision whether to accept will rest on an evaluation of the completed work which should include

- at least 100 pages of text;
- a table of contents;
- an informative introduction perhaps with some historical remarks which should be accessible to readers unfamiliar with the topic treated;
- a subject index.

3. Category iii). Conference proceedings will be considered for publication provided that they are both of exceptional interest and devoted to a single topic. One (or more) expert participants will act as the scientific editor(s) of the volume. They select the papers which are suitable for inclusion and have them individually refereed as for a journal. Papers not closely related to the central topic are to be excluded. Organizers should contact the Editor for CSE at Springer at the planning stage, see *Addresses* below.

In exceptional cases some other multi-author-volumes may be considered in this category.

4. Only works in English will be considered. For evaluation purposes, manuscripts may be submitted in print or electronic form, in the latter case, preferably as pdf- or zipped ps-files. Authors are requested to use the LaTeX style files available from Springer at http:// www. springer.com/authors/book+authors?SGWID=0-154102-12-417900-0.

For categories ii) and iii) we strongly recommend that all contributions in a volume be written in the same LaTeX version, preferably LaTeX2e. Electronic material can be included if appropriate. Please contact the publisher.

Careful preparation of the manuscripts will help keep production time short besides ensuring satisfactory appearance of the finished book in print and online.

5. The following terms and conditions hold. Categories i), ii) and iii):

Authors receive 50 free copies of their book. No royalty is paid.
Volume editors receive a total of 50 free copies of their volume to be shared with authors, but no royalties.

Authors and volume editors are entitled to a discount of 33.3 % on the price of Springer books purchased for their personal use, if ordering directly from Springer.

6. Commitment to publish is made by letter of intent rather than by signing a formal contract. Springer-Verlag secures the copyright for each volume.

Addresses:

Timothy J. Barth
NASA Ames Research Center
NAS Division
Moffett Field, CA 94035, USA
barth@nas.nasa.gov

Michael Griebel
Institut für Numerische Simulation
der Universität Bonn
Wegelerstr. 6
53115 Bonn, Germany
griebel@ins.uni-bonn.de

David E. Keyes
Mathematical and Computer Sciences
and Engineering
King Abdullah University of Science
and Technology
P.O. Box 55455
Jeddah 21534, Saudi Arabia
david.keyes@kaust.edu.sa

and

Department of Applied Physics
and Applied Mathematics
Columbia University
500 W. 120 th Street
New York, NY 10027, USA
kd2112@columbia.edu

Risto M. Nieminen
Department of Applied Physics
Aalto University School of Science
and Technology
00076 Aalto, Finland
risto.nieminen@tkk.fi

Dirk Roose
Department of Computer Science
Katholieke Universiteit Leuven
Celestijnenlaan 200A
3001 Leuven-Heverlee, Belgium
dirk.roose@cs.kuleuven.be

Tamar Schlick
Department of Chemistry
and Courant Institute
of Mathematical Sciences
New York University
251 Mercer Street
New York, NY 10012, USA
schlick@nyu.edu

Editor for Computational Science
and Engineering at Springer:
Martin Peters
Springer-Verlag
Mathematics Editorial IV
Tiergartenstrasse 17
69121 Heidelberg, Germany
martin.peters@springer.com

Lecture Notes in Computational Science and Engineering

1. D. Funaro, *Spectral Elements for Transport-Dominated Equations.*
2. H.P. Langtangen, *Computational Partial Differential Equations.* Numerical Methods and Diffpack Programming.
3. W. Hackbusch, G. Wittum (eds.), *Multigrid Methods V.*
4. P. Deuflhard, J. Hermans, B. Leimkuhler, A.E. Mark, S. Reich, R.D. Skeel (eds.), *Computational Molecular Dynamics: Challenges, Methods, Ideas.*
5. D. Kröner, M. Ohlberger, C. Rohde (eds.), *An Introduction to Recent Developments in Theory and Numerics for Conservation Laws.*
6. S. Turek, *Efficient Solvers for Incompressible Flow Problems.* An Algorithmic and Computational Approach.
7. R. von Schwerin, ***M**ulti **B**ody **S**ystem **SIM**ulation.* Numerical Methods, Algorithms, and Software.
8. H.-J. Bungartz, F. Durst, C. Zenger (eds.), *High Performance Scientific and Engineering Computing.*
9. T.J. Barth, H. Deconinck (eds.), *High-Order Methods for Computational Physics.*
10. H.P. Langtangen, A.M. Bruaset, E. Quak (eds.), *Advances in Software Tools for Scientific Computing.*
11. B. Cockburn, G.E. Karniadakis, C.-W. Shu (eds.), *Discontinuous Galerkin Methods.* Theory, Computation and Applications.
12. U. van Rienen, *Numerical Methods in Computational Electrodynamics.* Linear Systems in Practical Applications.
13. B. Engquist, L. Johnsson, M. Hammill, F. Short (eds.), *Simulation and Visualization on the Grid.*
14. E. Dick, K. Riemslagh, J. Vierendeels (eds.), *Multigrid Methods VI.*
15. A. Frommer, T. Lippert, B. Medeke, K. Schilling (eds.), *Numerical Challenges in Lattice Quantum Chromodynamics.*
16. J. Lang, *Adaptive Multilevel Solution of Nonlinear Parabolic PDE Systems.* Theory, Algorithm, and Applications.
17. B.I. Wohlmuth, *Discretization Methods and Iterative Solvers Based on Domain Decomposition.*
18. U. van Rienen, M. Günther, D. Hecht (eds.), *Scientific Computing in Electrical Engineering.*
19. I. Babuška, P.G. Ciarlet, T. Miyoshi (eds.), *Mathematical Modeling and Numerical Simulation in Continuum Mechanics.*
20. T.J. Barth, T. Chan, R. Haimes (eds.), *Multiscale and Multiresolution Methods.* Theory and Applications.
21. M. Breuer, F. Durst, C. Zenger (eds.), *High Performance Scientific and Engineering Computing.*
22. K. Urban, *Wavelets in Numerical Simulation.* Problem Adapted Construction and Applications.

23. L.F. Pavarino, A. Toselli (eds.), *Recent Developments in Domain Decomposition Methods.*

24. T. Schlick, H.H. Gan (eds.), *Computational Methods for Macromolecules: Challenges and Applications.*

25. T.J. Barth, H. Deconinck (eds.), *Error Estimation and Adaptive Discretization Methods in Computational Fluid Dynamics.*

26. M. Griebel, M.A. Schweitzer (eds.), *Meshfree Methods for Partial Differential Equations.*

27. S. Müller, *Adaptive Multiscale Schemes for Conservation Laws.*

28. C. Carstensen, S. Funken, W. Hackbusch, R.H.W. Hoppe, P. Monk (eds.), *Computational Electromagnetics.*

29. M.A. Schweitzer, *A Parallel Multilevel Partition of Unity Method for Elliptic Partial Differential Equations.*

30. T. Biegler, O. Ghattas, M. Heinkenschloss, B. van Bloemen Waanders (eds.), *Large-Scale PDE-Constrained Optimization.*

31. M. Ainsworth, P. Davies, D. Duncan, P. Martin, B. Rynne (eds.), *Topics in Computational Wave Propagation.* Direct and Inverse Problems.

32. H. Emmerich, B. Nestler, M. Schreckenberg (eds.), *Interface and Transport Dynamics.* Computational Modelling.

33. H.P. Langtangen, A. Tveito (eds.), *Advanced Topics in Computational Partial Differential Equations.* Numerical Methods and Diffpack Programming.

34. V. John, *Large Eddy Simulation of Turbulent Incompressible Flows.* Analytical and Numerical Results for a Class of LES Models.

35. E. Bänsch (ed.), *Challenges in Scientific Computing - CISC 2002.*

36. B.N. Khoromskij, G. Wittum, *Numerical Solution of Elliptic Differential Equations by Reduction to the Interface.*

37. A. Iske, *Multiresolution Methods in Scattered Data Modelling.*

38. S.-I. Niculescu, K. Gu (eds.), *Advances in Time-Delay Systems.*

39. S. Attinger, P. Koumoutsakos (eds.), *Multiscale Modelling and Simulation.*

40. R. Kornhuber, R. Hoppe, J. Périaux, O. Pironneau, O. Wildlund, J. Xu (eds.), *Domain Decomposition Methods in Science and Engineering.*

41. T. Plewa, T. Linde, V.G. Weirs (eds.), *Adaptive Mesh Refinement – Theory and Applications.*

42. A. Schmidt, K.G. Siebert, *Design of Adaptive Finite Element Software.* The Finite Element Toolbox ALBERTA.

43. M. Griebel, M.A. Schweitzer (eds.), *Meshfree Methods for Partial Differential Equations II.*

44. B. Engquist, P. Lötstedt, O. Runborg (eds.), *Multiscale Methods in Science and Engineering.*

45. P. Benner, V. Mehrmann, D.C. Sorensen (eds.), *Dimension Reduction of Large-Scale Systems.*

46. D. Kressner, *Numerical Methods for General and Structured Eigenvalue Problems.*

47. A. Boriçi, A. Frommer, B. Joó, A. Kennedy, B. Pendleton (eds.), *QCD and Numerical Analysis III.*

48. F. Graziani (ed.), *Computational Methods in Transport.*

49. B. Leimkuhler, C. Chipot, R. Elber, A. Laaksonen, A. Mark, T. Schlick, C. Schütte, R. Skeel (eds.), *New Algorithms for Macromolecular Simulation.*

50. M. Bücker, G. Corliss, P. Hovland, U. Naumann, B. Norris (eds.), *Automatic Differentiation: Applications, Theory, and Implementations.*

51. A.M. Bruaset, A. Tveito (eds.), *Numerical Solution of Partial Differential Equations on Parallel Computers.*

52. K.H. Hoffmann, A. Meyer (eds.), *Parallel Algorithms and Cluster Computing.*

53. H.-J. Bungartz, M. Schäfer (eds.), *Fluid-Structure Interaction.*

54. J. Behrens, *Adaptive Atmospheric Modeling.*

55. O. Widlund, D. Keyes (eds.), *Domain Decomposition Methods in Science and Engineering XVI.*

56. S. Kassinos, C. Langer, G. Iaccarino, P. Moin (eds.), *Complex Effects in Large Eddy Simulations.*

57. M. Griebel, M.A Schweitzer (eds.), *Meshfree Methods for Partial Differential Equations III.*

58. A.N. Gorban, B. Kégl, D.C. Wunsch, A. Zinovyev (eds.), *Principal Manifolds for Data Visualization and Dimension Reduction.*

59. H. Ammari (ed.), *Modeling and Computations in Electromagnetics: A Volume Dedicated to Jean-Claude Nédélec.*

60. U. Langer, M. Discacciati, D. Keyes, O. Widlund, W. Zulehner (eds.), *Domain Decomposition Methods in Science and Engineering XVII.*

61. T. Mathew, *Domain Decomposition Methods for the Numerical Solution of Partial Differential Equations.*

62. F. Graziani (ed.), *Computational Methods in Transport: Verification and Validation.*

63. M. Bebendorf, *Hierarchical Matrices.* A Means to Efficiently Solve Elliptic Boundary Value Problems.

64. C.H. Bischof, H.M. Bücker, P. Hovland, U. Naumann, J. Utke (eds.), *Advances in Automatic Differentiation.*

65. M. Griebel, M.A. Schweitzer (eds.), *Meshfree Methods for Partial Differential Equations IV.*

66. B. Engquist, P. Lötstedt, O. Runborg (eds.), *Multiscale Modeling and Simulation in Science.*

67. I.H. Tuncer, Ü. Gülcat, D.R. Emerson, K. Matsuno (eds.), *Parallel Computational Fluid Dynamics 2007.*

68. S. Yip, T. Diaz de la Rubia (eds.), *Scientific Modeling and Simulations.*

69. A. Hegarty, N. Kopteva, E. O'Riordan, M. Stynes (eds.), *BAIL* 2008 – *Boundary and Interior Layers.*

70. M. Bercovier, M.J. Gander, R. Kornhuber, O. Widlund (eds.), *Domain Decomposition Methods in Science and Engineering XVIII.*

71. B. Koren, C. Vuik (eds.), *Advanced Computational Methods in Science and Engineering.*

72. M. Peters (ed.), *Computational Fluid Dynamics for Sport Simulation.*

73. H.-J. Bungartz, M. Mehl, M. Schäfer (eds.), *Fluid Structure Interaction II - Modelling, Simulation, Optimization.*

74. D. Tromeur-Dervout, G. Brenner, D.R. Emerson, J. Erhel (eds.), *Parallel Computational Fluid Dynamics 2008.*

75. A.N. Gorban, D. Roose (eds.), *Coping with Complexity: Model Reduction and Data Analysis.*

76. J.S. Hesthaven, E.M. Rønquist (eds.), *Spectral and High Order Methods for Partial Differential Equations.*

77. M. Holtz, *Sparse Grid Quadrature in High Dimensions with Applications in Finance and Insurance.*

78. Y. Huang, R. Kornhuber, O.Widlund, J. Xu (eds.), *Domain Decomposition Methods in Science and Engineering XIX.*

79. M. Griebel, M.A. Schweitzer (eds.), *Meshfree Methods for Partial Differential Equations V.*

80. P.H. Lauritzen, C. Jablonowski, M.A. Taylor, R.D. Nair (eds.), *Numerical Techniques for Global Atmospheric Models.*

81. C. Clavero, J.L. Gracia, F.J. Lisbona (eds.), *BAIL 2010 – Boundary and Interior Layers, Computational and Asymptotic Methods.*

82. B. Engquist, O. Runborg, Y.R. Tsai (eds.), *Numerical Analysis and Multiscale Computations.*

For further information on these books please have a look at our mathematics catalogue at the following URL: `www.springer.com/series/3527`

Monographs in Computational Science and Engineering

1. J. Sundnes, G.T. Lines, X. Cai, B.F. Nielsen, K.-A. Mardal, A. Tveito, *Computing the Electrical Activity in the Heart.*

For further information on this book, please have a look at our mathematics catalogue at the following URL: www.springer.com/series/7417

Texts in Computational Science and Engineering

1. H. P. Langtangen, *Computational Partial Differential Equations.* Numerical Methods and Diffpack Programming. 2nd Edition
2. A. Quarteroni, F. Saleri, P. Gervasio, *Scientific Computing with MATLAB and Octave.* 3rd Edition
3. H. P. Langtangen, *Python Scripting for Computational Science.* 3rd Edition
4. H. Gardner, G. Manduchi, *Design Patterns for e-Science.*
5. M. Griebel, S. Knapek, G. Zumbusch, *Numerical Simulation in Molecular Dynamics.*
6. H. P. Langtangen, *A Primer on Scientific Programming with Python.*
7. A. Tveito, H. P. Langtangen, B. F. Nielsen, X. Cai, *Elements of Scientific Computing.* 2nd Edition
8. B. Gustafsson, *Fundamentals of Scientific Computing.*

For further information on these books please have a look at our mathematics catalogue at the following URL: www.springer.com/series/5151